DISCLAIMER OF WARRANTY

The technical descriptions, procedures, and computer programs in this book have been developed with the greatest of care and they have been useful to the author in a broad range of applications; however, they are provided as is, without warranty of any kind. Artech House, Inc. and the author and editors of the book titled *CMOS RFIC Design Principles* make no warranties, expressed or implied, that the equations, programs, and procedures in this book or its associated software are free of error, or are consistent with any particular standard of merchantability, or will meet your requirements for any particular application. They should not be relied upon for solving a problem whose incorrect solution could result in injury to a person or loss of property. Any use of the programs or procedures in such a manner is at the user's own risk. The editors, author, and publisher disclaim all liability for direct, incidental, or consequent damages resulting from use of the programs or procedures in this book or the associated software.

For a listing of recent titles in the *Artech House Microwave Library*,
turn to the back of this book.

CMOS RFIC Design Principles

CMOS RFIC Design Principles

Robert Caverly

BOSTON | LONDON
artechhouse.com

Library of Congress Cataloging-in-Publication Data
A catalog record for this book is available from the U. S. Library of Congress.

British Library Cataloguing in Publication Data
A catalogue record for this book is available from the British Library.

ISBN-13: 978-1-59693-132-9

Cover design by Igor Valdman

© 2007 ARTECH HOUSE, INC.
685 Canton Street
Norwood, MA 02062

All rights reserved. Printed and bound in the United States of America. No part of this book may be reproduced or utilized in any form or by any means, electronic or mechanical, including photocopying, recording, or by any information storage and retrieval system, without permission in writing from the publisher.

All terms mentioned in this book that are known to be trademarks or service marks have been appropriately capitalized. Artech House cannot attest to the accuracy of this information. Use of a term in this book should not be regarded as affecting the validity of any trademark or service mark.

10 9 8 7 6 5 4 3 2 1

To my parents, Bob and Janet:
for giving me a shortwave radio as a
Christmas present all those years ago.

Contents

Preface

CMOS IC technology is widely used today throughout industry and is the main technology for digital integrated circuits. In the last decade, the use of wireless techniques for transmission of data (mostly in digital form) has skyrocketed. Major progress has been made in integrating the rich variety of digital signal processors with RF and wireless circuitry so that single-chip solutions to wireless digital transmission can be achieved (the so-called "single-chip radio"). Already, a wide variety of wireless devices based on 802.11x, Bluetooth, WiMAX, or ZigBee standards, to name just a few, have been fabricated in CMOS technology. In addition to the numerous PCS applications, RFID and wireless sensors are being actively explored with CMOS being the preferred technology for both baseband and RF circuitry.

Another emerging technology based on software-defined radio (SDR) architectures, or its close relative, cognitive radio, promises to accelerate the need for highly integrated digital control and signal processing alternatives with rapidly reconfigurable RF circuitry. Historically, silicon CMOS has not been as robust an RF technology as gallium-arsenide because of a number of inherent limitations in MOS technology that limit frequency response. However, as progress in CMOS fabrication technology has continued to reduce the MOSFET feature size, the frequency response has likewise improved, and many RF integrated circuits using CMOS technology have been fabricated and are being used in a variety of products. The increasing use of CMOS for applications up to X-band has spawned a need for RF, wireless, and microwave designers familiar with the characteristics of RF/microwave CMOS and their influence on RF integrated circuit design.

The purpose of this book is to address the needs of RF, wireless, and microwave circuit designers, both professionals and students, to the design issues associated with CMOS RF integrated circuits. The book will be of utility to those RF and wireless designers who wish to begin working with the technology. The book will also be of use to those designers experienced in working with other technologies but who have a need to further investigate this technology. In addition, it should serve as a good "first look" for students interested in CMOS RFIC technology.

The book is divided into three major sections of roughly three chapters each. The first portion of the book is a review of RF system fundamentals and includes terminology and definitions widely used to describe these circuits. Chapters on CMOS active device and passive element fundamentals are included as a means to introduce the reader to the use of these design equations and their limitations for "first-pass" designs.

The second portion of the book covers elemental CMOS RF circuits such as low-noise amplifiers, general gain amplifiers, mixers, and oscillators. Ideal circuit topologies are discussed first and then merged with the nonideal CMOS circuit elements to provide insight into circuit modifications needed to achieve design specifications. Behavioral models for these elements are introduced for future use in later portions of the book.

The third portion of the book covers more advanced CMOS RF integrated circuits such as voltage-controlled oscillators, phase lock loops, frequency synthesizers, and power amplifier architectures. In addition to coverage of the fundamentals, numerous references are made to the current literature on these circuits.

All three sections of the book include a number of examples to aid in understanding of the principles discussed. Where feasible, simulations are performed (and illustrated) with CAD tools widely used in industry and academia. From the illustrations and accompanying simulation files, readers can look at various circuit scenarios to observe how variations of device parameters affect circuit performance. Because of the number of different RF circuit simulators available in the marketplace, the simulation files are primarily SPICE-based netlists that use standard BSIM parameters. Anyone having a basic familiarity with an RF CAD different simulator should be able to translate these easily for their own purposes. Links to various demonstration and student versions of CAD tools are included in an accompanying CD. CAD layouts and measured RF data on select RFIC circuits are also included so that further exploration of these concepts can be done by the reader.

Acknowledgments

I knew when I first started this book that there would be many people that I wanted to thank; little did I know, however, how many people would actually assist me in developing a book of this length. Therefore, I want to acknowledge a number of individuals and organizations, not necessarily in order of importance—they were all important.

I must first acknowledge the National Science Foundation for their support in funding a Combined Research and Curriculum Development project in an allied area that led to the idea for this book. This material is based upon work supported by the National Science Foundation under Grant No. 0203459. Any opinions, findings, and conclusions or recommendations expressed in this material are those of the author(s) and do not necessarily reflect the views of the National Science Foundation. I would also like to thank Dr. Jim Rautio and Mr. Yun Chase of Sonnett Software, Inc., for agreeing to provide a demo version of their popular Sonnet® Lite™ software for use in this book. Thanks also go to M. Englehart of Linear Technology for giving permission to include a copy of their SPICE simulator SwitcherCADIII on the CD. The MOSIS Service deserves my appreciation for allowing me to publish the SPICE BSIM parameters. These parameters are available on their public website, but it is convenient for the reader to see the detail in these MOSFET device descriptions as part of the text narrative.

I would like to thank my Artech House editors, Ms. Barbara Lovenvirth and Mr. Mark Walsh; Ms. Lovenvirth for her invaluable assistance in keeping the book on track and providing gentle reminders of upcoming deadlines and in general guidance through all aspects of the development and production process; and Mr. Mark Walsh for his assistance during the preliminary stages of development in helping me with understanding the entire publishing process. In addition to the editors, the technical reviewer gave me many suggestions for improving the overall manuscript; I used practically all of them. Two individuals gave me general but invaluable advice in the mechanics of book writing, advising me to stick to a strict and detailed outline and schedule, and to keep plugging away on a daily basis: Dr. Steve Maas and Dr. William Bushong (see, Bill, you made it into an engineering book). In addition, two general acknowledgments: the first to Mr. Gerry Hiller and Dr. Peter Rizzi, my mentors early in my career; and secondly, to the amateur radio community at large, for using RF and microwave technology as a hobby (amateur radio call WB4PWZ). I wish to thank my students in the Fall 2006 RFIC Design graduate class and its precursor, the ECE 9900 Special Topics–RFIC Design graduate class, at Villanova University in the preparation of the final manuscript. The students were quite proficient at finding errors in the manuscript, were constructively critical of unclear passages in the manuscript, and were invaluable in aiding with the "debugging" of the many equations and examples.

And last but not least, I'd like to thank my family for putting up with the long hours I spent at my computer during the course of the year. I want to thank my wife Maggie for her interest and continual support during the writing of the book; as a librarian and an overall lover of books, she had many constructive insights and ideas. I'd also like to thank my two sons, William and Matthew, for their interest and support during the preparation of this book

CHAPTER 1

Introduction

The computer will become the hub of a vast network of remote data stations and information banks feeding into the machine at a transmission rate of a billion or more bits of information a second.... Eventually, a global communications network handling voice, data and facsimile will instantly link man to machine—or machine to machine—by land, air, underwater, and space circuits.

—David Sarnoff, 1964 [1]

1.1 Historical Perspective and Background

1.1.1 A (Very) Brief History[1]

1.1.1.1 Portable Radio and Wireless Systems

During the Christmas season of 1906, radio operators at sea heard from their radio receivers not the usual Morse code[2] transmissions, but rather the sounds of Christmas music and a violin solo. This ground-breaking audio transmission scheme (and the violin solo) was the creation of Reginald Fessenden, a Canadian engineer who at one time had worked with Thomas Edison [2]. So began a century of rapid advances in radio and wireless transmission of voice information. A century ago, radio transmissions were based on a single transmitting and receiving architecture using "spark gap transmitters" and "coherer receivers." During the next 50 years, the codevelopment of vacuum tubes and various modulation schemes and system architectures vastly improved the efficiency and performance of radio communication links. The invention of the transistor as a replacement for vacuum tubes ushered in another 50 years of communication system development, not only in terms of performance but also in terms of the system's physical size, weight, and battery consumption. It is at this point in history that the portable wireless radio system really begins.

1. A number of books are available that outline in detail the early pioneers of radio, from Maxwell and Marconi to present-day heroes; a list of representative books on the history of radio and biographies of the early players (and by no means an exhaustive list) can be found in the Selected Bibliography at the end of this chapter.
2. The use of coding techniques for electronic transmission of information is not new; the Morse code had been used for decades in wired systems prior to 1906 and was used in the early days of radio. Morse code (named after Samuel F. B. Morse) is a four-state coding scheme defined as the transmission of a "dot," a "dash," and two time spacings, one between dots/dashes that defines a single letter of the alphabet and one that defines the spacing between the letters. For instance, the code for the phrase "it is" is ".. - .. ...". While not exactly a binary coded transmission, the Morse code has been widely used for decades to electronically transmit information, and is still used today by radio amateurs and experimenters [3].

The first truly portable wireless systems that were manufactured by the thousands were the SCR 300 and 536 "walkie-talkie" radios for the U.S. Army Signal Corps during World War II. These radios used either amplitude modulation (AM; SCR-300) or frequency modulation (FM; SCR-536), and were based on transmitter and receiver architectures developed by earlier engineers such as Armstrong, Carson, and de Forest [4]. The first commercial mobile telephone system (MTS) began right after the war in St. Louis, Missouri, and required operator assistance in connecting the call. The MTS was a cumbersome system that allowed only a few users and had a high overhead but was popular with its customers [5]. It wasn't until 1956 that the MTS became automated. During that same time frame, pagers made their debut and cellular-based radio was in its conceptual stage. By the late 1960s, the first cellular telephone systems were introduced, but it wasn't until 1978, with the introduction of the Advanced Mobile Phone Services (AMPS), that cellular radio took off in the United States. AMPS, sometimes referred to as a first generation (1G) system, provided analog voice service without any text messaging capability. Several years later, Scandanavia's Nordic MTS began, and with it, the worldwide deployment of cellular radio services. The year 1991 saw the development of the Global System for Mobiles or, in the original French, *Groupe Speciale Mobile*, (GSM). Since then, a new generation of cellular-based telephone products have been introduced every several years, each with increasing functionality for the mobile user. Digital second generation (2G) systems developed in Europe (GSM) and the United States (using the IS-95 standard) provided clearer voice systems as well as short messaging service (SMS) capability [6]. These systems did not communicate with each other since they were based on completely different digital protocols. Their relatively low data rate limited these 2G systems to simple text messaging and low data rate services. The next leap occurred with the introduction of 3G systems, which use a wider bandwidth than the earlier 2G systems, resulting in an increase in data rate (as high as 200 Kbps) that provides the capability for image and video services along with standard voice and text applications. The next generation, 4G, promises to provide tens of megabits per second throughput, providing for fast data downloads and smooth video and other streaming applications [7].

The use of radio and wireless systems is not confined, however, to just cellular-based systems for personal use. During the 1970s, computer-to-computer communications via local-area networking (LAN) debuted using wired connections with communications protocols such as Ethernet (and its standard, IEEE 802.3) [8]. The data rate for the early Ethernet protocols was 3.0 Mbps, increasing over time to 10 Mbps (10base-T), 100 Mbps (100base-TX), and then 1000 Mbps Gigabit Ethernet (1000base) [8]. These and other wired protocols (such as token rings and the fiber distributed data interface or FDDI) were expanded to the wireless realm in the late 1990s with the introduction of wireless local-area networks (WLANs) based on the IEEE 802.11 group of standards.[3] Currently, the most popular of the 802.11 standards are 802.11a, 802.11b, and 802.11g [9]. WLANs using the 802.11a standard

3. The term "IEEE 802" is the actual standards name for the "IEEE Standard for Local and Metropolitan Area Networks." The 802.11 standards make up the wireless section of the broad 802 standard, and the initials a, b, etc. (as in 802.11a, for example) are actually supplements and amendments to the original 802.11 standard developed in 1999. See the IEEE website for a detailed description of the 802 standard (http://www.ieee.org/getieee802/ about.html).

operate in the UNII (unlicensed national information infrastructure) band: 5.15 through 5.825 GHz. The other two protocols operate in the ISM (industrial, scientific, medical) band: 2.4 through 2.483 GHz. The 802.11g standard provides a higher data rate than 802.11b and is compatible with this protocol, although the data rate when operating in the "b" mode is lower than that of the "g" mode.

In recent years, a host of other wireless data transmission protocols have been developed. Protocols supporting low-power, short-range applications for wireless personal area networks (WPANs) include Bluetooth (using the IEEE 802.15.1 standard) [10, 11] and ZigBee (using the IEEE 802.15.4 standard) [12], whereas for wideband applications, WiMAX (World Interoperability for Microwave Access, based on the IEEE 802.16 standard) seems to have a bright future. There will undoubtedly be countless more protocols in the future. These low-power WPAN applications are especially amenable to complementary metal oxide semiconductor (CMOS) radio-frequency integrated circuit (RFIC) solutions. Both WLAN and WPAN applications often share the same frequency bands and so care must be taken in the design to ensure coexistence of the two different protocols. (The IEEE 802.15.2 standard helps address this issue.)

To ensure widespread acceptance with consumers, these portable cellular telephones and WLAN systems should be small and lightweight, exhibit long battery life, and be low cost, even as the actual functionality increases. Placing both digital and analog/RF functions on the same semiconductor die requires a high degree of integration for these so-called "systems on a chip" (SoC). Because the primary technology for digital electronics uses CMOS devices, it is only logical that to achieve the goal of a SoC, the analog/RF functions should also be fabricated using devices compatible with the CMOS processes. The next section outlines a short history of the development of CMOS technology and how RFIC development using this technology evolved.

1.1.1.2 MOS and MOS RFIC Development

The development of portable radio and wireless systems and the development of electronic devices, of course, go hand in hand. Radio engineers employed each new generation of devices to improve some (or several) aspects of the radio: the performance, size, weight, power consumption, or some combination.

The story of MOS actually starts in 1933 when Lilienfield was granted a patent for a "Device for conducting electrical current" [13]. At the time, that patent, along with another patent for a similar concept using metal [14], was considered merely a curiosity since the techniques to fabricate such devices were not available. It was almost another 30 years before the first MOS device was successfully fabricated. The first reported implementation of an MOS field effect transistor (MOSFET) as we currently recognize it occurred in 1960, almost simultaneously with Kilby's invention of the integrated circuit (IC) and Noyce's silicon IC implementation [15, 16]. The first MOSFET was an enhancement mode device, meaning that a path for electrons to flow had to be created (or "enhanced") by an applied voltage [17]. The complement to this MOSFET (hole carrier flow) was created a few years later [18], setting the foundation for the invention of *complementary* MOS circuits. Complete details of MOSFET operation will be covered in Chapter 2.

The classic MOSFET structure, which has not changed substantially since its inception, has two heavily doped semiconductor regions (D and S in Figure 1.1) separated by a semiconductor region of opposite (or complementary) doping.[4] Current cannot flow in this structure since the junction regions (JR) are of the PN type, representing back-to-back PN junction diodes. In the middle region, a thin insulating material (I) is placed between the semiconductor surface and another conductor placed on top of it, creating a third electrode (G). With the application of a sufficiently high voltage, a thin layer of charge carriers of the same type as the D and S regions is generated under the insulator, creating a channel that bypasses the back-to-back diodes and allows current to flow. This third electrode acts like a gate controlling the current, and hence is termed the *gate* electrode. If electron flow is assumed, electrons will flow from the S region (termed the *source*) and be collected in the D, or *drain*, region (the term *collector* was already taken for the bipolar junction transistor). The length of the gate, L_g or gate length, is an important parameter in MOSFET operation. In fact, when CMOS technologies are discussed, it is this gate length that specifies the technology, such as "0.25-micron CMOS technology" (meaning MOSFETs can be fabricated with the smallest gate length of 0.25 micrometers). Although details of the impact that this gate length has on CMOS device performance will be covered in Chapter 2, it is easy to see that a MOSFET will operate faster if the electrons have a shorter distance to travel. Faster operation implies fast switching speed (for digital electronics) and increased frequency response for analog/RF applications. Two figures of merit are used to describe the frequency response in MOSFETs (or other amplifying device). The first, f_T, is defined as the frequency where the short circuit current gain becomes unity. The second, f_{max}, is the maximum frequency at which the device can be used in an oscillator (also termed the maximum frequency of oscillation).

The results of MOS and CMOS technology development in the digital realm are well documented, with a huge increase in the number of MOSFETs that could be

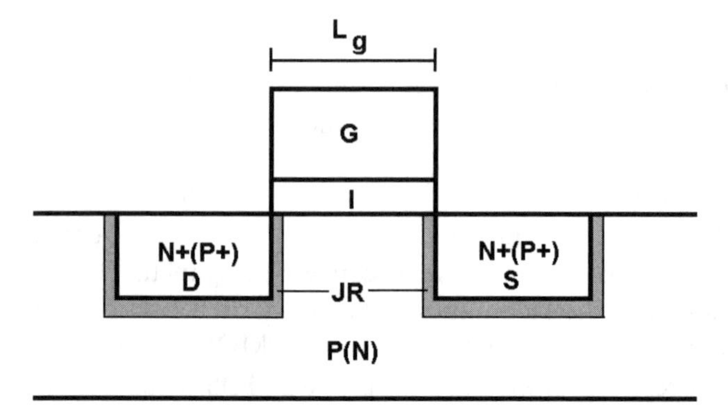

Figure 1.1 Cross section of a basic MOSFET structure defining the gate length, L_g, and the various regions of doped silicon material.

4. In Figure 1.1, the D and S regions are heavily doped with either donor or acceptor atoms (N+ or P+, respectively). The remaining region, termed the substrate, is doped with a reduced number of acceptor or donor atoms (P or N, respectively) to provide the complementary doping. Further details on MOSFET fabrication and operation will be covered in Chapter 2.

placed on a single silicon die. In 1965, Gordon Moore plotted five data points showing the "number of components per integrated function" versus time, and he extrapolated the now famous exponential increase in IC complexity that is now known as Moore's law [19]. What is so interesting about Moore's law is that, 40 years after it was first mentioned (not by that name, of course; the term *Moore's law* came later), this exponential increase in digital circuit complexity and density still holds. Table 1.1, for example, shows the trend over more than 40 years of the number of transistors in each new digital microprocessor generation. While bipolar transistors primarily made up ICs in the 1960s and 1970s, research continued on CMOS devices for ICs, with CMOS becoming the technology of choice for digital applications starting in the 1980s [20].

The increase in transistor density for analog/RF circuits, however, is not as remarkable as for digital circuits, a fact noted early on in IC development [19]. However, the increase in analog functionality achieved with CMOS is no less remarkable. The switched capacitor concept was a major driver in the use of CMOS analog circuit techniques for reasonably high-density functionality [21]. Until the early 1990s, however, the use of CMOS for RF applications was not considered feasible because of their poor high-frequency performance compared with bipolar and metal semiconductor field effect transistor (MESFET) technology, which uses compound semiconductor devices such as gallium arsenide (GaAs). That all changed with the publication of an article outlining the design of an integrated CMOS RF amplifier at 900 MHz. A decade-long study of the use of CMOS technology at frequency of 1000 MHz and higher began [21].

Table 1.1 The Number of Transistors on a Single Silicon Die Has Followed Moore's Law's Remarkable Exponential Increase with Time

Microprocessor	*Year of Introduction*	*Transistors*
4004	1971	2,300
8008	1972	2,500
8080	1974	4,500
8086	1978	29,000
Intel286	1982	134,000
Intel386™ processor	1985	275,000
Intel486™ processor	1989	1,200,000
Intel© Pentium© processor	1993	3,100,000
Intel© Pentium© II processor	1997	7,500,000
Intel© Pentium© III processor	1999	9,500,000
Intel© Pentium© 4 processor	2000	42,000,000
Intel© Itanium© processor	2001	25,000,000
Intel© Itanium© 2 processor	2003	220,000,000
Intel© Itanium© 2 processor (9-MB cache)	2004	592,000,000

© 2005 Intel Corporation.

1.1.1.3 What the Future May Hold

It is always interesting to try to pull out a crystal ball and predict the future. David Sarnoff's quote at the beginning of this chapter is a case in point and turned out to be quite prescient. Predictions such as these, however, are not based on pure speculation but rather are based on detailed knowledge of the present state of technology as well as business plans and technology road maps that help point a path to a future goal.[5] One such technology road map, the International Roadmap for Semiconductors (ITRS), is published every two years as a means to predict and benchmark improvements for all semiconductor technologies. The 2005 ITRS shows CMOS continuing to play a major role in the RF arena up to approximately 10 GHz, with steady improvements throughout the next 15 years (see Figures 1.2, 1.3, and 1.4). The 2005 ITRS covers two different CMOS performance measures: low-power RF ("performance RF") and high-power RF ("RF driver"). The steady decrease in gate length for these RF devices is directly linked to the continued reduction in feature size for the digital functions that make up the bulk of CMOS's use in IC technology [22]. With decreasing gate length, however, comes a corresponding decrease in power supply voltage, with 1.0V power supply voltages predicted for low-power CMOS RF devices by 2013 [23]. Somewhat higher power supply voltages are needed for high-power RF (1.8V by 2010). The biggest improvements in frequency response will occur for performance CMOS, with f_T and $f_{\max}$ values predicted to reach 800 and 1100 GHz, respectively, by the year 2020. The improvement in the frequency response for high-power RF CMOS is not expected to occur at the same rate as for low-power RF; instead, it is predicted to merely rise slowly over this same time period. It is this lower frequency response on the high-power RF side that is a major limiting factor for higher frequency deployment of CMOS RFICs.

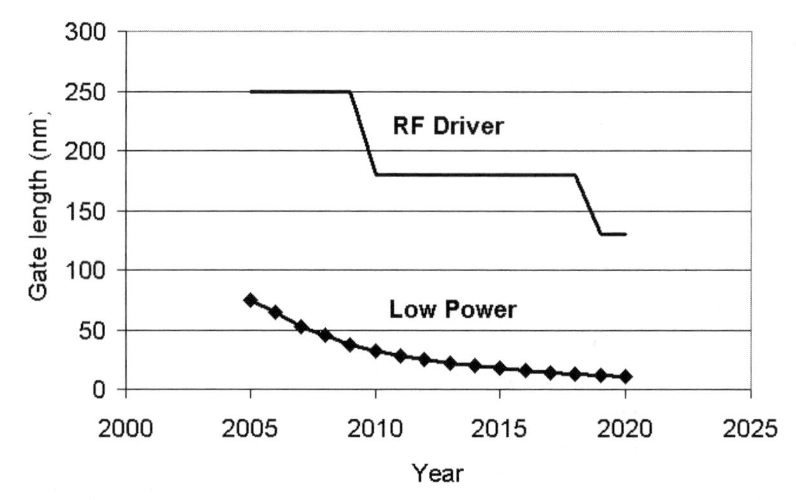

Figure 1.2 ITRS 2005 showing reduction in MOSFET gate length predictions out to the year 2020 for low- and high-power RF MOSFET devices.

5. Irrespective of U.S. baseball great Yogi Berra's famous quote: "If you don't know where you are going, you might wind up someplace else."

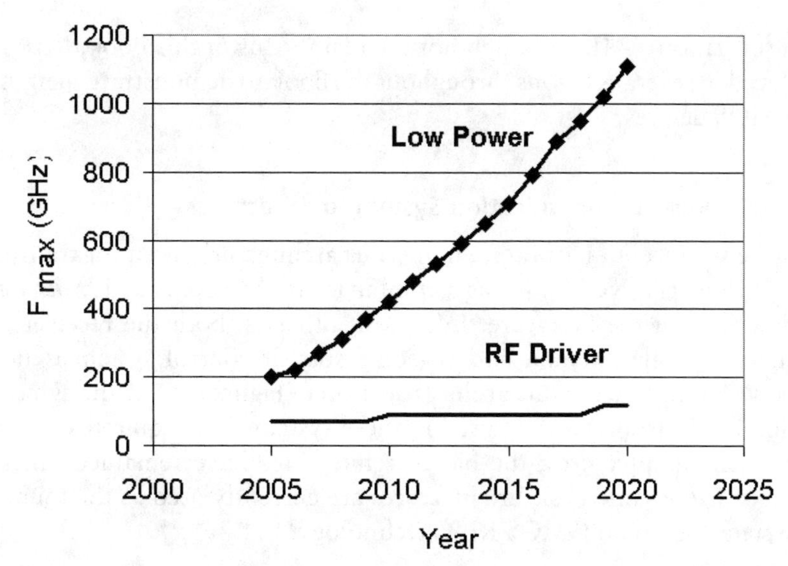

Figure 1.3 ITRS 2005 showing increase in f_T predictions out to the year 2020 for low- and high-power RF MOSFET devices.

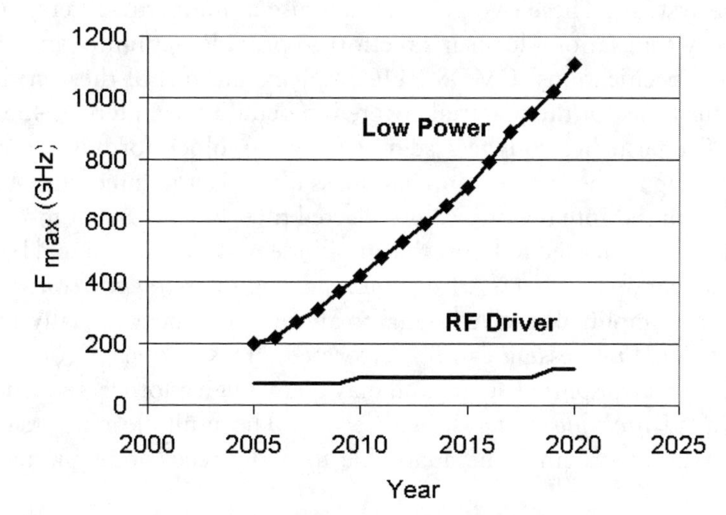

Figure 1.4 ITRS 2005 showing increase in f_{max} predictions out to the year 2020 for low- and high-power RF MOSFET devices.

The use of CMOS technology is not limited to active devices such as MOSFETs or even passive devices such as inductors, resistors, and capacitors. Micro electromechanical systems (MEMS) technology was first developed in the 1970s and units were first commercialized in the 1990s. Since that time, MEMS technology has made continued improvements and it is anticipated that this technology will be used for integrating large filters, switches, and resonator blocks in the future [23]. In fact, MEMS and CMOS devices are extremely compatible: CMOS devices are used for digital processing and communications tasks, while the MEMS components can be used for sensing (information input) and actuation (physical response to this information) [24]. Although not the main focus of this book, MEMS will be introduced in several sections throughout the book to demonstrate their interactions with CMOS devices.

1.1.2 Basic Communication System Architectures

Since the inception of radio, numerous architectures used for transmitters (XMTR) and receivers (RCVR) (or as a combination, they are called *transceivers*) have been described in the literature. In almost all cases, both the receiver and transmitter share a single antenna and so some sort of control system is needed to switch between the two circuits at the proper time (Figure 1.5). Although each new generation of electronic devices used in these systems has required some modification of existing architectures, the basic architectures have remained fairly constant over time. These same basic architectures are currently used as the foundations of radio systems based on CMOS RFIC technology.

1.1.2.1 Receivers

On the receiver side, the two basic architectures are the *heterodyne* and the *homodyne* systems. These two systems each use a number of filters, amplifiers, and frequency translation blocks in an effort to provide optimum performance for a given set of specifications. CMOS RFIC implementations of these critical system blocks are the basis for the material covered in detail in Chapters 3, 4, 5, and 6.

The basic heterodyne system is shown in block form in Figure 1.6. A signal is picked up by the antenna and bandpass filtered either inherently by the narrowband nature of the antenna or by a preselected filter block. One or more stages of amplification are then used to boost the amplitude of the desired signal band (which was set by the bandpass filter). A frequency shifting or *frequency conversion circuit* translates the amplified received signal to another frequency (usually a lower one) so efficient signal processing can be performed.[6] This frequency translated signal is then filtered and amplified again, and may go through another frequency translation (the so-called *multiple conversion* receiver). The multiple conversion process allows this type of system to be adaptable to many receiver requirements, although the

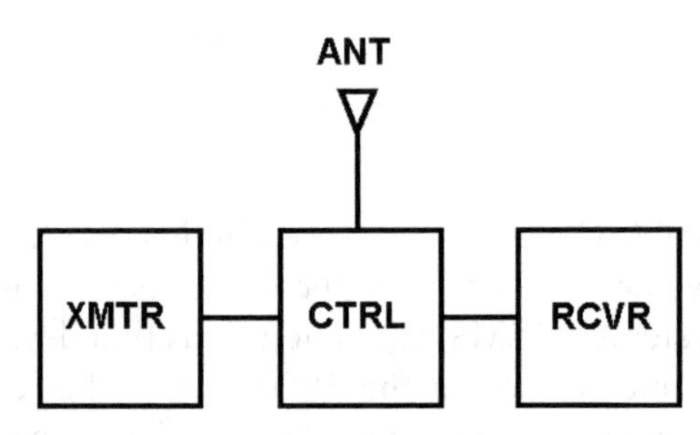

Figure 1.5 Block diagram of an RF transceiver showing the transmitter and receiver connected to an antenna control (CTRL) element.

6. This frequency translation or converter circuit is implemented using a combination of a mixer and a local oscillator. Details on both these topics are covered in Chapters 6 and 7.

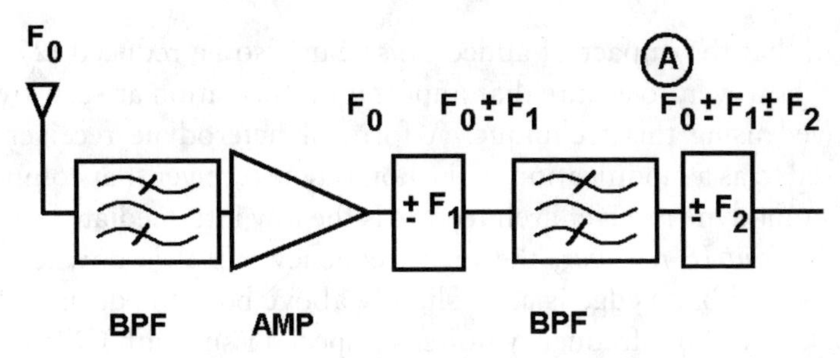

Figure 1.6 High-level block diagram of a heterodyne receiver architecture, showing bandpass filters (BPFs) and an amplifier (AMP), along with frequency conversion blocks $\pm F_1$ and $\pm F_2$.

complexity of the circuit and the need for a number of filtering and frequency translating components makes this a difficult architecture to implement in the highly integrated CMOS RFIC environment [25]. With a large number of circuits, power consumption also becomes an issue.

The basic homodyne system (Figure 1.7) shares block types with the heterodyne system, but instead of frequency converting to another frequency, the homodyne system downconverts the desired frequency band directly to dc or *baseband* (or at least a very low frequency); hence, these types of receivers are often termed *direct conversion receivers*. Compared to the heterodyne system, filtering requirements and parts/component counts are reduced with a corresponding reduction in power consumption. The relatively simple architecture makes this type of system a natural for the high levels of integration necessary for SoC applications. As a matter of fact, direct conversion receivers were initially developed to try to eliminate some of the drawbacks of the heterodyne receiver (most notably, the so-called "image" problem, discussed in further detail in Chapter 7). One of the major drawbacks of direct conversion receivers, however, is corruption of the desired downconverted signal by dc offset voltages (a problem since the band is directly converted down to zero frequency) and other low-frequency phenomenon, primarily noise (covered later in this chapter and in Chapter 7). Direct conversion receivers also require that the frequency translation circuits be highly stable. (Ways to increase the stability of these circuits will be covered in Chapter 8.) High-pass filtering, however, can be employed in certain system applications to reduce the effects of this dc offset voltage [26], and modifications to the basic architecture and improved CMOS devices have

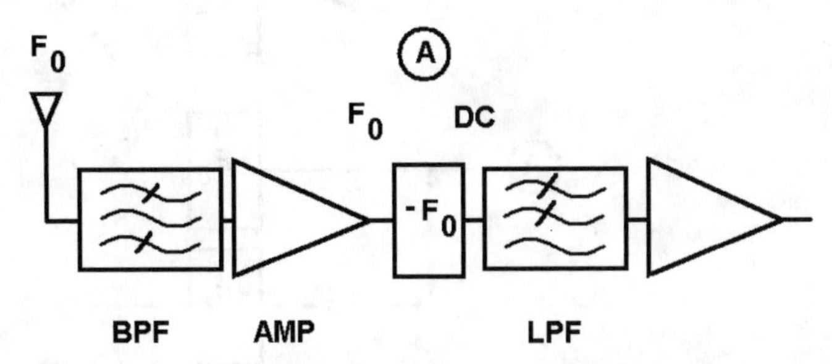

Figure 1.7 High-level block diagram of a homodyne receiver architecture, showing bandpass and low-pass filters (BPF and LPF) and an amplifier (AMP), along with a single frequency conversion block $-F_0$.

shown that the impact of added noise can also be reduced [27, 28]. Care must be taken, however, to ensure that important information at very low frequencies is not removed using this technique. A form of heterodyne receiver that is sometimes referred to as a modification to the homodyne receiver that compensates for the classical homodyne receiver architecture is the low intermediate frequency (IF) receiver. For the *low IF receiver*, the high-frequency signal is downconverted so that the lower signal band edge is at or slightly above both the dc as well as low-frequency noise levels. (Low-frequency noise is a special issue with CMOS devices and will be discussed in detail in Chapter 7.) Because the exact placement of the downconverted signal is strongly dependent on the receiver characteristics, a priori knowledge of the entire system requirements and performance is essential to a successful low IF receiver design.

Modern digital signal processing techniques use the concept of *in-phase* and *quadrature* signals (so-called *Cartesian I and Q signals*, respectively) to enhance performance of the entire communication system. These I and Q signals are by definition 90° out of phase with respect to each other, and can be derived in either heterodyne or homodyne receivers by introducing this phase shift during the frequency conversion process (Figure 1.8). I-Q signal creation and extraction is especially important in homodyne receivers since the downconversion process "folds" positive and negative frequencies together and important information could be lost (this is not the case with the heterodyne receiver). Because of the increasing complexity of communication systems, digital signal processing (DSP) techniques are taking on many of the processing chores previously performed by analog circuitry. Analog-to-digital converters (ADCs) are required at the back of the RF stage to perform the conversion to the digital domain. Because the conversion speed of the ADC is unlikely to provide direct conversion of the RF signal at the antenna terminals, heterodyne and homodyne receiver architectures will be necessary for the foreseeable future [29]. Some emerging systems based on software defined radio (SDR) concepts use direct ADC at the front end with no frequency downconversion

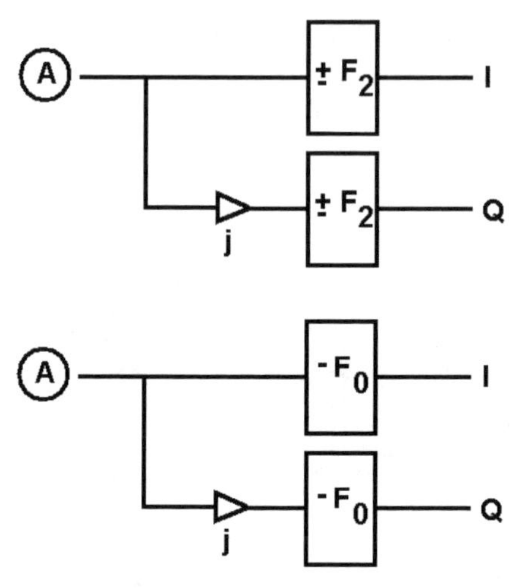

Figure 1.8 Cartesian I and Q signals can be generated by introducing a 90° phase shift (*j*) into the frequency translation process. Point A corresponds to the location indicated in Figures 1.6 and 1.7.

performed; this requires that the frequency range of interest be well within the conversion speed of the ADC.

1.1.2.2 Transmitters

The transmitter branch of the CMOS RFIC transceiver can be thought of, at least in block diagram form, as the inverse of the receiver architecture: a single channel (or I and Q channels) are frequency translated (usually upward) through one or more frequency conversion blocks. For digital-based signals, the single channel or I-Q channels may be directly taken from the output of a digital to analog converter (DAC). The phase relationships between the I and Q signals are introduced in the first frequency conversion block (Figure 1.9). Once the desired signal has been translated to the desired transmission frequency band, a series of amplifiers is employed to amplify the signal to the proper level for driving the antenna. The first stages of this amplifying chain are often referred to as *driver* stages, whereas the final stage is referred to as the *power amplifier* or *PA* stage (the term *final* or *final amplifier* is also used). Various filter stages are required at the output to band limit the signal to that of the desired frequency band. Design parameters and techniques for various PA architectures will be covered in Chapter 9.

1.1.3 Modulation Types

Where voice or digital information is to be transmitted, some means must be used to translate this data into a form that optimizes its transmission. The information signal is used to modify or *modulate* a *carrier signal* to effect the transmission of the information. A large number of these modulation techniques are used in modern communications systems, so this next section provides only the briefest of overviews of some of the techniques currently in use. This overview is useful, however, because system requirements may be specific to the modulation type. The reader is encouraged to refer to communications textbooks that delve into this interesting and important topic. (The Selected Bibliography at the end of this chapter contains a short listing of the vast number of books in print on the subject.)

Mathematically, the description of the transmitted waveform is relatively straightforward; (1.1) shows a time representation for a transmitted voltage:

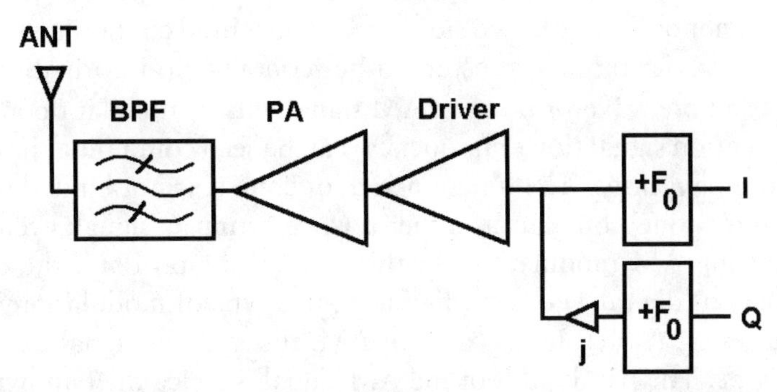

Figure 1.9 Simple transmitter architecture showing upconversion of the baseband I/Q signals, driver and PA amplifier chain, and an output bandpass filter connected to the antenna.

$$v(t) = A(t)\sin[\omega_c t + \phi(t)] \tag{1.1}$$

where ω_c is the carrier frequency, t is the time variable, $A(t)$ is a time-varying amplitude based on the information, and $\phi(t)$ is a time-varying phase term that may also be based on the information to be transmitted. The modulation types to be covered in the following section can be broadly divided into two forms: amplitude modulation and angle modulation.

1.1.3.1 Amplitude Modulation

The simplest form of amplitude modulation is termed, simply enough, amplitude modulation. Mathematically, the amplitude term $A(t)$ of an AM signal can be described by two terms, a modulation index m and an information signal $x(t)$:

$$A(t) = V_0[1 + m \cdot x(t)] \tag{1.2}$$

where both m and $x(t)$ have peak values of 1.0 and V_0 is the peak value of the voltage. The transmitted voltage term can then be written as (assuming $\phi(t) = 0$)

$$v(t) = V_0 \sin[\omega_c t] + mV_0 x(t)\sin[\omega_c t] \tag{1.3}$$

The first term of the transmitted signal carries no information but is often referred to as the carrier signal. The second term generates two sidebands on either side of the carrier frequency, both containing the same information. This can be easily illustrated by assuming a modulation index $m = 1.0$ and a single frequency sinusoid at frequency f_0 for $x(t)$. Using the standard trigonometric identities for the product of two sinusoids yields the following expression for the transmitted voltage waveform:

$$\begin{aligned} v(t) &= V_0 \sin[\omega_c t] + V_0 \sin[\omega_0 t]\sin[\omega_c t] = V_0 \sin[\omega_c t] \\ &+ \frac{V_0}{2}\sin[(\omega_c + \omega_0)t] + \frac{V_0}{2}\sin[(\omega_c - \omega_0)t] \end{aligned} \tag{1.4}$$

where the first term is the carrier and the second and third terms are the upper and lower sidebands (USB and LSB), respectively. This expression for the AM voltage waveform shows that this type of modulation is extremely inefficient in terms of power; not only are the two sidebands transmitted containing the same information, but the carrier term is transmitted but contains no information at all. Figure 1.10 shows a time waveform for an AM signal with a modulation index of $m = 0.5$. The information signal (lower frequency) can be seen containing the carrier signal within a certain *envelope*. The function $x(t)$, of course, is not limited to a single frequency signal (or tone), but can be either a voice or music signal; even digital data can be sent using AM modulation. In this case of digital data, the carrier is essentially turned on (digital 1) or off (digital 0); this type of modulation is termed amplitude shift keying (ASK). In all cases of AM, this type of signal exhibits a time-varying envelope. The bandwidth of the AM signal is twice the bandwidth of the information signal $x(t)$.

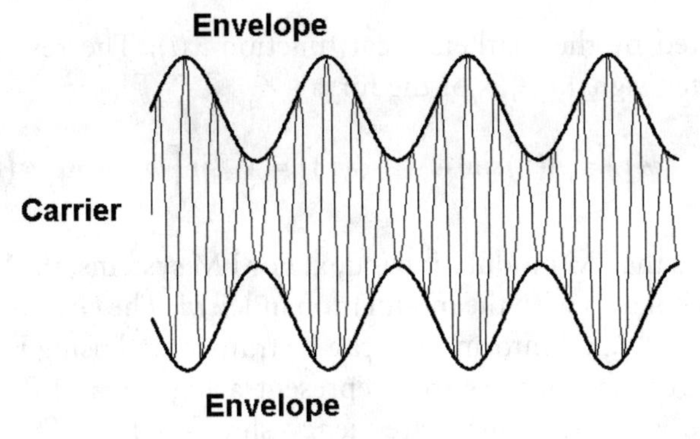

Figure 1.10 Single-tone AM signal, showing both the carrier signal and the information signal defining the transmitted envelope.

A large improvement in efficiency can be achieved if the carrier is removed (or suppressed), leaving only the USB and LSB terms. This form of amplitude modulation is termed double sideband suppressed carrier (DSB-SC). A further efficiency in transmission occurs if only one of the sidebands is transmitted, resulting in single sideband suppressed carrier (SSB-SC) or, simply, single sideband. Only the upper or the lower sideband need be transmitted using this technique; only one is required because they both contain the same information.

Amplitude modulation performance, as the name implies, is tightly bound to the amplitude of the signal. Any noise or fading on the channel will severely compromise the ability to received and detect the signal. In addition, because AM signals are so dependent on the purity of the signal amplitude for the transmission of information, linear amplification is required for these communication systems.[7] AM modulation is currently limited to commercial broadcasting applications because, even though the technique is inefficient in terms of power, simple receivers can be used. SSB is still in widespread use primarily for voice applications. ASK is not often used in RF channels because it is difficult to differentiate between the carrier signal being off due to a logic variable or through fading.

1.1.3.2 Angle Modulation

Far more angle modulation techniques are currently in use than AM techniques. Because only the phase angle is varied, angle modulation techniques are less susceptible to fading than AM; as long as the signal amplitude is large enough to detect the phase, a viable communications link can be established. The following discussion focuses on those angle modulation techniques frequently found in systems that are compatible with CMOS RFIC implementations.

1.1.3.2.1 Frequency Modulation

As the name implies, the FM signal changes or modulates the instantaneous frequency in accordance with the information to be transmitted, which can be

7. A discussion of the concept of linearity follows later in the chapter. A detailed discussion of the design of CMOS linear amplifiers is provided in Chapter 9.

represented by the mathematical function $x(t)$. The resulting constant amplitude transmitted signal $v(t)$ is of the form

$$v(t) = A(t)\sin\left[\omega_c t + \phi(t)\right] = V_0 \sin\left[\omega_c t + m\int x(t)dt\right] \tag{1.5}$$

where m is the FM modulation index. For FM systems, the bandwidth varies not just with $x(t)$ but also with the modulation index m. The envelope of the signal, however, is constant. Digital information can be transmitted using FM by transmitting one of two different frequencies that represent a logic 1 and 0. This digital modulation technique is termed binary frequency shift keying (BFSK) or simply FSK; Figure 1.11(a) shows an example of the time waveform of an FSK signal. Note in the figure the constant amplitude of the envelope. For FM and FSK systems, the linearity requirements for amplification are relaxed since only the frequency and not the amplitude of the signal is important. FM is widely used in the broadcast industry, whereas FSK was initially used for digital communications over telephone lines. A modified version of FSK is currently in use in modern digital systems.

1.1.3.2.2 Phase Modulation

Phase modulation (PM) is currently in wide use in digital communications. The simplest form of PM allows the time waveform $\phi(t)$ to toggle between two states that represent logic 1 and 0. Because the toggled waveform switches between two phases, 0 and 180°, this form of modulation is often referred to as binary phase shift keying; an example of a BPSK waveform is shown in Figure 1.11(b).

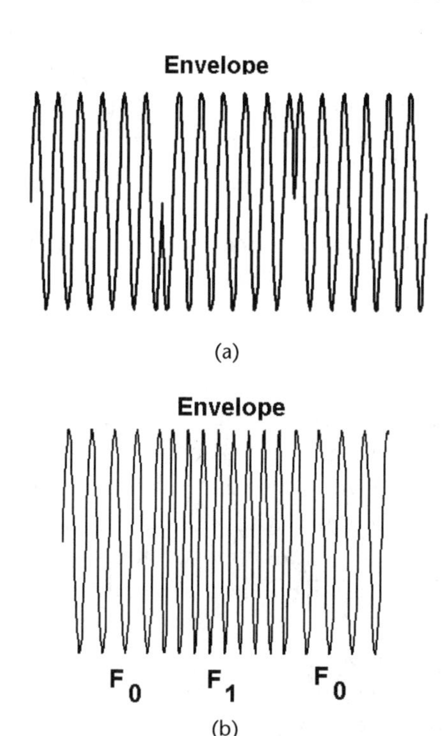

Figure 1.11 (a) Waveform of a FSK signal showing the two distinct frequencies representing the digital modulation. (b) Waveform of a BPSK signal showing the two distinct phases representing the digital modulation. The signal envelope is constant in simple FSK and BPSK.

Variations on the phase shift keying (PSK) theme are utilized in modern digital communications systems and detailed discussions of these PSK modulation techniques are covered in a variety of signals and systems texts [30]. These *M*-ary PSK modulation techniques increase the number of phase changes representing a certain symbol, thereby narrowing the overall bandwidth [31]. Table 1.2 summarizes a few of these PSK modulation techniques.

If we look at the modulation techniques shown in Table 1.2 on a real and imaginary axis, or more generally the Cartesian I and Q axis, the different PSK schemes show up in different locations in the I-Q space as shown in Figure 1.12. These I-Q plots are often referred to as *signal constellation* plots, or simply constellation plots. For a perfect receiver, the detected signals will appear exactly at the phase locations noted on the I-Q axis. However, the channel will adversely affect the signal, and so the resulting uncertainty causes the detected phases to cluster around the phase locations instead of being a distinct point in the I-Q space. As the signal becomes more degraded by the channel, the diameter of the cluster increases (Figure 1.13), increasing the probability of incorrectly decoding the proper phase

Quadrature amplitude modulation (QAM) combines both amplitude and phase modulation techniques to provide a higher data rate for digital communication systems. QAM operation can be described by looking at the signal constellation in Figure 1.12 for 16-QAM (16 states). Note that the data locations in I-Q space are there as in the previously described PM techniques, but now there is an amplitude component as well. (Data location "a" is further away from the origin than "b" and hence has a higher amplitude.) The QAM system must, therefore, be able to transmit both

Table 1.2 Summary of Some PSK Modulation Systems and Their Respective Phase Shifts

Name	*Abbreviation*	*M = 2N*	*Phase Shift (deg)*
Quadrature PSK	QPSK	4	90
8-ary PSK	8-PSK	8	45
16-ary PSK	16-PSK	16	22.5

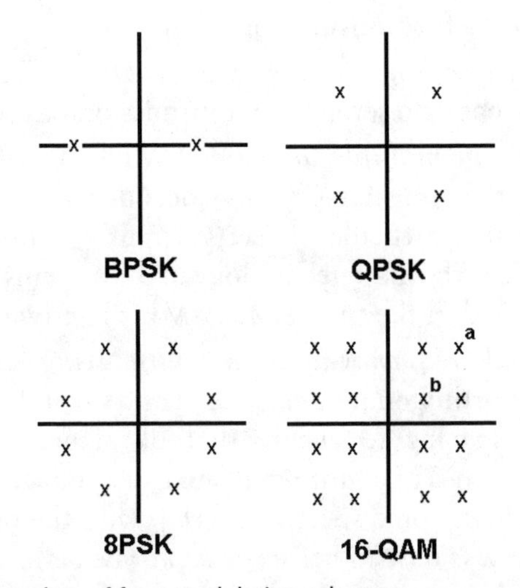

Figure 1.12 Constellation plots of four modulation schemes.

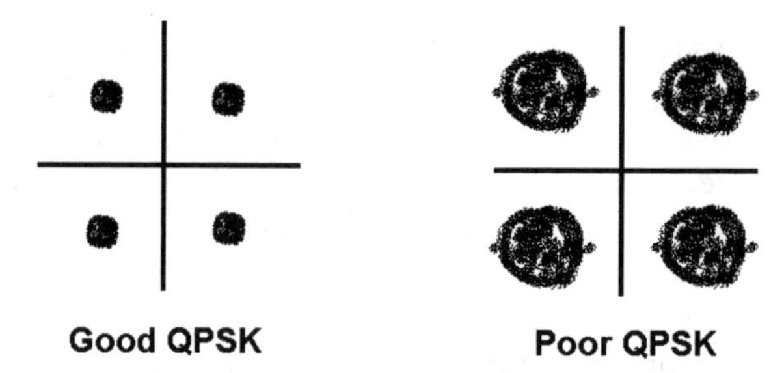

Figure 1.13 Detected signal constellations for a QPSK signal corrupted by the channel.

phase and amplitude information as well as decode the amplitude and phase differences at the receiver. This dual requirement sets strict linearity requirements on the system over that of purely phase modulation (linearity is discussed later in this chapter). Even higher bit rates may be transmitted (32-QAM, 64-QAM, 128-QAM, 256-QAM), although with each doubling of the number of data locations in the signal constellation, the data locations become more dense and therefore more susceptible to noise and other device and channel characteristics. From Figure 1.12, note that QPSK is a special case of QAM (4-QAM) with one data location per quadrant. (No amplitude modulation component is required for this modulation scheme.)

1.1.4 Multiple Users

The previous section described how the various bits that represent a particular piece of information are coded but not how a number of users, each with his or her own pieces of information, are accommodated in a particular system. Many techniques and variations on traditional techniques are in use; this section seeks to introduce the terms so the reader is familiar with the various technologies and their broad differences. The techniques that perform the user sharing operation are termed *multiple access techniques* and are broadly defined in one of four ways:

- *Time-division multiple access (TDMA):* The various users have specific time slots reserved for transmission of their data.
- *Frequency-division multiple access (FDMA):* The various users have specific frequency bands reserved for transmission of their data.
- *Code-division multiple access (CDMA):* All of the users' data is uniquely coded on the transmit side and decoded and separated on the receive side based on the particular code. Currently, wideband CDMA (WCDMA) is being used as the main technology in 3G systems and requires 5 MHz of bandwidth instead of the smaller 1.25-MHz bandwidth for traditional CDMA.
- *Orthogonal frequency-division multiplexing (OFDM):* OFDM extends the concept of orthogonal signals (i.e., I and Q) and applies it to FDMA. Instead of spectrum spacing to ensure that no signal spectra inadvertently overlap (through the use of *guard bands*) to cause adjacent channel interference (ACI), OFDM allows some spectral overlap with the orthogonality helping to minimize what was termed before as ACI. The main concept behind OFDM is that since low data rates tend to be less sensitive to changes in the communications

channel, it makes some sense to send a number of low data rate users in parallel rather than sending one high data rate signal that is more susceptible to the channel characteristics. The system bundles the various user data together and performs an inverse fast Fourier transform (IFFT) prior to transmission. Transmission is then performed using one of the classic modulation techniques (PSK, QPSK, etc.). The receiver then performs a fast Fourier transform (FFT) to recover the original user data. High-speed DSP is necessary for OFDM.

1.2 Review of System Fundamentals

The design of CMOS RFICs (or, for that matter, any RF system) is based on a thorough understanding of the various specifications the system must adhere to in order to achieve the desired performance. This section is a review of some of the fundamental system performance metrics often encountered when specifying RF systems with CMOS solutions.

1.2.1 System Gain

The gain of a single stage or a complex array of cascaded systems is usually referred to in terms of an output quantity (voltage, current, power) with respect to some input quantity (voltage, current, power). This array of output and input quantities gives rise to a number of definitions of system gain terms as shown in Table 1.3 and illustrated in Figure 1.14. The first three are the gain terms most often seen in RF systems, although the RF engineer should be conversant with all five. All gains except for the power gain may exhibit both magnitude and phase components and hence are complex quantities in general.

The remainder of the discussion will use the voltage gain for illustration purposes; the other terms can be derived by extending the analysis. Two or more circuit blocks in cascade have an overall gain that can be derived by noting that the output

Table 1.3 System Gain and Transfer Function Definitions

Term	*Gain/Definition*	*Units*
Voltage gain	V_{out}/V_{in}	V/V (or dimensionless)
Current gain	I_{out}/I_{in}	A/A (or dimensionless)
Power gain	P_{out}/P_{in}	W/W (or dimensionless)
Transimpedance	V_{out}/I_{in}	Ohms
Transconductance	I_{out}/V_{in}	Siemens

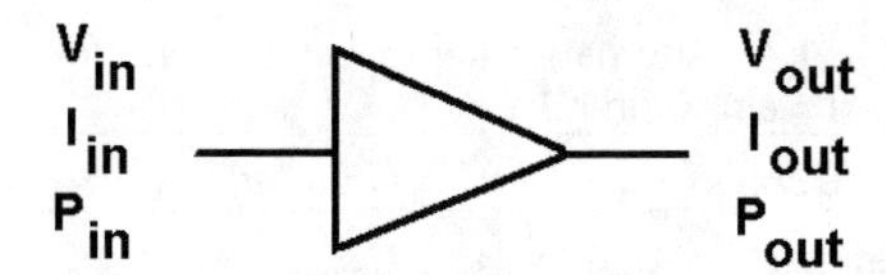

Figure 1.14 Input-output definitions for a general system block.

of the first stage is the input of the second, and so using chain mathematics, the total system gain is simply the product of the individual block gains:

$$G_{tot} = \frac{V_{out-2}}{V_{in-1}} = \frac{V_{out-2}}{V_{in-2}} \times \frac{V_{in-2\,(=out-1)}}{V_{in-1}} = G_1 G_2 \tag{1.6}$$

Additional stages merely require multiplication of the individual stage gains. The voltage gain may also be written in terms of decibels as

$$\begin{aligned} G_{tot-dB} &= 20\log(G_{tot}) = 20\log(G_1 G_2) \\ &= 20\log(G_1) + 20\log(G_2) = G_{1-dB} + G_{2-dB} \end{aligned} \tag{1.7}$$

indicating that the total gain (in dB) is simply the sum of the individual gains (also in dB). For power gain, the leading factor of 20 is replaced by 10 to account for the fact that the power is the square of the voltage (or current). The use of dB in estimating or describing stage gain allows for quick calculation of multiple stages of gain since only addition is required.

Implicit in (1.6) (or 1.7) is that the voltage gains are defined as a ratio of V/V. In RF systems and system design, voltage levels are frequently at the microvolt or millivolt level, and so the system gain parameter is then referenced to those voltages levels. The unit dB is then modified to show the voltage used as the reference: dBV (1V reference), dBmV (1 mV reference), and dBμV (1 μV reference) are possible variations of the dB unit. For example, a 2.0-mV signal can be equally represented as +6 dBmV, +66 dBμV, or –54 dBV. Table 1.4 shows the conversion between some more widely used measures of dB for voltage, current, and power.

In CMOS RFIC design, the on-chip circuits are designed primarily for specific voltages or currents, so the main focus will be on the voltage and current gain terms. The power gain form does come into use, however, when discussing noise characteristics and power at the input and output terminals.

1.2.2 System Noise Figure

1.2.2.1 Resistive Noise Model (Noise Figure, Temperature)

The random Brownian motion of charge carriers causes minute instantaneous fluctuations in the current flow in any current-carrying structure, even in dc supplies. In a resistor of value R, these current fluctuations can be mathematically described by some variation Δi about the mean current I. These current fluctuations, in turn, create voltage fluctuations across R and give rise to so-called *thermal resistive noise* or

Table 1.4 Alternative Representations of Voltage, Current, or Power Gain, G, in dB

Gain Term	*Base Unit*	*Gain (m)*	*Gain (μ)*
Voltage	G_{dBV}	$G_{dBmV} = 60 + G_{dBV}$	$G_{dB\mu V} = 120 + G_{dBV}$
Current	G_{dBA}	$G_{dBmA} = 60 + G_{dBA}$	$G_{dB\mu A} = 120 + G_{dBA}$
Power	G_{dBW}	$G_{dBm} = 30 + G_{dBW}$	$G_{dB\mu} = 60 + G_{dBW}$

just *resistive noise*. The noise power has been shown by Johnson and Nyquist to be linearly dependent on the physical temperature T of the resistor:

$$P = \alpha T = \Delta\bar{i}^2 R; \quad \Delta\bar{i}^2 = \alpha T/R = \alpha GT \tag{1.8}$$

where Δi is the rms value of the current fluctuations. The quantity α was found to be a function of the noise measurement bandwidth B and the Boltzmann constant k, so that the mean square noise current is written in what is now its traditional form:

$$\Delta\bar{i}^2 = 4kTGB = 4kTB/R \tag{1.9}$$

This form of noise current has been found to be very useful in circuit and system noise analysis because it provides an easy method of injecting a noise current into a circuit and calculating the circuit response to the noise input. The form of this noise current source is an ideal current source with a noiseless output resistance R as shown in Figure 1.15(a). The noise power available from this circuit is computed assuming the noise current source is driving a matched resistive load, that is,

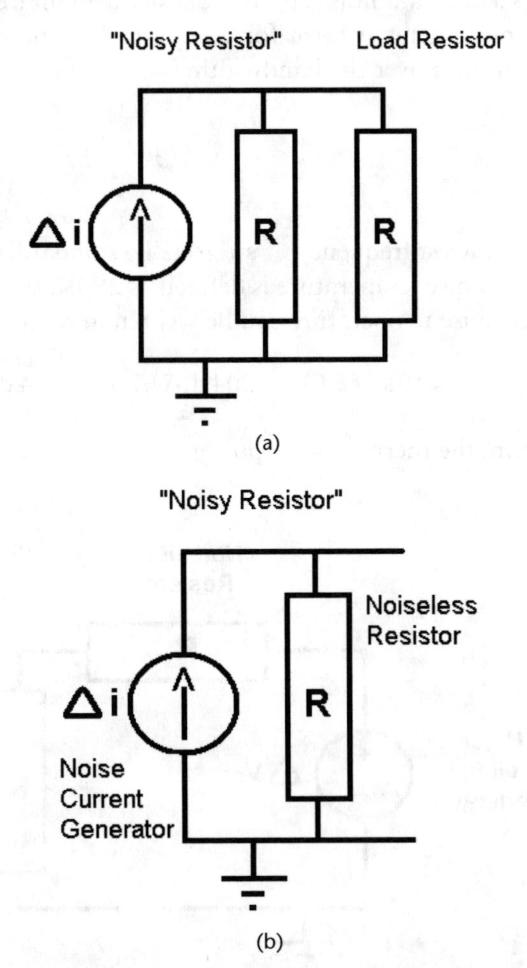

Figure 1.15 (a) The noisy resistor can be thought of as a noiseless resistor and a noise current generator. (b) The available power is computed using a matched load of a noiseless resistor of value *R*.

another noiseless resistor R [Figure 1.15(b)]. The noise current divides equally between the two resistors, and so the corresponding available noise power (in watts) in the resistive load is

$$P_N = \left(\frac{\Delta\bar{i}}{2}\right)^2 R = kTB \tag{1.10}$$

A Thevinen equivalent circuit for the noise source can also be drawn as shown in Figure 1.16, where $\Delta\bar{v}^2$ and the available noise power $P_N = kTB$ is the same as before. Both resistors R are considered noiseless in this model.

Equation (1.10) gives the total available thermal noise power over bandwidth B. A *noise power spectral density* S_N can be defined as the noise power per bandwidth (in dimensions of W/Hz):

$$S_N(f) = \frac{P_N}{B} = kT \tag{1.11}$$

which shows a constant noise spectral density over all frequencies at a specific temperature. A more general form for computing the noise power in terms of $S_N(f)$ requires integration over the bandwidth:

$$P_N = \int_{f_0}^{f_0+B} S_N(f)df \tag{1.12}$$

where f_0 is the lowest frequency of interest. To standardize noise measurements, the IEEE reference noise temperature is defined as 290K, so the noise spectral density at this reference noise temperature can be written in terms of dBm as

$$S_N(f) = 10\log(kT) = -204 \text{ dBW/Hz} = -174 \text{ dBm/Hz} \tag{1.13}$$

With this form, the thermal noise power over any bandwidth can be computed as

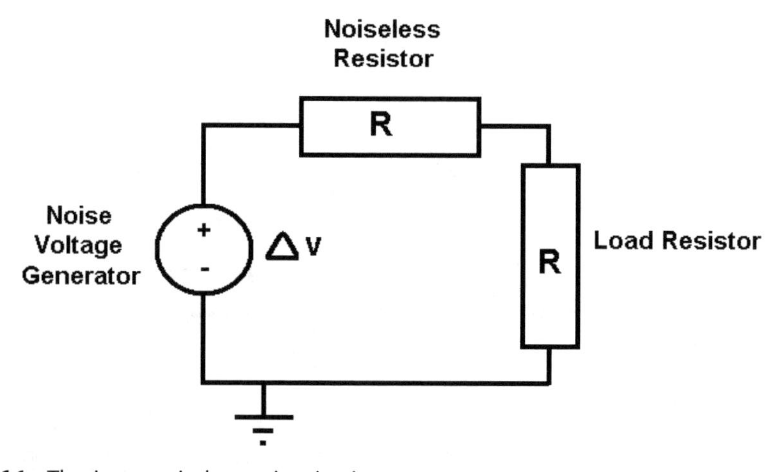

Figure 1.16 Thevinen equivalent noise circuit.

$$P_N(f) = 10\log B_{Hz} - 174\ \text{dBm} \tag{1.14}$$

For example, the thermal noise power measured over a 1.0-MHz bandwidth is –114 dBm or 4.07 *f*W. This noise power corresponds to a 0.63-μV thermal noise voltage if dropped across a 50-Ω resistor ($P = V2/2R$). Any desired signal received and dropped across this resistor will not be detectable (or is "in the mud") unless it is above this 0.63-μV thermal noise value.

1.2.2.2 Minimum Detectable (or Discernible) Signal

The thermal noise power P_N over a bandwidth B will be dissipated in a load even in the absence of any other signal applied to the system. Any signal applied to the system has to rise above this thermal noise floor to be observed or detected.[8] The thermal noise floor then gives the lower limit on the signal level or the minimum detectable signal (MDS). In the example in the previous section, the MDS level would have been just above the 0.63-μV thermal noise floor. As the input signal increases in value above the MDS, the ratio of the signal power P_{signal} with respect to the noise power P_{noise} increases (Figure 1.17). The *signal-to-noise ratio* (SNR) is defined as the ratio of these two powers, or the difference between those two powers if they are provided in dB terms:

$$\text{SNR} = \frac{P_{signal}}{P_{noise}};\ \text{SNR}_{dB} = P_{signal-dBx} - P_{noise-dBx} \tag{1.15}$$

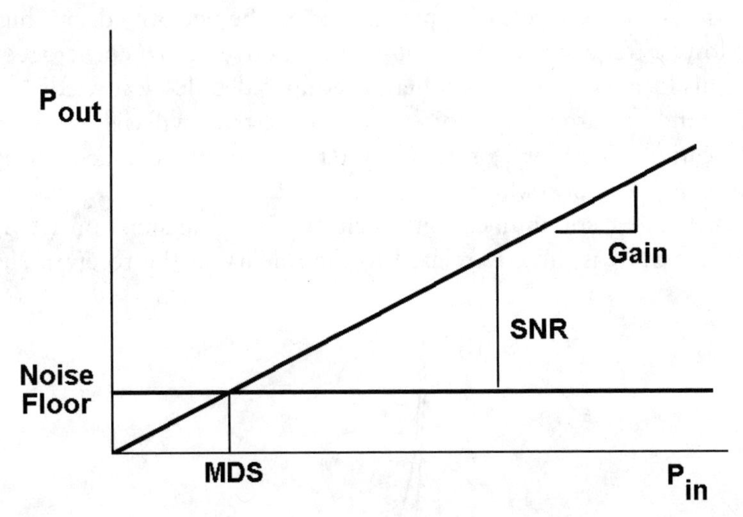

Figure 1.17 The MDS can be defined as the input signal level that provides an output equal to the noise level.

8. In some modulation techniques such as spread spectrum, the desired signal is detectable even though the received signal in the bandwidth B is below the noise floor. The spread spectrum technique uses advanced DSP techniques to effectively pull the signal above the noise floor limit and hence make the information retrievable.

Note that the same reference power must be used for the signal and the noise powers ($x = W, m, \mu$). The SNR for a signal whose power level equals the MDS is 1 or 0 dB.

An additional source of noise occurs in CMOS devices; *1/f* or flicker noise. This noise source comes from the relatively slow generation and recombination of carriers at the various interface layers in the MOS devices, primarily the insulator/semiconductor interface. As the term *1/f* implies, this noise source is highest at low frequencies and rolls off as *1/f*, eventually equaling the thermal noise value at the *1/f* corner frequency, $f_{1/f}$ (Figure 1.18).

Above this corner frequency, the thermal noise dominates the overall noise response. This corner frequency is on the order of a few megahertz for typical CMOS devices but is increasing as the feature size decreases because of increased surface state effects [22]. Because the thermal noise spectral density $S_N(f)$ is not constant with frequency, (1.12) is used to determine the overall thermal noise power if the system bandwidth contains the *1/f* corner frequency. Many CMOS RFIC systems operate at hundreds of megahertz or higher, and so at first glance it appears that *1/f* noise would have little bearing on these systems since they operate at frequencies much higher than the *1/f* corner frequency. The problem with *1/f* noise in RF systems occurs because most of these systems rarely process the signals in the hundreds of megahertz range; rather, these high-frequency signals are shifted or downconverted to lower frequencies for easier signal processing. The trade-off here, of course, is that now the downconverted signal (for easier signal processing) may appear at frequencies where *1/f* noise is significant, in which case the SNR is degraded since the overall noise is higher than the thermal noise value alone. The *1/f* noise contribution may be minimized in these systems by downconverting the signal to a frequency where easier processing can be performed, but high enough so that the lowest frequency in the bandwidth is above the *1/f* corner frequency. Determining this ideal conversion frequency requires detailed knowledge of the CMOS process (and its attendant noise properties) that will be used in the final RFIC fabrication. (In other words, the system and design engineers need to talk with the fabrication engineers.)

In analog communications such as voice transmission (AM, SSB, for example), the SNR is directly related to the fidelity of the received signal and how the

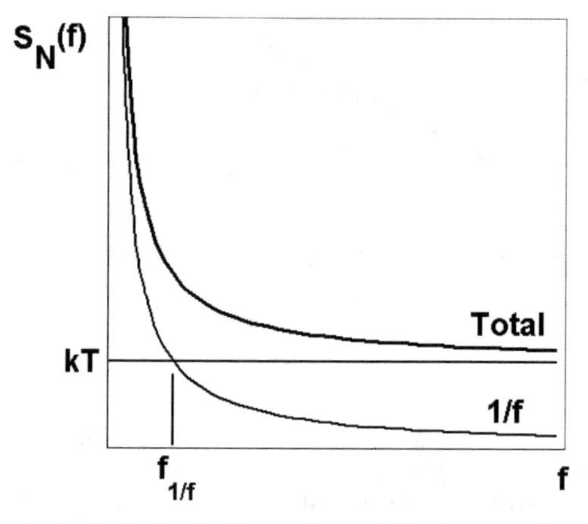

Figure 1.18 The total noise includes both thermal and *1/f* noise contributions.

information is perceived by the end user (a person or a speech recognition unit, for example). In digital communications, the SNR has a direct bearing on the rate at which data can be effectively transmitted. Shannon first formulated the *channel capacity*, or an upper bound on the number of bits occurring per unit time (the *bit rate*, BR), as a function of signal bandwidth B and SNR [32]:

$$\mathrm{BR} = B\log_2(1+\mathrm{SNR}) \tag{1.16}$$

The *symbol* or *baud rate* is a function of BR and the number of bits used to encode the symbol (N):

$$\mathrm{SR} = \frac{BR}{N} = \frac{B}{N}\log_2(1+\mathrm{SNR}) \tag{1.17}$$

These digital bits are not transmitted error free, and so the number of bits received in error divided by the total number of bits transmitted per unit time is referred to as the bit error rate (BER). The BER is strongly dependent on the type of modulation used but is still a function of the SNR. Mathematically, the BER is computed using the complementary error function (erfc). While an area of extensive research, Figure 1.19 shows representative BER values versus SNR for several PSK modulation techniques. Table 1.5 shows the approximations used in this calculation.

An improved SNR increases the bit rate and reduces the BER in digital modulation schemes for a given bandwidth. In terms of circuit design and the CMOS RFIC designer, improving the SNR requires either that the noise be reduced or the signal strength increased. For a given bandwidth, there is the thermal noise floor (kTB)

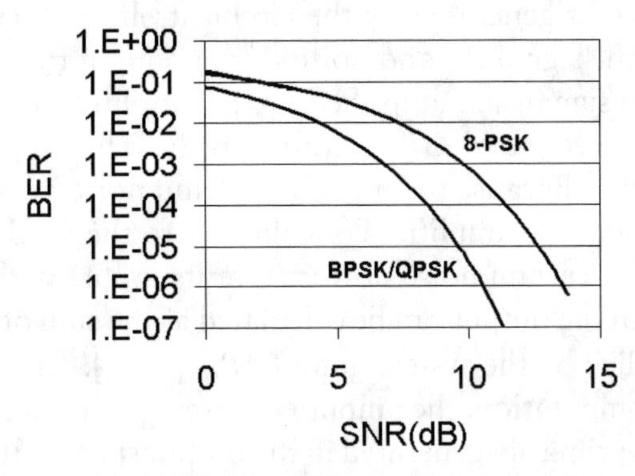

Figure 1.19 BER as a function of SNR for four different digital modulation schemes.

Table 1.5 Table of Various PSK BERs as a Function of SNR

Digital Modulation Scheme	*BER Equation*
BPSK	$0.5\mathrm{erfc}(\mathrm{SNR}^{1/2})$
QPSK	$0.5\mathrm{erfc}(\mathrm{SNR}^{1/2})$
8-ary PSK	$0.5\mathrm{erfc}[\mathrm{SNR}^{1/2}\cdot \sin(2\pi/8)]$

After: [32].

that introduces the minimum amount of noise in the system and that cannot be reduced; for increases in SNR, the received signal level then must be increased by increasing the transmitter power, reducing the channel distance, or improving the antenna efficiency. (See later discussion on link budget calculations for full details on these interactions.)

1.2.2.3 Amplifier Noise Model—Input Referenced Noise

The previous discussion focused on the noise generated by a resistor and divided this noise into an equivalent circuit composed of a noise generator and a noiseless resistor. Although incredibly interesting and powerful for introducing the concept of noise, the concept is a bit simplistic to be of much use in communication system analysis and design. Note, however, that the noise power and noise spectral density are independent of the value of the resistor, so these terms may be extended for use with any element or system block in an RFIC design. In this extended case, the noise temperature T is then defined as a general or equivalent noise temperature T_e and the IEEE noise reference temperature of 290K can be defined with the symbol T_{ref}:

$$P_N = kT_e B; \; P_{ref} = kT_{ref} B \tag{1.18}$$

Any RFIC—or for that matter, any circuit anywhere—will be constructed with elements that generate noise. Even theoretically noiseless reactive components such as inductors and capacitors actually have some losses associated with them, with the losses generating the noise. Because any circuit generates noise on its own, a signal with additive noise at the input will find its way to the output but with additional noise due to that generated by the circuit itself. This effect is illustrated in Figure 1.20, where the signal P_{sig} and noise $P_{N\text{-in}}$ are input to an amplifying state with gain G. These two signals are amplified by the amplifier so that at the output the signal and noise are GP_{sig} and $GP_{N\text{-in}}$, respectively. The amplifier also has its own self-generated noise. Because the input signal and noise are multiplied by G, the easiest way to compute the amplifier contribution to the total noise is to assume that the amplifier has a certain noise power P_A at its input. With this assumption, the total noise power at the output can be calculated as the sum of the two noise powers at the input multiplied by the system gain: $G[P_{N\text{-in}} + P_A]$.

In this computation, the amplifier noise P_A is termed *input referred noise*—even if the noise is primarily generated in the output stage of the amplifier. By referring all stage noise to the input, calculations of overall system noise are easily performed. With the addition of the amplifier noise contribution, two SNR values can be computed, one at the input and the other at the output:

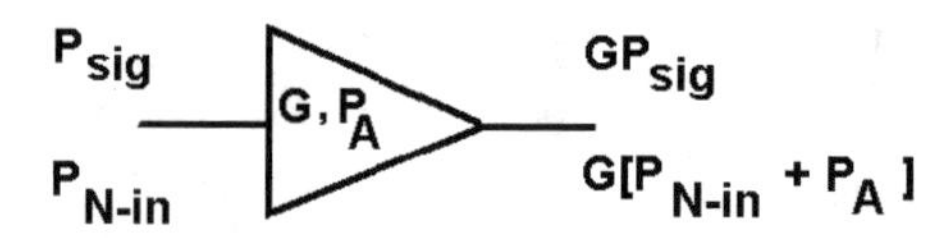

Figure 1.20 The amplifier multiplies both the signal and the noise by the gain and also adds self-generated noise.

$$\text{SNR}_{\text{in}} = \frac{P_{sig}}{P_{N\text{-in}}};\ \text{SNR}_{\text{out}} = \frac{GP_{sig}}{G[P_{N\text{-in}} + P_A]} = \frac{P_{sig}}{[P_{N\text{-in}} + P_A]} \tag{1.19}$$

The expression for output SNR, SNR_{out}, can be written in terms of input SNR, SNR_{in}, by rearranging terms and introducing F, the *noise figure*:

$$\text{SNR}_{\text{out}} = \frac{P_{sig}}{[P_{N\text{-in}} + P_A]} = \frac{P_{sig}/P_{N\text{-in}}}{1 + P_A/P_{N\text{-in}}} = \frac{\text{SNR}_{\text{in}}}{F};\ F = 1 + \frac{T_A}{T_{N\text{-in}}} \tag{1.20}$$

where T_A and $T_{N\text{-in}}$ are the *amplifier equivalent noise temperature* and *input noise temperature*, respectively. Further assuming that the input noise conforms to the IEEE standard allows F to be written as

$$F = 1 + \frac{T_A}{T_{ref}} \tag{1.21}$$

Equation (1.21) shows that noise figure is defined for a noise input temperature of 290K. In some applications where the antenna is looking at the sky, the sky noise temperature is very low and therefore noise figure F cannot be used for system noise analysis; however, equivalent noise temperature can still be used. Noise figure F is always greater than unity and shows that the output SNR is *always* degraded by the noise figure of the system. Even though the signals may be larger in amplitude (GP_{sig}), the amplifier also multiplies both the input noise and its own noise by its gain, thereby enhancing the noise at the output and reducing the SNR. The noise figure F may also be placed in dB terms, which shows the degradation of SNR in terms of dB:

$$F_{dB} = 10\log(F);\ \text{SNR}_{\text{out-}dB} = \text{SNR}_{\text{in-}dB} - F_{dB} \tag{1.22}$$

For example, a system with a 1.0-dB noise figure and an input SNR of 12 dB would exhibit an output SNR of 11 dB because of the noise figure. If only thermal noise is assumed, the output SNR in terms of the input signal (in dBm) can be written by combining (1.14) and (1.22):

$$\begin{aligned} \text{SNR}_{\text{out}} &= \frac{\text{SNR}_{\text{in}}}{F} \Rightarrow \text{SNR}_{\text{out}} = \text{SNR}_{\text{in}} - F_{dB} = P_{\text{in-}sig} - P_{N\text{-in}} - F_{dB}\ \text{dB} \\ \text{SNR}_{\text{out}} &= P_{\text{in-}sig} - F_{dB} - 10\log B_{Hz} + 174\ \text{dB} \end{aligned} \tag{1.23}$$

where the input signal power is in dBm. Expanding the SNR_{out} as the difference between the output signal and noise powers, and the gain G as the difference between the output and input powers (all in dB), (1.23) may be written as:

$$(P_{\text{out-}sig} - P_{\text{out-}noise}) - P_{\text{in-}sig} = -F_{dB} - 10\log B_{Hz} + 174\ \text{dBm} = G_{dB} - P_{\text{out-}noise} \tag{1.24}$$

The output noise term $P_{\text{out-}noise}$ is the output noise floor:

$$P_{\text{out-}noise} = P_{floor} = G_{dB} + F_{dB} + 10\log B_{Hz} - 174\ \text{dBm} \tag{1.25}$$

CMOS RFIC circuits are not always gain blocks but may actually have losses associated with them. Assuming that the loss is purely resistive loss, the resistive loss noise introduced is assumed to occur with an equivalent noise temperature equal to the reference temperature T_{ref}, so that the noise figure F is identical to the circuit loss L [33]:

$$F = L; \;\; F_{dB} = L_{dB} \tag{1.26}$$

Example 1.1 Determine the output noise floor for a 1.0-dB noise figure amplifier ($G = 10$ dB) with a bandwidth of 1 MHz. Determine the corresponding noise voltage dropped across a 50Ω load.

The noise floor can be calculated from (1.25):

$$\begin{aligned} P_{floor} &= G_{dB} + F_{dB} + 10\log B_{Hz} - 174\text{ dBm} = 10 + 1 + 60 - 174 = -103\text{ dBm} \\ &= -133\text{ dBW} = 5 \cdot 10^{-14}\text{ W} \end{aligned}$$

This thermal noise power, if dropped across a 50Ω resistor, gives rise to a noise voltage generated and dropped across the resistance:

$$P_{floor} = \frac{V^2}{2R} \Rightarrow \sqrt{100P_{floor}} = 2.24\ \mu V$$

The ability to provide input referenced noise sources for computation of an overall stage noise figure requires that the input thermal noise be described using both voltage and current sources, Δi or Δv (Figure 1.21). Using this more general form, the noise figure F can be written in terms of the expected value of the two noise sources [30]:

$$F = 1 + \frac{\overline{\left(\Delta v + R_g \Delta i\right)^2}}{4kTR_g} = 1 + \frac{\Delta\bar{v}^2 + R_g^2 \Delta\bar{i}^2 + 2R_g \overline{\Delta v \Delta i}}{4kTR_g} \tag{1.27}$$

where the overbar denotes the expected value and the third term denotes the correlation between the thermal voltage and current noise sources.

If the two noise sources are uncorrelated, this term is zero and the noise figure is simply the sum of the two noise power sources. The level of correlation between the

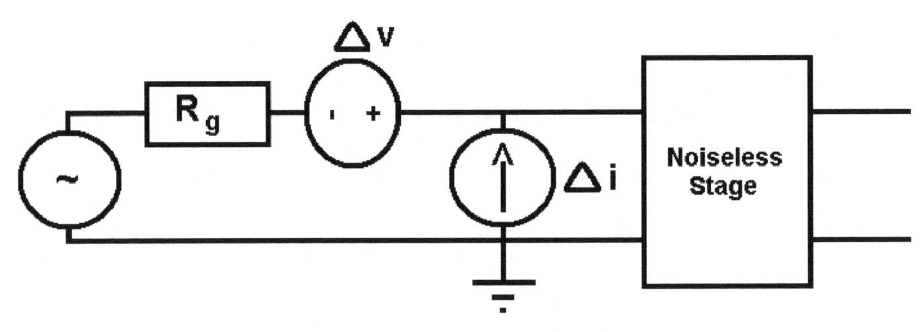

Figure 1.21 There are both voltage and current input referred noise sources.

Δi and Δv is determined after the system block noise contributions are input referred.

Example 1.2 Consider a simple MOSFET ac equivalent circuit where the noise contribution occurs at the output in the form of a drain thermal noise current Δi_d (Figure 1.22). Determine the input referred noise for this circuit.

The two input referred noise sources Δi and Δv are found by individually determining their relationships to Δi_d through the use of superposition. Setting $\Delta i = 0$ shows that Δv is identical to v_{gs}, and hence is related to Δi_d by the device transconductance, g_m:

$$\Delta i_d = g_m \Delta v \Rightarrow \Delta v = \frac{\Delta i_d}{g_m} \tag{1.28}$$

Setting $\Delta v = 0$ shows that Δi is identical to the current through capacitance C_{gs} and since Δ_{id} is related to v_{gs} through the transconductance, the input thermal noise current source value can be written as

$$\Delta i = j\omega C_{gs} v_{gs} = \frac{j\omega C_{gs}}{g_m} \Delta i_d \tag{1.29}$$

In this example, Δi and Δv are correlated noise sources since Δi can be written in terms of Δv:

$$\Delta i = \frac{j\omega C_{gs}}{g_m} \Delta i_d = j\omega C_{gs} \Delta v \tag{1.30}$$

with a correlation coefficient of $j\omega C_{gs}$. Both noise sources are required for a full accounting of the noise in this device.

1.2.2.4 Cascade System Noise

CMOS RFIC systems are, of course, not limited to a single stage or circuit, but rather have a number of stages, each providing gain (or loss) and their own noise

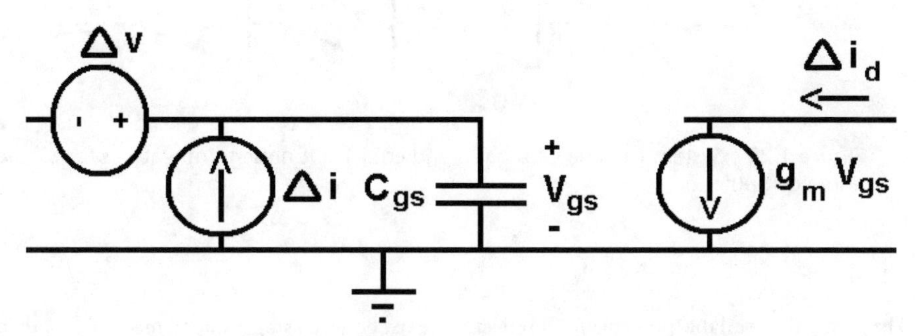

Figure 1.22 Schematic for calculating the input referred noise due to a drain thermal noise current Δi_d.

contributions. In each stage, the stage block is defined with a power gain[9] G_i and an equivalent input referred noise temperature T_i (Figure 1.23).

A thermal noise source is placed at the input, set to the reference noise temperature T_{ref}. At the output of the first stage, assuming a noise bandwidth B, the thermal noise power can be written as

$$P_{N-1} = kBG_1\left(T_1 + T_{ref}\right) \tag{1.31}$$

This noise then feeds the second stage, with the output of stage 2 multiplying P_{N-1} by G_2 and also contributing its own noise:

$$P_{N-2} = G_2 P_{N-1} + kBG_2 T_2 = kBG_2 G_1\left(T_1 + T_{ref}\right) + kBG_2 T_2 \tag{1.32}$$

In like fashion, the noise power at the output of the third stage can be written as

$$P_{N-3} = kBG_1 G_2 G_3\left(T_1 + T_{ref}\right) + kBG_2 G_3 T_2 + kBG_3 T_3 \tag{1.33}$$

In keeping with the concept of equivalent noise temperatures, the output noise power P_{N-3} can be written assuming a noise source with noise temperature T_{ref} is placed at the input stage of an amplifier with overall equivalent noise temperature T_e and gain $G_1G_2G_3$, in which case P_{N-3} can be written as

$$P_{N-3} = kBG_1 G_2 G_3\left(T_e + T_{ref}\right) \tag{1.34}$$

Equating (1.33) and (1.34) provides an expression for the equivalent noise temperature and noise figure F for the entire three-stage system:

$$T_e = T_1 + \frac{T_2}{G_1} + \frac{T_3}{G_1 G_2}; \quad F = 1 + \frac{T_e}{T_{ref}} \tag{1.35}$$

For more than three stages, the noise temperature of the succeeding stages is divided by the gains of the preceding stages. (1.35) shows that, regardless of the total

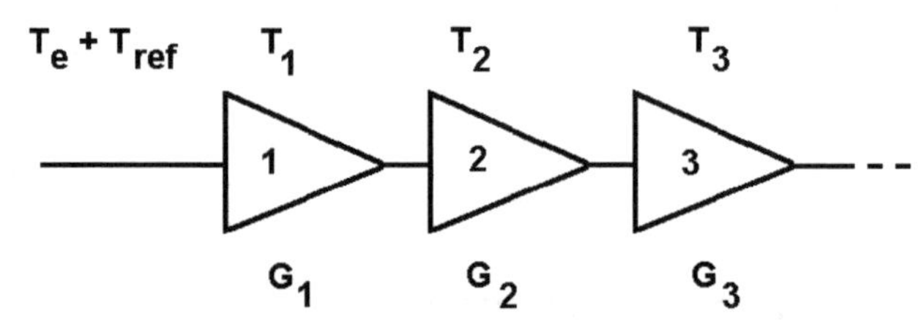

Figure 1.23 System cascade with gains and equivalent noise temperatures for cascade noise figure computation.

9. The gain is the available power gain. If reflections exist between stages due to mismatches, then the available gain should be used. The definition for the various gains is beyond the scope of this discussion, but the inquisitive reader may find detailed explanations elsewhere (for example, [33]).

number of system stages, it is the first stage and its noise properties that have the greatest impact on the overall system noise. Therefore, the first stage should be as low noise and high gain as possible to keep the noise figure of the system as low as possible. This explains why the first stage of RF systems is the low-noise amplifier (LNA). In some systems, filtering or control elements (band reject filters or antenna switches, for example) often precede the LNA; these components should exhibit low losses within the bandwidth of interest so that their noise contributions (which are directly related to the losses as discussed earlier) are minimized and do not unduly degrade the system noise figure.

1.2.2.5 Noise Figure Example

Determine the noise figure (in dB) of the three-stage system shown in Figure 1.24. Note that the first stage is a transmission line that exhibits a *loss* of 3 dB, and hence its noise figure is also 3 dB.

The first step in the calculation is to change all of the given information into absolute terms. Gain and noise information for the three stages is given in Table 1.6 along with corresponding computations. Note that noise may be given in equivalent noise temperature or noise figure and so the RF engineer should be able to handle both terms in any given problem.

Using (1.35), the overall system equivalent noise temperature and noise figure can then be computed:

$$T_e = T_1 + \frac{T_2}{G_1} + \frac{T_3}{G_1 G_2} = 290 + \frac{100}{0.5} + \frac{170}{0.5 \cdot 10} = 524\text{K}$$

$$F = 1 + \frac{T_e}{T_{ref}} = 1 + \frac{524}{290} = 2.8 \Rightarrow 4.48\text{ dB}$$

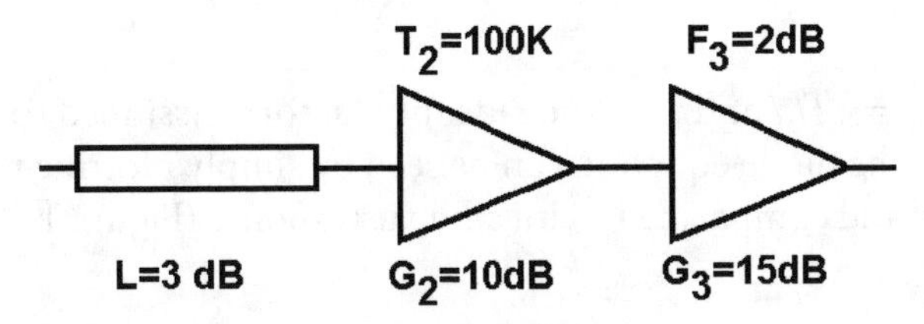

Figure 1.24 Block diagram for the noise figure example.

Table 1.6 Gain and Noise Terms for Computation of Overall System Noise Figure

Stage	*Gain Equation*	*Gain*	*Noise Equation*	*Noise Temperature (K)*
1	$10^{-L/10} = 10^{-3/10}$	0.5	$290\left(10^{F_{dB}/10} - 1\right) = 290\left(10^{3/10} - 1\right)$	290
2	$10^{G/10} = 10^{10/10}$	10	Given	100
3	$10^{G/10} = 10^{15/10}$	31.6	$290\left(10^{F_{dB}/10} - 1\right) = 290\left(10^{2/10} - 1\right)$	170

In this example, the noise figure is quite high because the first stage of the system is a lossy element, the transmission line. By interchanging the transmission line and first amplifier stage, the noise figure is reduced by approximately 2.5 dB, a significant improvement in SNR that shows the importance of putting the LNA ahead of lossy elements:

$$T_e = T_1 + \frac{T_2}{G_1} + \frac{T_3}{G_1 G_2} = 100 + \frac{290}{10} + \frac{170}{0.5 \cdot 10} = 163\text{K}$$

$$F = 1 + \frac{T_e}{T_{ref}} = 1 + \frac{163}{290} = 1.56 \Rightarrow 1.94 \text{ dB}$$

1.2.3 System Nonlinearities

In an ideal system, an input composed of the sum of two or more signals of the same or different frequency will yield an output that is composed of the same sum times the system response. Mathematically, this relationship may be expressed in the frequency domain as:

$$Y(f) = H(f)\left[X_1(f) + X_2(f) + \ldots\right] \tag{1.36}$$

where X_i is the input stimuli, Y is the output, and H is the system response. In all systems, CMOS RFIC and otherwise, most of the system blocks employ at least one active device. All active devices exhibit nonideal behavior in the form of *nonlinear* responses to the input stimulus. This nonlinear behavior generates additional signals in response to the input, with the new or *spurious* signals occurring at frequencies different than those of the stimulus. As an example, consider the simplest nonlinear active device, the diode. Here, the current through the diode is a function of the voltage across the terminals, given by the well-known Shockley diode equation:

$$I_d(V) = I_0\left(e^{V/V_T} - 1\right) \tag{1.37}$$

where $V_T = kT/q$ and the diode ideality factor is assumed to be unity. The nonlinear nature of the diode equation can be seen by simply plotting the current-voltage characteristic and comparing it with a linear response (Figure 1.25). An alternate form of

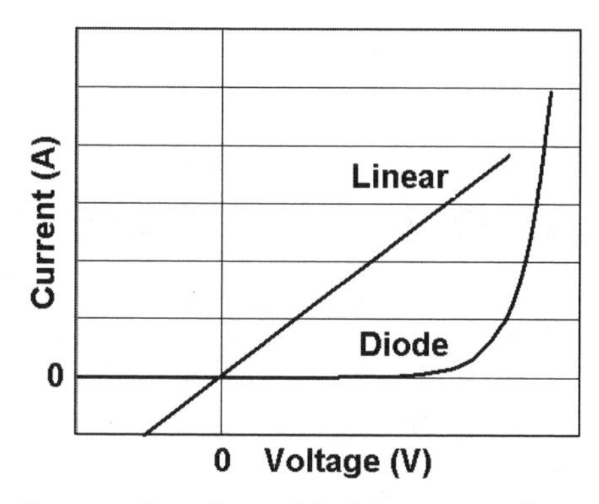

Figure 1.25 Example of a linear and nonlinear (diode) current voltage response.

the diode equation can be written by using a Taylor series expansion of the diode equation:

$$I_d(V) = \sum_{i=1}^{\infty} \frac{I_0}{i!(V_T^i)} V^i = \sum_{i=1}^{\infty} g_i V^i = g_1 V + g_2 V^2 + g_3 V^3 + \ldots \quad (1.38)$$

The first part of this equation is a general form that can be used to mathematically describe any signal, linear or nonlinear. For the linear case, $g_i = 0$ for $i > 2$. For the nonlinear case, all the powers of V and their corresponding coefficients must be included for an accurate representation of the signal.

1.2.3.1 Order of Nonlinearities

An estimate of the spurious signals generated by a nonlinear device described mathematically by (1.38) can be seen if a sum of two signals of different amplitude and frequency (the so-called *two-tone test*) is applied as voltage:

$$V = A\cos(\omega_a t) + B\cos(\omega_b t) \quad (1.39)$$

The current response is then written in powers of the applied voltage:

$$I_d(V) = \sum_{i=1}^{\infty} g_i \left[A\cos(\omega_a t) + B\cos(\omega_b t)\right]^i \quad (1.40)$$

Only orders up to $i = 3$ will be looked at in the next discussion, although the technique can be applied to any order. The resulting expression is quite lengthy, requiring a number of trigonometric substitutions, but can be simplified by looking at each order separately. The first-order expression ($i = 1$) is simply the linear response:

$$I_{d-1} = g_1\left[A\cos(\omega_a t) + B\cos(\omega_b t)\right] \quad (1.41)$$

where g_1 is the regular conductance relating voltage to current. The second-order response is related to the square of the voltage ($i = 2$) and can be written as

$$I_{d-2} = g_2 \begin{pmatrix} \dfrac{A^2}{2} + \dfrac{B^2}{2} + \dfrac{A^2}{2}\cos(2\omega_a t) + \dfrac{B^2}{2}\cos(2\omega_b t) \\ +AB\cos\left[(\omega_a + \omega_b)t\right] + AB\cos\left[(\omega_a - \omega_b)t\right] \end{pmatrix} \quad (1.42)$$

The first two terms are dc components and represent a change in the dc level that is a function of the square of the RF amplitudes.[10] The second two terms represent second-order harmonics (H) or signals that occur at twice the input frequency or *fundamentals*. The third set of terms represents second-order intermodulation

10. This is the mathematical origin describing the operation of the square law diode detector, where the diode's dc current is related to the square of the applied voltage (which essentially measures power since the square of the voltage is related to power).

(IM) or the *sum and difference beat frequencies*. Note that for closely spaced frequencies, the IM products are at nearly twice the fundamental frequency (actually, in between ω_a and ω_b) and at very low frequencies (but not quite dc). During an RF two-tone test, the fundamental amplitudes A and B are usually set equal to one another, with the result being that the IM products are always twice the level of the harmonics (6 dB). In addition, because the second-order products increase as the square of the fundamental drive, for every 1-dB increase in the fundamental amplitude, the second-order products will increase by 2 dB.

The third-order response ($i = 3$), unfortunately, is even more mathematically complicated than the second-order response (higher orders become progressively even more complex), but is actually of more importance to the RF designer and so will be considered in its full form:

$$I_{d-3} = g_3 \begin{bmatrix} \left(\frac{3A^3}{4} + \frac{3}{2}AB^2\right)\cos(\omega_a t) + \left(\frac{3B^3}{4} + \frac{3}{2}BA^2\right)\cos(\omega_b t) \\ + \frac{A^3}{4}\cos(3\omega_a t) + \frac{B^3}{4}\cos(3\omega_b t) + \frac{3}{4}A^2 B\cos\left[(2\omega_a \pm \omega_b)t\right] \\ + \frac{3}{4}AB^2 \cos\left[(2\omega_b \pm \omega_a)t\right] \end{bmatrix} \tag{1.43}$$

Equation (1.43) shows that the third-order response in a two-tone test is rich in signals. The first two terms are spurious components that have been converted (or beat) down to the level of the fundamentals and tend to *compress* or reduce the fundamental signal levels at high drive levels (usually $g_3 << g_1$). The second set of terms shows third-order harmonic signals at three times the fundamental frequency. The last set of signals is especially important because the two IM difference terms ($2\omega_a - \omega_b$) and ($2\omega_b - \omega_a$) actually convert *down* to frequencies very close to the original signal fundamental frequencies. During an RF two-tone test, equal power drive signals imply that the third-order IM terms are three times (9.54 dB) higher than the third-order harmonic terms. In addition, the third-order IM terms increase at three times the rate of the fundamentals (in dB terms), so that for every 1-dB increase in fundamental signal power increase, the third-order IM products increase by 3 dB. The two spurious IM signals (and indeed all odd-order IM signals) tend to be the most problematic to deal with from a circuit perspective. All other signals (second-order H, IM and third-order H, and the sum IM products) can be easily removed through filtering. The difference IM products, however, residing so close to the fundamentals, are difficult if not impossible to adequately filter without serious degradation of the desired signals.

Figure 1.26 shows a frequency domain plot of the second- and third-order H and IM products for a two-tone test. During a two-tone test, as the fundamental drive power is increased, the H and IM products increase at two and three times the rate for second and third order, respectively. If these signal powers are plotted on a graph of output power versus input drive power, the difference in slopes of the various output components is easily seen (Figure 1.27). By extrapolating the second- and third-order power curves up to the level of the fundamental, it is seen that the curves intersect the fundamental power curve at distinct power levels. These intersection or

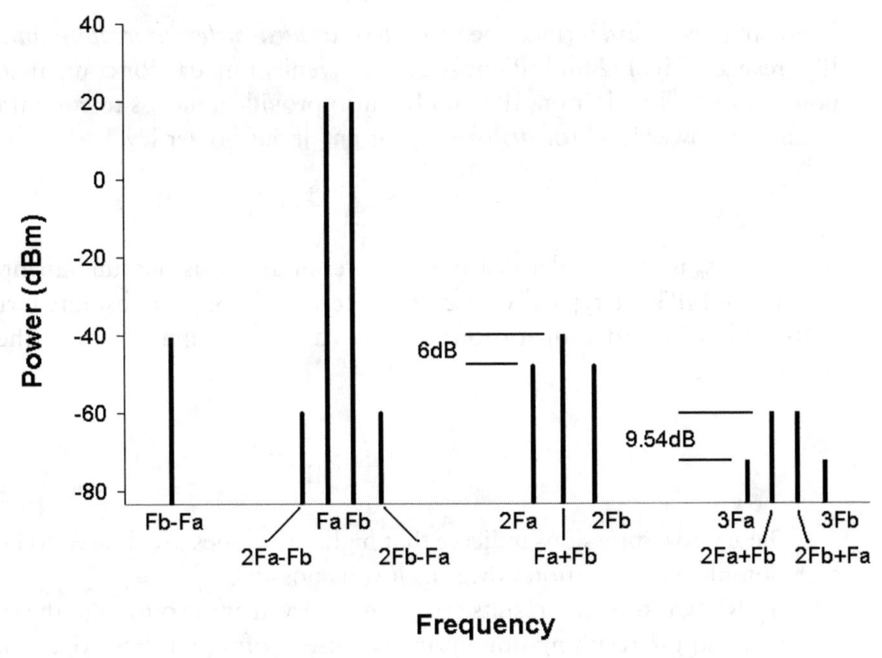

Figure 1.26 Spectrum plot showing fundamental and second- and third-harmonic and intermodulation products. (*After*: [34].)

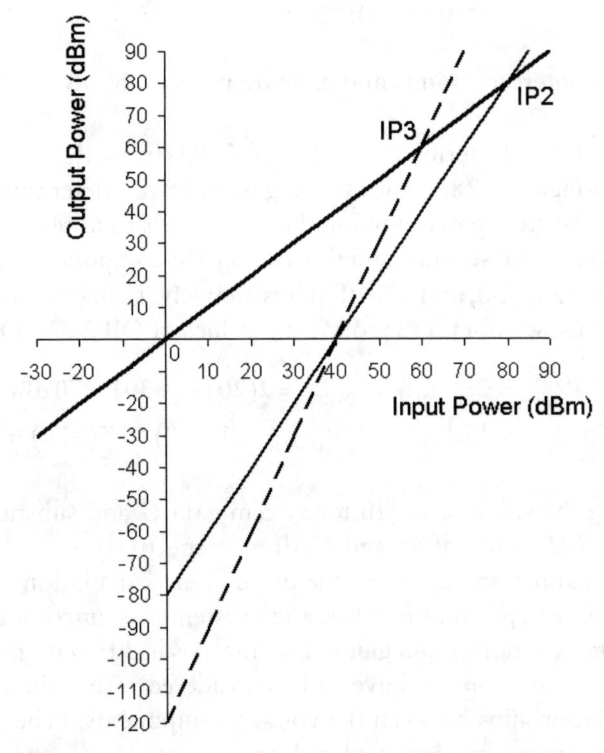

Figure 1.27 The intercept point can be computed given measured fundamental and distortion powers. (*After*: [34].)

intercept powers are termed the *second-* and *third-order intercept points* or IP2 and IP3, respectively. IP2 and IP3 provide a convenient metric for comparison of device nonlinearities. In addition, IP2 and IP3 also provide a means to compute the actual nonlinear power level (or *distortion*) for any input power level:

$$P_{2D-dB} = 2P_{F-dB} - \text{IP2}_{dB} \quad P_{3D-dB} = 3P_{F-dB} - 2 \cdot \text{IP3}_{dB} \tag{1.44}$$

where P_{xD-dB} is the x-order distortion power and P_{F-dB} is the fundamental power in dB. IP2 and IP3 are typically specified in terms of dBm. In absolute terms, the second- and third-order distortion powers can be computed using the following relationships:

$$P_{2D} = \frac{P_F^2}{\text{IP2}}; \; P_{3D} = \frac{P_F^3}{\text{IP3}^2} \tag{1.45}$$

These two expressions indicate that higher IP values are desired to keep the generated nonlinear distortion power as low as possible.

The RF two-tone test results are often observed at the output of the system block or circuit, so the resulting nonlinear response is often referred to as *output* IP2 or *output* IP3 or OIP2 and OIP3, respectively. An alternate measure is to refer the intercept points to the input of the system block or circuit (*input referred intercept points*, IIP2 and IIP3). The relationship between OIP3 and IIP3 is simply the gain of the system or circuit:

$$\text{IIP2} = \text{OIP2} - G \quad \text{IIP3} = \text{OIP3} - G \tag{1.46}$$

where the intercept points and gain are in terms of dB.

Example 1.3 Determine OIP2 and OIP3 from the spectrum analyzer output shown in Figure 1.28. If the system gain is 25 dB, determine IIP2 and IIP3.

The first step in determining the IP values is to note the power levels in the fundamental and the second- and third-order IM components. From Figure 1.28, these values are +20, –30, and –70 dBm, respectively. Using (1.44) and rearranging to find the IP terms, we obtain the following values of OIP2 and OIP3:

$$\text{OIP2}_{dBx} = 2P_{F-dBx} - P_{2D-dBx} = 2(20) - (-30) = 70 \text{ dBm}$$
$$\text{OIP3}_{dBx} = 1.5P_{F-dBx} - 0.5P_{3D-dBx} = 1.5(20) - 0.5(-70) = 65 \text{ dBm}$$

Using the value of 25 dB for system gain G and substituting into (1.46) yields IIP2 and IIP3 values of 45 and 40 dBm, respectively.

In a manner analogous to the noise figure calculations in the previous section, the total intercept point in a cascaded system is a function of the individual circuit block intercept points and gains. The analysis is different, however, in that the various voltage components have to be considered rather than powers because of the phase relationships between the voltage components. (The powers in the noise figure calculations were considered uncorrelated and therefore could be added directly.) Consider an input/output (I/O)-matched (Z_0), two-stage amplifier with

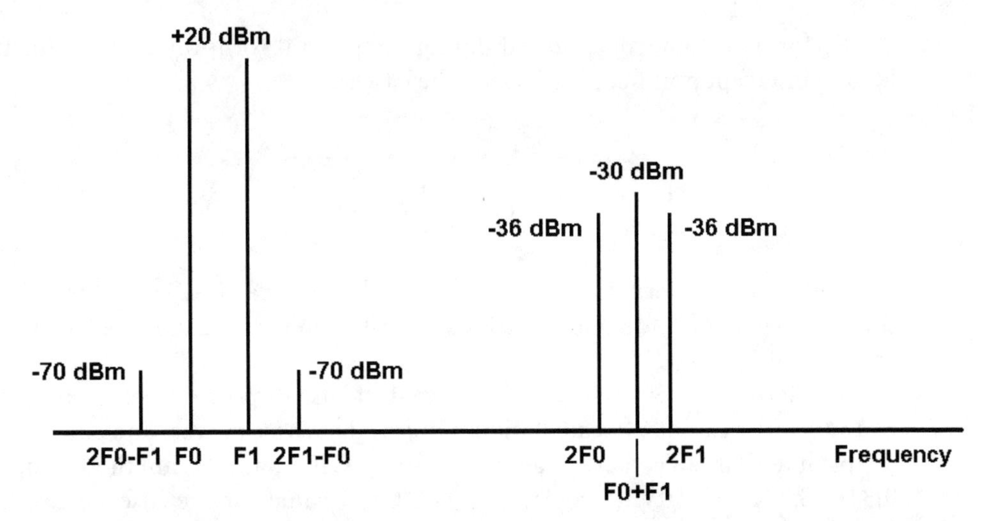

Figure 1.28 Spectrum analyzer output for Example 1.3, showing power levels up to third order.

power gains and OIP3 values as shown in Figure 1.29. Because voltage calculations are required, the absolute values of the gains and IP are needed (not dB). The fundamental signal power at the output is easily calculated through chain multiplication as follows (the square root comes from the original definition of power gains):

$$V_{\text{out}-F} = \sqrt{G_1 G_2}\, V_{\text{in}} \tag{1.47}$$

The third-order distortion voltage at the output of the first stage is computed using the voltage equivalent of (1.45):

$$V_{1-3D} = \frac{\left(G_1 V_{\text{in}}^2\right)^{3/2}}{\text{OIP3}_1} \frac{1}{2Z_0} \tag{1.48}$$

The third-order distortion voltage at the output of the second stage is a function of both the input third-order voltage presented at the input as well as the second-stage self-generated third-order power. Assume these two nonlinear voltages add in phase, which provides a worst case distortion scenario:

$$V_{\text{out}-3D} = \left[\frac{\sqrt{G_2}\left(G_1 V_{\text{in}}^2\right)^{3/2}}{\text{OIP3}_1} + \frac{\left(G_1 G_2 V_{\text{in}}^2\right)^{3/2}}{\text{OIP3}_2}\right] \frac{1}{2Z_0} \tag{1.49}$$

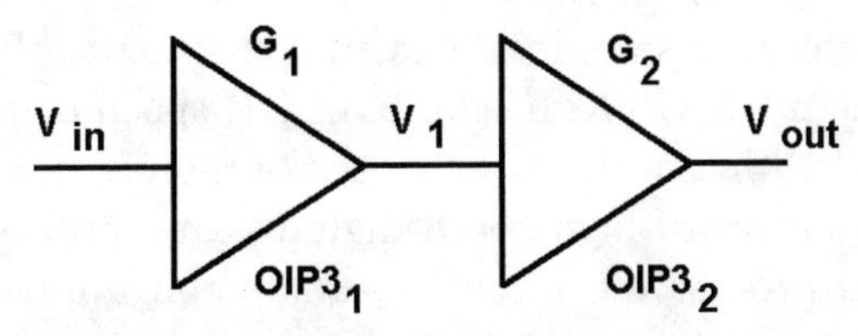

Figure 1.29 Two-stage amplifier cascade used for cascade OIP3 calculation.

Performing some algebra and noting the general form for OIP3 yields the following intercept point expression for the cascade:

$$\mathrm{OIP3}_{cas} = \frac{1}{\dfrac{1}{\mathrm{OIP3}_2} + \dfrac{1}{G_2 \cdot \mathrm{OIP3}_1}} \tag{1.50}$$

Example 1.4 Determine the system IP3 of the cascade of two amplifiers if the first stage has a gain of 15 dB and an OIP3 of 25 dBm, and stage 2 has a gain of 10 dB and an OIP3 of 35 dBm.

While the previous IP example allowed the use of power levels to obtain OIP2 and OIP3, the cascade IP calculation requires that these powers and gains be placed in absolute terms. The key parameters, $\mathrm{OIP3}_1$, $\mathrm{OIP3}_2$, and G_2 are then computed as 0.316, 3.16, and 10W, respectively. The OIP3 value for the cascade can then be found using (1.50):

$$\mathrm{OIP3}_{cas} = \frac{1}{\dfrac{1}{\mathrm{OIP3}_2} + \dfrac{1}{G_2 \cdot \mathrm{OIP3}_1}} = \frac{1}{\dfrac{1}{3.16} + \dfrac{1}{10 \cdot 0.316}} = 1.58\mathrm{W} \Rightarrow 32 \text{ dBm}$$

Because the overall system gain is 25 dB, the cascade IIP3 = 7 dBm.

1.2.3.2 Dynamic Range

In the previous section, the nonlinear system response provides both first-order and third-order components at the fundamental frequencies. As the fundamental power increases, the third-order terms increase at three times the rate and start to become observable at higher power levels, with the result that the fundamental output signal does not continue to increase at the same rate but will actually start to saturate or compress above a certain power level. The power level where this compression causes a 1-dB deviation from the ideal linear response is termed the *1-dB compression point* or P_{1dB}. P_{1dB} provides a measure of the upper power limit for the system to behave in a linear fashion.

With the thermal noise floor as defined in the previous section and with knowledge of the 1-dB compression point P_{1dB}, the lower and upper limits of the usable linear range of the system have now been defined as shown in Figure 1.30. The difference between these two boundaries defines the *dynamic range* (DR) of the system:

$$DR = P_{1dB} - P_{floor} \tag{1.51}$$

where the power terms are in a logarithmic format (dBW, dBm, etc.). Typical systems designed in CMOS RFIC can have DR specifications of 90 dB or better. P_{1dB} may be defined either at the input or the output.

With the advent of high-speed digital signal processors, much can be done to condition and process the received signals and improve overall system SNR (by noise shaping and translation) through these processing techniques. The signal must

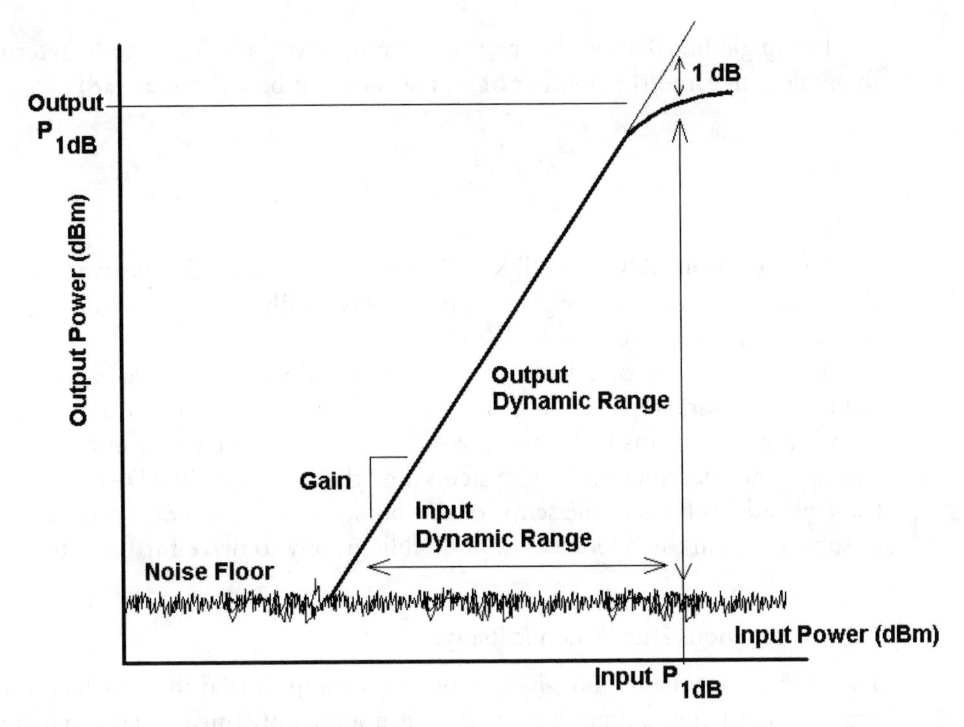

Figure 1.30 Definition of system dynamic range.

be digitized for the on-chip DSPs to operate, however, and so the DR of the system directly influences the number of bits required in the ADC stage. The ADC operates on the voltage waveform at the output of the RF subsystem, and an increase of 1 bit (2^1) essentially doubles the voltage range, or adds another 6.02 dB to the ADC range (20 log2). Using this information, the total number of bits that the ADC must possess is related directly to the DR:

$$\mathrm{DR} = 6.02N \text{ or } N = \mathrm{INT}(\mathrm{DR}/6.02) \tag{1.52}$$

where the INT operator rounds up the results to the next integer value. For a 90-dB dynamic range system, $N = 15$ bits or the equivalent of 32,768 voltage steps. Assuming a 70% voltage overhead for a 1.8V power supply system, this 15-bit digitization corresponds to about 38.5 μV per bit resolution and is approaching the upper limit for current techniques. Modern ADCs use techniques such as oversampled[11] delta-sigma ($\Delta\Sigma$) modulators to improve interference and digitization noise generation to achieve the large bit resolution [35–37]. For a single-bit $\Delta\Sigma$ modulator-based ADC, the DR is slightly modified from (1.52) [38]:

$$\mathrm{DR} = 6.02N + 1.76 \text{ dB} \tag{1.53}$$

11. The Nyquist rate indicates that data sampling need only be done at twice the highest frequency of the signal. Sampling at this rate, however, requires so called *antialiasing* filters with steep roll-off rates, and so sampling at a rate considerably above the Nyquist rate (hence, the term oversampling) reduces the need (and therefore parts count and overall cost) for such high roll-off rate filters.

For single-bit ΔΣ modulators, the minimum oversampling rate M in terms of the dynamic range and the number of bits needed can be written as [38]

$$M \cong \left[\frac{2\pi^2}{9} \frac{\mathrm{DR}^2}{2^N - 1} \right]^{1/3} \tag{1.54}$$

For the 15-bit ADC and a DR of 90 dB ($10^{90/20} = 31{,}623$), the oversampling rate should be approximately 39; since sampling typically is done at multiples of 2, 64× oversampling should be used.

The previous set of calculations was intended to give the reader an idea of the number of bits and the sampling rate needed for the DSP side of a CMOS RFIC and is intended by no means to be anything more than the briefest of introductions. The design of these fascinating ADC systems (and the corresponding DAC systems for the transmit side) is beyond the scope of this book, but the reader is encouraged to look at suggestions in the reference list and bibliography to delve further into this topic.

1.2.3.3 Spurious Free Dynamic Range

The 1-dB compression point P_{1-dB} indicates the impact that third-order (and higher) products have on the fundamental. This is not the only third-order product that has an influence on the system response. The third-order IM products are increasing at three times the rate of the fundamental power increase as well. For low fundamental power levels, the third-order IM products are still well below the noise floor. However, as the fundamental power increases, these third-order IM products with then begin to appear above the noise floor and increase at a 3:1 rate with the fundamental. Another system metric, the *spurious free dynamic range* (SFDR), is defined as the range between the MDS and the fundamental power that yields a third-order IM power equal to the MDS [30, 39]. This definition is shown graphically in Figure 1.31, where the linear and third-order power levels are indicated along with the

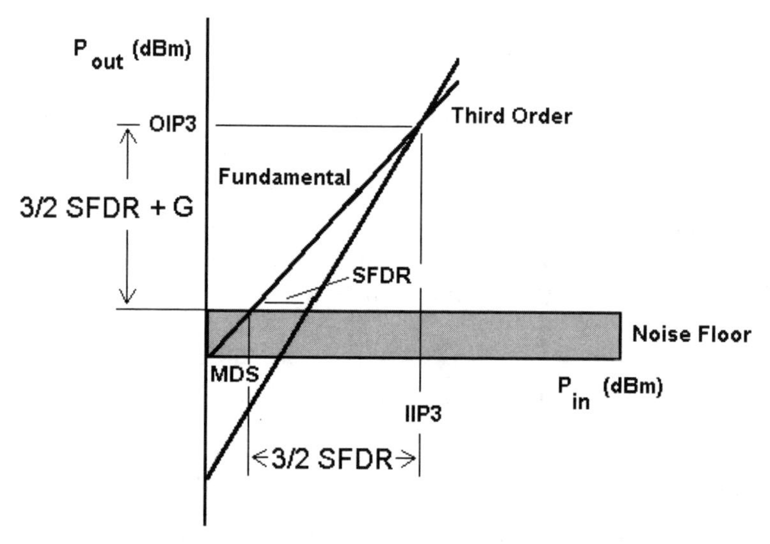

Figure 1.31 Fundamental and third-order power levels for computation of SFDR.

SFDR. The third-order power that yields the MDS for a given input power can be computed from the IIP3 expression, which is then input to the SFDR definition:

$$P_{3D} = 3P_{\text{in}} - 2 \cdot \text{IIP3} = \text{MDS} \Rightarrow P_{\text{in}} = \frac{2}{3}\text{IIP3} + \frac{1}{3}\text{MDS} \tag{1.55}$$
$$\text{SFDR} = P_{\text{in}} - P_{3D} = P_{\text{in}} - \text{MDS}$$

Employing a bit of algebra and including the relationship between OIP3 and IIP3 yields the SFDR_{dB} expression:

$$\text{SFDR}_{dB} = \frac{2}{3}(\text{IIP3} - \text{MDS}) = \frac{2}{3}(\text{OIP3} - G - \text{MDS})\ \text{dB} \tag{1.56a}$$

$$\text{SFDR}_{dB} = \frac{2}{3}(\text{IIP3} - \text{MDS}) - \text{SNR}_{\text{min}} = \frac{2}{3}(\text{OIP3} - G - \text{MDS}) - \text{SNR}_{\text{min}}\ \text{dB} \tag{1.56b}$$

where all the gain and power levels are in their logarithmic format (dB, dBm, etc.). If the system has a minimum SNR requirement (SNR_{min}), then the SFDR expression in (1.56a) may be modified to include this requirement (1.56b).

Example 1.5 For the cascade system of Example 1.4, determine the SFDR for a 1-MHz bandwidth signal. Assume the overall amplifier noise figure is 2 dB with a gain of 25 dB.

The amplifier cascade noise figure of 2 dB corresponds to an effective noise temperature of 169.6K, with a corresponding noise power of

$$P_N = kB(T_e + T_{ref}) = 1.38 \cdot 10^{-23} \cdot 10^6 \cdot (169.6 + 290) = 6.35\ \text{fW} \Rightarrow -112\ \text{dBm}$$

which can also be assumed to be the minimum detectable signal. With an OIP3 of 32 dBm and a total gain of 316.7 (25 dB), the SFDR can be computed as

$$\text{SFDR}_{dB} = \frac{2}{3}(\text{OIP3} - G - \text{MDS}) = \frac{2}{3}(32 - 25 - (-112))\ \text{dB} = 79.3\ \text{dB}$$

1.2.4 Link Budget

The major applications of CMOS RFICs are at the front end of modern integrated communication systems. Between the two end points of the communications link, the signals suffer a number of effects (path loss, interference, fading) that conspire to reduce the signal strength and purity. These so-called *link budget* calculations are very useful for the CMOS RFIC designer to know because these calculations can aid in the prediction of system parameters such as noise figure, system gain, dynamic range, and transmitter power that the RF front end of the communication system must be designed to accomplish.

Of these effects, the *path loss* is the most easily quantifiable, with a $1/R^2$ decay of the power density from the point of origin (i.e., the transmitter) reducing the signal amplitude at the receive end. The various communications protocols mentioned earlier in the chapter each have certain specifications on the power level at the receiver terminals for a given value of system performance. For example, the Bluetooth WLAN specification states that the minimum receiver sensitivity is –70 dBm (100 pW) for a bit error rate of 10^{-4} [11]. With a Class 2 transmitter output of +4 dBm, the total loss from transmitter to receiver must be less than 74 dB. An additional 10 dB of received signal strength is required to enhance BER to 10^{-5}.

1.2.4.1 Friis Equation

The starting point for communication link prediction is the *Friis equation*, which relates the received power to the transmitted power between the two communication points in terms of the antenna gains, distance, and wavelength:

$$P_{RCVR} = \frac{G_{RCVR}G_{XMTR}}{(4\pi)^2}\frac{1}{(R/\lambda)^2}P_{XMTR} \tag{1.57}$$

where $P_{XMTR(RCVR)}$ is the transmitted (received) power, $G_{XMTR(RCVR)}$ is the transmitted (received) antenna gain, R is the link distance, and λ is the wavelength of the transmitted signal (Figure 1.32).

If propagation is assumed to be through the atmosphere, then the wavelength λ is its free space value ($3 \times 10^8/F_{Hz}$). The Friis equation can be put in logarithmic terms for easier calculation:

$$\begin{aligned} P_{RCVR-dBx} &= P_{XMTR-dBx} + G_{RCVR} + G_{XMTR} \\ &\quad -\left[20\log(R_{km}) + 20\log(F_{GHz}) + 92.44\right] \text{dBx} \end{aligned} \tag{1.58}$$

where the power levels are in dBx (dBx = dBW, dBm, etc.). The term in brackets in (1.58) is often termed the *path loss*, PL:

$$\text{PL} = 20\log(R_{km}) + 20\log(F_{GHz}) + 92.44\ \text{dB} \tag{1.59}$$

Figure 1.33 shows a plot of $P_{RCVR} - P_{XMTR}$ as a function of distance for four different frequencies assuming unity gain antennas; essentially the path loss. For a 1.0-GHz signal, the PL at 1.0 km is 92.4 dB and increases by 20 dB per decade increase in distance. As the frequency increases, the PL increases as well since the normalized distance (R/λ) increases.

Example 1.6 Determine the maximum transmitter-receiver separation for a Bluetooth link assuming a PL of 74 dB.

Figure 1.32 Schematic diagram of a communications link showing terms described in the Friis equation.

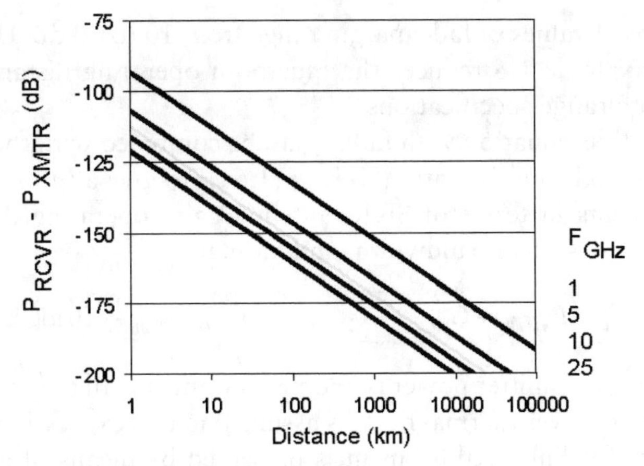

Figure 1.33 Path loss as a function of link distance and frequency.

Bluetooth WLAN components operate in the ISM frequency band at 2.4 GHz. Combining this frequency information with the link requirement of 74 dB allows the link distance R_{km} to be computed by rearranging (1.58):

$$\text{PL} = 20\log(R_{km}) + 20\log(F_{GHz}) + 92.44\ \text{dB}$$

$$R_{km} = 10^{\frac{[PL-20\log(F_{GHz})-92.44]}{20}} = 10^{\frac{[74-20\log(2.4)-92.44]}{20}} = 49.5\text{m}$$

The maximum operating distance for this example would be approximately 50m. This particular operating distance is higher than that noted in the Bluetooth technology; however, in this example, the link was assumed "ideal" with no loss mechanisms other than the $1/R^2$ path loss. The next section will present another loss mechanism that must be included for link budget analysis.

1.2.4.2 Fading and Fade Margin

The basic Friis path loss expression assumes that the RF signal follows a single path between the transmitter and the receiver with no other loss mechanisms except for the $1/R^2$ path loss. In actual RF communication links, the RF signal will take a number of paths because of reflections off of nearby objects (walls, ground, ceilings, and trees). These *multipath* signals will be slightly delayed, which causes a phase shift in the signals when received. The sum of all of these multipath signals, some in phase and some out of phase, can cause a time-varying increase and/or decrease in the signal strength; this effect is termed *channel fading*, or simply *fading*. The effects of fading and the corresponding reduction in received signal strength can be introduced into the path loss expression by use of a term called the *fade margin* (FM). This makes sense because the reduction in signal strength does "appear" to be in addition to that of the normal path loss mechanism. Equation (1.59) can then be easily modified to include the fade margin:

$$\text{PL}_{fade} = 20\log(R_{km}) + 20\log(F_{GHz}) + \text{FM}_{dB} + 92.44\ \text{dB} \tag{1.60}$$

Typical values of fade margin range from 10 to 30 dB. Using a 20-dB fade margin in Example 1.6 reduces the Bluetooth operating distance to 5m, in line with Bluetooth range specifications.

The Friis equation with fading can be combined with the previous equation for output signal-to-noise ratio (SNR_{out}; (1.61)) to give a full accounting of the system requirements in terms of SNR, antenna gain, operating distance and frequency, noise figure, system bandwidth, and fading:

$$SNR_{out} = P_{XMTR} + G_{RCVR} + G_{XMTR} - PL_{fade} - F_{dB} - 10\log B_{Hz} + 174 \text{ dB} \qquad (1.61)$$

where the transmitter power is in units of dBm and all the other terms are in units of dB. Only receiver thermal noise is assumed in this expression. The use of these relationships for link establishment is presented by means of an example in the next section.

1.2.4.3 Link Budget Example

A WLAN link requires a 10-MHz bandwidth at 2.4 GHz and uses unity gain antennas on both the transmitter and receiver. The data rate of this system is dependent on the received power and corresponding SNR as shown in Table 1.7. In the table, the "Received Power" and "SNR" columns are measured at the receiver antenna terminals. If the transmitter is limited to a maximum output power of 10 mW (10 dBm), find the maximum operating range for each of the data rates listed. Assume a fade margin of 20 dB.

The data rate in this example is a function of two system requirements, the received power and the SNR. Both parameters must be satisfied if the link is to be established at the specified data rate; therefore, calculations of both are required, namely (1.58) and (1.61) (to include fade margin):

$$\begin{aligned} SNR_{out} = {} & P_{XMTR} + G_{RCVR} + G_{XMTR} - 20\log(R_{km}) - 20\log(F_{GHz}) \\ & -92.44 - 10\log B_{Hz} - FM + 174 \text{ dB} \end{aligned}$$

and

$$\begin{aligned} P_{RCVR-dBx} = {} & P_{XMTR-dBx} + G_{RCVR} + G_{XMTR} \\ & -\left[20\log(R_{km}) + 20\log(F_{GHz}) + 92.44 + FM\right] \text{dBx} \end{aligned}$$

Table 1.7 WLAN Specifications for the Link Budget Example

Data Rate (Mbps)	*Received Power (dBm)*	*SNR (dB)*
5.0	–84	15
2.55	–88	10
1.0	–92	6
0.5	–94	4

Since the calculations are at the receiver antenna terminals and not the receiver output, the noise figure term F in (1.61) has been neglected.

Table 1.8 shows the values of the various terms to be included in the link budget calculation assuming that the system bandwidth is twice the maximum data rate (10 MHz). Substituting the parameter values in Table 1.8 into the previously listed equations provides the maximum operating distances for the received power and output SNR values specified in Table 1.9.

In all cases, the received power level governs the maximum operating range, with the resulting data rate as a function of operating distance shown in Table 1.10. For example, if the operating distance is 40m, then the link conditions are such that data can be transferred at the maximum rate. If the operating distance increases to 125m, then the data rate must be reduced to 0.5 Mbps. If the transmitter can adaptively increase its output power by 2.0 dB, then the data rate can be doubled to 1.0 Mbps and still be within the system specifications for received power. Beyond 152m for the 10-dBm transmitted power level, a link cannot be established at any data rate; hence, the link fails at this distance. An increase in transmitted power level, of course, will increase the maximum operating range of the system. (If this is a battery operated device, the increase in power level will also drain the batteries at a faster rate.)

1.3 Introduction to the Book

With this chapter as an introduction, Chapter 2 provides a detailed look at the MOSFET and the fabrication technology used in its implementation. The chapter begins with a review of some general characteristics of the CMOS fabrication

Table 1.8 Parameter Values for the Link Budget Example Calculations

Parameter	*Value*
P_{XMTR}	10 dBm
G_{XMTR}, G_{RCVR}	0 dB, 0 dB
FM	20 dB
$10 \log B_{Hz}$	70
$20 \log F_{GHz}$	7.6

Table 1.9 Maximum Operating Range Calculated Separately for the Received Power Requirement and the SNR Requirement

Received Power (dBm)	R_{max} *(m)*	*SNR (dB)*	R_{max} *(m)*
–84	50	15	89
–88	79	10	158
–92	126	6	251
–94	158	4	316

Table 1.10 Maximum Operating Range for a Specific Transmission Data Rate in the Link Budget Example

Data Rate (Mbps)	*Maximum R (m)*
5.0	50
2.55	79
1.0	126
0.5	158

process. The various modes of MOSFET operation are then reviewed, with an emphasis on the development of equivalent circuit models that are beneficial for RF MOSFET simulation. Several simple RF MOSFET circuits are introduced. A number of simulation examples are performed throughout this chapter (and other chapters as well) based on a review of the SPICE simulation package.

Chapter 3 focuses on passive structures in the CMOS process. While some of these passive structures are necessary for RF circuit design (resistors, capacitors, inductors, interconnections), this chapter also looks at the origin of loss mechanisms that are a special problem at high frequencies. An introduction to RF MEMS is followed by a discussion of basic packaging issues and grounding. Several examples of passive element equivalent circuit prediction are performed using a commercial electromagnetic field solver package.

Chapter 4 is a broadly based discussion on CMOS RFIC amplifiers (but not power amplifiers). Basic concepts of amplifiers such as equivalent circuits, biasing, and impedance matching are presented. A thorough discussion of LNAs is presented to show how the MOSFET's construction can be used to reduce the noise introduced into the system. A number of single-ended and differential amplifier circuit topologies are introduced and analyzed.

Chapter 5 builds on the concepts introduced in Chapters 2, 3, and 4 with discussions on important CMOS RFIC ancillary circuits, such as negative resistance circuits, automatic gain control circuits, active reactance circuits, and simple filtering circuits. A discussion on measurements and *S*-parameter deembedding techniques is also presented.

Chapter 6 presents the use of the MOSFET in oscillator applications. The chapter begins with an overview of general oscillation criteria in terms of design via gain and phase margins. Fixed and variable reactance oscillators are covered as are resonator-based oscillators. Modern communications systems CMOS RFICs are primarily digital in nature, and so oscillators based on digital technologies, in the form of ring oscillators, are also covered in this chapter. The chapter concludes with a discussion of oscillator phase noise and how this important noise source can be reduced through circuit techniques.

Chapter 7 covers the essentials of frequency translation through the use of mixer technology. An introduction to general mixer concepts begins the chapter, advancing to a presentation of various mixer types often used in CMOS RFICs. These fundamental mixer circuits are then used as foundational blocks in a discussion of more advanced mixer structures that are used to improve image rejection.

Chapter 8 presents a discussion on improving the controllability of the oscillators introduced in Chapter 6 through the use of phase-locked loops (PLLs). The chapter begins with an introduction to the theory of PLLs using both analog and digital loop techniques. Various loop filters are discussed, primarily those seen in modern CMOS RFICs. Simplified behavioral modeling of PLLs is also introduced. The chapter concludes with a discussion of frequency synthesizers based on both PLL and direct digital synthesis techniques.

The final chapter, Chapter 9, builds on the amplifier material in Chapter 4 but with a focus on the high-power amplifiers used in CMOS RFICs. The chapter begins with an overview of the various amplifier classes (A through F) and then focuses on the amplifier classes most often seen in CMOS RFICs. The focus of this chapter is on power amplifiers integrated with the other system elements for complete CMOS SoCs; discrete MOS power amplifier devices such as LDMOS (an interesting topic in and of itself) are beyond the scope of the intended discussion.

The text contains a number of appendices as well. The first two contain sample CMOS fabrication and MOS transistor parameters that are used throughout the text for estimation and simulation of active and passive circuit performance. Appendix C shows a detailed RF equivalent circuit of the RF MOSFET and its associated admittance or *Y*-parameters. The corresponding *S*-parameters for this RF equivalent circuit can be obtained using the *Y*-to-*S*-parameter conversion relationships presented in Appendix D. Appendix E contains a summary of important physical constants applicable to the silicon CMOS environment.

The accompanying CD-ROM contains a number of items that may be of interest to the reader. Many examples in the text are based on SPICE simulations. The SPICE netlist files are available on the CD if the reader wishes to continue further study into variations of the material covered in the examples. For the examples where SPICE netlists are available, the file name is listed in parentheses; for example, (ch2-1.txt). A vast number of circuit simulators are available for the RF engineering student or professional. Rather than attempt to provide simulation examples for all of these simulators (with their attendant differences in input definitions), SPICE netlists were chosen for the widest possible applicability with readily available simulators. A number of SPICE programs are available, all of which may have graphical input interfaces but can be usually set up to use a netlist as input. One such SPICE package, SwitcherCAD III from Linear Technologies, is included on the CD along with installation instructions. The number of components in each SPICE netlist file has been intentionally kept to a minimum so that student or demonstration versions of SPICE (which may be active device or component number limited) can be used. The reader is strongly encouraged to modify the files to investigate the variations in circuit performance and frequency response with changes in device or component values.

The CD also contains a version of the Sonnet® Lite™ electromagnetic field solver software. This software package is used extensively in Chapter 3 in estimation of the performance of passive structures (primarily capacitors and inductors) fabricated in the CMOS process. Design files for the various examples in Chapter 3 are included on the CD. Installation instructions for Sonnet Lite are also contained on the CD. As with the SPICE examples, the reader is strongly encouraged to modify the various passive component layouts to investigate the variations in circuit performance and frequency response.

References

[1] Lyons, E., *David Sarnoff: a Biography*, New York: Harper Row, 1966.

[2] Bellaver, R. F., "Wireless: From Marconi to McCaw," *2000 IEEE Int. Symp. Tech. Society*, Sept. 2000, p. 197.

[3] American Radio Relay League, http://www.arrl.org (amateur radio organization).

[4] Sarkar, T., et al., *History of Wireless*, Wiley Series in Microwave and Optical Engineering, New York: John Wiley–IEEE Press, 2006.

[5] Lee, T. H., *The Design of CMOS Radio Frequency Integrated Circuits*, 2nd ed., Cambridge, UK: Cambridge University Press, 2004.

[6] Hussain, A., *Advanced RF Engineering for Wireless Systems and Networks*, Hoboken, NJ: John Wiley and Sons, 2005.

[7] Rouffet, D., et al., "4G Mobile," Technical Paper, *Alcatel Telecommun. Rev.*, 2005.

[8] Cisco Systems, "Internetworking Technology Handbook: Chapter 7: Ethernet," 2002, http://www.cisco.com/univercd/home/home.htm.

[9] Zargari, M., S. Mehta, and D. Su, "Challenges in the Design of CMOS Transceivers for the IEEE 802.11 Wireless LANs: Past, Present and Future," *Proc. 2005 RFIC Symp.*, 2005, p. 353.

[10] Bluetooth, "Specification v. 2.0 + EDR [Vol. 3]: Radio Specification," 2004, http://www.bluetooth.com.

[11] Bluetooth, 2006, http://www.bluetooth.com/bluetooth.

[12] ZigBee Alliance, 2006, http://www.zigbee.org.

[13] Lilienfeld, J., "Device for Controlling Electrical Current," U.S. Patent 2,900,018, filed March 28, 1928, granted March 7, 1933.

[14] Heil, O., "Improvements in or Relating to Electrical Amplifiers and Other Control Arrangements and Devices," British Patent 439,457, filed March 4, 1935, granted Dec. 6, 1935.

[15] Kilby, J., "Invention of the Integrated Circuit," *IEEE Trans. Electron Devices*, Vol. ED-23, 1976, p. 648.

[16] Noyce, R., "Semiconductor Device and Lead Structure," U.S. Patent 2,981,877, filed July 30, 1959, granted April 25, 1961.

[17] Kahang, K., and M. Atalla, "Silicon–Silicon Dioxide Field Induced Surface Devices," *IRE-AIEEE Solid State Devices Research Conf.*, 1960.

[18] Wanlass, F., and C. Sah, "Nanowatt Logic Using Field Effect Metal Oxide Semiconductor Triodes," *Proc. 1963 IEEE Int. Solid State Circuits Conf.*, Feb. 1963, p. 32.

[19] Moore, G., "Cramming More Components onto Integrated Circuits," *Electronics*, Vol. 38, No. 8, April 19, 1965, p. 114.

[20] Liou, J., and F. Schweirz, "RF MOSFET: Recent Advances and Future Trends," *Proc. 2003 IEEE Conf. on Electron Devices and Solid State Circuits*, Dec. 2003, p. 185.

[21] Abidi, A., "RF CMOS Comes of Age," *IEEE Microwave Mag.*, Dec. 2003, p. 47.

[22] Bennett, H. S., et al., "Device and Technology Evolution for Si-Based RF Integrated Circuits," *IEEE Trans. Electron Devices*, Vol. 52, No. 7, July 2005, p. 1235.

[23] "International Technology Roadmap for Semiconductors: Radio Frequency and Analog/Mixed-Signal Technologies for Wireless Communications," 2005 ed., http://www.itrs.net/Common/2005ITRS/Home2005.htm (accessed 2006).

[24] All About MEMS, "What Is MEMS? MEMS Introduction," 2006, http://www.allaboutmems.com/whatismems.html.

[25] Maurer, L., K. Chabrak, and R. Weigel, "Design of Mobile Radio Transceiver RFICs: Current Status and Future Trends," *Proc. 2004 1st Int. Symp. on Control, Communications, and Signal Processing*, 2004, p. 53.

[26] Mikkelsen, J., et al., "Feasibility Study of DC Offset Filtering for UTRA-FDD/WCDMA Applications," *Proc. 17th IEEE NORCHIP Conf.*, Oslo, Norway, Nov. 1999, p. 34.

[27] Lee, T. H., and A. Hajimiri, "Oscillator Phase Noise: A Tutorial," *IEEE J. Solid State Circ.*, Vol. 35, No. 4, March 2000, p. 326.

[28] Hajimiri, A., and T. H. Lee, "Design Issues in CMOS Differential LC Oscillators," *IEEE J. Solid State Circ.*, Vol. 34, No. 5, May 1999, p. 717.

[29] Razavi, B., "RF CMOS Transceivers for Cellular Telephony," *IEEE Commun. Mag.*, Aug. 2003, p. 144.

[30] Proakis, J. G., *Digital Communications,* Singapore: McGraw-Hill, 1995.

[31] Razavi, B., *RF Microelectronics*, Upper Saddle River, NJ: Prentice Hall, 1998.

[32] Shannon, C. E., "A Mathematical Theory of Communication," *Bell Syst. Tech. J.,* Vol. 27, July 1948, p. 379, and Oct. 1948, p. 623.

[33] Pozar, D., *Microwave Engineering*, 2nd ed., New York: John Wiley, 1998.

[34] Caverly, R., and G. Hiller, "Distortion in p-i-n Diode Control Circuits," *IEEE Trans. Microwave Theory Techniques*, Vol. MTT-35, No. 5, May 1987, p. 492.

[35] Boser, B., and B. Wooley, "The Design of Sigma Delta Modulation Analog-to-Digital Converters," *IEEE J. Solid State Circ.*, Vol. SC-23, Dec. 1988, p. 1298.

[36] Candy, J., and G. Temes, *Oversampling Delta Sigma Data Converters*, New York: IEEE Press, 1997.

[37] Norsworthy, S., R. Schreier, and G. Temes, *Delta Sigma Data Converters*, New York: IEEE Press, 1992.

[38] Allen, P., and D. Holberg, *CMOS Analog Circuit Design*, 2nd ed., New York: Oxford University Press, 2002.

[39] Lerdworatawee, J., and W. Namgoong, "Revisiting Spurious-Free Dynamic Range of Communication Receivers," *IEEE Trans. Circ. Syst. I*, Vol. 53, No. 4, April 2006, p. 937.

Selected Bibliography

Coe, L., *Wireless Radio: A History*, Jefferson, NC: McFarland & Company, 2006.

Garratt, G. R. M., *The Early History of Radio: From Faraday to Marconi,* IEE History of Technology Series, London: IEE Press, 1994.

Holt, K., "Wireless LAN: Past, Present, and Future," *Proc. 2005 Design, Automation and Test in Europe*, 2005, p. 92.

Hong, S., *Wireless: From Marconi's Black-Box to the Audion,* Transformations: Studies in the History of Science and Technology Series, Boston: MIT Press, 2001.

Sarkar, T., R. Mailloux, A. Oliner, M. Salazar-Palma, and D. Sengupta, *History of Wireless*, Wiley Series in Microwave and Optical Engineering, New York: John Wiley-IEEE Press, 2006.

Saxby, S., *The Age of Information: The Past Development and Future Significance of Computing and Communications*, New York: New York University Press, 1990.

U.S. Federal Communications Commission, 2006, http://www.fcc.gov/omd/history/radio/ bibliography.html (extensive bibliography on the history of radio as well as biographies of some of the early pioneers).

CHAPTER 2

CMOS Integrated Circuit Fundamentals

The use of CMOS for microwave and RF design was broached to the microwave community in the early 1990s when GaAs was the technology of choice for designs in the UHF region and above [1]. Being able to design a CMOS RFIC (or any other RFIC technology device for that matter) is not simply a matter of getting time on a CAD workstation, inputting a certain circuit topology, and then letting a circuit optimization routine run for hours. An understanding of the underlying device operation is crucial to understanding the entire circuit operation. Time is well spent in the initial stages of an RFIC design by doing "back-of-the-envelope" calculations to determine if such a circuit is feasible before computer simulations verifying its operation are even started. In addition, these early-stage designs also aid in determining if unrealistic component or device values might be needed and what component or device variables need to be adjusted to provide a remedy to the problem. Once the computer simulation stage has been reached, intelligent adjustment of device parameters will ultimately yield functioning RFICs.

This chapter contains an overview of CMOS technology from both a fabrication as well as a device perspective. CMOS fabrication parameters and accompanying device layouts provide the basis for the estimation of RF equivalent circuit models for FETs and for their ultimate performance. This chapter focuses on linear small-signal equivalent circuit models for both *n*-channel and *p*-channel MOSFET devices. Numerous complex MOSFET models have been discussed in the literature, but this chapter focuses on simple models to aid in understanding some of the trade-offs in the design. Equivalent circuit models are also introduced for the two major sources of MOSFET noise. A look at the operation of the MOSFET in its linear or resistive mode of operation is presented with an RF switch application discussed.

2.1 Review of CMOS Technology

2.1.1 The CMOS Physical Structure

Numerous examples in the literature, both research and educational, review the processes involved in fabricating a CMOS integrated circuit (see, for example, [2–6]). They all share, however, a degree of commonality that allows for some generalizations in the CMOS design process. The MOSFETs are fabricated using highly doped drain and source diffused regions separated by a gate material grown over a thin insulating layer (Figure 2.1).

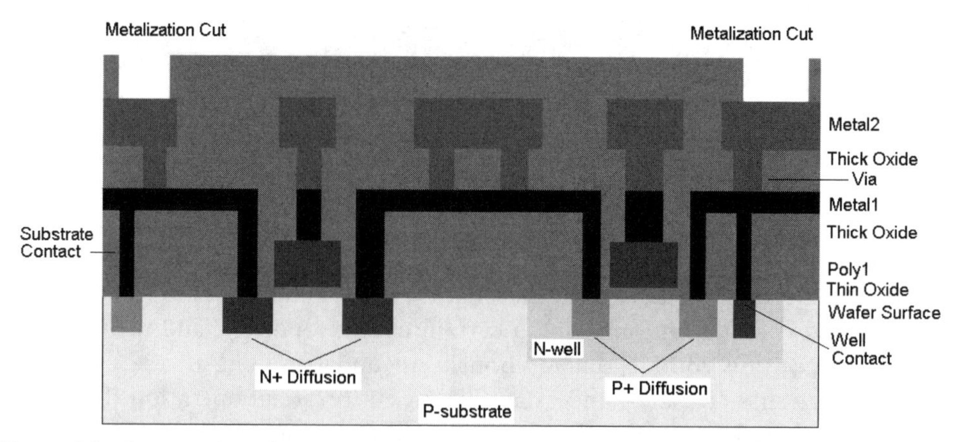

Figure 2.1 Cross section of conventional bulk CMOS process with a single poly layer (two or more may be available in CMOS processes) and two levels of metal (seven or more metal layers may be available in CMOS processes).

This gate material was originally aluminum (Al) in early MOSFETs. As geometries were reduced in size, however, fabrication processes changed and the higher temperatures required for these processes prohibited the use of low melting point metals such as Al. Replacing the metal gate material with highly doped (and therefore highly conducting) polycrystalline silicon provided good long-term reliability as well as tolerance to high-temperature processing[1] [7]. The term *polysilicon* (or *poly*) is traditionally used when discussing the polycrystalline silicon gate material. The insulating layer can be silicon dioxide or other insulating material with good material matching properties, but the term *oxide* is typically used when discussing this gate insulating material.

Some CMOS fabrication technologies provide the capability for a second poly layer over the first (with a thin oxide layer in between). The thin oxide between the two poly layers provides the RFIC designer with a high capacitance per unit area and hence the possibility of high values of on-chip capacitance. There may be up to seven layers of metal above the wafer surface in advanced CMOS processes (aluminum and copper) with thicker lines on the higher layers. Each metal layer is separated by a thick layer of insulating oxide (on the order of 1 μm); therefore, each metal layer has no DC connection (only two metal layers shown in Figure 2.1). These metal layers can be used for signal interconnections and power supply connections as well as various RF components.

The dc current-carrying capability of these metal layers depends on the metal layer cross section, with dc current densities on the order of 2 mA/μm^2 typically assumed for reliability [8]. For RF signals, studies have shown that the maximum current density is approximately 25 to 30 times higher [9]. Duty cycles will also affect this maximum current density, with low duty cycles allowing higher current

1. To this day, the term metal is still used to describe the MOSFET, although a more accurate term may be SOS (semiconductor oxide semiconductor) or SIS (semiconductor insulator semiconductor); these terms are sometimes seen in the literature.
2. A conjugate match can be created by matching the source with one that has the identical resistive component but a reactive component value of the opposite sign.

densities. Vias are used to connect adjoining metal layers, with contacts used to connect semiconductor layers or metal layers to other semiconductor layers. Manufacturers publish or otherwise make available to RFIC designers process parameters that describe electrical parameters such as threshold voltage, resistance, and capacitance, or physical parameters such as oxide and other layer thicknesses. Representative process parameters for a 0.5-μm CMOS process are illustrated in Table 2.1. Knowledge of these process parameters is crucial to successful designs. More detailed process parameters can often be obtained from the IC manufacturer but in many cases nondisclosure agreements are involved because of proprietary issues and protection of trade secrets.

The primary fabrication technology for digital CMOS relies on a bulk technology; the majority of the silicon used is purely for mechanical and thermal purposes. This bulk CMOS technology unfortunately adds unwanted parasitic characteristics termed *substrate effects* that dramatically influence the RF behavior of both active and passive devices. Silicon-on-insulator (SOI) technology utilizes a thin epilayer of silicon grown on an insulating substrate (glass or sapphire is usually used) that reduces many of the unwanted bulk effects (see Table 2.1) that dramatically impact bulk CMOS.

A variation on bulk CMOS technology, laterally diffused MOS (LDMOS), was developed for high-power applications. Although similar in many respects to the classical MOS technology, LDMOS devices include a long laterally diffused *N*-doped region that allows the drain to be physically removed from the gate, increasing breakdown voltage, improving drain source resistance, and reducing feedback capacitance (Figure 2.2 [10]). The heavily doped sinker layer under the source node provides a low resistance path to the backside metallization.

2.1.2 Technology Scaling

The phrase "0.18-micron CMOS technology," often heard when referring to an IC, is the minimum feature size associated with the fabrication process and is almost always synonymous with the MOSFET gate length, L. The primary indicator of MOSFET frequency (and for that matter speed) response is also this gate length (Figure 2.3). The time required for a charge carrier to transit this length L is dependent on the mobility μ and the applied electric field E:

$$\tau = \frac{L}{\mu E} = \frac{L}{v_{sat}} \tag{2.1}$$

However, the transit time cannot be decreased indefinitely by increasing the electric field; in silicon, the velocity saturates (v_{sat}) at approximately 10^5 m/s, requiring reductions in L to improve transit time response. A quick calculation of frequency response would look at the inverse of the transit time (at least dimensionally):

$$f = \frac{1}{2\pi\tau} = \frac{v_{sat}}{2\pi L} = 15.9 \text{ GHz}/\mu\text{m} \tag{2.2}$$

Table 2.1 Sample Set of Process and Other MOS Fabrication Parameters for a 0.5-μm IC Process

Process Parameters

	N+	*P+*	*POLY*	*POLY2*	*M1*	*M2*	*M3*	*N\PLY*	*N_Well*	*Units*
R_S	82.6	104	23.4	43.4	0.09	0.10	0.05	844	836	Ω/square
R_C	61.2	158	17.0	28.4		0.83	0.78			Ω

Capacity Parameters

	N+	*P+*	*POLY*	*POLY2*	*M1*	*M2*	*M3*	*N_W*	*UNITS*
Area (substrate)	430	729	87		32	17	11	40	aF/μm^2
Area (*N*+active)			2496		37	17	12		aF/μm^2
Area (*P*+active)			2411						
Area (poly)				920	56	16	10		aF/μm^2
Area (poly2)					50				aF/μm^2
Area (metal1)						34	14		aF/μm^2
Area (metal2)							36		aF/μm^2
Fringe (substrate)	315	266			75	58	39		aF/μm
Fringe (poly)					62	38	29		aF/μm
Fringe (metal1)						57	35		aF/μm
Fringe (metal2)							50		aF/μm
Overlap (*N*+active)			218						aF/μm
Overlap (*P*+active)			238						aF/μm
R_S—sheet resistance of layer									
R_C—contact resistance per square of via, contact									
Gate oxide thickness		138							Å

Transfer Parameters

	N-Channel	*P-Channel*	*Units*
Threshold voltage *VT*	0.66	–0.91	volts
Instrinsic transconductance, KP	56.3	19.1	μA/V2
Low-field mobility	450	152	cm^2/V-s

Source: MOSIS, http://www.mosis.org. Used with permission.

Although a crude measure of frequency response, (2.2) indicates that significant improvements are garnered by reducing the gate length. Reducing L, however, also means varying many of the fabrication process parameters such as layer thicknesses, layer separations, and doping concentrations. Details of the adjustments required as CMOS feature sizes shrink are discussed later in the chapter.

2.2 The MOSFET

2.2.1 The Basic *n*-Channel MOSFET

The basic *n*-channel MOSFET (nMOSFET) device structure is shown in Figure 2.4 with standard current and voltage polarities indicated. When V_{GS} exceeds the threshold voltage, V_T, the region under the gate material "inverts" by having more

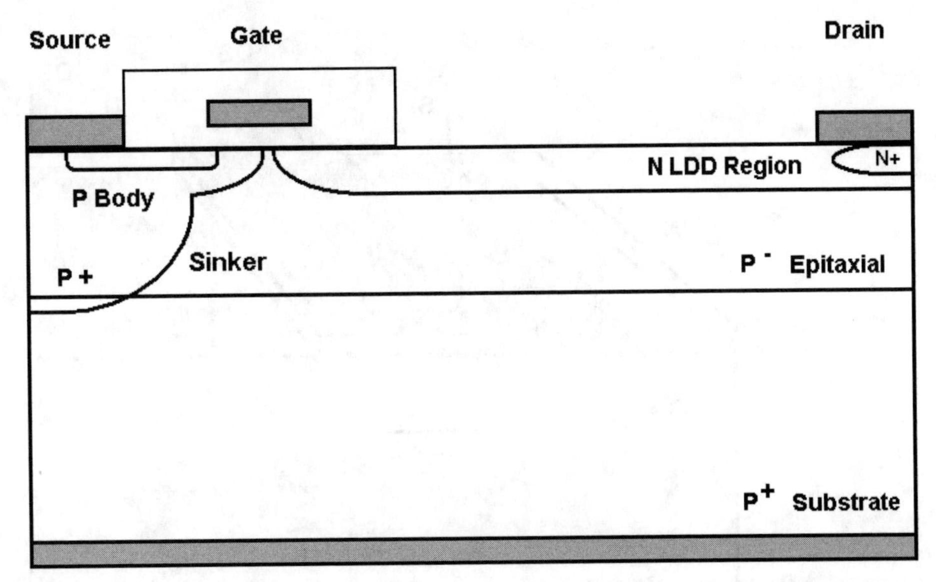

Figure 2.2 Cross section of conventional LDMOS for use at RF. (*Source:* [10]. © 2003 IEEE.)

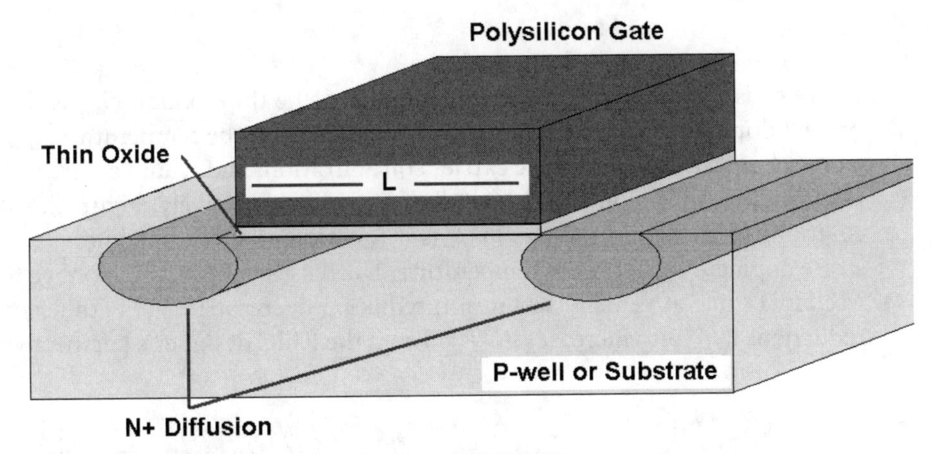

Figure 2.3 Simplified cross section of *n*-channel MOSFET.

mobile electrons than holes, creating a conducting channel that allows a dc current I_{DS} to flow when a drain-source voltage, V_{DS}, is applied. The threshold voltage V_T is determined during the MOS fabrication process and can be estimated using the expression [11]:

$$V_T = \frac{t_{ox}}{\varepsilon_{ox}} \sqrt{2\varepsilon_{Si} q N_{sub} (2\phi_B)} + 2\phi_B \tag{2.3}$$

where

$$\phi_B = \frac{kT}{q} \ln\left(\frac{N_{sub}}{n_i}\right) \tag{2.4}$$

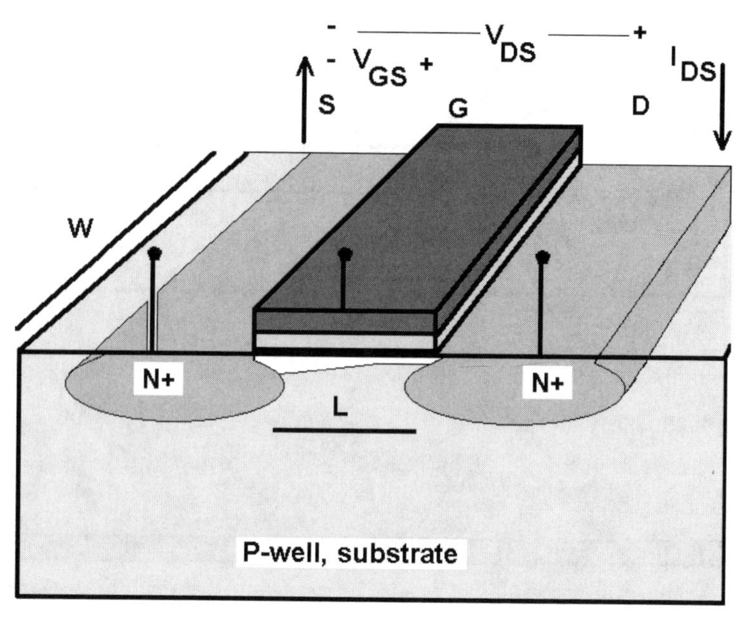

Figure 2.4 Cross section of an *n*MOSFET showing channel region, device width *W*, and dc bias voltage polarities.

where t_{ox} is the oxide thickness under the gate (the thin oxide), N_{sub} is the substrate or well doping (m^{-3}), k is Boltzmann's constant, T is the temperature, q is the electronic charge, n_i is the intrinsic carrier concentration, and ε_{Si} and ε_{ox} are the dielectric permittivities of silicon and the insulating oxide, respectively. Figure 2.5 shows calculations of threshold voltage (V_T) as a function of oxide thickness (t_{ox}) and substrate doping (N_{sub}) at room temperature. Decreases in operating voltage to keep the MOSFETs out of velocity saturation require a decrease in V_T, with accompanying reductions in t_{ox} and increases in N_{sub}. From the RFIC designer's perspective, V_T is set

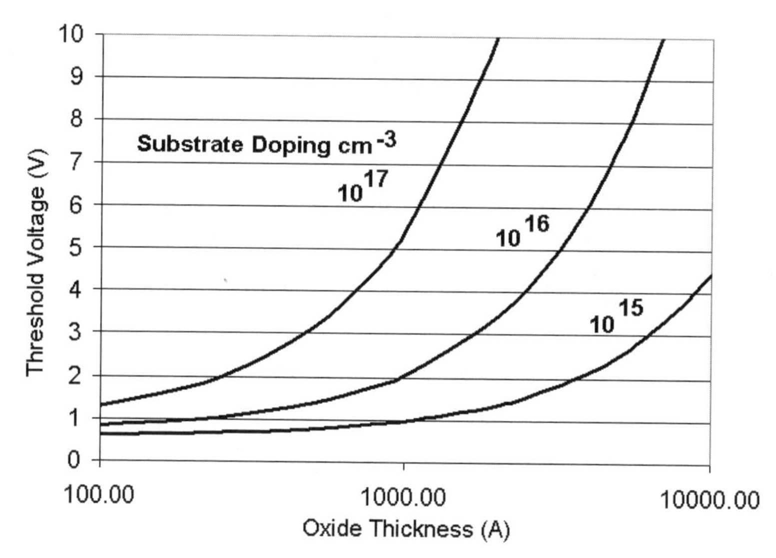

Figure 2.5 Room temperature threshold voltage versus oxide thickness for three different substrate dopings.

by the process and is therefore specified by the IC fabricator (Table 2.1) and is typically 10% to 20% of the target operating voltage. For small values of V_{DS}, I_{DS} increases linearly (the so-called "linear" region of operation) and the current-voltage expression may be written as [12]

$$I_{DS} = \mathrm{KP}\frac{W}{L}V_{DS}\left(V_{GS} - V_T - 0.5V_{DS}\right) \quad \mathrm{KP} = \mu C_{ox} = \frac{\mu\varepsilon_{ox}}{t_{ox}} \tag{2.5}$$

where μ is the mobility of the dominant charge carrier in the induced channel (electrons in *n*MOSFETs, holes in *p*MOSFETs), and KP is the intrinsic transconductance, another parameter set by the fabrication process and not under the designer's direct control (other than choosing a particular fabrication process). The factor *W*/*L* in (2.5) occurs very frequently in CMOS design (RFIC, analog and digital) and is often referred to as the MOSFET *aspect ratio*; this factor will be an important design parameter as will be seen throughout the book.

As V_{DS} increases, the voltage drop in the channel eventually equals the effective gate-to-channel voltage at the drain, at which point the current becomes relatively constant even if V_{DS} increases. This *saturation voltage* (V_{Dsat}) equals the difference between the gate-source voltage and V_T:

$$V_{Dsat} = \left(V_{GS} - V_T\right) \tag{2.6}$$

and the device is operating in the *saturation region* with I_{DS} described as

$$I_{DS} = \frac{1}{2}\mathrm{KP}\left(\frac{W}{L}\right)\left(V_{GS} - V_T\right)^2\left(1 + \lambda V_{DS}\right) \tag{2.7}$$

The additional term in (2.7), λ, is termed the *channel length modulation factor* and is used to quantify the nonzero slope of I_{DS} in the saturation region [Figure 2.6(a)]. Equation (2.7) shows that not every parameter governing the MOSFET I-V behavior is controllable by the designer; the operating voltage V_{GS} and the gate width *W* are the only design variables for a given CMOS process (and V_{DS} to a lesser extent since λ is typically much less than one).

2.2.2 The Basic *p*-Channel MOSFET

The *p*-channel MOSFET (*p*MOSFET) operates in a fashion similar to that of the *n*MOSFET device except that the current and voltage polarities (including the threshold voltage) are reversed [Figure 2.6(b)] as a direct consequence of the *N*- and *P*-type material comprising the *p*MOS devices (Figure 2.7). In addition, the primary current carriers in the channel region are holes. The channel mobility μ in *p*MOSFETs is therefore that of the holes; the channel mobility for holes is approximately three times less than that of electrons in silicon and so the current-handling capability of a *p*MOSFET is less than that of an identically sized *n*MOSFET for a given gate overdrive voltage ($V_{GS} - V_T$).

The standard voltage and current polarities for both *n*MOSFETs and *p*MOSFETs are shown in Figure 2.8. This figure shows a fourth node on both the *n*MOSFET and *p*MOSFET schematic diagrams labeled "B" in addition to the

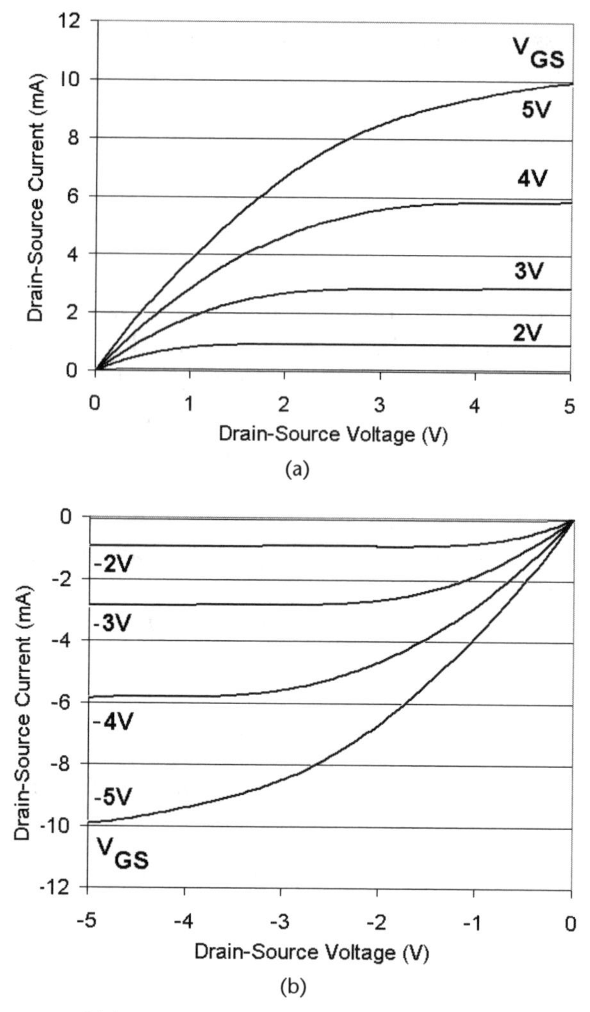

Figure 2.6 $I_{DS} - V_{DS}$ curve of (a) an *n*MOSFET with 0.66V threshold voltage and (b) a *p*MOSFET with –0.66V threshold voltage.

previously discussed gate, drain, and source nodes (G, D, and S, respectively). This B node is often referred to as the *bulk* or *substrate* node and is required for accurate modeling of the influence of the substrate on both the dc and RF characteristics of the MOSFET. The origin of this node can be found by looking closely at the cross-sectional view of the *n*MOSFET shown earlier in Figure 2.4. The drain and source connections on the surface of the N^+ layers define the corresponding nodes (D, S) on the *n*MOSFET schematic. The side and bottom of the N^+ layers, however, are in contact with the P-type substrate (or well) and are one connection to this bulk node B on the schematic. In addition, while the top of the inverted channel region is located at the oxide semiconductor interface, the bottom of the channel is in contact with the P-type substrate, again connected to the B node. To keep these PN junctions reverse biased (which is always the case unless a PN junction device is specifically required), the P substrate is usually biased at the lowest voltage available (usually ground). The converse is true for the N substrate (or well) in *p*MOSFET devices; the

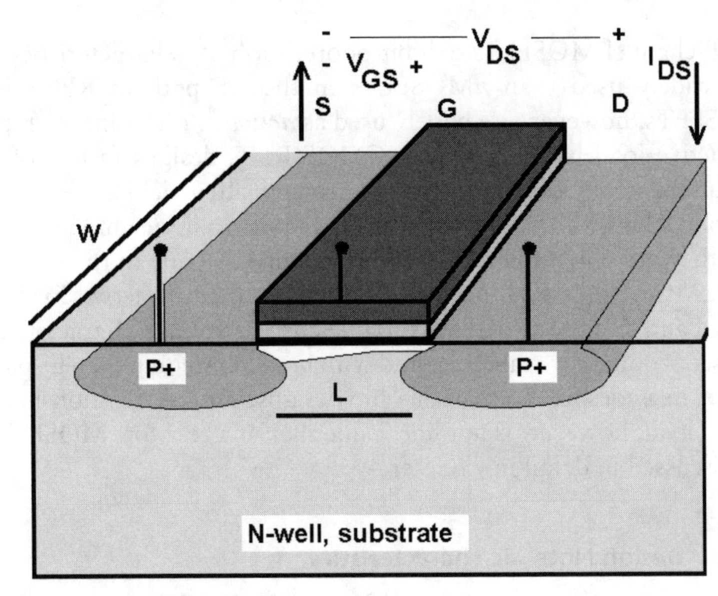

Figure 2.7 Cross section of *p*MOSFET showing channel region, device width *W*, and dc bias voltage polarities.

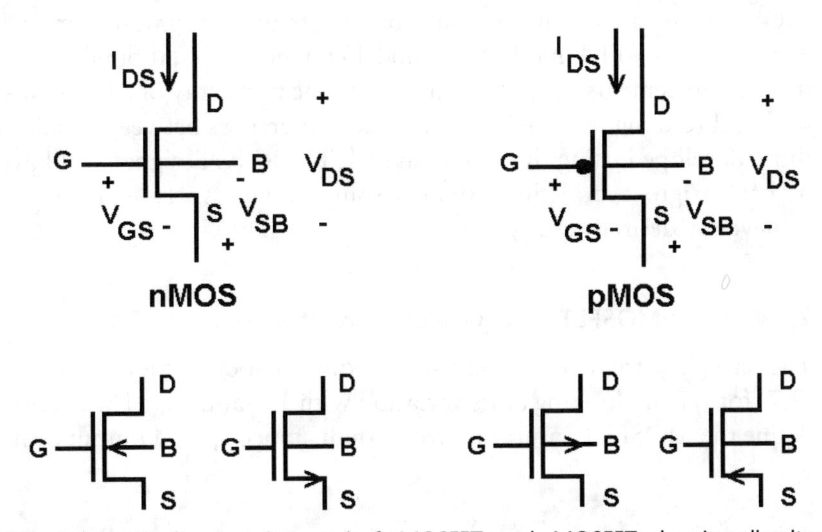

Figure 2.8 Schematic diagrams (top row) of *n*MOSFETs and *p*MOSFETs showing all voltage and current polarities used throughout the text. Alternative schematic diagrams for the *n*MOSFET and the *p*MOSFET are shown on the bottom row.

substrate node is biased at the highest potential (usually V_{DD}). If one or more of these PN junctions (and the NPN/PNP structures) become forward biased, excessive current will flow and the CMOS circuit can *latch up* [12]. This latch-up condition is to be avoided at all costs because only a power cycling of the system can ensure a complete "unlatching" of the CMOS circuit. The effects of the reverse biased drain and source regions are introduced later in this chapter when MOSFET equivalent circuit capacitances are discussed.

P-channel MOSFETs exhibit poorer mobility characteristics and are therefore less widely used than *n*MOSFETs in the RF path in RFIC designs. *P*-channel MOSFETs, however, are widely used as *active loads* or in on-chip biasing schemes. An active load is often used in CMOS RFIC designs in lieu of a passive resistor because, in many cases, design requirements often yield large resistance values. Fabricating a high-value resistor typically requires a large amount of silicon real estate, and they are only accurate to within about 10% due to the inevitable variations in the CMOS process. A MOSFET configured as an active load, however, can be biased and sized to provide a suitable current voltage characteristic but with much less semiconductor real estate and with better matching than its passive counterpart; hence, its widespread use in on-chip designs. The I-V relationship for the MOSFET active load, however, is not linear like that of a resistor; MOSFET active loads will be discussed in detail in Chapter 4.

2.2.3 Design Note: dc characteristics

The preceding I-V analyses for both the *n*MOSFET and the *p*MOSFET are based on the simplistic *square-law I-V model*. The power of this model is that it provides a tractable means of performing a circuit design process that provides reasonable ballpark results that can be used as a starting point for circuit simulation. However, MOSFET operation is much more complex than the square-law model indicates and numerous MOSFET model and model enhancements abound [2–6, 13]. Many of these enhancements are programmed into the vast array of RF circuit simulators and so the RFIC designer should expect to see differences between simulation results and those developed using the simple model. The RFIC designer will have to iterate his or her design several times (which some designers say is an understatement) to achieve the desired results.

2.2.4 Basic MOSFET RF Equivalent Circuit Model

The basic low-frequency, small-signal equivalent circuit model can be derived from (2.7) for I_{DS} and looking at its variation with V_{GS} and V_{DS}. The resulting expressions define the MOSFET saturation transconductance g_m and output drain conductance g_{ds}:

$$g_m = \frac{\partial I_{DS}}{\partial V_{GS}} = \mathrm{KP}\left(\frac{W}{L}\right)(V_{GS} - V_T)(1 + \lambda V_{DS}) \cong \sqrt{2\mathrm{KP}\frac{W}{L} I_{DS}} \tag{2.8a}$$

$$g_{ds} = \frac{1}{r_{ds}} = \frac{\partial I_{DS}}{\partial V_{DS}} = 0.5\mathrm{KP}\lambda\left(\frac{W}{L}\right)(V_{GS} - V_T)^2 \cong \lambda I_{DS} \tag{2.8b}$$

The MOSFET g_m increases as the dc current increases but at the expense of reducing r_{ds}. The parameters I_{DS} and g_m are under the RFIC designer's control through the dc bias voltage on the gate (V_{GS}) and the MOSFET aspect ratio *W*/*L*. The simple equivalent circuit diagram for the MOSFET can then be drawn as shown in Figure 2.9(a). The *p*MOSFET has an identical equivalent circuit diagram, including current and voltage polarities. Care should be taken, however, when creating the

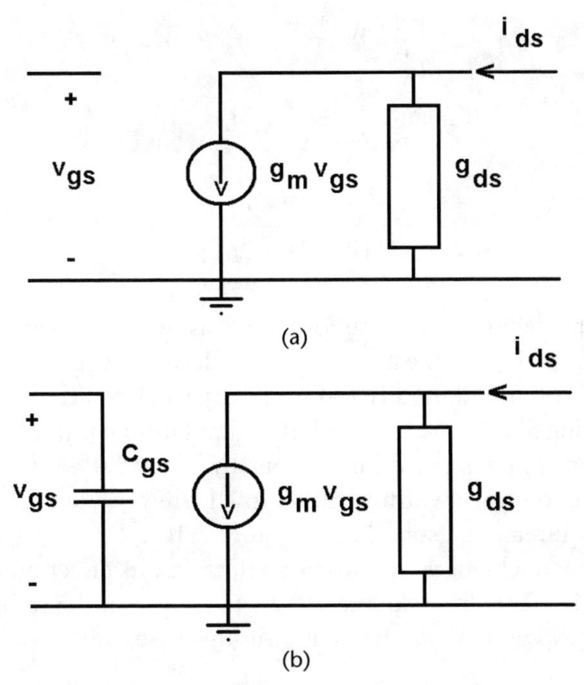

Figure 2.9 Basic RF equivalent circuit model (a) for the MOSFET and (b) for the MOSFET with C_{gs} included.

RF equivalent circuit diagram in a full CMOS circuit by recalling that the *p*MOSFET has its source at a *higher* potential (that is, closer to the V_{DD} rail) than the drain. Figure 2.9(b) shows a simple RF equivalent circuit model that includes a gate input capacitance, C_{gs}. Neglecting the output drain conductance g_{ds} for the moment, the current gain can be derived from the equivalent circuit diagram in Figure 2.9(b) as

$$|A_i| = \left|\frac{i_{ds}}{i_{gs}}\right| = \frac{g_m v_{gs}}{\omega C_{gs} v_{gs}} = \frac{g_m}{\omega C_{gs}} \tag{2.9}$$

Based on (2.9), the *unity current gain frequency*, f_T, is defined as the frequency at which $|A_i| = 1$:

$$f_T = \frac{g_m}{2\pi C_{gs}} \tag{2.10}$$

Example 2.1: Determine the *n*MOSFET width *W* that yields a constant current in saturation of 5 mA at $V_{DS} = 3.3$V with 2.0V on the gate. Use the 0.5-μm parameters listed in Table 2.1 for the design (ch2-1.txt).

Using (2.7) and performing some algebra gives the required *n*MOSFET gate width for 5 mA of current:

$$I_{DS} = \frac{1}{2} KP \frac{W}{L} (V_{GS} - V_T)^2$$

$$W = \frac{2I_{DS}}{KP(V_{GS} - V_T)^2} L$$

$$W = \frac{2(5 \cdot 10^{-3})}{56.3 \cdot 10^{-6}(2.0 - 0.66)^2} 0.5\mu\text{m} = 49.5\mu\text{m}$$

The preceding result was obtained using an approximate design formula for *n*MOSFET sizing, so the next step in the design process is to perform a simulation to verify the result and modify it if necessary to obtain the desired current. Simulation results using SPICE and 0.5-μm BSIM simulation parameters indicate the simple formula overestimates the drain current by about 20% (dashed line in Figure 2.10). Decreasing the gate width by 20% to 41 μm provides a much closer match to the design requirements (solid line in Figure 2.10).

The channel length modulation effect and its effect on g_{ds} can be readily seen in Figure 2.10. The drain-source conductance g_{ds} can be estimated by calculating the change in I_{DS} with V_{DS} at fixed V_{GS} (in this case, 2.0V):

$$g_{ds} = \frac{\Delta I_{DS}}{\Delta V_{DS}} = \frac{(5.23 - 5.09)\text{ mA}}{(3.0 - 2.5)V} = 0.28 \text{ mS}$$

$$r_{ds} = \frac{1}{g_{ds}} = 3.57 \text{ K}\Omega$$

Example 2.2: Estimate and verify the transconductance in saturation for the previous *n*MOSFET in Example 2.1 (ch2-2.txt).

Using the optimized value of *W* from Example 2.1, the transconductance g_m can be estimated from (2.8):

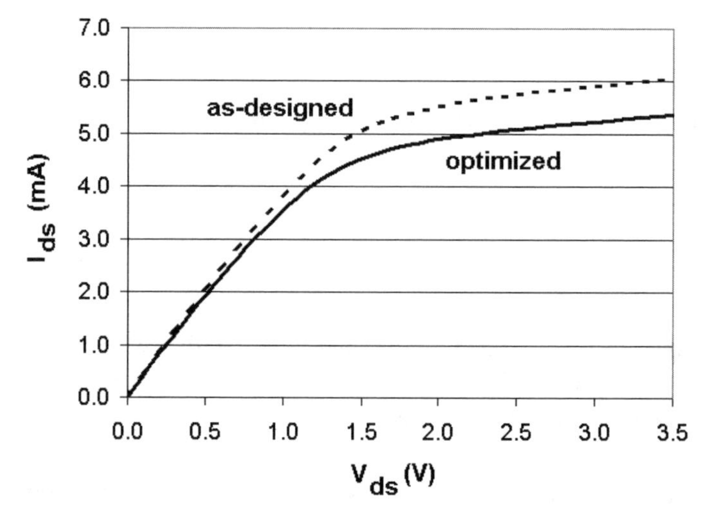

Figure 2.10 Simple dc bias circuit for an *n*MOSFET (V_{GS} = 2.0V) with designed and optimized I_{DS}–V_{DS} characteristics.

$$g_m = \mathrm{KP}\left(\frac{W}{L}\right)(V_{GS} - V_T) = 56.3 \cdot 10^{-6}\,\frac{41}{0.5}(2 - 0.66) = 6.18\ \mathrm{mS}$$

SPICE simulations can be used to verify this result. Figure 2.11 shows the I-V characteristics for the *n*MOSFET ($W/L = 41/0.5$) at 1.9, 2.0, and 2.1V on the gate. The g_m can be estimated by looking at the change in I_{DS} with respect to V_{GS} at constant V_{DS} (for this calculation, $V_{DS} = 3.0$V is assumed):

$$g_m = \frac{\Delta I_{DS}}{\Delta V_{GS}} = \frac{(5.58 - 4.88)\ \mathrm{mA}}{(2.1 - 1.9)V} = 3.5\ \mathrm{mS}$$

2.2.5 Advanced MOSFET RF Equivalent Circuit Model

2.2.5.1 Drain, Source Capacitances, CV, Area, and Sidewall

Among the important factors limiting MOSFET RF performance are the capacitances that are an inherent part of the device structure (Figure 2.12). These MOSFET capacitances are functions of the actual device size, with the gate width W being a consistent parameter in all capacitance estimation. The gate oxide (C_{gc}), gate overlap (C_{ovr}), and bulk channel (C_{cb}) capacitances are the components that make up the MOSFET gate source capacitance, C_{gs}, and are dependent on fabricator process parameters (which the designer again does not have control over for a given process) and the gate width W (which the designer does have control over):

$$\begin{aligned}
C_{ovr} &= C_{gx0} \cdot W \\
C_{gc} &= \frac{\varepsilon_{ox}}{t_{ox}} \cdot WL \\
C_{gs} &= C_{ovr} + \frac{C_{gs} C_{cb}}{C_{gc} + C_{cb}} \cong C_{ovr} + C_{gc} \\
C_{gd} &= C_{gd0} \cdot W
\end{aligned} \tag{2.11}$$

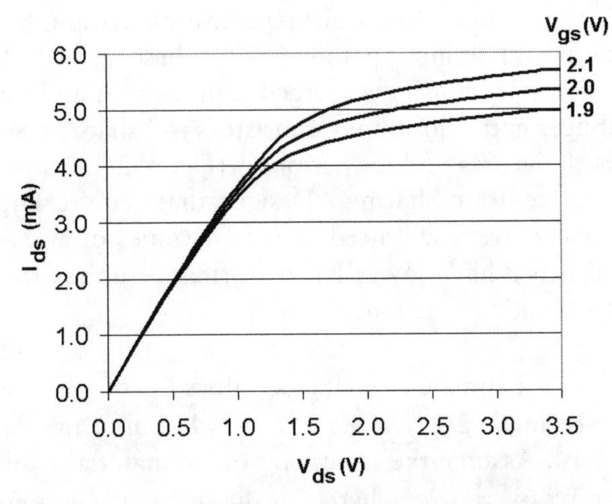

Figure 2.11 I_{DS}–V_{DS} curves for three different dc bias voltages for computation of transconductance g_m.

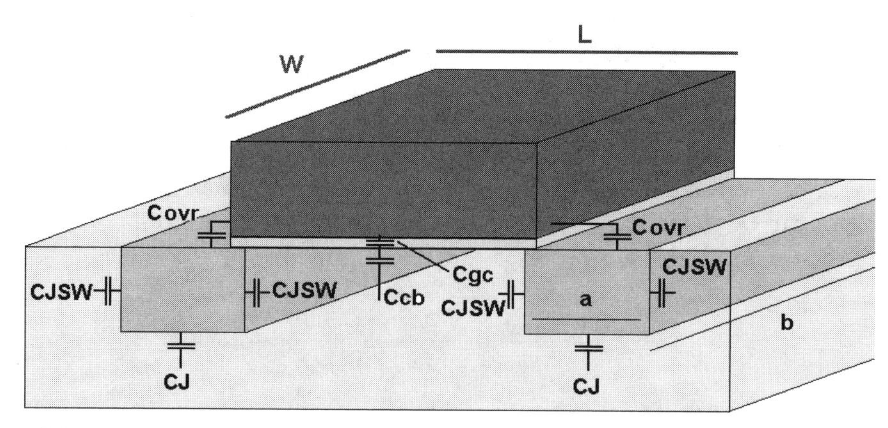

Figure 2.12 Capacitances in the MOS structure.

where C_{gd0} is the gate-drain feedback capacitance and C_{gx0} overlap capacitance (X is either drain or source) per unit length. Capacitance C_{cb} in Figure 2.12 is primarily a depletion region capacitance and is typically much greater than C_{gc}; hence the parallel combination of the two is simply C_{gc}.

The drain and source regions are reverse biased PN junctions under typical operating conditions and define the drain-bulk and source-bulk capacitances (C_{db} and C_{ds}, respectively). The PN junction surrounds the diffused region on three sides, yielding a capacitance dependent not only on the area of the diffused region [(a)–(b) in Figure 2.12] but also the perimeter ($2a + 2b$). The area (perimeter) term combines with the process area C_J (sidewall C_{JSW}) capacitance per unit area (length) parameter to form the drain-bulk C_{db} and source-bulk C_{sb} diffusion capacitance contributions (Figure 2.12):

$$\begin{aligned} C_{db(sb)} &= \text{Area} \cdot C_J + \text{Periphery} \cdot C_{JSW} \\ C_{db(sb)} &= (ab) \cdot C_J + 2(a+b) \cdot C_{JSW} \end{aligned} \tag{2.12}$$

The area and perimeter terms are direct input in circuit simulators and must be included for an accurate frequency response simulation. (For example, AS, AD, PS, and PD are the SPICE input parameters for these terms.) These capacitances can be reduced somewhat by using a merged source-drain and multifingered gate layout. The multifinger gate allows two transistors to "share" a source and drain, thereby reducing both the area and the periphery (Figure 2.13 shows a two-finger MOSFET with this source-drain sharing.) Designs that require large transconductance g_m and/or dc bias current will need increased values of W and exhibit larger capacitances, and hence have lower RF performance unless the layout is optimized to reduce the parasitic capacitance.

Example 2.3: Estimate capacitance values C_{gs}, C_{gd}, C_{sb}, and C_{db} for the nMOSFET designed in Example 2.1 ($L = 0.5\ \mu$m, $W = 41\ \mu$m) using the top view layout shown in Figure 2.14. Assume the insulating oxide material exhibits a relative dielectric constant of 3.9 ($\varepsilon_{ox} = 3.9\varepsilon_0$ where ε_0 is the free space permittivity, 8.85 pF/m).

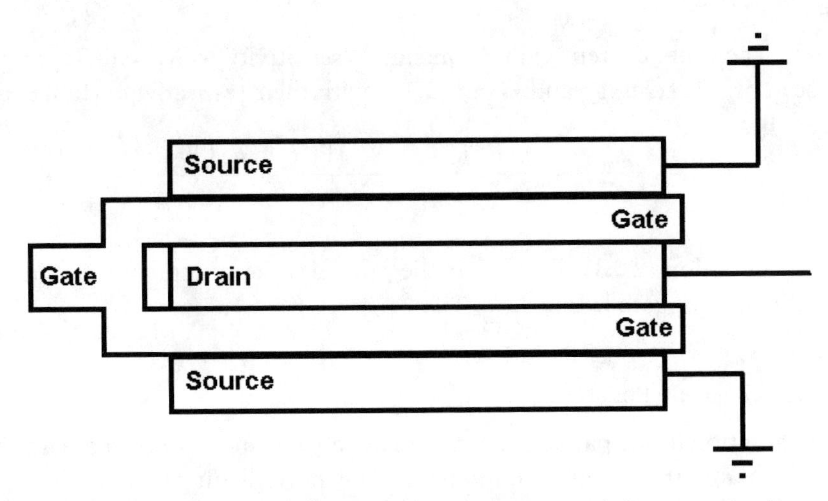

Figure 2.13 MOSFET with two gate fingers and a merged source-drain connection for capacitance reduction.

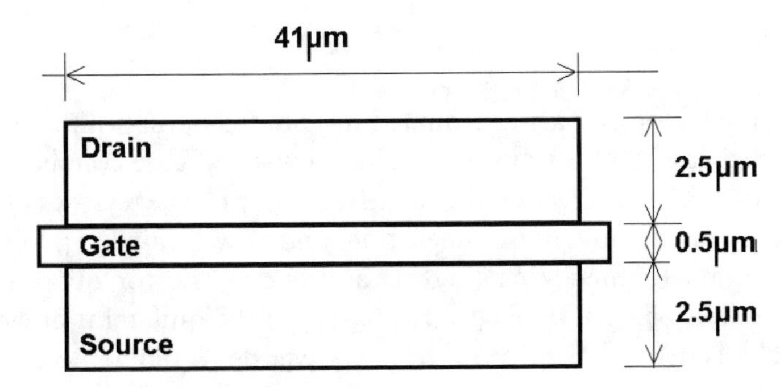

Figure 2.14 Example layout of a MOSFET.

From the figure, the area of the drain and source are $41 \times 2.5\ \mu\text{m}^2 = 102.5\ \mu\text{m}^2$, and the perimeter of the drain and source diffused regions is $2(2.5\ \mu\text{m} + 41\ \mu\text{m}) = 87$ μm. Using the sample parameters listed in Appendix A, each capacitance can be computed:

$$C_{ovr} = C_{gx0} \cdot W = 2.18 \cdot 10^{-10} \cdot 41 \cdot 10^{-6} = 8.94 \text{ fF}$$

$$C_{gc} = \frac{\varepsilon_{ox}}{t_{ox}} \cdot WL = \frac{3.9\varepsilon_0}{138 \cdot 10^{-10}} \cdot 41 \cdot 10^{-6} \cdot 0.5 \cdot 10^{-6} = 51.3 \text{ fF}$$

$$C_{gs} = C_{ovr} + C_{gc} = 60.24 \text{ fF}$$

$$C_{gd} = C_{gd0} \cdot W = 2.18 \cdot 10^{-10} \cdot 41 \cdot 10^{-6} = 8.94 \text{ fF}$$

$$C_{db(sb)} = AC_J + PC_{JSW} = 102.5 \cdot 10^{-12} \cdot 4.27 \cdot 10^{-4} + 87 \cdot 10^{-6} \cdot 3.05 \cdot 10^{-10}$$

$$C_{db(sb)} = 43.7 \text{ fF} + 26.6 \text{ fF} = 70.3 \text{ fF}$$

The unity current gain frequency f_T sensitivity to MOSFET parameters can be seen by replacing g_m and C_{gs} in (2.10) with their respective formulas to give

$$f_T = \frac{g_m}{2\pi C_{gs}} = \frac{1}{2\pi}\frac{KP(V_{GS} - V_T)W/L}{\varepsilon_{ox}/t_{ox}\cdot WL} = \frac{\mu_n(V_{GS} - V_T)}{2\pi}\frac{1}{L^2} \tag{2.13}$$

Equation (2.13) shows that the most dramatic improvement in f_T comes from reduction of gate length L.

2.2.5.2 Gate Resistance

In addition to the parasitic capacitances, a resistance associated with the polysilicon gate material must also be modeled. This polysilicon gate exhibits a nonzero sheet resistance R_S and therefore contributes to a series gate resistance R_G. Gate resistance R_G is partly a distributed resistance and is estimated using the following expression for multifinger (N fingers) gate MOSFETs:

$$R_G \cong \frac{1}{k}\frac{1}{N}\frac{W}{L}R_S \tag{2.14}$$

where the factor of k arises from the distributed nature of the resistance and is 3 for one-sided gate connections and 12 for two-sided gate connections [14]. This resistance has a large impact on the high-frequency characteristics of MOSFETs, such as impedance matching, noise generation, and power gain [15]. In addition to the unity current gain frequency f_T, there is another metric for quantifying the upper frequency limit of a MOSFET that is based on the unilateral power gain, G_{TU}. Power gain G_{TU} is defined as the ratio of the power delivered to a conjugate matched[2] load $P_{L\text{-}C}$ with respect to the available input power from the conjugate matched generation $P_{avs\text{-}C}$, and is the most gain that can be possibly be expected from the device [16]. The *maximum frequency*, $f_{\max}$, is defined as the frequency at which the unilateral power gain $|G_{TU}|$ equals one and is given in (2.15) [15, 16]:

$$f_{\max} \cong \sqrt{\frac{f_T}{8\pi R_G C_{gd}}} \tag{2.15}$$

where R_G is the resistance in the gate lead and C_{gd} is the gate-drain feedback capacitance.

The $f_{\max}$ dependence on MOSFET sizing can be seen by noting that f_T scales as $1/L^2$, R_G scales as W/L, and C_{gd} scales as W:

$$f_{\max} \propto \sqrt{\frac{1}{W^2 L}} \tag{2.16}$$

This dependence indicates that smaller MOSFETs have higher $f_{\max}$ values.

Example 2.4: Estimate f_T and $f_{\max}$ for the nMOSFET design in Examples 2.1 through 2.3. Assume a single-finger MOSFET gate layout ($N = 1$).

The only term left to estimate is the gate resistance, R_G. The example parameters in Table 2.1 show that the sheet resistance for the gate is 23.4 Ω/sq; (2.14) can now be used to estimate this parameter:

$$R_G = \frac{1}{3}\frac{W}{L}R_S = \frac{1}{3}\frac{41}{0.5}23.4 = 639\Omega$$

The unity current gain frequency f_T can be calculated from (2.10) as

$$f_T = \frac{g_m}{2\pi C_{gs}} = \frac{6.18 \cdot 10^{-3}}{2\pi \cdot 60.2 \cdot 10^{-15}} = 16.3 \text{ GHz}$$

and the maximum frequency $f_{\max}$ can be calculated from (2.15):

$$f_{\max} \cong \sqrt{\frac{f_T}{8\pi R_G C_{gd}}} = \sqrt{\frac{16.3 \cdot 10^9}{8\pi \cdot 639 \cdot 8.94 \cdot 10^{-15}}} = 10.7 \text{ GHz}$$

2.2.5.3 Bulk, RF Substrate Effects, and Threshold Voltage

Bulk CMOS technology exhibits a heavily doped substrate directly beneath both the active and passive devices and must be considered for accurate RFIC designs in MOS technology. For MOSFETs, the region under the channel is assumed to reverse bias the depletion regions (negative potential or ground for *n*MOSFETs, positive potential for *p*MOSFETs). Any voltage in the bulk region under the gate will tend to increase V_T and therefore change both the dc operating point of the MOSFET as well as the RF equivalent circuit parameters (g_m and g_{ds}). For a general source-bulk voltage V_{SB}, the modified threshold voltage becomes

$$V_T = \frac{t_{ox}}{\varepsilon_{ox}}\sqrt{2\varepsilon_{Si} q N_{sub}\left(2\phi_B + |V_{SB}|\right)} + 2\phi_B \tag{2.17}$$

This so-called *bulk effect* is primarily seen in "stacked" transistors where the MOSFET sources are not directly connected to their respective rails (where the source bulk V_{SB} is zero) but have active or passive devices between the source and power rail.

Because of the heavily doped substrate, the path from the bulk-substrate node to RF ground is influenced by the skin depth and is not directly under the MOSFET. This "effective distance" into the substrate for RF ground is a frequency-dependent parameter, which can be approximated by [17]

$$h_{sub} = \frac{1-j}{2}\sqrt{\frac{1}{\pi f \mu_0 \sigma_{Si}}}\tanh\left[(1+j)h_{bulk}\sqrt{\pi f \mu_0 \sigma_{Si}}\right] \tag{2.18}$$

where h_{bulk} is the distance from the oxide-substrate boundary to the physical RF ground connection and σ_{Si} is the conductivity of the silicon substrate or well. Figure 2.15 shows the relationship between the physical distance to the ground plane (normalized to the substrate skin depth δ) and the normalized distance to the RF ground

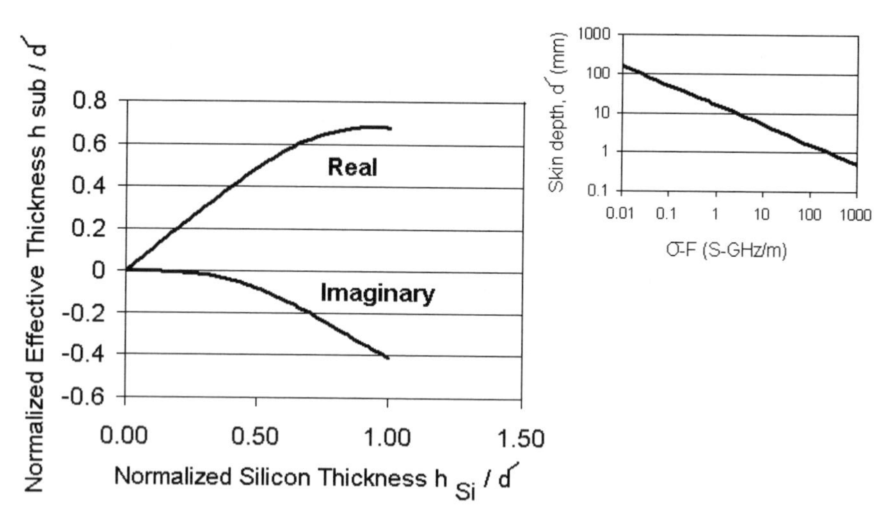

Figure 2.15 Effective distance to RF ground as a function of physical distance to RF ground, normalized to the substrate skin depth. (*After*: [18].)

[18]. Figure 2.15 shows little difference between the effective and actual RF thicknesses for large skin depths. As the physical distance to the RF ground approaches the skin depth, however, the effective distance to RF ground parameter h_{SUB} begins to display complex behavior. This form of h_{SUB} can be used in deriving an RF equivalent circuit model for substrate effects if a parallel RF circuit based on a lossy parallel plate structure is assumed. The parasitic substrate resistance (R_{bulk}) and capacitance (C_{bulk}) can be estimated assuming a parallel plate lossy capacitor model. (2.18) shows that the substrate or well doping and the distance from the bottom of the channel to the effective ground is related to the real or imaginary part of h_{SUB}, yielding the following equivalent circuit parameters:

$$R_{bulk} = \frac{\mathrm{RE}\{h_{SUB}\}}{\sigma_{Si} WL} \quad C_{bulk} = \frac{\varepsilon_{SUB} WL}{\mathrm{IM}\{h_{SUB}\}} \tag{2.19}$$

Combining all of the parasitic capacitances, resistances, and bulk effects yields an RF equivalent circuit model that is widely used in RFIC design (Figure 2.16). This model does not include other parasitic effects such as contact resistance, but these extra elements can be added during simulation setup.

2.2.5.4 MOSFET Noise Model

The MOSFET RF equivalent circuit (Figure 2.16) shows two resistive elements in the direct RF path, R_G and r_{ds} (neglecting substrate effects), which are the main sources of thermal noise in MOSFETs. A third noise source, the induced gate noise, which models the noise coupled from the channel to the gate, is typically much small than that due to r_{ds} and hence will be neglected [15]. The output drain channel thermal noise can be modeled as a current source from the drain to source with mean square value given by

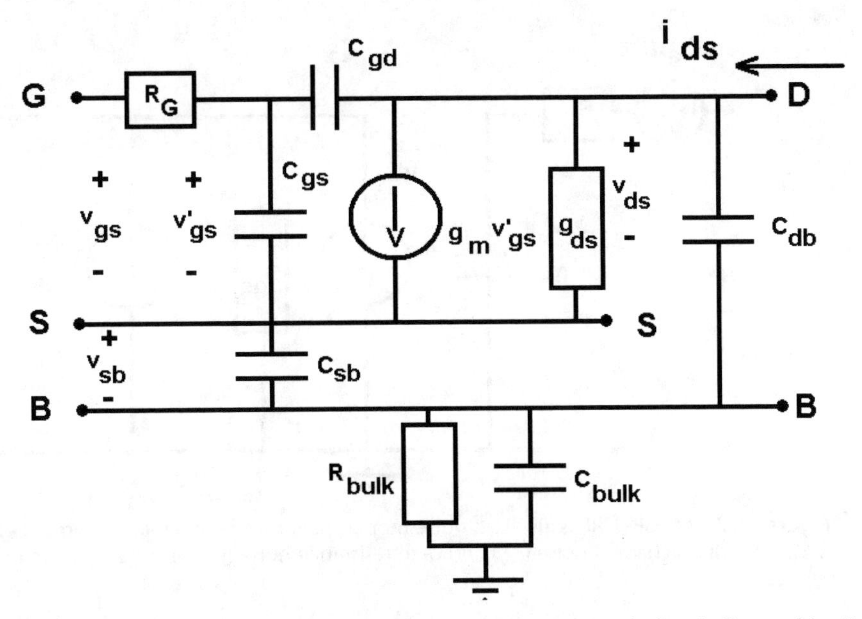

Figure 2.16 Small-signal RF equivalent circuit model of MOSFET including bulk parasitic effects. The equivalent circuit model is greatly simplified if the source and the bulk are at the same potential (C_{sb}; R_{bulk} and C_{bulk} are shorted out and effectively removed).

$$\bar{i}^2_{dN} = 4kT\gamma g_{d0} B \tag{2.20}$$

where γ is typically between 0.5 and 1.0, and g_{d0} is the effective channel conductance and is typically calculated using the expression for g_m. The thermal noise generated by the gate resistance is modeled as a noise voltage source whose mean square value is given by

$$\bar{v}^2_{gN} = 4kTR_G B \cong 4kT \frac{1}{kN} \frac{W}{L} R_S B \tag{2.21}$$

where R_S is the gate sheet resistance. Figure 2.17 shows the equivalent noise model for the MOSFET (with substrate effects neglected for clarity).

From the general discussion in Chapter 1 on system noise, the noise characteristics of an RF stage are calculated with noise sources assumed to be at the input (so-called *input referenced noise*). In keeping with this assumption, the output drain channel noise represented by (2.20) must be referred back to the input of the MOSFET so that, along with the gate resistance noise, meaningful noise computations can be performed. The general technique for input referencing output noise sources is to assume both voltage and current noise sources at the input and equate their noise contributions at the output with the drain channel noise [19]. By neglecting g_{ds} and all capacitances except C_{gs}, simplified noise equivalent circuits that must equate are shown in Figure 2.18. From Figure 2.18(a), the output noise current i_{dN} originates from the feed-forward term $g_m v'_{gs}$:

$$\bar{i}_{dN} = g_m \bar{v}'_{gs} \tag{2.22}$$

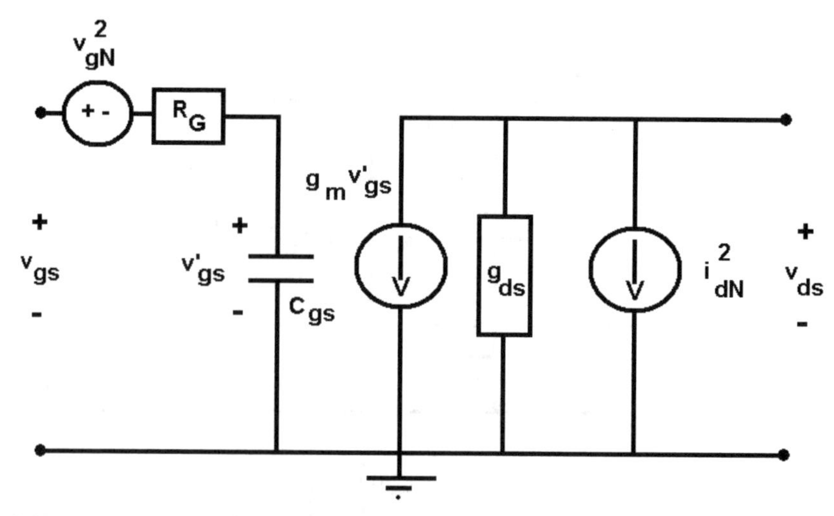

Figure 2.17 MOSFET RF equivalent noise model showing the two major sources of noise in MOSFETs: drain channel noise (i_{dN}^2) and gate resistance noise (v_{gN}^2). Bulk effects are neglected in the figure.

Assuming for a moment that there is no input gate noise voltage ($V_{gN}{}^2 = 0$), the mean square value of the input noise current source $\bar{i}_{dN_{\text{in}}}^2$ can be written in terms of the output mean square drain noise current $\bar{i}_{dN_{\text{in}}}^2$ as

$$\bar{i}_{dN_{\text{in}}}^2 = \omega^2 C_{gs}^2 \bar{v}_{gs}'^2 = \omega^2 C_{gs}^2 \frac{\bar{i}_{dN}^2}{g_m^2} = \frac{f^2}{f_T^2} \bar{i}_{dN}^2 = 4kT\gamma g_{d0} B\left(\frac{f}{f_T}\right)^2 \tag{2.23}$$

Similarly, by only considering the voltage noise source V_{dN} at the input, $v'_{gs} = V_{dN}$ and the mean square value of the input noise voltage source may be written as

$$\bar{V}_{dN}^2 = \bar{v}_{gs}'^2 = \frac{\bar{i}_{dN}^2}{g_m^2} \tag{2.24}$$

Note that the input referenced noise voltage ($\bar{V}_{dN}^2$) and noise current ($\bar{i}_{dN_{\text{in}}}^2$) sources have common terms and are therefore correlated. Equations (2.23) and (2.24) show that the input referenced channel noise contribution to the overall device noise may be minimized using high transconductance g_m and f_T MOSFETs. The gate resistance noise source V_{gN} is uncorrelated with the input referenced drain channel noise sources; therefore, the two mean square values of the noise voltage sources are additive:

$$\bar{V}_{TOT_N}^2 = \bar{V}_{dN}^2 + \bar{V}_{gN}^2 \tag{2.25}$$

The full impact of MOSFET noise will be discussed in further detail in Chapter 4 where this phenomenon plays an important role in estimating the noise figure in LNA circuits.

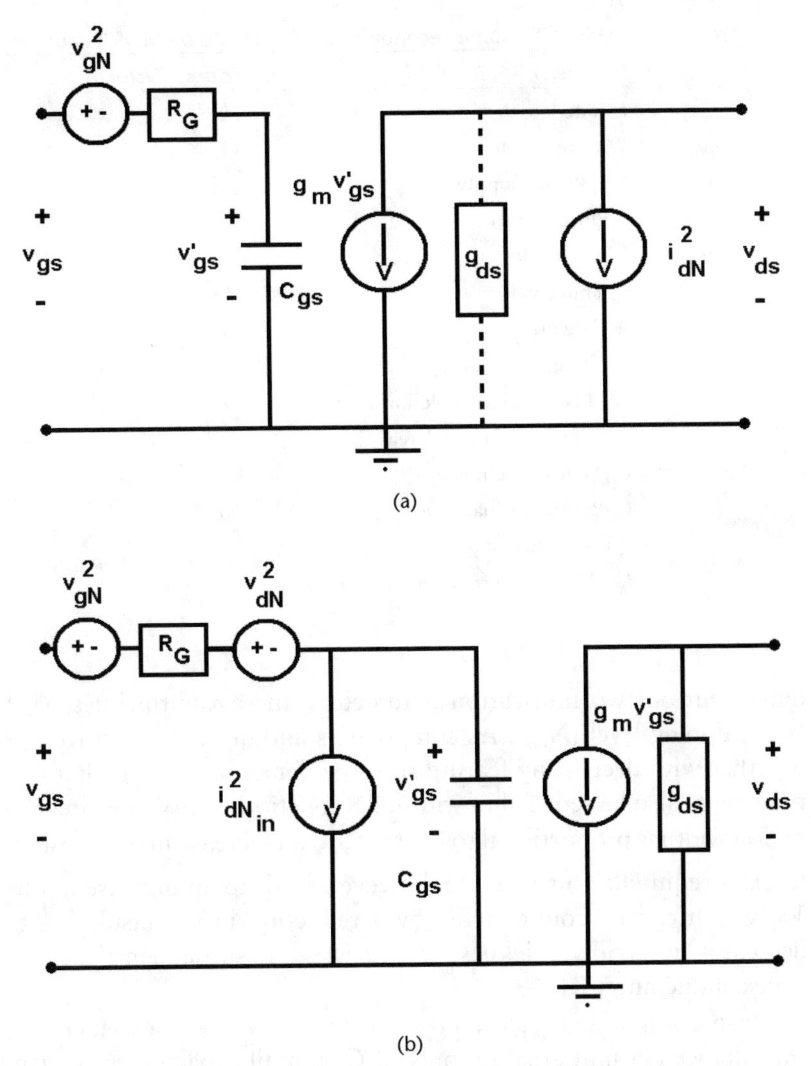

Figure 2.18 (a) Noise equivalent circuit for the MOSFET showing both input gate resistance and output channel noise sources. (b) Input referenced noise equivalent circuit. The output response of this noise equivalent circuit is identical to the noise equivalent circuit shown in part (a).

2.2.5.5 Impact of Technology Scaling on MOS Parameters

A set of *scaling rules* has been developed to show improvements and degradations due to the reduction in minimum feature size L. In modern CMOS technology, scaling of gate length L also requires scaling of many of the other fabrication parameters associated with the technology. If L is scaled by a factor α (α greater than unity), the other MOS parameters must vary by this factor, as shown in Table 2.2. The origin of these can be seen by close inspection of (2.3) through (2.5) and (2.11) as well as knowledge of the electric field in the MOSFET and surrounding structures.

To keep the electric field constant (and hence in the low field regime), the voltages in the circuit must also scale by this same factor, yielding a reduction in the overall power supply voltage, V_{DD}. Because V_{DD} reduces by the factor α, the threshold voltage V_T must also be reduced. Note, however, that the threshold voltage V_T is

Table 2.2 Some Technology Scaling Effects on MOS Parameters

Parameter	*Scaling Factor*
L Gate length	$1/\alpha$
W Gate width	$1/\alpha$
V_{DD} Power supply	$1/\alpha$
E Electric field	1
J Current density	α^2
t_{ox} Thin oxide	$1/\alpha$
A Gate area	$1/\alpha^2$
N_{sub} Substrate doping	α
W_D PN junction depletion width	$1/\alpha$
C_{ox} Gate capacitance/area	α
C_G Gate capacitance	$1/\alpha$
C_x Parasitic capacitance	$1/\alpha$
f_T Unity current gain frequency	α
f_{max} Maximum frequency	α

dependent on two fabrication parameters, the oxide thickness (t_{ox}) and substrate (well) doping level (N_{sub}). As components and devices are moved physically closer together with decreasing feature size, the depletion regions that occur at PN junctions must also be reduced in width. PN junction theory indicates that the depletion region width is proportional to $1/\sqrt{N_{sub}}$, so an increase in the substrate (well) doping levels is required. This increase, however, leads to an increase in threshold voltage, V_T, which can be counteracted by a reduction in the insulating thickness, t_{ox}. A decrease in oxide thickness t_{ox} causes a subsequent increase in intrinsic transconductance, KP.

Inspection of (2.11) shows that the term ε_{ox}/t_{ox} will also increase, increasing the capacitance per unit area not only of C_{gc} but all capacitances in general. However, because both W and L are each reduced by the factor α, the overall capacitance scales as $1/\alpha$. Because all vertical and horizontal components reduce by α, the current density in conductors increases by α^2. Although these technology scaling rules are not hard and fast and have a great deal of variability from one process to another, they do give a broad estimate of the improvements or degradations that can be expected by a reduction in minimum IC feature size.

2.2.6 Linear Operation of the MOSFET

2.2.6.1 Basic Operation

When operated in the linear or so called *triode* region,[3] the MOSFET acts as a voltage-controlled resistor with the gate being the controlling voltage. Looking at the change

3. The term triode region is a holdover from vacuum tube terminology. In a vacuum tube, the output current (from the high-voltage plate) was controlled by a voltage applied to the two other terminals: the cathode (where the electrons originated) and the grid. The resulting curves were linear in nature and since three terminals were used, the term triode was created [20].

of I_{DS} with respect to V_{DS} (2.5) shows the dependence of fabrication parameters on the MOSFET resistance:

$$R_{ON} = \left(\frac{\partial I_{DS}}{\partial V_{DS}}\right)^{-1} = \frac{1}{\frac{W}{L}\text{KP}(V_{GS} - V_T)} \tag{2.26}$$

When operated in series with the RF path, bulk effects can cause an increase in the threshold voltage, requiring a somewhat larger W for a given value of resistance. Shunt connected MOSFETs (in which the source is tied directly to a rail) do not suffer from this bulk phenomenon. MOSFET switches interface well with digital control signals and can be used to switch various circuit components in and out of RF tuned circuits. (An example of MOSFETs used to switch in capacitors is shown in Figure 2.19.)

Example 2.5: Design an *n*MOSFET to exhibit 1000Ω at V_{GS} = 2.0V. Show the range of I_{DS} and V_{DS} where this *n*MOSFET can be considered a resistor. Assume L = 0.5 μm and V_T = 0.66V (ch2-3.txt).

Using the expression for the on-state resistance in the triode region, the *n*MOSFET gate width W can be derived as

$$R_{ON} = \frac{1}{\frac{W}{L}\text{KP}(V_{GS} - V_T)}$$

$$W = \frac{L}{R_{ON}\text{KP}(V_{GS} - V_T)} = \frac{0.5 \cdot 10^{-6}}{1000 \cdot 56.3 \cdot 10^{-6}(2.0 - 0.66)} = 6.6\ \mu\text{m}$$

Figure 2.20 shows the on-state resistance dependence on V_{DS} over the full power supply range. Note that the triode region assumption is only valid for V_{DS} less than approximately 0.5V (on the curve labeled "Transistor"), and R_{ON} varies between approximately 700 and 800Ω within the triode region. Decreasing W to 5.0 μm will bring R_{ON} to 1000Ω at approximately V_{DS} = 0.25V.

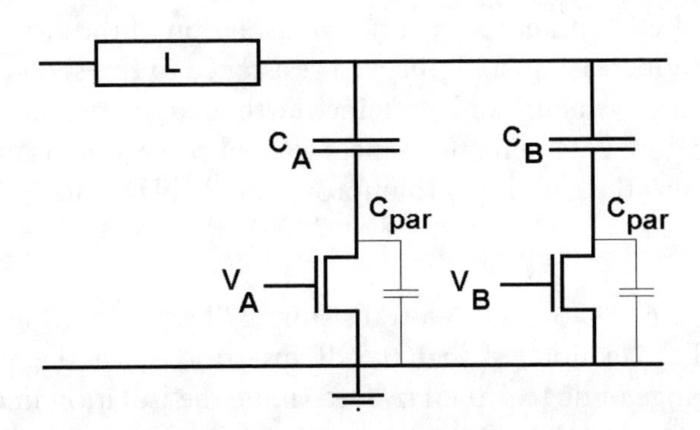

Figure 2.19 Example of MOSFET switching of lumped element components.

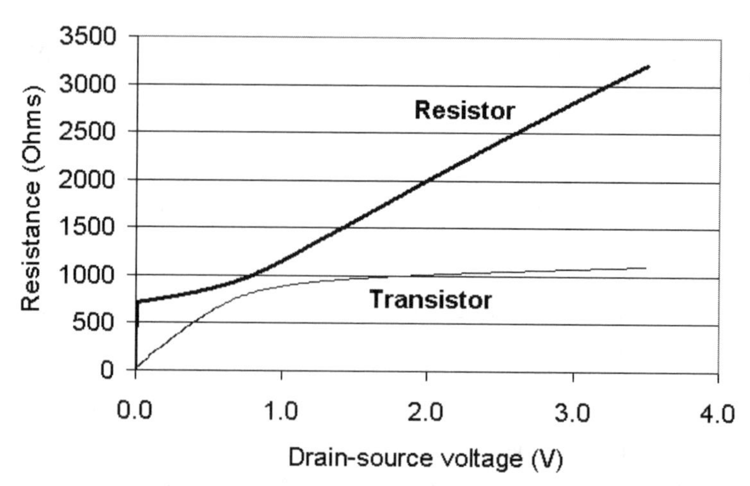

Figure 2.20 Drain-source resistance for the MOSFET in the linear (resistor curve) and saturation (transistor curve) regions. Of interest is the resistance below approximately 0.5V (linear region).

2.2.6.2 T/R and Element Switching

Figure 2.19 shows an example of the use of one such MOSFET switch application: a switch-tuned LC circuit. When V_A or V_B are biased above threshold, their respective capacitances C_A and C_B are connected to ground. This allows digital control of the total capacitance, allowing step variations of C_A, C_B, and $C_A + C_B$. In the MOSFET off-state, C_A and C_B are in series with the small parasitic capacitances of the switch (C_{par}). Detailed simulations are necessary to determine the influence of the MOSFET on-state resistance and parasitic capacitance on desired circuit behavior.

MOSFETs with multiple gate fingers can exhibit small on-state resistance with enough reduced parasitic capacitance as transmit-receive (T/R) switches for use in 50Ω RF systems where a receiver and transmitter share a common antenna. A common T/R single-pole, double-throw (SPDT) switch circuit is shown in Figure 2.21 where two control voltages are also indicated. A switch table showing applied control voltages and the resulting RF connection is shown in Table 2.3. Device aspect ratios as high as several thousand (depending on KP) are required for insertion loss values less than 0.5 dB [21, 22]. With these large MOSFET values for W, parasitic capacitances are also large for 50Ω switch applications and so SOI-type technology is often employed to improve frequency response. The off-state capacitance (primarily C_{gs} and C_{gd}) influences the off-state isolation of the switch; these also increase in value with increasing W. Improved resistance to threshold voltage effects and general RF improvement can be attained with a large number of substrate connections [22]. In Figure 2.21, gate resistances (R_G) of several thousand ohms are deliberately inserted into the gate leads to improve the dc bias isolation over a wide frequency range.

Example 2.6: For a two-transistor T/R switch (Figure 2.22), design the MOSFETs to exhibit less than 0.5-dB insertion loss in the on-state arm over the frequency range of dc to 1.0 GHz. Determine the isolation in off-state arm.

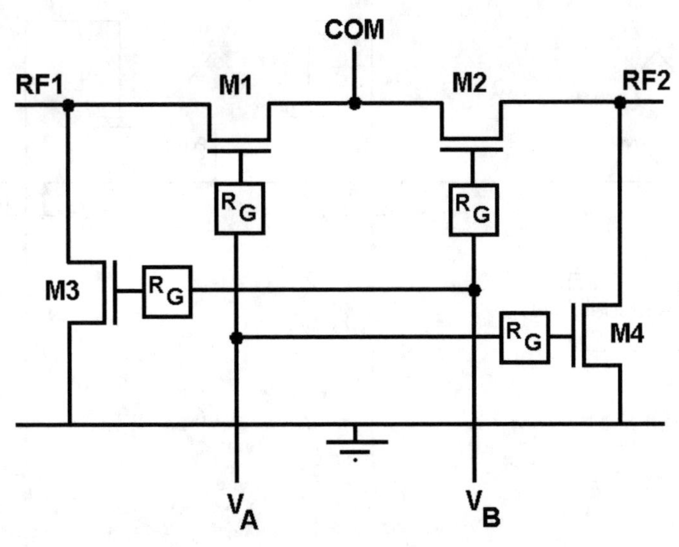

Figure 2.21 SPDT T/R switch using MOSFETs. (*After*: [21].)

Table 2.3 Switching Table for the SPDT T/R Switch Shown in Figure 2.21

V_A	V_B	*ON-MOSFETs*	*COM*
*V*1	*V*2	*M*1/*M*4	RF1
*V*2	*V*1	*M*2/*M*3	RF2

Note: *V*1 – gate voltage to turn the MOSFET on; *V*2 – gate voltage to turn the MOSFET off.

An insertion loss of 0.5 dB requires that the on-state resistance of the MOSFETs be on the order of 5.0Ω. For a 0.5-μm CMOS, this requires the following MOSFET gate width *W*:

$$R_{ON} = \frac{1}{\frac{W}{L}\text{KP}(V_{GS} - V_T)}$$

$$W = \frac{L}{R_{ON}\text{KP}(V_{GS} - V_T)} = \frac{0.5 \cdot 10^{-6}}{5 \cdot 56.3 \cdot 10^{-6}(3.3 - 0.66)} = 673\ \mu\text{m}$$

Figure 2.23 shows simulation results indicating that the 0.5-dB insertion loss specification has been met. The corresponding isolation is better than 40 dB at 1.0 GHz, a respectable isolation for this type of switch. This simulation represents an "ideal" simulation in that any RF pad capacitance, bulk effects, or substrate coupling between the RF ports (1 and 2) and the common or antenna port (3) have not been included. However, even this simple simulation compares favorably with published measurement data over a similar frequency range (Figure 2.24 [23]). If a part like this is to be a stand-alone packaged product, then package parasitics (bond wires, pad and package capacitance) have to be included. These packaging issues are reviewed in the next chapter.

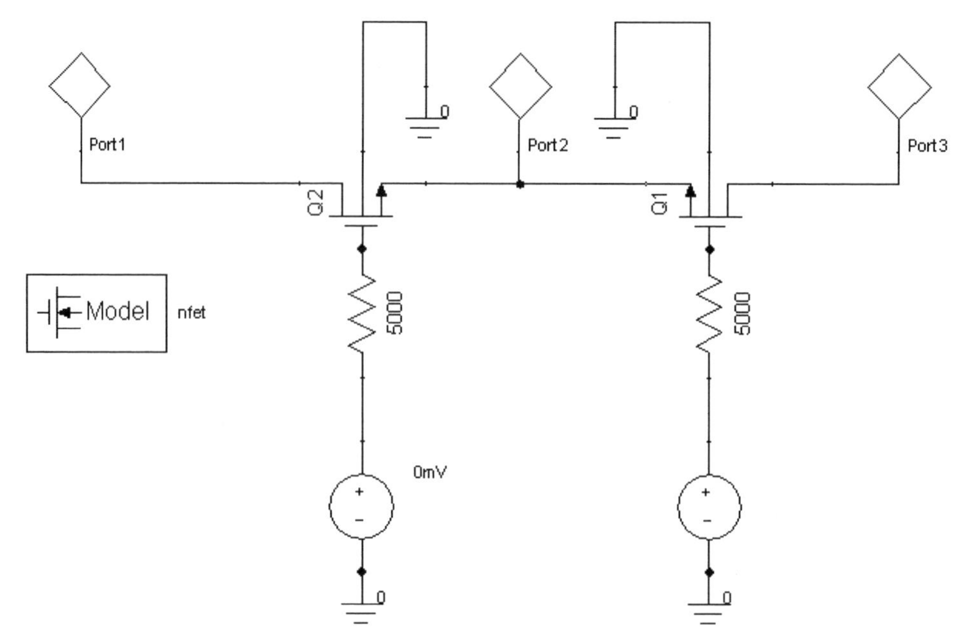

Figure 2.22 Simulation example: two-transistor MOSFET T/R switch for insertion loss and isolation characterization.

2.3 MOSFET Weak Inversion and Accumulation Operation

The simple square-law model for *n*MOSFET operation described earlier assumed that the channel inversion layer was only induced for gate voltages V_{GS} greater than or equal to the threshold voltage, V_T. In actuality, the channel does not invert abruptly at V_T but rather reaches an electron density in the channel twice that of the background acceptor doping. There are still electrons in the channel region even for $V_{GS} < V_T$, and hence a small I_{DS} can flow. This region of *n*MOSFET operation is often termed *subthreshold* or *weak inversion*. For very small V_{GS} or $V_{GS} < 0$, the channel region takes on the characteristics of the *P* substrate (or well). For V_{GS} significantly less than zero, the hole density is actually increased in the channel region, yielding a mode of operation termed *accumulation* (the holes, rather than electrons, accumulate under the gate). Subthreshold and accumulation modes of *n*MOSFET operation are discussed in this section. The *p*MOSFET operation in these regions below inversion is similar to that of the *n*MOSFET except for differences in carrier type and voltage polarities. These modes of operation can be exploited to create novel variable-voltage capacitors based on the MOSFET structure; these are also covered in this section.

2.3.1 Accumulation Mode

In the *n*MOS structure without any applied bias, the *P* substrate-well under the gate is the anode of two PN junctions whose cathodes are the drain and source diffusions (Figure 2.25). In this configuration, no matter what the voltage polarity across the drain to the source is, one of the diodes will be reverse biased, so no current will flow

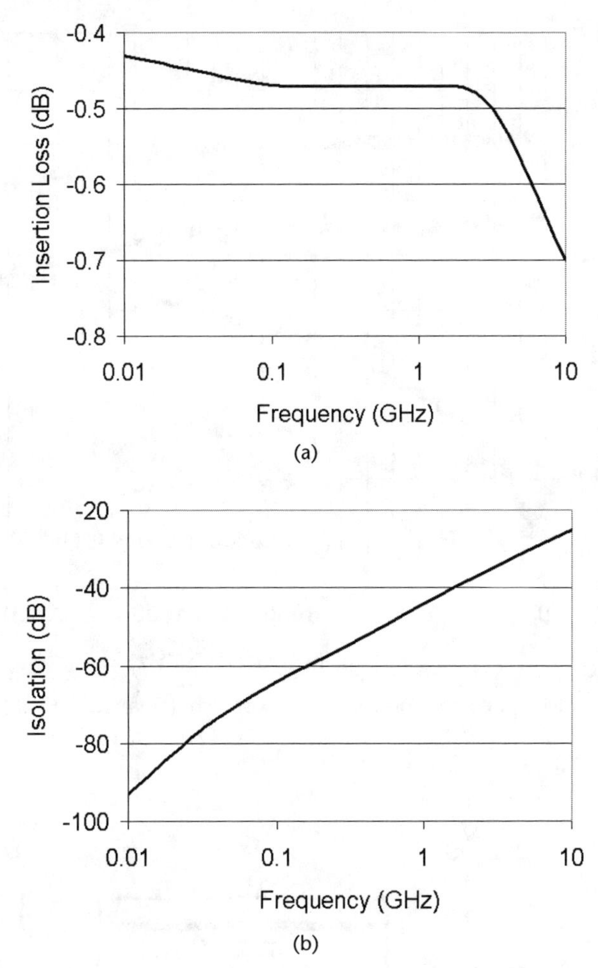

Figure 2.23 (a) Insertion loss and (b) isolation of the two-transistor MOSFET T/R switch. The insertion loss is approximately 0.48 dB up to 1.0 GHz, whereas the ideal isolation is better than 40 dB at 1.0 GHz.

(other than some small reverse current that will be neglected in this discussion). There is also a depletion region associated with the PN junctions that extends into the P material, with an associated depletion capacitance. Decreasing V_G causes the energy bands at the semiconductor oxide interface to bend to higher energies, causing even more holes to accumulate at this interface. This is termed the *accumulation mode* of MOSFET operation. The back-to-back diodes are still intact and no current flows regardless of the drain-source voltage polarity.

2.3.1.1 Capacitances

The capacitance between the gate and interface region is simply that due to the gate oxide, C_{ox}. The capacitance between the oxide semiconductor interface and drain and source is the series combination of C_{ox} and PN junction capacitances. The capacitance associated with the PN junctions is much larger than C_{ox} in accumulation due to a narrow junction region, and so both C_{gs} and C_{gd} in this mode are C_{ox}.

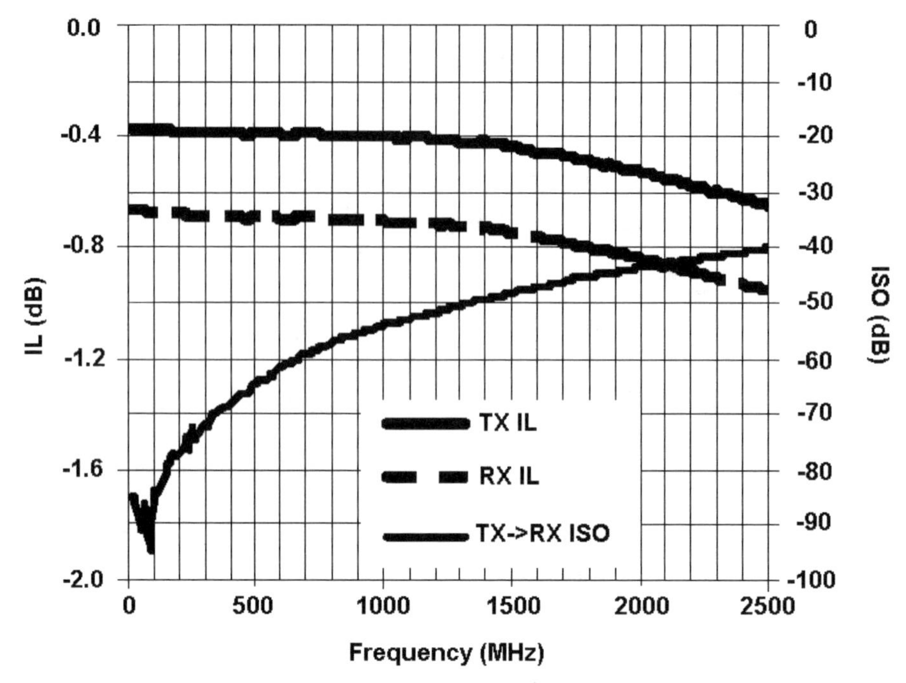

Figure 2.24 Measured data on a CMOS T/R switch. (*From:* [23]. © 2005 IEEE).

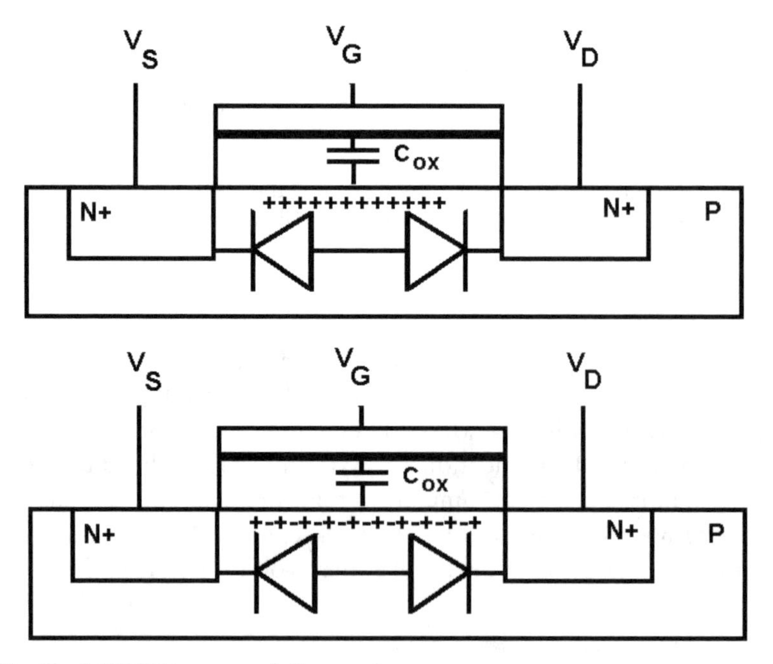

Figure 2.25 The MOSFET in accumulation mode.

2.3.2 Weak Inversion–Subthreshold Operation

Increasing V_G above 0V starts to bend the energy bands at the oxide semiconductor interface downward, causing electrons to migrate to this interface. The electrons are stopped by the insulating oxide and are therefore trapped in a thin interface layer. For small V_G, however, the number of electrons does not exceed the number of holes in the substrate-well under the gate; however, the number of holes there is being reduced or depleted and so this operational mode of the MOSFET is termed *depletion mode*. Little current flows since the number of holes exceeds the number of electrons and the PN junctions are still there. However, the depletion region associated with the PN junctions is changing and the associated capacitance is changing as well and becoming smaller as the charge stored in the region is reduced. The smaller capacitances in the depletion mode manifest themselves as a reduction in capacitance seen by the MOSFET gate.

Increasing V_G still further brings more electrons to the oxide semiconductor interface and eventually the number of electrons equals the number of holes supplied by the substrate-well doping, causing the MOSFET to enter the *weak inversion* or *subthreshold* mode of operation. There are now no longer PN junctions that can prevent the flow of current for an applied drain or source voltage of any polarity. However, the inverted region under the gate has so few electrons that any applied potential can inject electrons from the heavily doped drain-source regions into this channel region. As a voltage from drain to source is applied (V_{DS}), the potential on the drain side is higher than the source, making the effective potential across the inverted region actually less than the applied V_G; therefore, there is a reduction in electron number near the drain. However, the source side is unchanged (Figure 2.26). This asymmetric electron density under the gate causes a diffusion current to flow as the electrons from the higher concentration area (the source) move toward the area of lower concentration (the drain). While not a PN junction diode, the two do share this diffusion current concept and so the current flow from drain to source in the MOSFET will have an exponential variation with voltage.

2.3.2.1 The dc Characteristics (Subthreshold Current)

If the bulk node is grounded, the subthreshold current that flows from drain to source as a function of V_{GS} and V_{DS} can be written as [2]

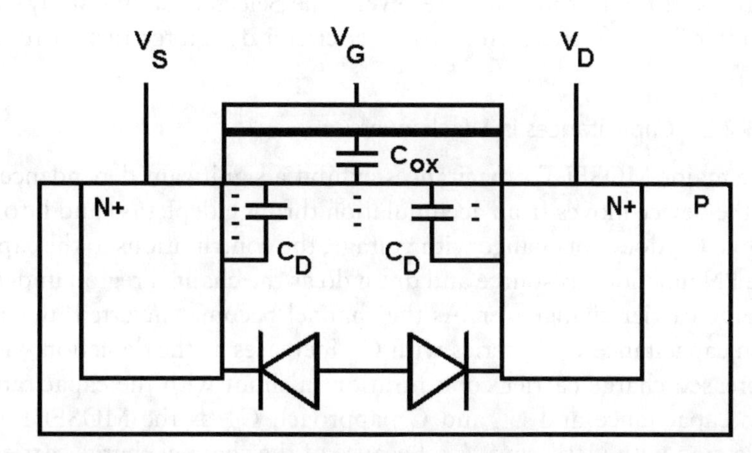

Figure 2.26 The MOS structure in weak inversion showing an electron density gradient in the region under the gate.

$$I_{DS} = \frac{W}{L}\frac{\mu}{2}\sqrt{\frac{q^2 \varepsilon_{silicon} N_{SUB}}{kT \ln\left(\frac{N_{SUB}}{n_i}\right)}}\left(\frac{kT}{q}\right)^2 e^{\frac{V_{GS}-V_T}{nkT/q}}\left(1-e^{\frac{V_{DS}}{kT/q}}\right) = \frac{W}{L} I_M e^{\frac{V_{GS}-V_T}{nkT/q}}\left(1-e^{\frac{V_{DS}}{kT/q}}\right) \quad (2.27)$$

where the factor n is the *subthreshold voltage shape factor* [24] and is given by

$$n = 1 + \frac{\gamma}{2\sqrt{2(kT/q)\ln\left(\frac{N_{SUB}}{n_i}\right)}} \quad (2.28)$$

The subthreshold region of operation can span several decades of I_{DS}, and with the exponential variation of I_{DS} with V_{GS}, there is a high sensitivity to signal variations on V_{GS}. Equation (2.29) can be used to find the gate voltage swing S needed to reduce I_{DS} by a factor of 10 [12]:

$$S = \ln 10 \frac{\partial V_{GS}}{\partial \ln I_{DS}} = \frac{\ln 10}{\partial \ln I_{DS} / \partial V_{GS}} = 2.3n\frac{kT}{q} \quad (2.29)$$

which shows that only a few kT/q ($kT/q = 25.8$ mV at 30°C) are needed for a decade change in I_{DS}. This sensitivity is also borne out in the expression for the *subthreshold transconductance* g_m, found using (2.27) and taking the derivative of I_{DS} with respect to V_{GS}:

$$g_m = \frac{\partial I_{DS}}{\partial V_{GS}} = \frac{W}{L}\frac{I_M}{kT/q} e^{\frac{V_{GS}-V_T}{nkT/q}}\left(1-e^{\frac{V_{DS}}{kT/q}}\right) = \frac{I_{DS}}{kT/q} \quad (2.30)$$

Operation of MOSFET devices in the low-current subthreshold region gives rise to a number of exciting low-power circuit possibilities. The low current these circuits provide, however, gives rise to slow rates of capacitance charge and discharge, thus limiting their use at very high frequencies but being ideal for lower frequencies such as in the IF range of a receiver. The Selected Bibliography at the end of this chapter lists references on low-power circuit design for further reading.

2.3.2.2 Capacitances in Weak Inversion

The major MOSFET capacitances exhibit a significant dependence on gate voltage as the device moves from accumulation through depletion and into weak inversion. While C_{ox} does not change with voltage, the contributions to the capacitances due to the PN junctions as source and drain do as the channel region undergoes the type of charge carrier changeover. As the channel becomes inverted, the associated depletion capacitance C_D in series with C_{ox} increases as the depletion width shrinks. The increased charge carrier concentration in shunt with the capacitance "shorts out" this capacitance and C_{gs} and C_{gd} approach C_{ox} as the MOSFET moves to strong inversion with increasing V_{GS}. Because of the channel charge carrier gradient, there is more charge at the source end than drain end, resulting in larger voltage variations in C_{gs} than in C_{gd} by virtue of its larger capacitance (see Figure 2.31 later in this chapter).

2.3.3 MOSFET Variable-Voltage Capacitors

In RFIC circuits there is often a need for a variable-voltage capacitor (VVC) or *varactor*. In discrete circuits, the varactor diode is a widely used element that works as a VVC in its reverse bias mode. In CMOS technology, there are a number of ways to exploit the fabrication structure to obtain a varactor. The PN junctions that are everywhere throughout the CMOS RFIC can be used to implement a varactor; however, the range of the capacitance variation of the PN junction is limited to about 10% over the relatively narrow range of voltages, and so they are not widely used [25]. The MOSFET, however, can be configured as a VVC. The simplest way to do so is by tying the drain, source, and bulk together and obtaining the VVC from the gate and common drain connections (Figure 2.27).

This configuration of VVC has been shown to exhibit an approximate 30% tuning range [25]. The VVC operation for the *n*MOSFET varactor may be explained over the various regions of operation as follows (*p*MOSFET varactor operation is similar except for the opposite voltage polarity and carrier type). For $V_G < 0$, the *n*MOSFET is operating in accumulation mode. The large number of charge carriers (substrate holes in this case) yields a highly conductive region, and along with the good conducting properties of the gate, the capacitance is simply that of the oxide, C_{ox}. For $0 < V_G < V_T$, the MOSFET operates in depletion mode and the capacitance is the series combination of the oxide capacitance C_{ox} and the *P–N+* depletion region capacitances that make up the drain and source diffused regions (Figure 2.27). Both capacitance terms are functions of the MOSFET size, and so the MOSFET can be sized for the required capacitance range.

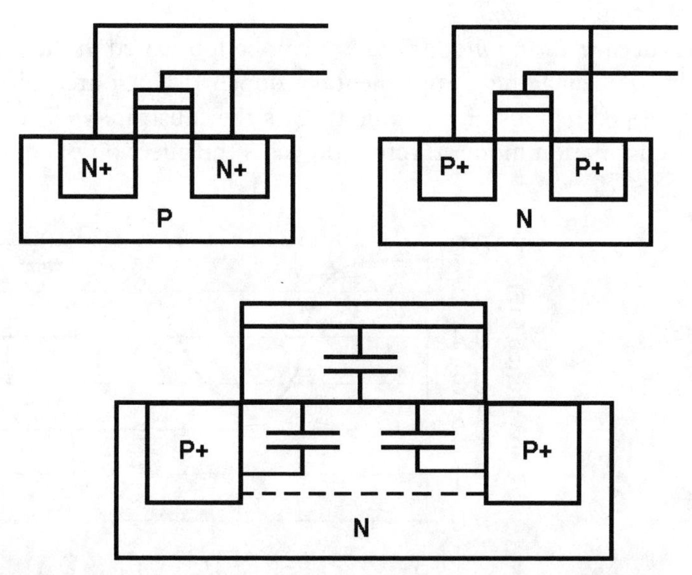

Figure 2.27 MOSFET varactors can be configured by tying the drain and source regions to the same potential. The variable capacitance is between the gate and this connection. The bulk connection has a large influence on varactor operation. (*From:* [25]. © 2000 IEEE.)

At the other voltage extreme, $V_G > V_T$, the generated channel inversion layer under the gate provides a good conducting path to the drain-source connections, leaving only C_{ox} contributing the total capacitance. The transition region between depletion and strong inversion shows a smoothly varying capacitance that varies with the applied gate voltage (Figure 2.28). There are also parasitic resistances in the MOSFET varactor: a channel resistance in shunt with the desired capacitance and a series resistance term due to gate resistance R_g. The ideal VVC just exhibits capacitive reactance. However, the parasitic resistances add a real component to the overall varactor impedance that is described using the varactor quality factor or simply the *varactor Q*:

$$Q = \frac{|X_C|}{R} = \frac{1}{\omega RC} \tag{2.31}$$

The lower the parasitic resistance R, the higher the Q. Varactor Q should be as high as practically possible since a low-Q varactor will significantly lower the overall Q of a resonant circuit; this effect of low Q on resonant circuits will be discussed later in Chapter 6, the chapter on oscillators. Although this simple MOSFET varactor configuration gives a large variation in capacitance, the bulk of the capacitance change occurs over a very small voltage range. A wider voltage tuning range varactor can be configured with the MOSFET structure by tying the bulk node to a rail potential (ground for *n*MOS, V_{dd} for *p*MOS) instead of the drain-source, creating the *inversion mode MOSFET varactor*; this connection prevents the MOSFET from entering the accumulation mode, thereby flattening the CV curve for $V_G > 0$ [26]. The C-V characteristic for a $W/L = 500/0.5$ MOSFET operating as an inversion mode varactor at 1000 MHz is shown in Figure 2.29, showing a smooth capacitance change for gate voltages greater than approximately –1.0V. The simulation of Q for this varactor ($R_g = 0$) is also shown in the figure and shows good Q (greater than 15) over the tuning range.

An *accumulation mode varactor* can be fabricated in the standard CMOS process by replacing the complementary doped drain-source diffusion regions with highly doped regions of the same type as the substrate-well (Figure 2.30) [25, 26]. The accumulation mode structure prevents the injection of complementary carriers

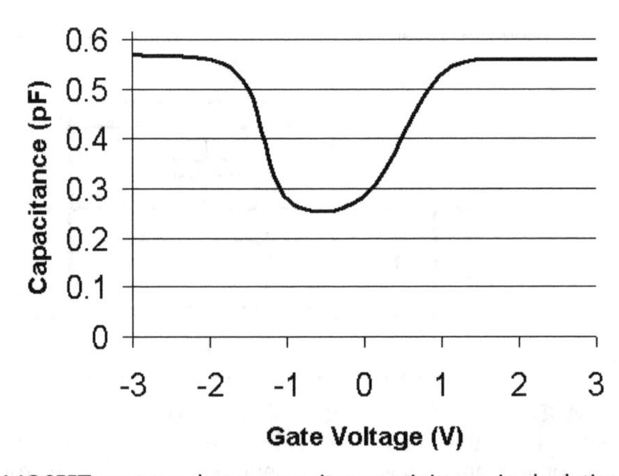

Figure 2.28 The *n*MOSFET varactor shows capacitance minimum in depletion and weak inversion.

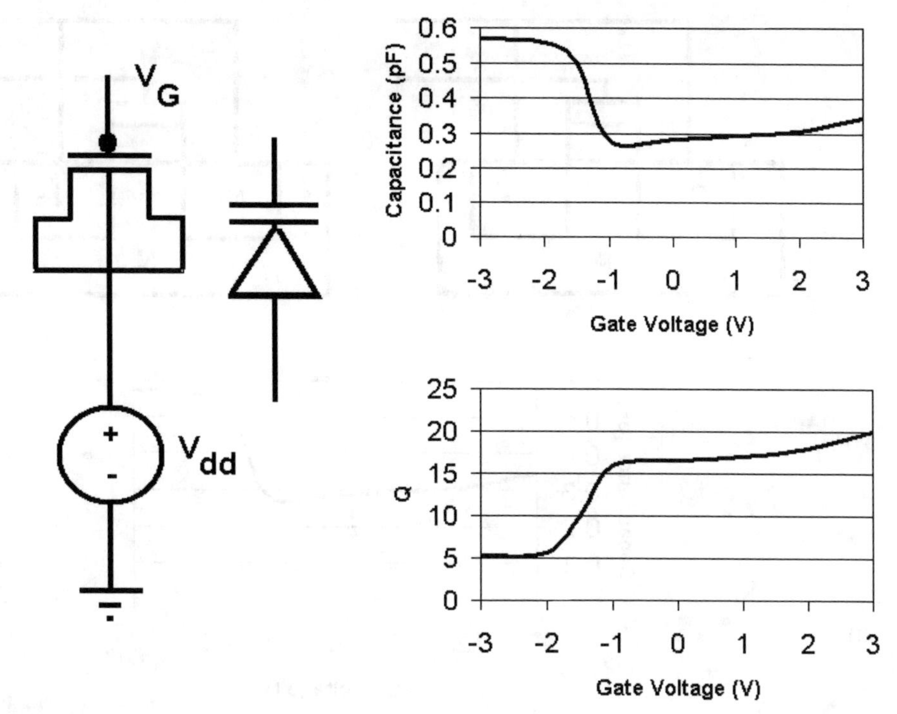

Figure 2.29 A *p*MOS varactor operating in inversion mode provides a wider tuning voltage range and good Q for $V_G > 0$.

into the substrate and therefore suppresses any channel inversion, thereby removing the PN junction capacitance associated with the drain-source diffusion regions. The resulting C-V characteristic shows a wide voltage tuning range into the previously described inversion regime (Figure 2.31). The structure has the added benefit of minimizing parasitic resistances associated with the substrate, although resistances are associated with the accumulation layer [27]. Note that this is not a traditional MOSFET structure and so the modeling of the accumulation mode varactor is more difficult than its inversion counterpart [25]. In addition, the layout is not a traditional CMOS layout and so may cause design rule errors in CAD layout programs.

2.4 Review of *S*-Parameters

No review of technology fundamentals involving RF and microwave circuitry would be complete without a discussion of the preferred method of describing an RF network, the scattering or so-called *S*-parameters. Like the other various network parameters, *S*-parameters are used to completely specify the network's response in terms of voltage (or current) waves both incident *on* and reflected (or scattered) *from* the various ports of an arbitrary network. For a network with I input ports and J output ports, the total number of ports is the sum of the input and output ports, $N = I + J$, and the network is then termed an N-port network. *S*-parameters differ dramatically from other descriptions of N-port networks, such as impedance (Z) or admittance (Y) parameters, in that the manufacture of open or

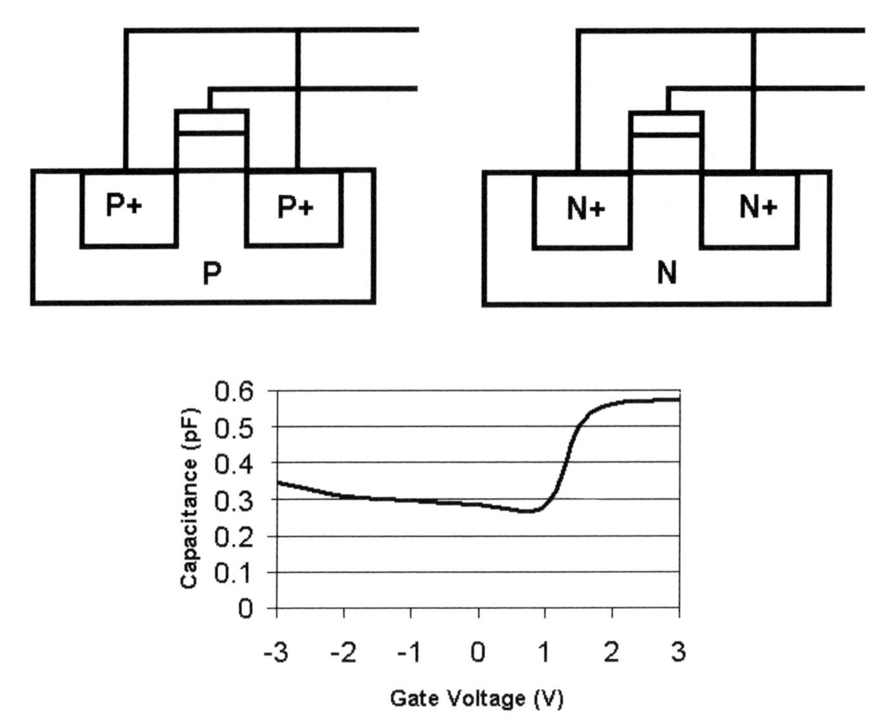

Figure 2.30 The accumulation mode varactor can be fabricated in a traditional CMOS process but can be difficult to model.

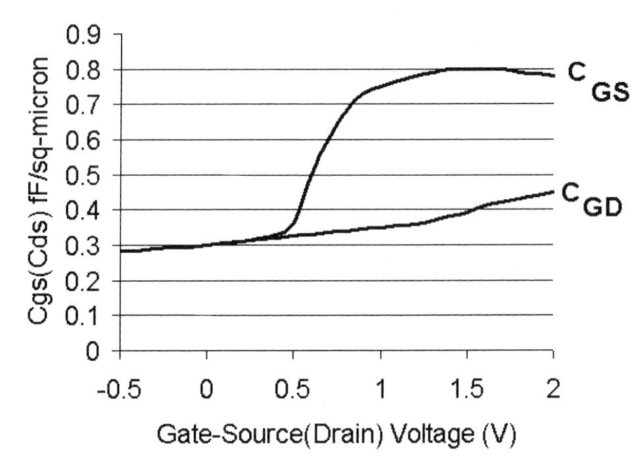

Figure 2.31 The capacitance in the MOSFET is governed by the gate voltage, V_G. (*After*: [28].)

short circuits is not required[4]; rather, the ports are terminated with loads of the same impedance as the characteristic impedance of the system, Z_0. The determination of the individual S-parameters of the network then consists of a series of measurements at the network ports under these matched conditions. The following discussion is limited to a simple two-port network such as that shown in Figure 2.32 but can be

4. This is in addition to possible device failure occurring during a measurement involving an open or short circuit at the device input or output.

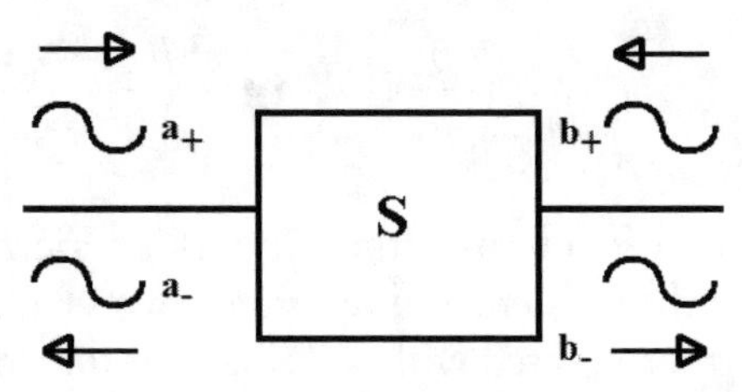

Figure 2.32 General two-port network in terms of incident and reflected waves on both ports.

easily extended to a larger number of ports (with N^2 terms needed for the N-port network description). Voltage or current waves incident on the two-port network are defined by symbols a_+ and b_+, whereas those waves scattered from the two ports are defined as a_- and b_-.

The scattered output waves a_- and b_- will in general have contributions from both incident waves a_+ and b_+ with these contributions weighted by the corresponding S-parameter:

$$\begin{aligned} a_- &= S_{11}a_+ + S_{12}b_+ \\ b_- &= S_{21}a_+ + S_{22}b_+ \end{aligned} \quad \text{or} \quad \begin{pmatrix} a_- \\ b_- \end{pmatrix} = \begin{bmatrix} S_{11} & S_{12} \\ S_{21} & S_{22} \end{bmatrix} \begin{pmatrix} a_+ \\ b_+ \end{pmatrix} = [\mathrm{S}] \begin{pmatrix} a_+ \\ b_+ \end{pmatrix} \tag{2.32}$$

Measurement of the individual parameters S_{ij} requires that the incident wave contribution from one of the ports be eliminated; this can be done by terminating the proper port with a matched impedance Z_0. To measure S_{11}, for example, the incident wave b_+ can be eliminated by placing a matched load Z_0 at the output (Figure 2.33). The measurement of S_{11} is then made by simply measuring the ratio of the reflected to incident waves on the "a" port:

$$S_{11} = \frac{a_-}{a_+}\bigg|_{b_+=0} \tag{2.33}$$

The remaining S-parameters (2.34) can be obtained in a similar fashion by placing a matched load at the proper port:

$$S_{11} = \frac{a_-}{a_+}\bigg|_{b_+=0} \quad S_{12} = \frac{a_-}{b_+}\bigg|_{a_+=0}$$

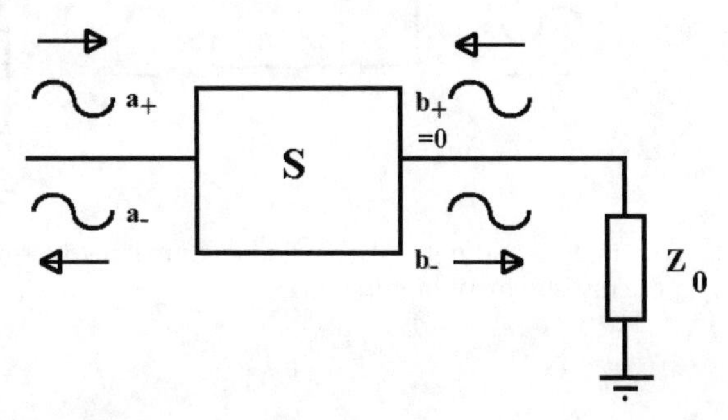

Figure 2.33 General two-port network, showing a matched termination and subsequent elimination of wave b_+ in measurement of S_{11}.

$$S_{21} = \frac{b_-}{a_+}\Big|_{b_+=0} \quad S_{22} = \frac{b_-}{b_+}\Big|_{a_+=0} \tag{2.34}$$

S_{11} and S_{22} are often termed the network's *reflection coefficients,* since they are measurements of the incident and reflected waves from the network (with, of course, the opposite port terminated with a Z_0 load). If the "b" port is considered an output port, then S_{21} describes the exiting wave with respect to that incident on the input "a" port; hence, S_{21} is often termed the network's *feed-forward* characteristic. Similarly, because S_{12} describes the wave exiting the input port "a" with respect to that incident on the output "b" port; this term is often referred to as the *feedback* term.

If the "b" measurement port in the measurement of S_{11} is not terminated in a matched load by rather an arbitrary impedance Z_b, a reflection from the port termination will occur and provide a nonzero incident wave b_+ that may eventually find its way to the "a" port (Figure 2.34). By defining the reflection coefficient[5] from Z_b as Ãb, the incident wave on the "b" port can be defined in terms of the scattered wave b_-: b_+ = Ãb b_ . Combining this relationship with (2.35) shows the relationship between b_- and a_+:

$$b_- = S_{21}a_+ + S_{22}b_+ = S_{21}a_+ + S_{22}\Gamma_b b_- \Rightarrow b_- = \frac{S_{21}a_+}{1 - S_{22}\Gamma_b} \tag{2.35}$$

The resulting output from port "a" then becomes

$$a_- = S_{11}a_+ + S_{12}b_+ = S_{11}a_+ + S_{12}\Gamma_b b_- = S_{11}a_+ + \frac{S_{12}S_{21}\Gamma_b a_+}{1 - S_{22}\Gamma_b} \Rightarrow \frac{a_-}{a_+} = S_{11} + \frac{S_{12}S_{21}\Gamma_b}{1 - S_{22}\Gamma_b} \tag{2.36}$$

For unilateral devices, $S_{12} = 0$ and the preceding equation reduces to the form given in (2.34). Relationships for the other S-parameters in the presence of unmatched port terminations can be determined in a similar fashion.

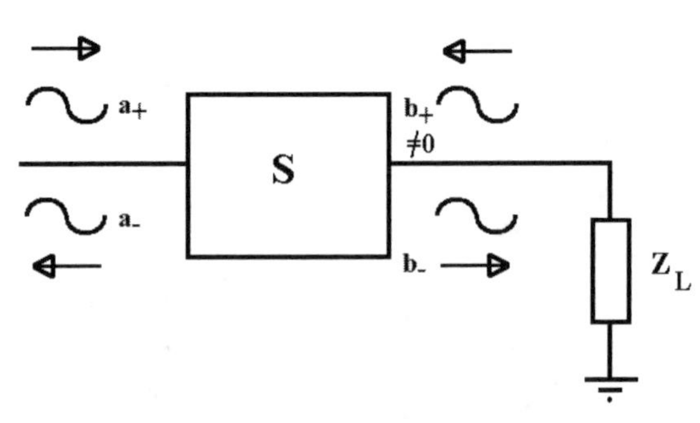

Figure 2.34 General two-port network showing an unmatched termination and subsequent nonzero wave b_+ in the measurement of a_-/a_+.

5. The reflection coefficient is calculated as $\Gamma_b = \frac{Z_b - Z_0}{Z_b - Z}$.

Several circuit properties are easily defined by the use of the circuit's *S*-parameters. An *N*-port network is considered to be a reciprocal network if the **S** matrix is symmetrical:

$$[\mathrm{S}] = [\mathrm{S}]^T \text{ or } S_{ij} = S_{ji} \tag{2.37}$$

where the superscript *T* denotes a matrix transpose. A lossless *N*-port network exhibits the property that the sum of all powers entering the network leaves the network [29]. Because the *S*-parameters relate the input and scattered waves to each other, the lossless condition can be specified strictly in terms of the network's [S] matrix:

$$[\mathrm{S}]^* = \left([\mathrm{S}]^T\right)^{-1} \tag{2.38}$$

where the asterisk (*) indicates a complex conjugate of all the **S** matrix elements.

S-parameters play such a large role in the RF and microwave industry that numerous papers and texts have been written describing their properties, so the reader is advised to consult some of those texts for further details on this method of describing the input/output behavior of arbitrary RF and microwave networks.

2.5 SPICE Modeling of CMOS RF Circuits

Circuit simulation is an art unto itself. A widely "observed to be true" equation gives a relationship between the amount of time expended on simulation and technical expertise [24]:

$$[\text{Time spent on simulation}] \times [\text{Common technical sense}] = \text{Constant}$$

The goal of any RFIC designed then is to try to minimize the first term in this equation while maximizing the second. The designer should first use the simple design equations to develop a general circuit topology and then perform a dc simulation to test proper biasing. If the circuit is not properly dc biased, it certainly is not going to work when an RF signal is applied. If the simple dc simulation results look promising, then the "full" model would be used for a more refined solution. The same process is repeated for AC/RF simulations, but this time circuit and device capacitances are included. Adjustment of MOSFET and other circuit parameters will certainly need to be performed, keeping in mind the preceding relationship. This procedure is still valid in the initial design stages, even for submicron technologies. Always start simple and then work to higher levels of complexity.

2.5.1 SPICE Level 3

Many different simulators are available for the RFIC designer but a widely used "standard" simulator is SPICE [13]. In fact, almost all simulators (free, shared, and proprietary ones) use the same set of parameters for MOSFET modeling as more generic SPICE simulators, and so in keeping with the idea of a generic simulator, this discussion will focus on the generic SPICE definitions.

Previous technology generations (gate lengths of approximately 1 μm and greater) used the SPICE Level 3 model (SPICE-3). IC fabricators would provide test MOSFET data with extracted SPICE-3 of previous runs to RFIC designers for use in future fabrication. SPICE-3 has approximately 25 parameters [30], but only a small subset is actually needed for first-pass feasibility studies. In addition, the SPICE-3 model is a physics-based model that can be seen by looking at the design equation forms earlier in this chapter. (Appendix A contains a sample listing of SPICE-3 parameters.) A minimum set of SPICE-3 parameters for dc modeling includes those listed in Table 2.4.

For AC/RF modeling, additional capacitance terms are required for accurate modeling. SPICE-3 uses the definitions listed in Table 2.5 for the model capacitances.

For AC/RF noise modeling, both thermal (kTB) and flicker ($1/f$) noise parameters are required. SPICE-3 uses the definitions given in Table 2.6 for noise parameters:

2.5.2 BSIM Parameters

As MOSFET device geometries entered the submicron realm, the SPICE-3 model was found to be not accurate enough in modeling MOSFET operation. The Berkeley Short-Channel IGFET Model for MOS Transistors (BSIM) was developed in the 1980s to provide for more accurate submicron MOSFET simulation [31]. Since this initial new modeling approach, BSIM has gone through numerous versions and has

Table 2.4 Minimum Set of SPICE-3 Parameters for dc Modeling

Parameter	*Description*	*Units*
VTO	Threshold voltage	V
UO	Surface mobility	$cm^2/V\text{–}s$
KP	Intrinsic transconductance	A/V^2
LAMBDA	Channel length modulation factor	V^{-1}
GAMMA	Body effect parameter	$V^{1/2}$
NSUB	Substrate or well doping	cm^{-3}

Table 2.5 SPICE-3 Model Capacitance Terms

Parameter	*Description*	*Units*
CBD	Zero bias bulk drain capacitance	F
CBS	Zero bias bulk source capacitance	F
CGBO	Gate bulk overlap capacitance	F/m
CGDO	Gate drain overlap capacitance	F/m
CGSO	Gate source overlap capacitance	F/m
CJ	Bulk PN junction capacitance/area	F/m^2
CJSW	Bulk PN junction sidewall capacitance	F/m

Table 2.6 SPICE-3 Thermal and Flicker Noise Parameters

Parameter	*Description*	*Units*
KF	Flicker noise coefficient	None
AF	Flicker noise exponent	None
GDSNOI	Channel shot noise coefficient	None

branched out from the original BSIM3 version to BSIM4 (for sub–100-nm technologies) and BSIMSOI (for silicon-on-insulator technology). Full details and information can be found on the BSIM website (http://www-device.eecs.berkeley.edu/~bsim3/).

More recent BSIM models contain significantly more parameters than SPICE-3 in an effort to describe the more complex behavior of submicron MOSFET operation [32]. For example, BSIMSOI contains such enhancements as gate tunneling current (an issue for deep submicron MOSFETs), improved noise modeling, better descriptions of the bulk-substrate network (important to the RF designer), improved modeling in the weak and moderate inversion regions of operation, and better modeling of multiple-fingered MOSFETs [33].

One of the significant differences between SPICE-3 and early BSIM models was that, although many (but not all) of the SPICE-3 parameters were physically based, many of the BSIM model parameters were not physically based but had instead been determined from extensive measurements on fabricated MOSFET test structures. Curve-fitting routines were then used with the measured data to determine the BSIM model parameters. This provided better accuracy but at the expense of some physical intuition, primarily because of the large number of parameters required in the BSIM model. More recent versions of BSIM have returned to the physics-based approach with the BSIM models being industry standards.

The number of BSIM parameters fills up 34 pages in the BSIM4.5.0 Users' Manual appendix [32], so the full set will not be repeated here. Even the basic model parameters number more than 65 and so will also not be repeated here. The users manual, however, is readily available on the web (http://www-device.eecs.berkeley.edu/~bsim3/) and contains in-depth discussions of the model equations and the measurement techniques used in extracting the model parameters. A sample listing of BSIM parameters for a double poly triple metal process is shown in Appendix B.

References

[1] Camilleri, N., et al., "Silicon MOSFETs, the Microwave Device Technology for the 1990s," *Proc. 1993 IEEE MTT-S Int. Microwave Symp.*, 1993, Vol. 2, pp. 545–548.

[2] Tsividis, Y. P., *Operation and Modeling of the MOS Transistor*, New York: McGraw-Hill, 1987.

[3] Enz, C., "An MOS Transistor Model for RF IC Design Valid in All Regions of Operation," *IEEE Trans. Microwave Theory Techniques*, Vol. 50, No. 1, Jan. 2002, pp. 342–359.

[4] Enz, C., and Y. Cheng, "MOS Transistor Modeling for RF IC Design," *IEEE Trans. Solid State Circ.*, Vol. 35, No. 2, Feb. 2000, pp. 186–201.

[5] Ou, J., et al., "CMOS RF Modeling for GHz Communication IC's," *1998 Symp. on VLSI Technology*, 1998.

[6] Das, M., "High Frequency Network Properties of MOS Transistors Including the Substrate Resistivity Effect," *IEEE Trans. Electron Devices*, Vol. ED-16, 1969, pp. 1049–1069.

[7] Reinhard, D., *Introduction to Integrated Circuit Engineering*, Boston: Houghton Mifflin, 1987.

[8] Banerjee, K., "Trends for ULSI Interconnections and Their Implications for Thermal, Reliability and Performance Issues," *Proc. 7th Int. Symp. on Dielectrics and Conductors for ULSI Multilevel Interconnection (DCMIC)*, 2001, pp. 38–50.

[9] Ducarouge, B., et al., "Power Capabilities of RF MEMS," *Proc. 24th Int. Conf. on Microelectronics (MIEL 2004)*, May 2004, Vol. 1, p. 65.

[10] Wilson, P., "A Trench Gate LDMOS for RF Applications," *Proc. 2003 IEEE Int. Symp. on Electron Devices and Microwave Optoelectronic Applications,* 2003, pp. 43–47.

[11] Sze, S., *Physics of Semiconductor Devices,* New York: Prentice Hall, 1982.

[12] Pucknell, D., and K. Eshraghian, *Basic VLSI Design*, New York: Prentice Hall, 1994.

[13] UC Berkeley BSIM3/BSIM4 Device Group, http://www-device.eecs.berkeley.edu/~bsim3.

[14] Abou-Allam, E., and T. Manku, "A Small Signal MOSFET Model for Radio Frequency IC Applications," *IEEE. Trans. Computer-Aided Design*, May 1997, pp. 437–447.

[15] Manku, T., "Microwave CMOS—Device Physics and Design," *IEEE. J. Solid State Circ.*, Vol. 34, No. 3, March 1999, pp. 277–285.

[16] Vendelin, G., A. Pavio, and U. Rohde, *Microwave Circuit Design Using Linear and Nonlinear Techniques*, New York: John Wiley and Sons, 1990.

[17] Melendy, D., and A. Weisshaar, "A New Scalable Model for Spiral Inductors on Lossy Silicon Substrate," *2003 IEEE MTT-S Digest*, 2003, pp. 1007–1010.

[18] Caverly, R., et al., "Modeling and Characterization of Integrated Inductors and Transformers for Amplifier Applications in Silicon High Frequency Systems," *Proc. 5th Topical Meeting on Silicon Monolithic Integrated Circuits for RF Systems (SiRF04)*, Sept. 2004.

[19] Razavi, B., *RF Microelectronics*, New York: Prentice Hall, 1997.

[20] American Radio Relay League, *The ARRL Handbook*, Newington, CT: ARRL, 1986.

[21] Caverly, R., "Linear and Nonlinear Characteristics of the Silicon CMOS Monolithic 50 ohm Microwave and RF Control Element," *J. Solid State Circ.*, Vol. 34, No. 1, Jan. 1999, pp. 124–126.

[22] Li, Z., et al., "5.8-GHz CMOS T/R Switches with High and Low Substrate Resistances in a 0.18-μm CMOS Process," *IEEE Microwave Wireless Components Letts.*, Vol. 13, No. 1, 2003, pp. 1–3.

[23] Kelly, D., et al., "The State of the Art of Silicon-on-Sapphire CMOS RF Switches," *CSIC 2005 Digest*, 2005, pp. 200–203.

[24] Allen, P., and D. Holberg, *CMOS Analog Circuit Design*, 2nd ed., New York: Oxford University Press, 2002.

[25] Hui, Z., et al., "A Wide Tuning Range Gated Varactor," *IEEE J. Solid State Circ.*, Vol. 35, No. 5, May 2000.

[26] Andreani, P., and S. Mattisson, "On the Use of MOS Varactors in RF VCO's," *IEEE J. Solid State Circ.*, Vol. 35, No. 6, June 2000.

[27] Soorapanth, T., et al., "Analysis and Optimization of Accumulation-Mode Varactor for RF ICs," *1998 Symp. VLSI Circuits Digest of Technical Papers*, June 1998, pp. 32–33.

[28] Wu, D., et al., "Notched Sub-100 nm Gate MOSFETs for Analog Applications," *Proc. 6th Int. Conf. on Solid State Integrated Circuit Technology*, Vol. 1, 2001, pp. 539-542.

[29] Collin, R. E., *Foundations for Microwave Engineering*, 2nd ed., New York: McGraw-Hill, 1992.

[30] MicroSim Corp., *MicroSim PSPICE A/D Reference Manual*, Version 8, 1997.

[31] Sheu, B., et al., "BSIM: Berkeley Short-Channel IGFET Model for MOS Transistors," *IEEE J. Solid State Circ.,* Vol. SC-22, 1987, pp. 558–566.

[32] Xi, X., et al., *BSIM4.5.0 MOSFET Model Users Manual,* Berkeley: Regents of the University of California, 2004.

[33] BSIMSOI4.0 News Release, Nov. 30, 2005.

Selected Bibliography

Lee, T. H., *The Design of CMOS Radio-Frequency Integrated Circuits*, 2nd ed., Cambridge, UK: Cambridge University Press, 2004.

Rogers, J., and C. Plett, *Radio Frequency Integrated Circuit Design*, Norwood, MA: Artech House, 2003.

CHAPTER 3

The Passive Components

In low-frequency RF design, it is taken for granted that high-quality discrete components are available. Discrete lumped element components such as chip capacitors, inductors, and resistors can be manufactured to exhibit high Q values well into the gigahertz range. The CMOS RFIC designer, however, does not have this same luxury in assuming that high-Q components are available. In fact, without significant postprocessing or other fabrication techniques, inductor Q values of 10 are difficult to obtain. In addition, with the heavily doped silicon substrate (compared with semi-insulating substrates used in technologies such as GaAs), parasitic effects can be almost of the same order as the desired component value, and so careful consideration of the layout and design of RFIC passive elements is critical to circuit performance. This chapter takes a look at passive components, their nonidealities, as well as layout and simulation issues that provide details on their overall performance in CMOS RFIC designs. An introduction to transmission line and interconnect issues is covered since the equivalent circuit models for these components can be derived (at least to first order) from knowledge of the layout capacitance, inductance, and resistance. RF MEMS are finding increased use in CMOS RFIC designs, so an introduction to this technology is provided in this chapter. Finally, a primer on packaging, thermal, and grounding issues completes the chapter.

3.1 Capacitors

3.1.1 Metal-Insulator-Metal Capacitors

Passive capacitors used in RFIC designs are primarily overlays of the various layers available in the CMOS process. Each layer in the process is separated from another layer by an insulator, and hence the general term *metal-insulator-metal* (MIM) is used to describe the overall capacitor structure (although some layers may be silicon in nature, the letter *M* is used to describe the high-conductivity layer, similar to the way *M* in *MOSFET* refers to a metallized gate even though polysilicon is actually used). The three primary overlay capacitors that can be created in the CMOS process are metal-metal, metal-polysilicon, and polysilicon-polysilicon (termed poly-poly and available only in a multiple-poly layer process). Of these three, metal-metal and poly-poly are the more widely used.

The layer overlap capacitor is initially estimated using the simple parallel plate capacitor approximation:

$$C_{layer} = \frac{\varepsilon_r \varepsilon_0 WL}{t_{ox}} = \frac{\varepsilon_r \varepsilon_0}{t_{ox}} WL = C_\square WL \tag{3.1}$$

where L (W) is the gate length (width), t_{ox} is the insulating layer thickness, and ε_r is the relative dielectric permittivity of the insulating material. Since the insulating layer material and the layer thickness are specified by the process, IC fabricators provide the designer a parameter termed the *overlap capacitance per unit area* $C_\square$ to aid in estimating the capacitance; the desired capacitance is then simply the product of this term and the overlapping layer area. Table 3.1 shows representative values of $C_\square$ for a 0.5-μm CMOS process. Of interest is the poly-poly overlap capacitance $C_\square$, which is at least an order of magnitude greater than the overlap capacitance $C_\square$ for adjacent metal-metal layers (such as metal1-metal2), a consequence of the thinner insulating material (so-called *thin oxide*) between the two poly layers; capacitors fabricated with this layer exhibit significantly higher capacitances for a given area. For the 0.5-μm CMOS process, the thin oxide is approximately 140 to 150 Å, reducing[1] to approximately 80 (40) Å in a 0.35- (0.18)-μm CMOS process. The remaining

Table 3.1 Parameters for Determining the Simplified Capacitor RF Equivalent Circuit

Parameter	*Value*	*Unit*
Poly1 R_{SH}	23.4	Ω/sq
Poly2 R_{SH}	43.4	Ω/sq
Metal1 R_{SH}	0.09	Ω/sq
Metal2 R_{SH}	0.10	Ω/sq
Metal3 R_{SH}	0.05	Ω/sq
$C_\square$ (poly-poly2)	920	aF/μm^2
$C_\square$ (poly-metal1)	56	aF/μm^2
$C_\square$ (poly-metal2)	16	aF/μm^2
$C_\square$ (poly-metal3)	10	aF/μm^2
$C_\square$ (poly2-metal1)	50	aF/μm^2
$C_\square$ (metal1-metal2)	34	aF/μm^2
$C_\square$ (metal1-metal3)	14	aF/μm^2
$C_\square$ (metal2-metal3)	36	aF/μm^2
C_{ox} (substrate)	87	aF/μm^2
C_F (poly-poly)	88	aF/μm
C_F (poly-metal1)	62	aF/μm
C_F (poly-metal2)	38	aF/μm
C_F (poly-metal3)	29	aF/μm
C_F (metal1-metal2)	57	aF/μm
C_F (metal1-metal3)	35	aF/μm
C_F (metal2-metal3)	50	aF/μm
$C_{ox\text{-}F}$ (poly-substrate)	20	aF/μm

1. This reduction is a direct consequence of technology scaling described in Chapter 2 and summarized in Table 2.2.

layers are separated by a significantly thicker insulating material (so-called *thick oxide*). The thick oxide layers are typically 20 to 30 times thicker than thin oxide layers, as evidenced by the significant reduction in the capacitance per unit area for these layers. A high capacitor value can be accomplished in a particular area with poly-poly capacitors; however, due to the thin oxide, care must be taken to ensure that the electric field between the poly-poly layers does not exceed the breakdown field in the oxide material (approximately 100 V/μm). Table 3.1 also shows the so-called *fringing capacitance per unit length* C_F; this term addresses the fringing capacitance caused by the fringing electric fields off the sides. This fringing capacitance is additive to the overall capacitance and can be a significant contributor to C_{TOT} (Figure 3.1).

The total capacitance is the sum of the area and fringing capacitance contributions:

$$C_{TOT} = WLC_{\square} + 2(W + L)C_F \tag{3.2a}$$

For square capacitors, (3.2a) reduces to

$$C_{TOT} = W^2 C_{\square} + 4WC_F = W(WC_{\square} + 4C_F) \tag{3.2b}$$

Layer N of the MIM capacitor is also suspended above another conductor, the low-resistivity silicon substrate and its effective RF ground. The *layer-to-substrate capacitance per unit area* C_{ox} and *layer-to-substrate fringing capacitance* $C_{ox\text{-}F}$, area and length capacitance terms, respectively, are parasitic capacitances that must be considered as part of the overall RFIC design process. These capacitances can be minimized by using MIM capacitors on higher layers since $C_{ox(ox\text{-}F)}$ is smaller due to higher t_{sub}; Table 3.1 shows these capacitances per unit area decreasing with higher

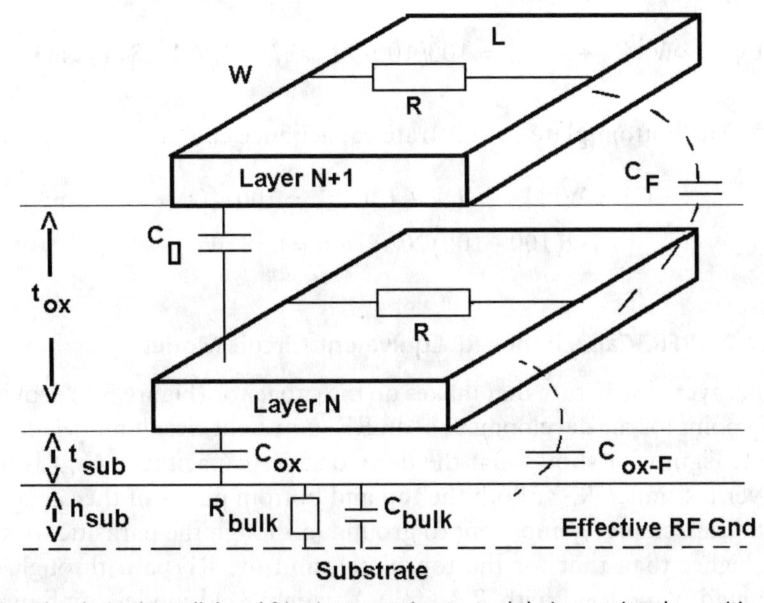

Figure 3.1 The desired parallel and fringing capacitance and their associated parasitic elements are functions of the layout.

metal layers. Equation (3.3) can be used to estimate the bottom plate capacitance to substrate C_B from fabrication parameters:

$$C_B = WLC_{ox} + 2(W + L)C_{ox-F} \tag{3.3}$$

In addition to the capacitance associated with these layers, the plates themselves exhibit nonzero resistances. For square capacitors ($W = L$), the resistance of each plate is simply the sheet resistance of the associated layer:

$$R_{layer} = R_{SH} \frac{L}{W} = R_{SH} \tag{3.4}$$

For metal plate capacitors, R_{layer} can safely be ignored; for capacitors with at least one polysilicon plate, however, the relatively high sheet resistance of this layer requires that it be considered as a design parasitic. Asymmetrical capacitors (W, L unequal) require knowledge of the RF displacement current direction to estimate the capacitance; the variable L in (3.4) will then be in the direction of the RF current.

Example 3.1: Determine the layer resistance, plate-to-plate capacitance, and bottom plate capacitance to ground for a 100- by 100-μm^2 poly-poly capacitor.

Using (3.4) and the sample fabrication parameters in Table 2.1, the resistance for each square plate ($W = L$) is simply the layer sheet resistance:

$$R_{layer} = R_{SH} \begin{cases} 23.4\Omega & \text{poly1} \\ 43.4\Omega & \text{poly2} \end{cases}$$

Equation (3.2b) and the sample fabrication parameters in Table 2.1 provide the plate-to-plate capacitance for the two-poly capacitor:

$$C_{TOT} = W^2 C_{\square} + 4WC_F = 100\left(100 \cdot 920 \text{ aF}/\mu\text{m}^2 + 4 \cdot 88 \text{ aF}/\mu\text{m}\right) = 9.23 \text{ pF}$$

The bottom plate-to-substrate capacitance can be estimated using (3.3):

$$\begin{aligned} C_B &= WLC_{ox} + 2(W + L)C_{ox-F} = 100 \cdot 100 \cdot 87 \text{ aF}/\mu\text{m}^2 \\ &+2(100 + 100)20 \text{ aF}/\mu\text{m} = 0.88 \text{ pF} \end{aligned}$$

3.1.2 RFIC Capacitance RF Equivalent Circuit Model

The layered structure that makes up the capacitor (Figure 3.1) provides a good starting point for the development of an RF equivalent circuit model of the passive capacitor. Figure 3.1 shows that the desired series capacitance C_{TOT} is in series with the layer resistance R_{layer}. Both the top and bottom plates of the capacitor have a parasitic capacitance component to ground (although the parasitic for the bottom plate is higher than that for the top plate), and the RF path through the substrate to ground is modeled with $R_{bulk}C_{bulk}$. A widely used model combining all of the elements, desired as well as parasitic, is shown in Figure 3.2. Although widely used, the

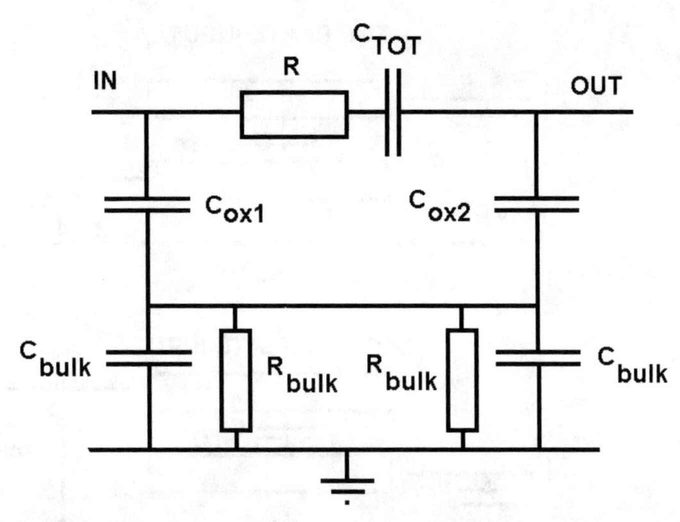

Figure 3.2 RF equivalent circuit for an MIM capacitor in a CMOS process.

model has some limitations; namely, if R_{layer} is included (and is needed for poly-poly capacitors because of their larger sheet resistance), then the R_{layer}–C_{TOT} element should be modeled as a distributed RC network [1]. A distributed RC model, however, complicates the analysis, and good results can still be obtained with the simple model. Estimates of each element in the model are made using (3.2) through (3.4) and fabrication parameters such as those listed in Table 3.1.

An alternate method of determining the MIM capacitor's RF behavior is through the use of electromagnetic simulators (for example, one of the Sonnet® software suites (http://www.sonnetusa.com/); there are several others). This method requires detailed knowledge of process parameters such as layer and insulator thickness and doping levels for the best accuracy. RFIC designers not employed by the IC fabrication house may need to sign nondisclosure or other agreements to obtain this data, however. The payoff is an accurate description of the MIM capacitor over a wide frequency range. The downside to the full EM simulation is its lengthy computation time. A middle ground approach is to perform the EM simulation over a relatively narrow frequency range and use that data to extract RF equivalent circuit parameters (Figure 3.2). Although this approach misses possible frequency dependencies of the RF equivalent circuit elements themselves [$R_{layer}(f)$, $C_{bulk}(f)$, and so forth], the advantage is that simulation times are dramatically reduced.

3.1.3 Concept of Top/Bottom Plate

The MIM capacitor layout shown in Figure 3.1 can be used as a series capacitor of value C_{TOT}. The capacitor's RF behavior, however, is somewhat dependent on which of the two layers or plates acts as the RF input and output (Figure 3.3). Using layer N (bottom plate) as the input of the series capacitor places a larger parasitic capacitance (C_{ox1}) in shunt with the driver circuit; layer $N + 1$ (top plate) at the output has a smaller parasitic capacitance (C_{ox2}) because this layer is farther from the substrate. The larger shunt capacitance will reduce the frequency response of the driver circuit because of the capacitive loading at the output. Using layer $N + 1$ (top

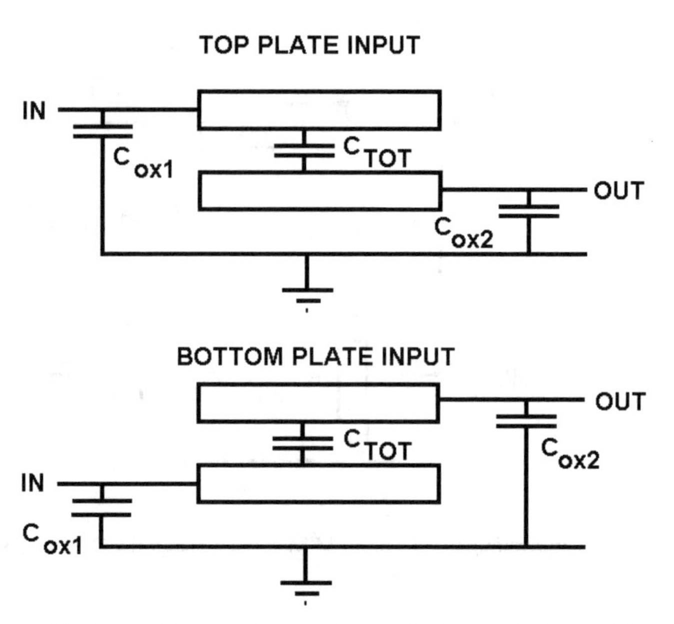

Figure 3.3 Top and bottom plate capacitor inputs and their associated parasitics.

plate) as the input of the series capacitor places the smaller parasitic (C_{ox2}) in shunt with the driver circuit; the larger parasitic C_{ox1}, comparable in magnitude to C_{TOT}, creates a voltage/charge divider circuit, reducing the output voltage level to the next stage. The frequency response of the driver circuit is better, but the voltage is reduced by the capacitive divider circuit. Therefore, inputs to series capacitors should usually be made on the bottom plate.

3.1.4 Modeling Example

The determination of the equivalent circuit parameters (neglecting bulk effects) for a series-connected, two-layer polysilicon-metal1 capacitor will be shown in this example for a 100- by 100-μm overlay capacitor. The important parameters to estimate the simple equivalent circuit parameters (Figure 3.4) are indicated in Table 3.1 for a 0.5-μm CMOS process.

The layer resistance can be calculated using (3.4):

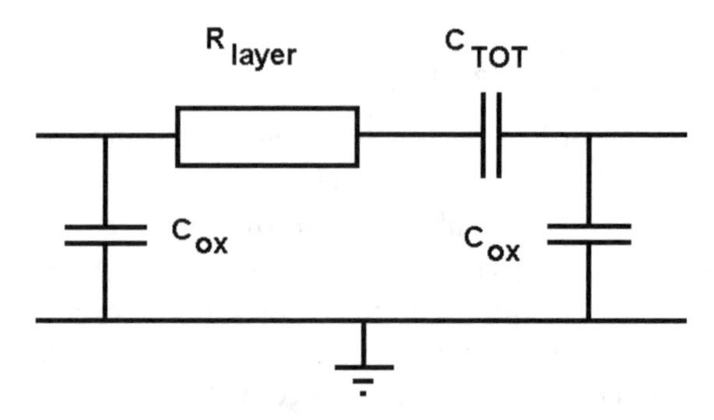

Figure 3.4 Simplified capacitor RF equivalent circuit (neglecting substrate effects).

$$R_{layer} = R_{SH}\frac{L}{W} = R_{SH} = \begin{Bmatrix} 23\Omega & \text{poly} \\ 0.09\Omega & \text{metal1} \end{Bmatrix} \quad (3.5)$$

The poly-metal and poly-substrate (ground) capacitances can be computed using (3.2):

$$C_{TOT} = W(WC_{\square} + 4C_F) = 100\left(100 * 54 \text{ af}/\mu\text{m}^2 + 4 * 62 \text{ af}/\mu\text{m}\right) = 0.58 \text{ pF} \quad (3.6)$$

$$C_{ox} = WLC_{sub} = 100 * 100 * 87\text{Faf}/\mu\text{m}^2 = 0.87 \text{ pF}$$

A 2D EM field solver can be employed to estimate the various RF equivalent circuits if the insulating layer thicknesses are known. If these layer thicknesses are not directly known, an estimate of the interlayer oxide (t_{ox}) and layer-to-substrate (t_{sub}) thicknesses can be made from the parameters listed in Table 3.1 (assuming that the dielectric is known; here, assume silicon dioxide with $\varepsilon_r = 3.9$):

$$t_{ox} = \frac{\varepsilon_r \varepsilon_0}{C_{\square}} = \frac{3.9 * 8.85 \cdot 10^{-12}}{56 \cdot 10^{-6} \ F/m^2} = 0.62 \ \mu\text{m} \quad t_{sub} = t_{ox}\frac{C_{\square}}{C_{sub}} = 0.62\frac{56}{87} = 0.40 \ \mu\text{m} \quad (3.7)$$

The layout in a 2D EM field solver, Sonnet Lite™, for the overlay capacitor in this example is shown in Figure 3.5.[2] The boundary box defining the simulation region is 300 by 300 μm. The results of the 2D simulation (S-parameters and R_{layer} and C_{TOT} shown below) indicate that at 1000 MHz, R_{layer} is 37Ω, C_{TOT} is 0.56 pF,

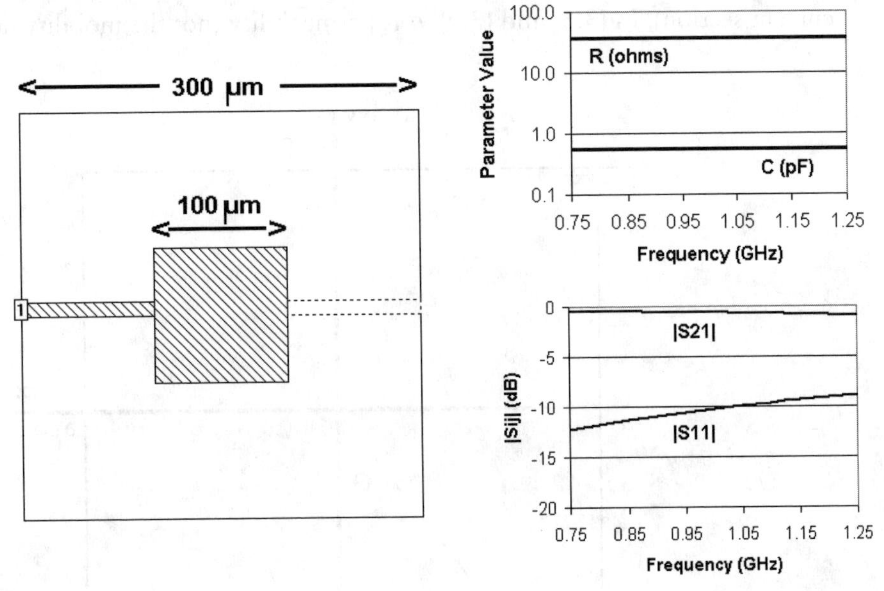

Figure 3.5 Sonnet Lite layout of a capacitor and the resulting S-parameters and equivalent circuit parameters (capacitor_example.son).

2. A version of Sonnet Lite is available on the accompanying CD. Installation information is located on the CD. Layout files for examples in this chapter can also be found on the CD.

and each capacitance $C_{ox} = 0.39$ pF (both in parallel yield a 0.78-pF total capacitance). The estimated results of the first calculation are in reasonable agreement with the 2D simulation results and indicate that the estimates defined by (3.1) to (3.7) provide a good first-pass approach to capacitance modeling.

3.2 Inductors

3.2.1 On-Chip Inductor Types

Inductive reactance on an RFIC can be created using planar inductors (planar since the majority of the windings are on a single metal layer or plane). The inductance of the windings is a complex function of the number of turns, the width of th[illegible] line, the spacing between windings, and the area cross section of th[illegible] ple estimate of the inductance of a single rectangular loop ca[illegible] the Biot-Savart law

$$dB = \mu dH = \frac{\mu I dl \sin \Theta}{4\pi r^2}$$

and the classical definition of inductance

$$L = \frac{N}{I} \int\int \vec{B} \cdot d\hat{S} \quad (3.9)$$

to the rectangular current loop shown in Figure 3.6 (where $Idl = Idx$ is an elemental current section). In (3.8) and (3.9), μ is permeability (not the mobility here), r is the

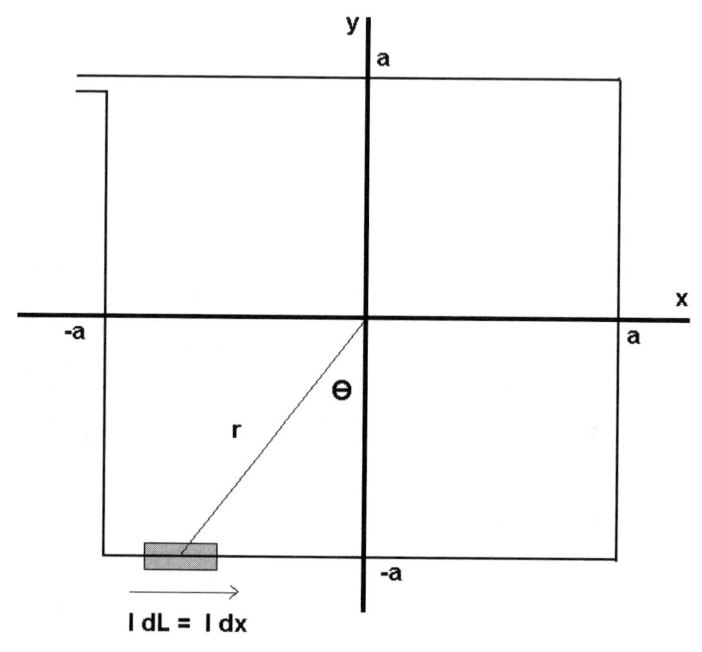

Figure 3.6 Single rectangular conductor loop for calculation of inductance using the Biot-Savart law.

distance from the current element, B is magnetic flux, H is the magnetic field, N is the number of turns of the loop, S is the loop differential area, and L is the inductance.

Using the geometry of the problem as shown in Figure 3.6, $r^2 = x^2 + a^2$ and $a = r \sin(\Theta)$, so the total magnetic flux out of the center of the loop can be written as

$$B_{TOT} = \mu H_{TOT} = 4\frac{a\mu I}{4\pi}\int_{-a}^{a}\frac{dx}{\left(x^2 + a^2\right)^{3/2}} = \left.\frac{\sqrt{2}\mu I}{\pi a}\right| N\text{ turns}: B_{TOT} = \frac{\sqrt{2}\mu I N}{\pi a} \tag{3.10}$$

and the inductance then becomes

$$\begin{aligned} L &= \frac{N}{I}\iint B \cdot d\hat{S} \approx B_{TOT}\, 4a^2 = \left.\frac{N}{I}\frac{4Ia\sqrt{2}\mu}{\pi}\right| N\text{ turns} \\ L &= \frac{2(2a)\sqrt{2}\mu N^2}{\pi} \approx 2.26 a_{\mu m} N^2\,\text{pH} \end{aligned} \tag{3.11}$$

where $a_{\mu m}$ is the side dimension a in microns. Admittedly a simple analysis, a number of important characteristics of planar on-chip inductors become apparent by looking at (3.11). The inductance is strongly dependent on the side length (and hence area cross section) of the inductor as well as the number of turns that make up the inductor. For example, a 1.13-nH planar inductor with a single turn requires 1.0 mm of side length, a considerable amount of silicon real estate for a single passive element. It is also important to note that the inductance does *not* scale as a smaller gate length technology is used. A smaller line width technology may provide higher frequency use with a certain reactance able to be obtained with a smaller inductance, but for a given value of inductance, the inductance does not scale. The side length can be reduced by increasing the number of current loops (i.e., turns N), but as will be seen, this increases the length of the metal used in the inductor, and consequently its parasitic resistance and capacitance will also increase. An often cited study of various planar inductors (studied geometries in Figure 3.7) derived more detailed expressions for the inductance for square, rectangular, octagonal, and circular inductors [2] based on an extensive numerical study using the ASITIC (Analysis and Simulation of Inductors and Transformers for ICs) program [3].

Using a current sheet approach, the inductance can be computed from layout parameters using the fit data shown in Table 3.2 and the inductance expression in (3.12):

$$\begin{aligned} L &= \frac{\mu N^2 d_{avg} c_1}{2}\left[\ln\left(\frac{c_2}{\rho}\right) + c_3\rho + c_4\rho^2\right] \\ &= 0.628 d_{avg-\mu m} N^2 c_1\left[\ln\left(\frac{c_2}{\rho}\right) + c_2\rho + c_4\rho^2\right]\text{pH}/\mu\text{m} \end{aligned} \tag{3.12}$$

where $d_{avg\text{-}\mu m}$ is the average length of the inductor (in microns) and ρ is the inductor fill factor:

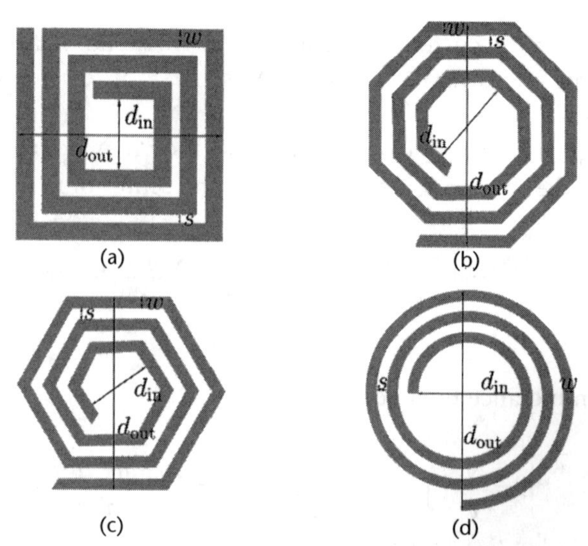

Figure 3.7 Layouts for the (a) square, (b) octagonal, (c) hexagonal, and (d) circular spiral inductors. (*From:* [2]. © 1999 IEEE.)

$$d_{avg} = \frac{d_{out} + d_{in}}{2}; \ \rho = \frac{d_{out} - d_{in}}{d_{out} + d_{in}}; \ d_{in} \cong d_{out} - 2[N(W+S) - S] \quad (3.13)$$

These closed-form expressions for the inductors should provide design accuracy to within about 10% but, more important, should give the designer a straightforward starting point for inductor design. The closed-form expressions allow the designer to quickly see how various shapes, turns, line widths, and spacings can yield a given inductance. The eventual layout of planar inductors, however, will give rise to a number of parasitic elements (Figure 3.8).

The windings of the inductor are of finite width W, and so a distributed capacitance is formed by each trace between the winding and the substrate (C_{ox}). While the inductor windings are metal with a very small sheet resistance, the overall metal runs are extremely long for a reasonable inductance, with a corresponding large L/W and resistance R generated. The parasitic capacitance and resistance can be approximated assuming a square inductor with N closely spaced W thickness lines and with d_{avg} side length:

$$C_{ox} = \sum_{\text{all sections}} C_A WL \approx C_A WN4d_{avg} \quad R = \sum_{\text{all sections}} R_S \frac{L}{W} \approx 4N\frac{d_{avg}}{W} \quad (3.14)$$

Table 3.2 Coefficients for Inductance Expression

Layout	c_1	c_2	c_3	c_4
Square	1.27	2.07	0.18	0.13
Hexagonal	1.09	2.23	0.00	0.17
Octagonal	1.07	2.29	0.00	0.19
Circle	1.00	2.46	0.00	0.20

Source: [2]. © 1999 IEEE.

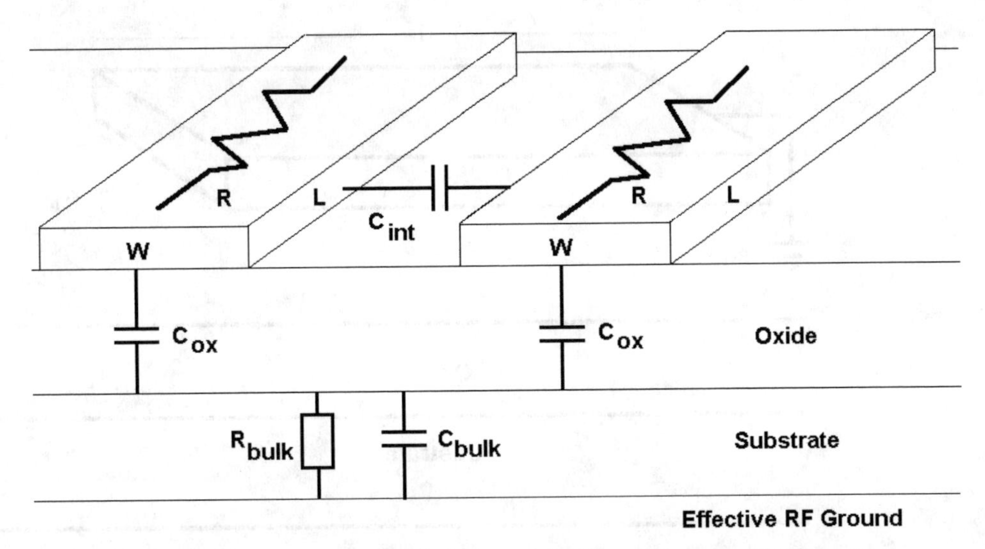

Figure 3.8 Layout of parallel inductor windings and their associated parasitic elements.

The parasitic capacitance can be reduced by using narrow inductor windings (small W); however, the series resistance of the windings will increase by a similar amount. There is an interwinding capacitance C_{int} and the substrate/bulk RC network from the substrate oxide interface to the effective RF ground. The parasitic capacitance affects the frequency response of the inductor through the natural resonant frequency of the element ($\omega^2 = 1/LC$), whereas the series resistance of the inductor affects its overall Q:

$$Q_{series} = \frac{\text{Magnetic energy stored}}{\text{Energy dissipated}} = \frac{\text{Inductor reactance}}{\text{Inductor series resistance}} = \frac{\omega L}{R} \tag{3.15}$$

The inductor layouts shown in Figure 3.7 all have windings that terminate in the center of the inductor. Connections to this end require a duck-under that will cross all the windings; there will be a small amount of overlap capacitance between the input and output leads of the inductor:

$$C_p = NW^2 \frac{\varepsilon_{r-ox}\varepsilon_0}{t_{ox}} \tag{3.16}$$

where t_{ox} is the spacing between the two metal layers. The inductor metal layer and the associated capacitances are grounded along paths through the low-resistivity silicon substrate (effective RF ground). The effective RF ground through the substrate is modeled as a parallel RC network as in the capacitance case.

Another type of inductor termed the *slab inductor* can be fabricated. The slab inductor's characteristics are based on a transmission line equivalent with one end short-circuited [4] (Figure 3.9):

$$Z_{in} = Z_0 \tanh(\beta d) \tag{3.17}$$

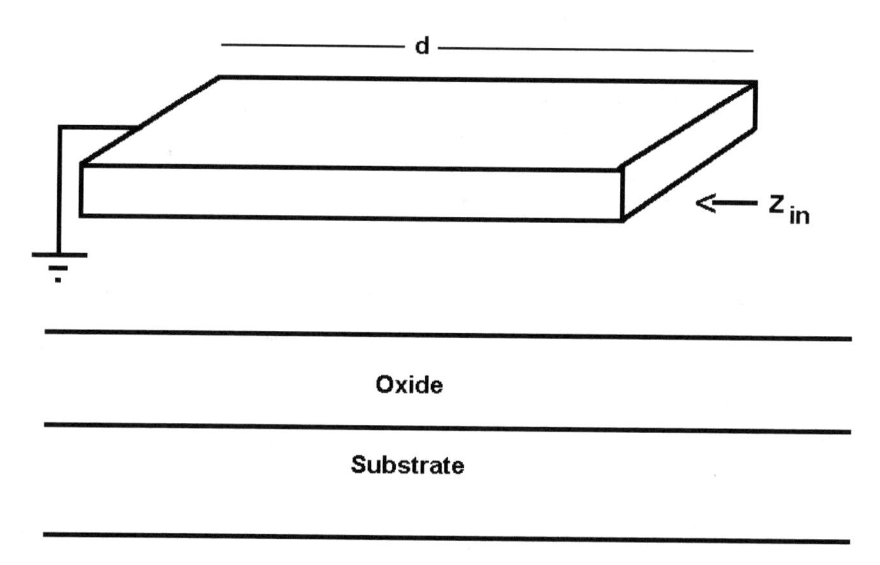

Figure 3.9 Slab inductor fabricated from a grounded upper level metal trace.

where β is the complex propagation constant of the transmission line ($\beta_r + j\beta_i$) and d is the length of the slab inductor. For a short section of line, the input impedance Z_{in} can be written as

$$Z_{\text{in}} = Z_0 \tanh(\beta d) \approx Z_0 \beta_r d + jZ_0 \beta_i d = R_{eq} + j\omega L_{eq} \tag{3.18}$$

The Q of the slab inductor is simply a function of the ratios of the complex propagation constant where the transmission line model equivalent circuit parameters are indicated:

$$\text{Q} = \frac{\omega L_{eq}}{R_{eq}} = \frac{\beta_i}{\beta_r} = \frac{\text{IM}\left\{\sqrt{(R + j\omega L)(G + j\omega C)}\right\}}{\text{RE}\left\{\sqrt{(R + j\omega L)(G + j\omega C)}\right\}} \tag{3.19}$$

If the loss terms (R, $1/G$) are negligible, the Q is infinite since the denominator goes to zero. If the loss terms are large, then the denominator will be large and the Q will be low; this is the case with low-resistivity silicon substrates.

If the transmission line is modeled as microstrip over a lossy ground plane, variations of inductance and Q with line length d and width of the strip W can be determined. The microstrip parameters per unit length (R, L, C, G) are complex for the MOS microstrip structure. Analyses have shown, however, that there is an optimal width and length of the slab inductor that will maximize the Q for a given inductance, such as that shown in Figure 3.10.

3.2.2 Planar Spiral Inductor RF Equivalent Circuit Model

The individual elements that comprise the planar spiral inductor are combined into an equivalent circuit model as shown in Figure 3.11. From the inductor layout in

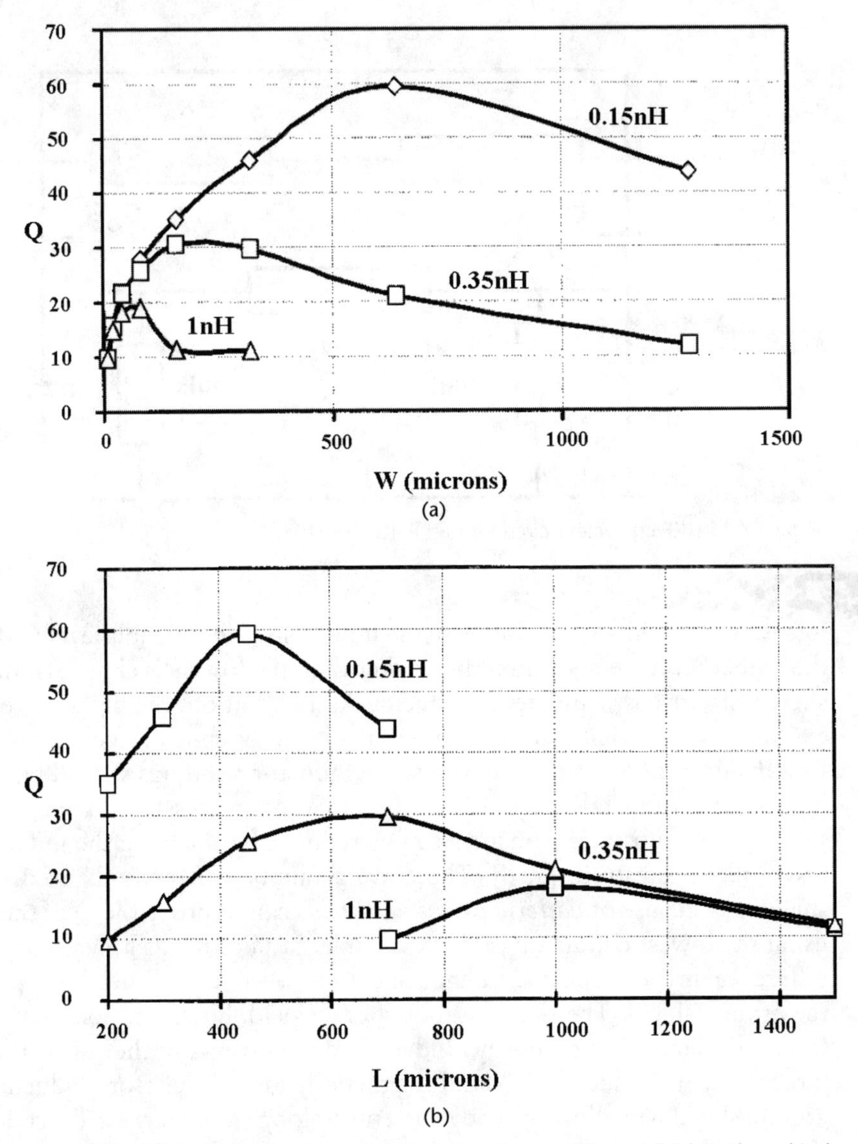

Figure 3.10 The slab inductor has its best Q for (a) a given width and (b) length and inductance. (*From:* [4]. © 2002 IEEE.)

Figure 3.8, the resistance R_s is the series resistance of the long line length required for a given inductance L. The area of the windings gives rise to a parasitic capacitance to the substrate (C_{ox}). The bulk parameters $R_{bulk}C_{bulk}$ model the influence of the substrate and its effective distance to RF ground. The capacitance C_p describes the duck-under capacitance of the inductor layout and could also include interwinding capacitance C_{int}.

3.2.3 Reduction of Inductor Parasitics

The main parasitic elements in the planar inductor are the R_s and C_{ox}. Minimizing C_{ox} will minimize the amount of coupling to the substrate and hence reduce the bulk

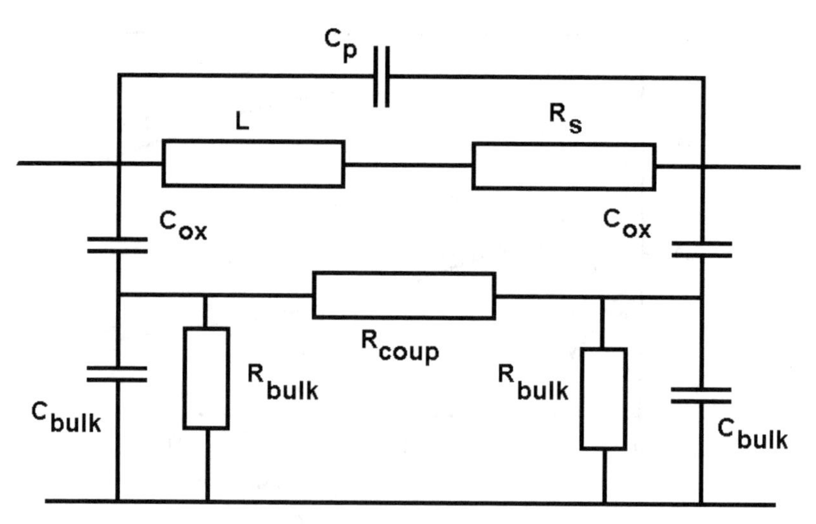

Figure 3.11 RF equivalent circuit for an RFIC inductor.

losses. Minimizing R_s will increase the inductor Q. In the multilevel CMOS process, the higher level metals tend to be thicker than the lower level metals, reducing their sheet resistance R_{SH} and hence reducing R_s. In addition, the higher level metals will exhibit lower capacitance per unit area (t_{ox} will be much larger), thereby reducing C_{ox} (Figure 3.12). For these reasons, the inductor windings should be placed at the highest level of metal available in the fabrication process.

Inductor substrate parasitics can be reduced by shielding the inductor from the lossy silicon substrate [5, 6]. There are a number of methods used to effect this shielding. The use of *patterned ground shields* can improve the Q of the inductor by using the lowest conducting layer available. Studies have shown that there is little difference in performance enhancement between using metal or polysilicon as the ground shield. The shape of the ground shield, however, does play a role in the level of inductor improvement. Induced eddy currents in the shield flow in such a direction as to reduce the overall magnetic field and therefore inductance. A solid ground shield will allow the eddy currents to loop over a large area, creating a larger

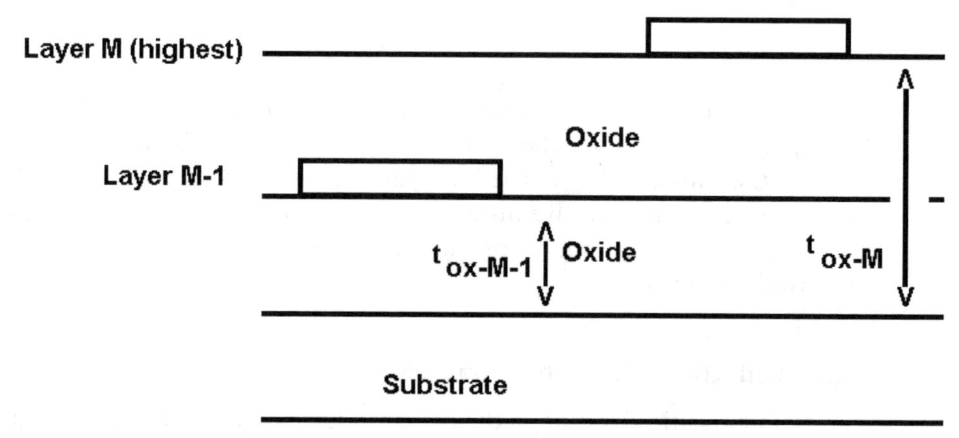

Figure 3.12 Higher level metals are farther from the conducting substrate, and they reduce parasitic capacitances.

magnetic field cancellation. Small cuts placed in the ground shield reduce these eddy current loops to smaller area, reducing the magnetic flux density associated with each loop and thereby lowering the cancellation effect. In addition, there is less magnetic field penetration into the substrate with the associated reduction in substrate losses. Figure 3.13 shows an example of a patterned ground shield that fits directly under the inductor windings; the periphery of the ground shield of course should be grounded.

Reduction in substrate parasitic effects can be done by reducing the value of C_{ox} and its associated coupling to the substrate. Suspending the inductors in air by removing the underlying oxide can reduce C_{ox} by approximately a factor of 4 [7]. A drawback to this method is that postprocessing is required to remove the oxide underneath the inductor. Other techniques include moving the inductor winding plane so that the magnetic field is parallel to the substrate instead of perpendicular [8]. This reduces eddy currents and their associated substrate losses, improving both the inductance and the Q of the inductor. The drawbacks to these methods are the specialized layout and postprocessing techniques required for their implementation; however, this increased effort and cost may be worthwhile for the level of inductor Q improvement in some applications.

3.2.4 Modeling Example

A 10-nH inductor is needed for a particular application. The top level metal is 3 μm above an ideal 1,000-μm-thick silicon substrate (the bottom of the silicon substrate is assumed grounded), and the insulating layer is silicon dioxide ($\varepsilon_r = 3.9$). Determine a set of layout parameters that will achieve this inductor using the simple equations and verify using an EM simulator.

The spreadsheet in the accompanying CD contains a file to aid in this design. One set of parameters that achieves the specifications is a three-turn inductor with 20-μm-wide windings spaced 3 μm apart and 570 μm on a side. The design tool gives an inductance of 10.06 nH, a series resistance of 9.53Ω, a shunting capacitance of 0.24 pF (C_{ox} in parallel with C_{sub}), and an overlap capacitance of 83 fF. At

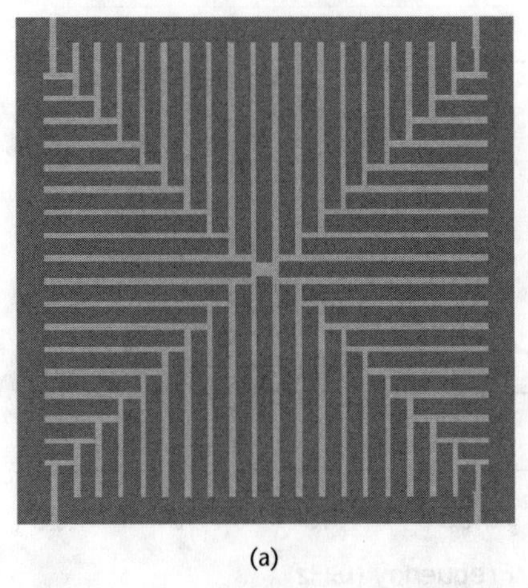

(a)

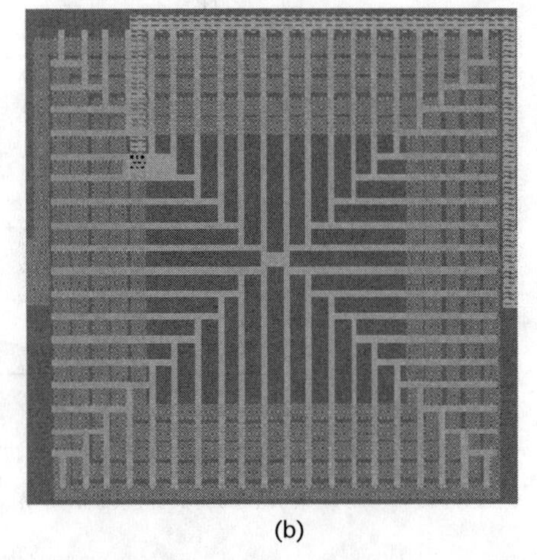

(b)

Figure 3.13 (a) A polysilicon ground shield for inductor improvement. (b) The inductor is wound over the ground shield and the shield is grounded.

900 MHz, the Q of this inductor is approximately 6. Using the simple RF equivalent circuit (Figure 3.14) with these derived equivalent circuit parameters, Figure 3.15 shows S_{11} and S_{21} over the frequency range 750 to 1250 MHz.

Using the same layout and substrate characteristics with an EM solver (in this case, Sonnet Lite), the equivalent circuit parameters can be extracted from the simulated S-parameters (Figure 3.16). In the initial EM simulation, the parameters in Table 3.3 were used.

The resulting EM simulation results were used to extract the RF equivalent circuit parameters, with the results showing that both R_s and the inductance are weak functions of frequency (Figure 3.17).

The extracted series resistance is 10.2Ω, an inductance of 8.5 nH, and an oxide capacitance of 0.2 pF, in reasonable agreement with the simple design tool. The Q is somewhat lower (4.7) due to the lower inductance value. The inductance and capacitances are sensitive to the substrate thickness. Little change in extracted equivalent circuit parameters is observed with a 0.1 S/m substrate, but with a 1000 S/m substrate, the extracted parameters change significantly and become strong functions of frequency. The series resistance at 900 MHz drops to approximately 8Ω, the inductance is reduced by about 0.5 to 8 nH, but the oxide capacitance rises dramatically to 0.72 pF, indicating much more substrate interaction. The transfer parameter $|S_{21}|$ is approximately –6 dB, indicating much higher losses in the circuit (Figure 3.18).

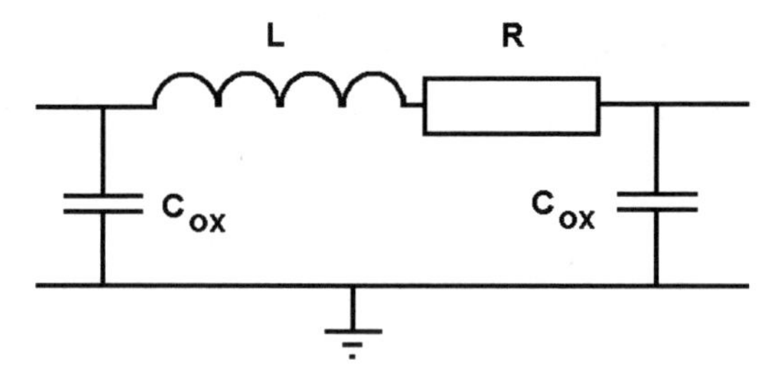

Figure 3.14 RF equivalent circuit for inductor (bulk effects neglected).

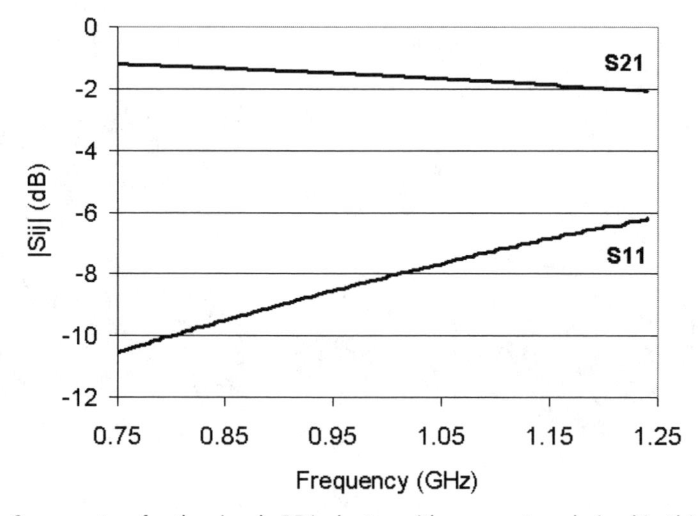

Figure 3.15 *S* parameters for the simple RF inductor with parameters derived in this example.

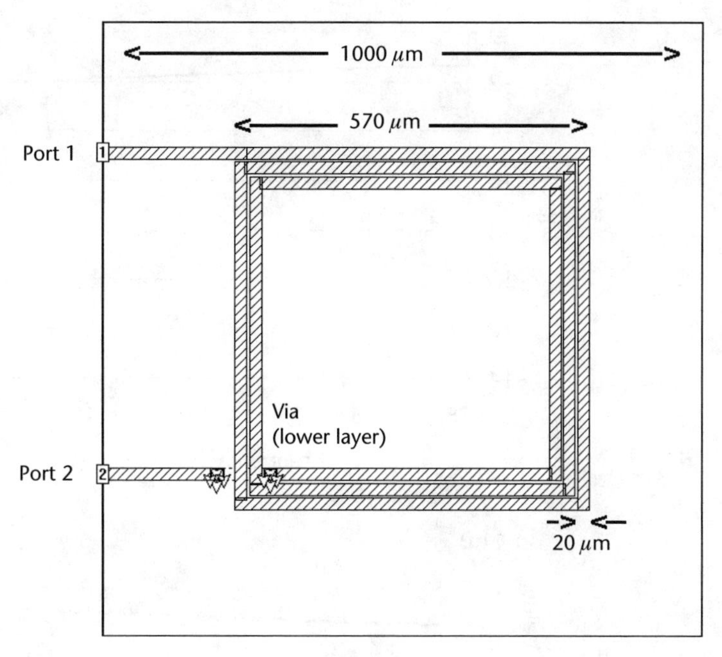

Figure 3.16 Three-turn inductor layout for simulation in Sonnet Lite (inductor_example.son).

Table 3.3 Inductor Simulation Parameters

Parameter	*Value*
Work space	1,000 × 1,000 μm
Space above top metal	100 μm
Metal width	20 μm
Metal-to-metal spacing	3 μm
Substrate thickness	1000 μm
Insulator ε_r	3.9
Substrate ε_r	11.8
Metal sheet resistance	0.03 Ω/sq
Substrate conductivity	0 S/m

While the methods described above provide reasonable "ballpark" estimates of inductor parameters, the frequency dependent nature of inductor equivalent circuit parameters shows that either full EM simulations or simulations using other methods are crucial for successful inductor designs [3].

3.2.5 Transformers

3.2.5.1 On-Chip Transformer Types

If placed in proximity, two or more inductors will be linked by mutual inductance and will exhibit transformer properties. Planar inductors may be fabricated by closely spaced parallel and concentric windings [9] on a single level of metal or by

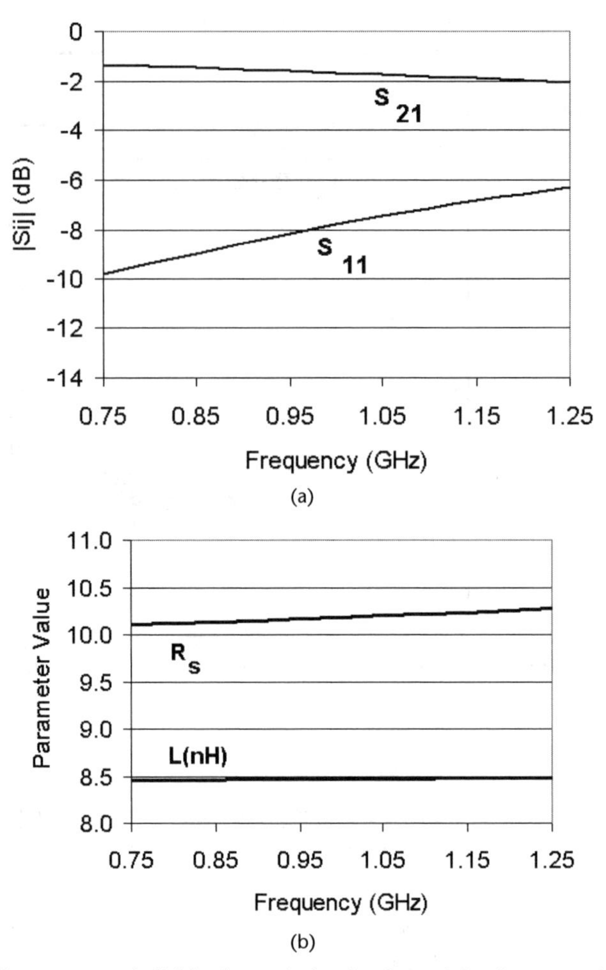

Figure 3.17 (a) Electromagnetic field solver solution for S_{11} and S_{21} of rectangular spiral inductor and (b) the extracted series resistance (in ohms) and inductance (in nanohenries).

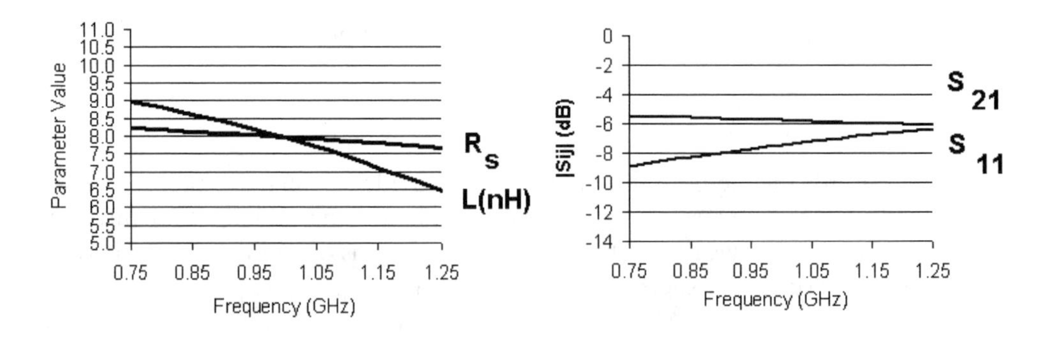

Figure 3.18 High substrate conductivity increases the losses in on-chip inductors and makes the parameters more strongly dependent on frequency.

stacking the inductors on adjacent metal layers [10] (some layout examples are shown in Figure 3.19).

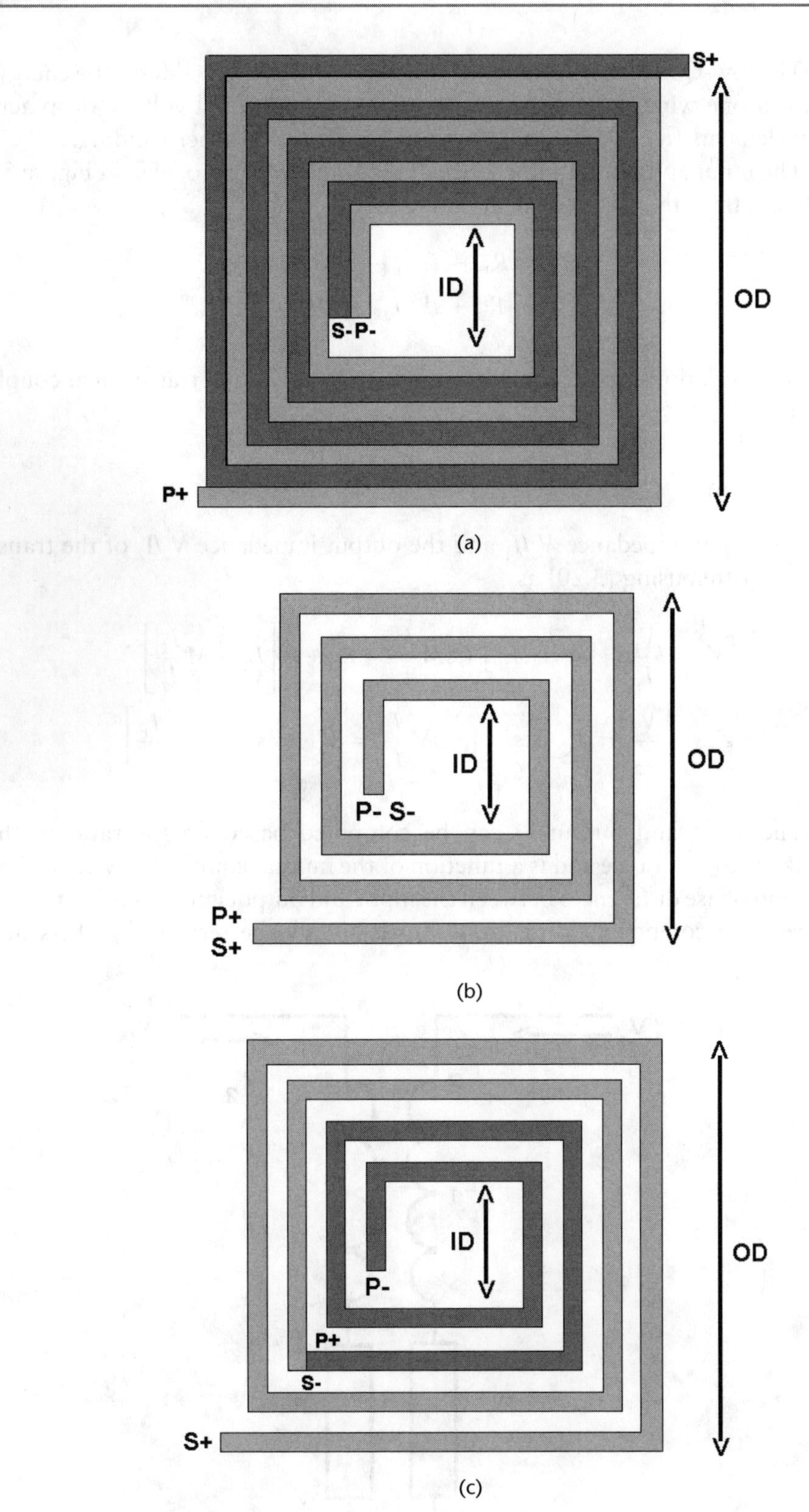

Figure 3.19 Layouts of planar transformers: (a) parallel winding, (b) overlay (only top layer shown), and (c) concentric. *P* and *S* indicate primary and secondary connections. (*After:* [9].)

The low-Q inductors that make up planar inductors will limit the energy linked between one winding and the other due to an additional voltage drop across the lossy elements (R_1 and R_2) that does not couple to the other winding.

The input and output impedance of the transformer modeled in Figure 3.20 can be found from the RF equivalent circuit:

$$\begin{aligned} V_1 &= [R_1 + sL_1 I_1] + sMI_2 \\ V_2 &= [R_2 + sL_2 I_2] + sMI_1 \end{aligned} \tag{3.20}$$

where the relationship between the mutual inductance and transformer coupling k is given by

$$M = k\sqrt{L_1 L_2}$$

The input impedance V_1/I_1 and the output impedance V_2/I_2 of the transformer can be written using (3.20) as

$$\begin{aligned} Z_{in} &= \frac{V_1}{I_1} = [R_1 + sL_1] + sM\frac{I_2}{I_1} = R_1 + j\omega\left[L_1 + M\frac{I_2}{I_1}\right] \\ Z_{out} &= \frac{V_2}{I_2} = [R_2 + sL_2] + sM\frac{I_1}{I_2} = R_2 + j\omega\left[L_2 + M\frac{I_1}{I_2}\right] \end{aligned} \tag{3.21}$$

The input and output Q can be computed based on the ratio of the total reactance to resistance and is a function of the mutual coupling as well as the amplitude and phase differences between the input and output currents [6]. At first glance, the reactive component of Z_{in} and Z_{out} is simply the term in brackets in (3.21).

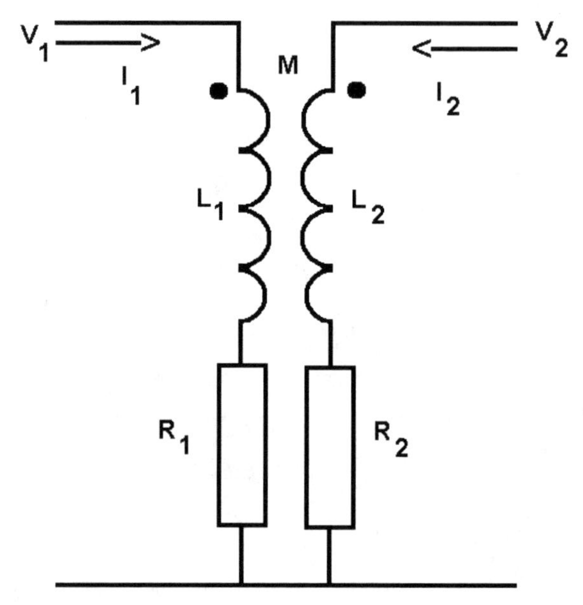

Figure 3.20 Simple RF equivalent circuit model of a transformer with resistive losses in both windings.

However, if I_1 and I_2 are of quadrature phase (90°), then the mutual coupling term actually becomes a resistive component and not reactive. For example, if the amplitude of the ratio of the two currents I_2/I_1 is A and the phase difference is +90°, then the two impedances become

$$\begin{aligned} Z_{\text{in}} &= R_1 + j\omega\left[L_1 + MjA\right] = R_1 - \omega AM + j\omega L_1 \\ Z_{\text{out}} &= R_2 + j\omega\left[L_2 - jM\frac{1}{A}\right] = R_2 + \omega\frac{M}{A} + j\omega L_2 \end{aligned} \tag{3.22}$$

By proper design of A and M, the input Q may be enhanced:

$$Q_{\text{in}} = \frac{X}{R} = \frac{\omega L_1}{R_1 - \omega AM} = \frac{\omega L_1}{R_1}\frac{1}{1 - \dfrac{\omega AM}{R_1}} = Q_1\frac{1}{1 - \dfrac{\omega AM}{R_1}} \tag{3.23}$$

The output Q, however, may be degraded since the current relationship provides additional loss at the output:

$$Q_{\text{out}} = \frac{X}{R} = \frac{\omega L_2}{R_2 + \omega\dfrac{M}{A}} = \frac{\omega L_2}{R_2}\frac{1}{1 + \omega\dfrac{M}{R_2 A}} = Q_2\frac{1}{1 + \omega\dfrac{M}{R_2 A}} \tag{3.24}$$

In addition to the mutual inductance created by the magnetic field linkages, there will also be capacitive coupling between the inductors as well as the usual inductor parasitic capacitance. Typical magnetic coupling factors for closely spaced windings are $0.75 < k < 0.9$. The k factors can be reduced by increasing the winding spacing. Layout techniques that improve stand-alone inductor Q such as line width, high metal layers, and substrate shielding also improve the performance of planar inductors.

An RF equivalent circuit model that includes the magnetic field coupling is shown in Figure 3.21 and combines the usual transformer action with the parasitic capacitances associated with the two inductors. The model shows an ideal transformer surrounded by the series and shunt parasitic elements, including intra-winding capacitance (C_{IW}).

A transformer based on the use of a slab inductor has been developed and is termed the *distributed active transformer* (DAT) [4]. The DAT consists of two stacked slab inductors on adjacent metal layers (Figure 3.22). Unlike in some of the planar spiral transformers, the two DAT connections can be physically far apart, a layout topology that can be useful in differential circuitry where the primary and secondary connections may be reasonably far apart. Differentially driving the DAT means that the center of either side of the transformer is an effective RF ground where dc bias may be applied [4].

3.2.5.2 Magnetic Coupling

An approximate expression for the mutual inductance M based on a series of measurements for a 0.25-μm CMOS process [11] was found to be

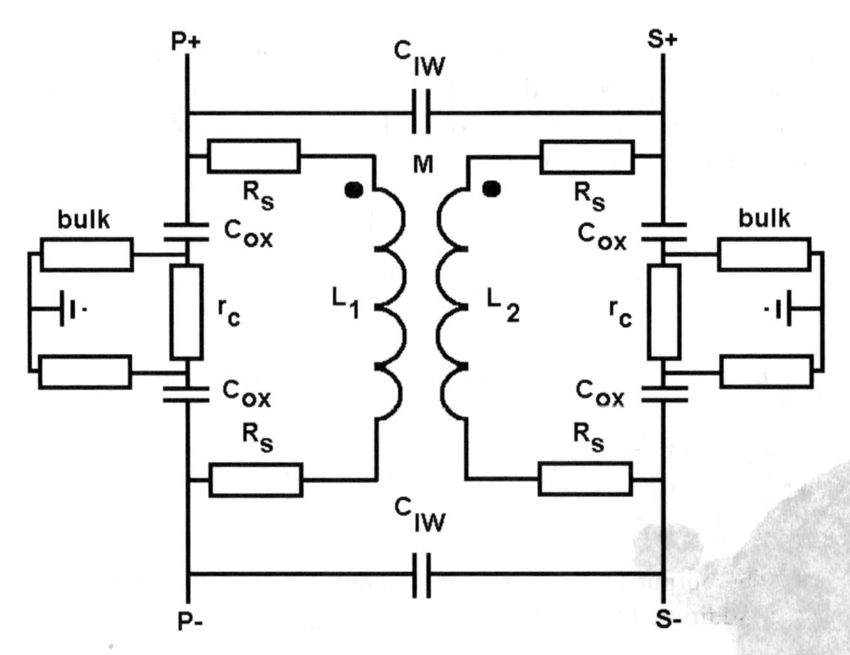

Figure 3.21 Detailed RF equivalent circuit model for an on-chip transformer, showing each inductor's parasitic elements as well as desired inductive (M) and undesired capacitive (C_{IW}) coupling.

$$M = 1.33 \cdot 10^{-6} D_{\mu m}^{1.15} e^{-0.0794 d_{\mu m}^{0.6} + 0.439 OD_{\mu m}^{1/2}} \text{ nH} \tag{3.25}$$

where D is the trace length, d is the spacing between windings, and OD is the outer diameter of the coil (all in microns). The expression indicates that there is a strong dependence on the spacing and size of the inductors making up the transformer (d, OD). For example, the metal inductance between two parallel wound square transformers of $OD = 240\ \mu$m, spacing of 5 μm, and trace length D of 240 μm would be 0.5 nH. For primary and secondary inductances of 5.0 nH, the transformer coupling would then be 0.1:

$$k = \frac{M}{\sqrt{L_1 L_2}} = \frac{0.5 \text{ nH}}{5.0 \text{ nH}} = 0.1 \tag{3.26}$$

3.2.5.3 Layout Examples

Transformers can be wired in other configurations to take advantage of their inherent voltage and current transformation as well as dc isolation. The layouts shown in Figure 3.19 indicate that both primary and secondary windings have similar numbers of turns, creating approximately 1:1 transformers. The transformer can be used to transform an unbalanced signal (such as from an antenna) to a balanced signal that can be fed directly to a differential amplifier. This so-called transformer *balun* (for *balanced to unbalanced*) is configured with one side (say, the primary) having its $P-$ winding grounded, and the differential signal appears at $S+$ and $S-$ [12] (Figure 3.23).

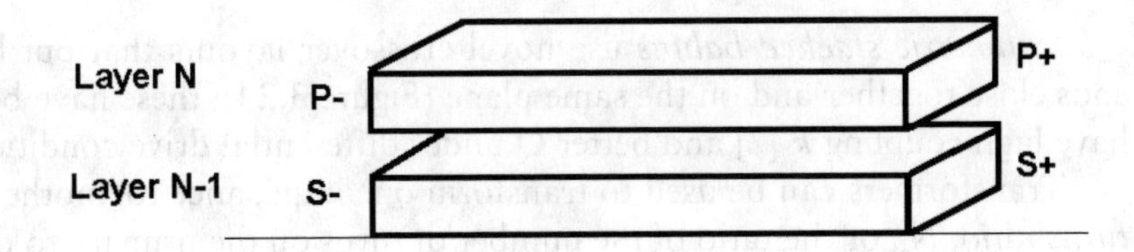

(a)

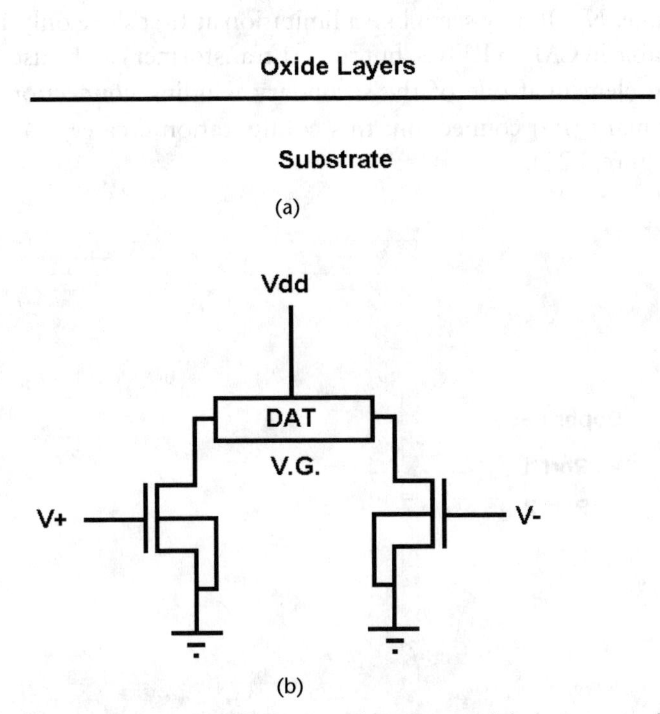

(b)

Figure 3.22 (a) The distributed active transformer consists of two slab inductors on adjacent metal layers. (b) The DAT exhibits an effective RF ground (*VG*) when used in a differential circuit.

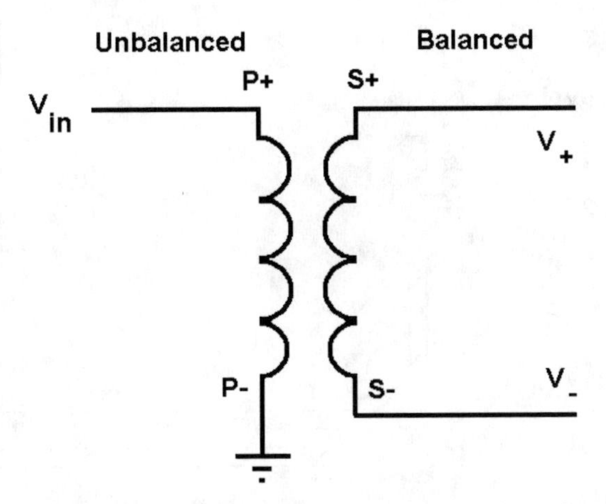

Figure 3.23 Transformers can be used to transform an unbalanced signal to a balanced signal (*balun*).

Symmetric stacked baluns use novel crossover layouts that put both winding ends close together and on the same plane (Figure 3.24); these have been shown to have high coupling *k* [2] and better Q under differential drive conditions [13].

Transformers can be used to transform one impedance to another through the *turns ratio*, N_r, or the ratio of the number of turns on the primary to the number of turns on the secondary. The impedance transformation goes as the square of this value, N_r^2. It may seem like a limitation at first since only 1:1 transformers are preferable in CMOS RFICs, but the 1:1 transformer can be used as an impedance matching element if one of the secondary winding connections (*S*–) is fed back to the primary (*P*+) connection; this configuration creates a 4:1 impedance transformer (Figure 3.25).

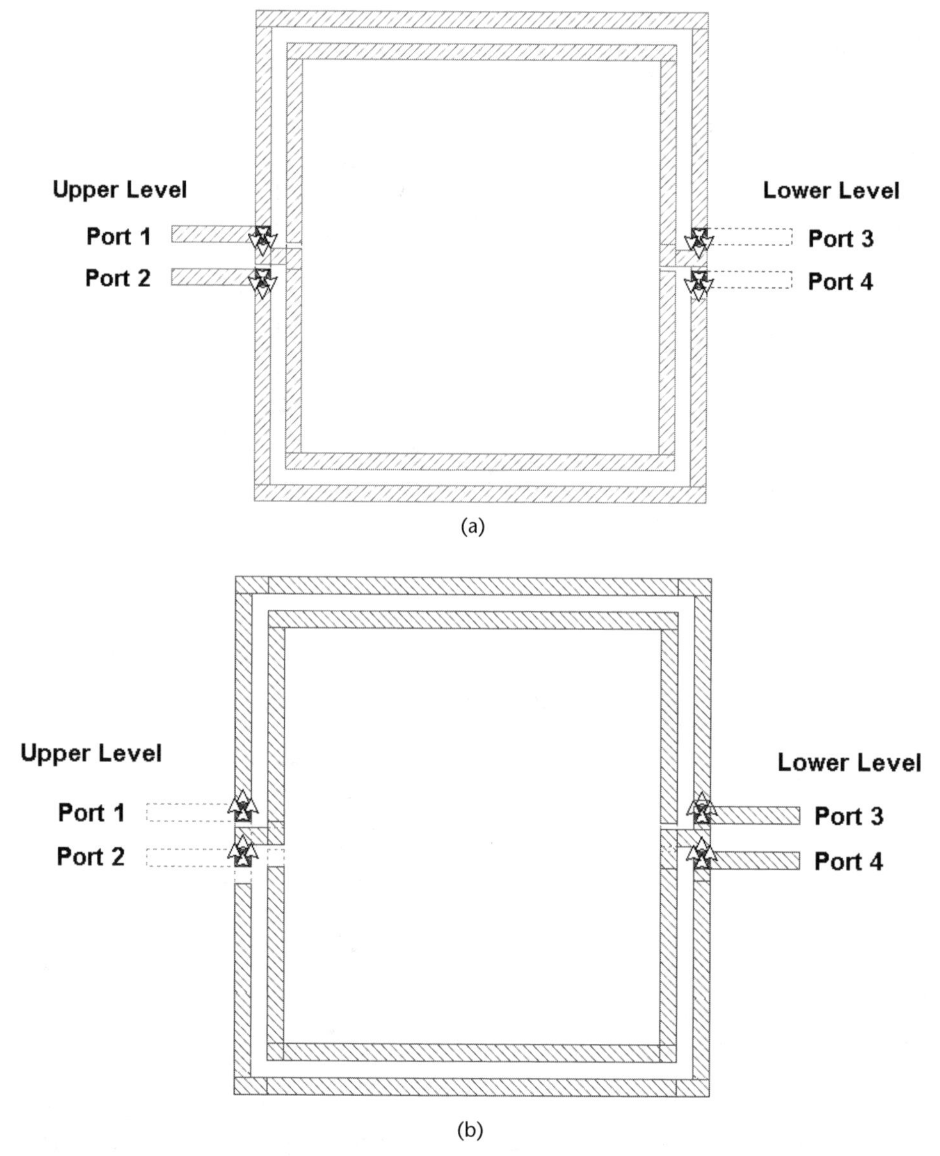

Figure 3.24 Symmetric stacked inductor layouts can be used as baluns: (a) top level and (b) lower level metallization. (*After:* [13].)

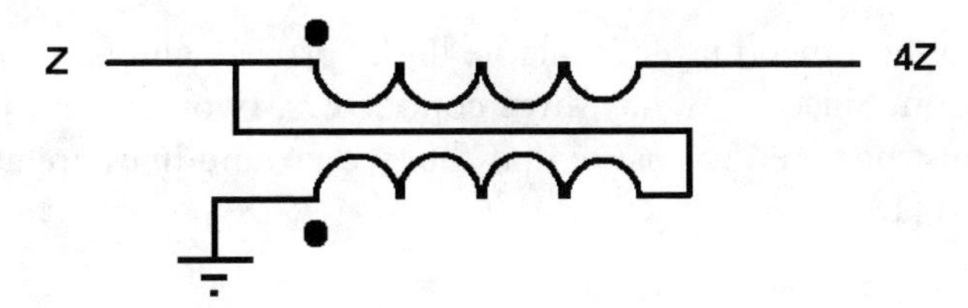

Figure 3.25 A 4:1 impedance transformer can be made with a 1:1 transformer.

3.2.6 Transmission Lines and Equivalents

The two most widely used transmission lines for RFICs are microstrip and coplanar waveguide. Conventional microstrip lines (single trace over a grounded substrate) are very lossy because of the low-resistivity substrates commonly found in the CMOS process. A modification to traditional structure for microstrip is to utilize the number of metal layers at the designer's disposal and use one metal layer as the "true" ground plane and a higher level metal as the signal trace [14, 15]. This configuration keeps the electric fields out of the silicon substrate, providing a lower loss structure. The exact location of the RF ground is also known, so traditional microstrip design equations can be used [16–18]. A drawback to this form of microstrip for use at frequencies above a few gigahertz is that the inductance per unit length is quite small, causing a corresponding low transmission line Q [19]. A widely used alternative transmission line structure has all the metal portions on the same metal layer (coplanar) and is termed *coplanar waveguide*.

3.2.6.1 Coplanar Waveguide

A coplanar waveguide (CPW) is a three-trace structure. Two outside traces are at RF ground and the center trace carries the signal (so-called GSG, or *ground-signal-ground,* structure). The width of the center trace *W* and the spacing *S* govern the characteristic impedances of coplanar waveguide (along with the dielectric constant of the surrounding material) (Figure 3.26).

CPW can be designed to have significantly higher Q than conventional microstrip by adjusting the width to reduce the losses. In addition, placing the CPW

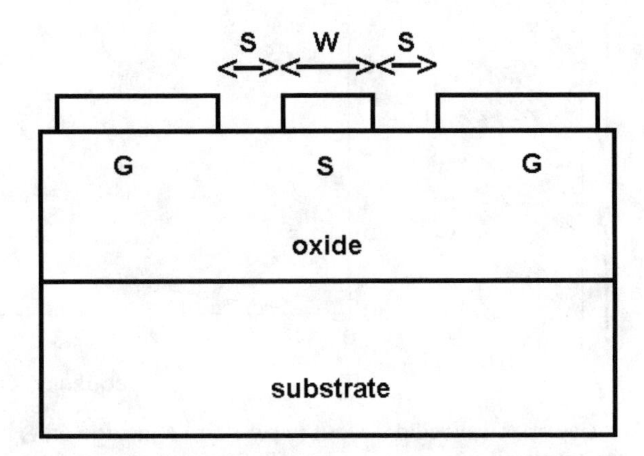

Figure 3.26 A coplanar waveguide structure resides on an oxide layer above the silicon substrate.

on the highest metal layer available in the process will further reduce any substrate interaction. Since CPW has three conductors, two different modes can be excited. Care must be taken to ensure that the two ground lines are at the same RF ground potential [15].

3.2.6.2 Modeling Example

Typical characteristic impedances of CPW on silicon range from approximately 25 to 75Ω. As an example, a 10-μm-wide line with 2-μm ground spacing at a height of 6 μm above the effective RF substrate ground yields a Z_0 of approximately 47Ω [20]. A 2D simulation using Sonnet Lite was performed to verify this impedance with the layout shown in Figure 3.27. The figure shows a close-up view of only one side of the connection; the line was simulated with a length of 1000 μm and a 100-μm width. The 2D simulation setup used the CMOS process parameters listed in Table 3.4. The 2D simulation results of the CPW structure on lossy substrate shows a near match to 50Ω input and output loads over a very wide frequency range with less than 0.3-dB insertion loss ($|S_{21}|$) over the entire frequency band.

3.3 Interconnections

The connection between various circuits and subsystems on an RFIC can have a major impact on the frequency response, time delay, and phase shift between the circuit blocks [21]. Modeling of these *interconnects* is crucial for obtaining an estimate of the overall system level of performance. In general, the longer the interconnection, the more impact on circuit performance. Two models are often used for

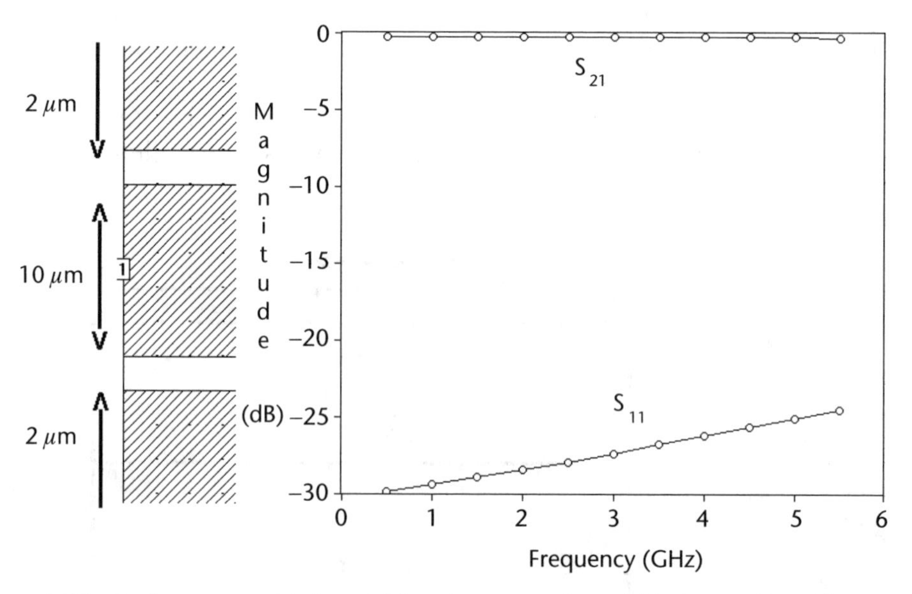

Figure 3.27 Coplanar waveguide layout (view along one edge only) in Sonnet Lite with the return loss and transfer *S* parameter for the 1000-μm-long transmission line indicated (cpw_example.son).

Table 3.4 Parameters Used in CPW Simulation

Parameter	*Value*
Work space	1,000 × 100 μm
Space above top metal	100 μm
Metal width	10 μm
CPW spacing (each side)	2 μm
Substrate thickness	1,000 μm
Insulator ε_r	3.9
Substrate ε_r	11.8
Metal sheet resistance	0.03 Ω/sq
Substrate conductivity	100 S/m

estimates of the interconnect behavior: those based on simple interconnect resistance and capacitance and those based on distributed or transmission lines.

3.3.1 Simple RC Models

The simplest RC model describes the interconnect in terms of the layer resistance and capacitance. The model elements are usually computed using the following equations:

$$R_{IC} = \frac{L}{W} R_{SH}; \; C_{IC} = \frac{\varepsilon_r \varepsilon_0}{t_{ox}} WL = WLC_{\square} \tag{3.27}$$

To keep the resistance associated with the interconnect to a minimum, metal interconnects are employed. In addition, the highest level of metal yields the lowest interconnect capacitance. However, the designer may not have the luxury of placing all the interconnects on the highest metal layer; in this case, the increased interconnect capacitance will need to be considered in the overall circuit design and performance specifications. Polysilicon interconnects should be avoided since they exhibit higher resistance (higher sheet resistance) and capacitance (closer to the substrate and effective RF ground) than the metal layers.

Moving vertically from one metal layer to another requires the use of a *via* or *via contact*. This via exhibits a resistance as well and should be considered as part of the entire interconnect system. The via resistances are usually provided assuming a square via contact, with values of 0.5 to 1Ω not uncommon. The via contact resistance between metal layers can be further reduced by using a number of via contacts in parallel (the so-called *stitched contact,* Figure 3.28); the overall resistance is reduced by $1/N$, where N is the number of parallel contacts. Multiple vias are needed when jumping two or more metal layers; each via will then add to the contact resistance.

The simple RC model routinely overestimates the actual delay associated with an interconnect since the interconnect should ideally be modeled as a distributed RC line. From time delay calculations [1] it has been shown that the two elements are

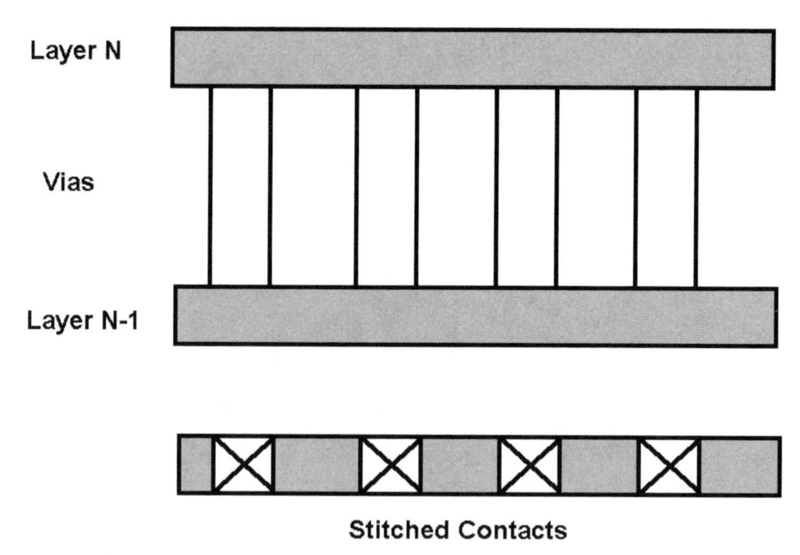

Figure 3.28 Parallel or stitched contacts can reduce the interconnect resistance by placing a number of via contacts in parallel.

about 30% too high, resulting in a twofold increase in RC time constant for the structure:

$$R_{IC} = \frac{1}{\sqrt{2}} \frac{L}{W} R_{SH}; \; C_{IC} = \frac{1}{\sqrt{2}} \frac{\varepsilon_r \varepsilon_0}{t_{ox}} WL = \frac{1}{\sqrt{2}} WLC_{\square} \tag{3.28}$$

3.3.2 Transmission Line Models

Transmission line models exploit further the distributed nature of the interconnect and look at all four parameters: the resistance, inductance, capacitance, and conductance per unit length. The complex layout of the RFIC virtually ensures that there will be inductance effects due to the current loops created by the ground return path(s). While EM numerical modeling will yield the most accurate results, this method is also time consuming and is prohibitive for chipwide simulation. Using a partial element equivalent circuit (PEEC) methods and the concept of the effective distance to ground, closed-form estimates for the resistance and inductance per unit length can be made. A so-called *complex effective inductance per unit length* can be derived using a complex image approximation using the closed-form microstrip inductance formula with the substrate height being related to the effective ground location h_{eff} [22]:

$$L^*(\omega) = \frac{\mu_0}{4\pi} \ln\left[1 + 32\left(\frac{W}{h_{eff}(\omega)}\right)^{-2}\left(1 + \sqrt{1 + \left(\frac{\pi W}{8h_{eff}(\omega)}\right)^2}\right)\right] \tag{3.29a}$$

$$h_{eff}(\omega) = \frac{\mu_0}{W}\left(t_{ox} + \frac{1-j}{2}\delta_{Si} \tanh\frac{(1+j)t_{Si}}{\delta_{Si}}\right)$$

where W is the trace width, t_{ox} (t_{Si}) is the oxide (silicon) thickness, δ_{Si} is the skin depth in the silicon, and h_{eff} is the effective distance to RF ground [22]. The inductance and resistance per unit length are then functions of this effective inductance L^*:

$$R(\omega) = -\omega \mathrm{IM}\left(L^*\right) \quad L(\omega) = \mathrm{RE}\left(L^*\right) \tag{3.29b}$$

As an example to show the frequency dependence of these two model elements, consider the following structure: a 10-μm-wide interconnect trace over a 3-μm-thick insulating oxide and a 500-μm silicon substrate with a 100-S/m conductivity (Figure 3.29). The application of (3.29) shows that the inductance and resistance per unit area are only weakly dependent on frequency. However, above approximately 3 GHz, the distributed elements show a strong frequency dependence, with the loss term rising significantly.

3.4 RF Microelectrical Mechanical Systems

MEMS, and more specifically RF MEMS, are finding increased use in a number of switching and control functions at RF and microwave frequencies [23]. In addition, the structures are used as variable capacitors (*varactors*) and resonating structures [24, 25]. This section covers material on understanding basic RF MEMS operation and applications as a way to see how the technology fits within the framework of RF analog circuit design. RF MEMS switches have been shown to work up to 70 GHz with good insertion loss and isolation characteristics [23]. Resonator circuits using MEMS have been designed for use beyond the X-band [26]. The RF MEMS literature is quite vast, so these are just a sampling of the current state of technology.

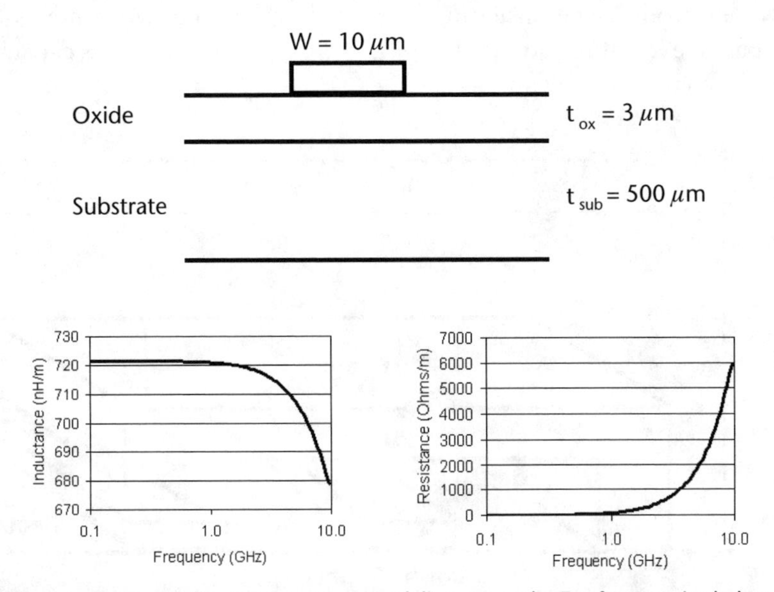

Figure 3.29 A transmission line interconnect modeling example. For frequencies below approximately 2.0 GHz, the interconnect exhibits low losses and constant transmission line equivalent circuit parameters.

A number of good books and articles are available on the subject [27, 28], and the reader is encouraged to investigate these references for more detailed information.

3.4.1 Basic Types and Operation

The fundamental physics of RF MEMS component operation is relatively simple: To create movement, mechanical forces need to be overcome by an applied electrostatic force. In the previous discussion of CMOS fabrication, all layers and materials were embedded firmly in an insulating matrix, with no movement possible. Some of this insulator must be selectively removed so that particular layers and materials can be free to move and provide the MEMS operation. During layout, fabrication, and postprocessing, insulating layers that will eventually be removed are often referred to as *sacrificial layers*. The concept of the sacrificial layer can be easily seen by looking at a widely used MEMS structure, the *cantilever beam*, or *diving board*, structure (Figure 3.30). Here, a metal layer is anchored on one side and embedded in the oxide layer. In this form, the beam cannot be moved; however, if the oxide in the region indicated by the dotted line were removed, then the beam would be free to move.

If the metal layer were anchored on both sides and the oxide layer removed between the two anchors, then a *MEMS membrane* would result (Figure 3.31). In both beam and membrane structures, the MEM exhibits a certain range of motion, and if enough electrostatic force is applied, the gap length g will go to zero. This bottom contact can be either bare metal for a *dc contact* or a thin insulating layer of high dielectric constant for a *capacitive contact.*

3.4.2 Actuation Voltage

Referring to Figure 3.31, a *control* or *actuation voltage* source placed between the beam or membrane and the pull-down electrode sets up an electric field between the two electrodes. The insulating material between the two contacts (typically air) is capacitive in nature, so the change in the energy stored in this capacitor E_C is related to the change in the gap:

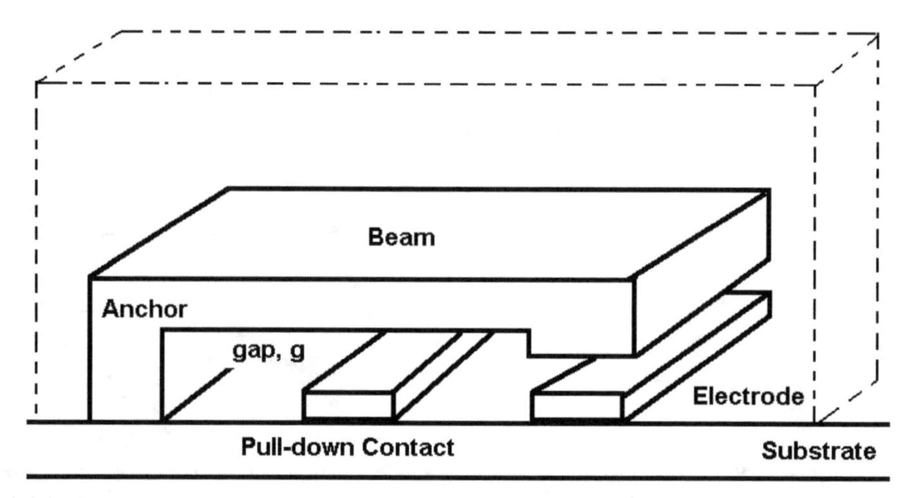

Figure 3.30 MEMS cantilever beam structure with region of oxide to be removed via postprocessing.

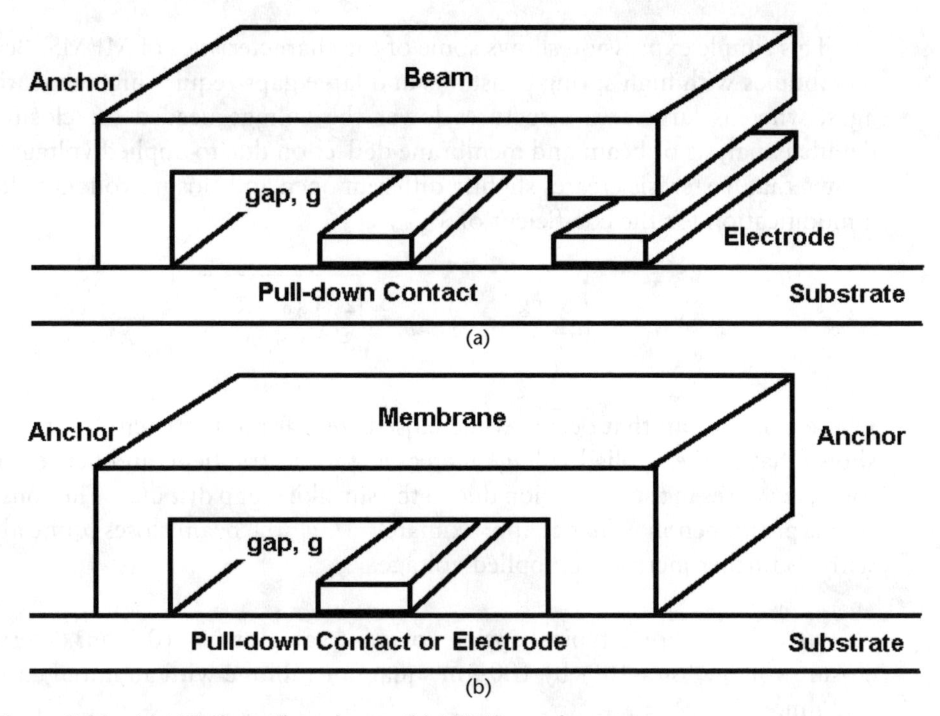

Figure 3.31 (a) MEMS cantilever beam. (b) MEMS membrane.

$$E_C = \frac{1}{2}CV^2 = \frac{1}{2}\frac{\varepsilon_0 A}{\Delta g}V^2 \tag{3.30}$$

where Δ_g is the gap change or deflection. The force created by the application of voltage V is countered by the mechanical energy of the moving beam or membrane. This kinetic energy expression can be described in terms of the beam or membrane *spring constant k* and the change in the gap:

$$E_{MEMS} = \frac{1}{2}k(\Delta g)^2 \tag{3.31}$$

Equating the two energy terms and rearranging yields a simple relationship between the gap deflection and the applied voltage:

$$V = \sqrt{\frac{k(\Delta g)^3}{\varepsilon_0 A}} \tag{3.32}$$

To completely close the gap, $\Delta g = g_0$, so the actuation or closure voltage can be written as

$$V_{close} = \sqrt{\frac{k g_0^3}{\varepsilon_0 A}} \tag{3.33}$$

This simple expression shows some of the characteristics of MEMS. Beams and membranes with high spring constants and large gaps require high activation voltages, whereas large-area structures lower the voltage needed for closure. More detailed analysis of beam and membrane deflection due to applied voltage [28] has shown that hysteresis creates slightly different open and closure voltages along with a modification for the coefficient of V_{close}:

$$V_{close} = \sqrt{\frac{8}{27}\frac{kg_0^3}{\varepsilon_0 A}}; \quad V_{open} = \sqrt{\frac{2}{3}\frac{k(g_0 - g_c)^3}{\varepsilon_0 A}} \tag{3.34}$$

where g_c is the gap that occurs when closed (zero for a dc switch). Further analysis shows that as the applied voltage increases, the electric field, and hence pull-down force, increases in the gap region due to the shrinking gap distance. The consequence of this phenomenon is that at approximately $1/3g_0$, the beam closes (quite abruptly), with no further increase in applied voltage.

Example 3.2: For a typical spring constant value of $k = 10$ N/m, determine the closure voltage for a 100- by 100-μm square membrane with an initial gap spacing of 2 μm.

Using (3.34), the closure voltage is computed as

$$V_{close} = \sqrt{\frac{8}{27}\frac{kg_0^3}{\varepsilon_0 A}} = \sqrt{\frac{8}{27}\frac{10(2\cdot 10^{-6})^3}{8.85\cdot 10^{-12}\cdot 10^4\cdot 10^{-12}}} = 16.3\text{V}$$

Note that this voltage is significantly higher than the typical voltage levels needed for active device operation (3.3V and lower for small-feature technologies).

3.4.3 MEMS Switches

The two main types of MEMS switches are dc contact and capacitive switches. One of the more widely used types is the membrane switch used in conjunction with coplanar waveguide [23, 24]. The membrane is anchored to the CPW grounds with the freely moving portion suspended over the signal line of the CPW (Figure 3.32). The control voltage can be applied either to the signal line or to pull-down electrodes. The primary type of switch using this structure is capacitive, although there have been dc contact switches studied as well. The capacitive switch structure will often have a thin layer of high dielectric constant material coating the signal line of the CPW. The membrane then moves from a low capacitance-to-ground state (membrane up) to a high capacitance-to-ground state (membrane down). The two capacitance states are then dependent on the active area of the membrane, the changing gap distance, and the dielectric layer coating the signal line:

$$C_{down} = \frac{\varepsilon_r \varepsilon_0 A}{t_K}; \quad \frac{1}{C_{up}} = \frac{t_{air}}{\varepsilon_0 A} + \frac{1}{C_{down}} \Rightarrow C_{up} = \frac{\varepsilon_0 A}{t_{air}} \tag{3.35}$$

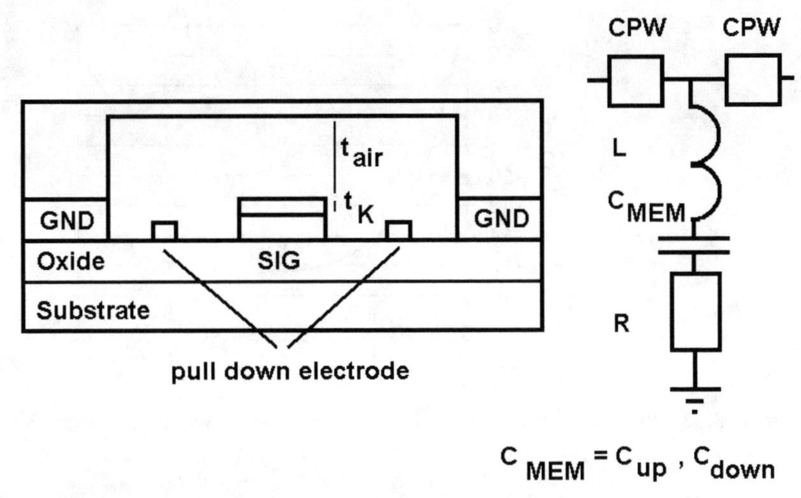

Figure 3.32 Coplanar waveguide MEMS switch with RF equivalent circuit. (*After:* [23].)

Ratios of C_{down}/C_{up} can be 100 or more. In a CPW application, a low insertion loss occurs during the membrane up state, corresponding to small C_{up}. When the membrane is pulled down, the low shunting reactance of C_{down} reflects the majority of the microwave or RF signal, resulting in high isolation.

The RF equivalent circuit for the MEMS capacitive membrane switch includes an inductive and resistive term modeling the length of the membrane between the capacitance and the RF ground. Typical values of L and R for shunt membrane switches are 10 pH and 0.5Ω [23].

Example 3.3: Determine the insertion loss and isolation for a capacitive membrane switch if $C_{up} = 50$ fF and $C_{down} = 2$ pF over the frequency range 1.0 to 50 GHz.

For a shunt impedance, the insertion loss and isolation expression can be written as

$$IL/ISO = 20\log\left|1 + \frac{1}{2}\frac{Z_0}{Z_{MEM}}\right|;\ Z_{MEM} = R + j\omega L + \frac{1}{j\omega C_{MEM}} \tag{3.36}$$

The resulting calculations shows very low insertion loss (less than 0.2 dB) up to 50 GHz, with isolation better than 16 dB at 30 GHz (Figure 3.33). The resonance of the L and C_{MEM} in the circuit provides the improved isolation at 30 GHz, a phenomenon that may be used as part of the design process [23].

Series and shunt dc contact switches can be made with either cantilever beams or membranes. Series switches using beams are straightforward in operation, with the beam acting as the signal line. Closing the switch allows the signal to flow to the other electrode. Shunt switches have been designed by modifying the capacitive membrane switch structure. The high-dielectric coating is removed from the CPW signal line and a "dimple" is placed in the membrane (Figure 3.34). This dimple makes dc contact with the CPW signal line when the switch is activated.

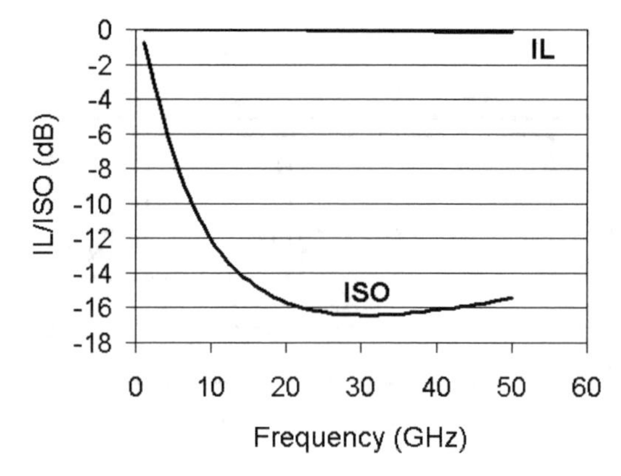

Figure 3.33 Insertion loss and isolation for a CPW MEMS switch in Example 3.3.

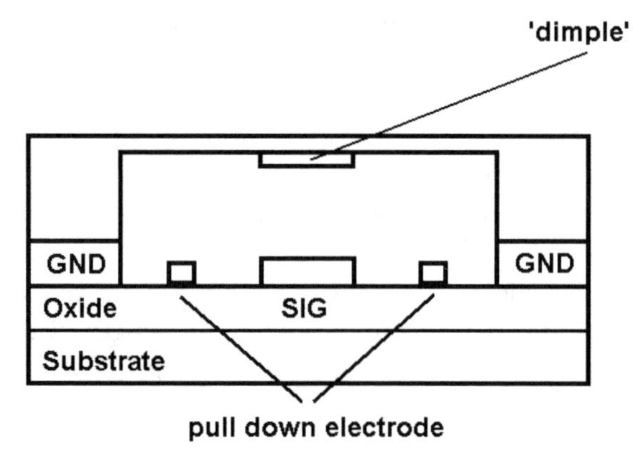

Figure 3.34 The dc contact MEMS CPW switch.

3.4.4 MEMS Resonators

MEMS structures can be used as mechanical resonators in on-chip applications. The beam and membrane structures mentioned previously, like all mechanical systems, have natural resonant frequencies that are dependent on the structure's geometry and physical properties. In its simplest form, the mechanical resonance frequency of a cantilever beam can be written as

$$f = \frac{1}{2\pi}\sqrt{\frac{k}{m}} \tag{3.37}$$

where k is the spring constant describing the beam stiffness, and m is the mass of the beam. A high mechanical resonance frequency occurs with high spring stiffness and a low beam mass. In terms of the beam dimensions, for a narrow beam, the resonant frequency can be modified [27]:

$$f = 1.03\kappa\sqrt{\frac{E}{\rho}}\frac{h}{L_r^2} \tag{3.38}$$

where E is Young's modulus, ρ is the density of the beam, h is the beam thickness, and L_r is the length of the beam. The factor κ is a scaling factor describing the surface conditions. For polysilicon beams, E is approximately 150×10^9 Pa and the beam density is approximately 2000 kg/m^3 [29]. For example, a 25-μm-long, 1.5-μm-thick polysilicon beam has a natural resonant frequency of 21.4 MHz (assuming unity κ). The Q of MEMS resonators is also quite high, in direct contrast to the relatively low Q of on-chip resonating structures using inductors and capacitors. Q's of 10,000 or higher have been achieved, with even higher values in vacuum, where atmospheric damping is minimized [29–31]. These high values of Q are reminiscent of another mechanical resonating structure: the quartz crystal resonator. MEMS resonator equivalent circuits are often described in terms of *motional parameters*, similar to quartz crystal resonators. The motional parameters, however, are RF equivalent circuit parameters only and do not represent actual physical component values.

Early MEMS resonators were limited to resonant frequencies under 1.0 MHz. Recent advances in both technology and novel resonator structures (with higher order mechanical resonance modes excited) have managed to push into the low gigahertz range [32]. A unique structure has been proposed that provides on-chip high-Q resonating structures at frequencies as high as K band and beyond [26]. The back plane of this structure is metallized, forming one side of a traditional dielectric loaded (high-resistivity silicon) microwave cavity structure. The cavity side walls are created with a "picket fence" structure of closely spaced vias from the silicon top surface to the back plane metallization. The final plate is fed by a CPW. Q's greater than 70 and frequencies as high as 33 GHz have been demonstrated.

Until recently, the only way to obtain required high-Q resonating structures was with off-chip components (SAW filters, for example); high-Q MEMS structures promise to provide this functionality on chip and be completely integrated with active CMOS circuit elements. An example of such a MEMS resonating structure is shown in Figure 3.35 [33]. For a 100-MHz high-Q resonator, typical dimensions are $W = h = 2$ μm and $L = 12.9$ μm.

3.4.5 MEMS Reliability and Packaging

Three related factors that affect the rollout of more RF MEMS applications are packaging, reliability, and their power handling capability. The MEMS package must allow freedom of movement of the MEMS structure, with a hermetically or near-hermetically sealed package using low-pressure and/or inert gas. Low-pressure gas inside the package reduces switching delays and vibrational damping due to atmospheric drag, while the inert gas (such as nitrogen) prevents aging and corrosion of the fragile MEMS structures. Frequently, elevated temperatures are used in the sealing process, which may cause changes to the MEMS structure. Both circuit and package designers for RF MEMS should be involved with the design of the product from the outset and work in parallel rather than in series to ensure a successful design [34].

Another issue unique to RF MEMS switches and resonators is their long-term reliability. Because of the mechanical nature and movement of the actuators, the structures will degrade over time, leading to failure after a certain number of cycles

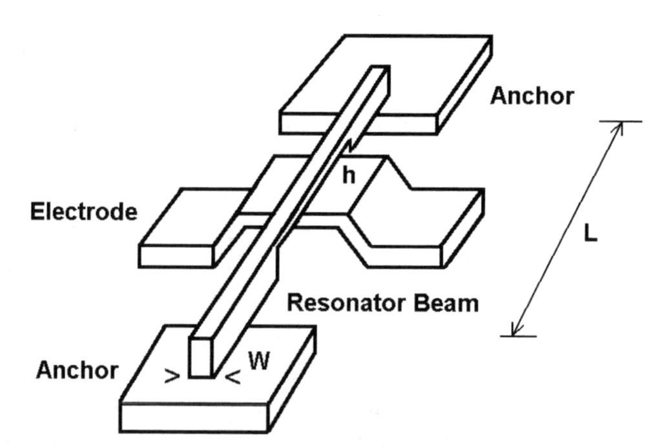

Figure 3.35 Example of a MEMS resonator that can replace a SAW filter but requires much less silicon real estate. (*After:* [33].)

or operations. For example, dc contact switches are subject to mechanical damage from impact or "welding" of the contact in the presence of high currents [23]. Capacitive switches can fail prematurely due to dielectric breakdown in high-power environments or by stiction (stuck contact) caused by dielectric charging [23, 34, 35]. In spite of these important issues, a major effort in MEMS research has improved the lifetime and reliability of MEMS for RF switching and circuit applications. The number of operations has been improved to greater than 10 billion cycles in some commercial devices.[3] Design methodologies to improve the power handling capability of RF MEMS and their high-power reliability are also under way [36, 37].

3.5 Basic Packaging

3.5.1 Anatomy of an RF Package

There are many different types of RF packages in use, with a variety of abbreviations. There are traditional in-line packages with pins on one or both sides of the package (DIP, for example). For surface mounting, there are small-outline packages (SOICs such as SOP, SOT, TSOP, and QSOP, for example), flat pack and leadless chip carriers (FP and LCC, respectively), no-lead carriers (QFN, for example), and micro-P; in short, a real "alphabet soup" of packages, each with its own advantages in terms of RF, real estate, and thermal performance. Plastic and ceramic packages are available, with the plastic package frequently chosen for low cost and high-volume usage; ceramic packages are used for very high frequencies because of their lower loss. However, all these packages, regardless of their abbreviations, share a number of characteristics that are reviewed in this section.

At the center of the package is the *paddle,* where the silicon RFIC die is attached [Figure 3.36(a)]. This paddle is metallic in nature and helps in dissipating the heat produced by the RFIC as well as providing a sturdy mechanical base for the die. Depending on how much heat must be dissipated, the paddle may have a connection to the exterior of the package for external heat sinking. The die is bonded to the

3. See for example http://www.radantmems.com/ and http://www.memtronics.com/.

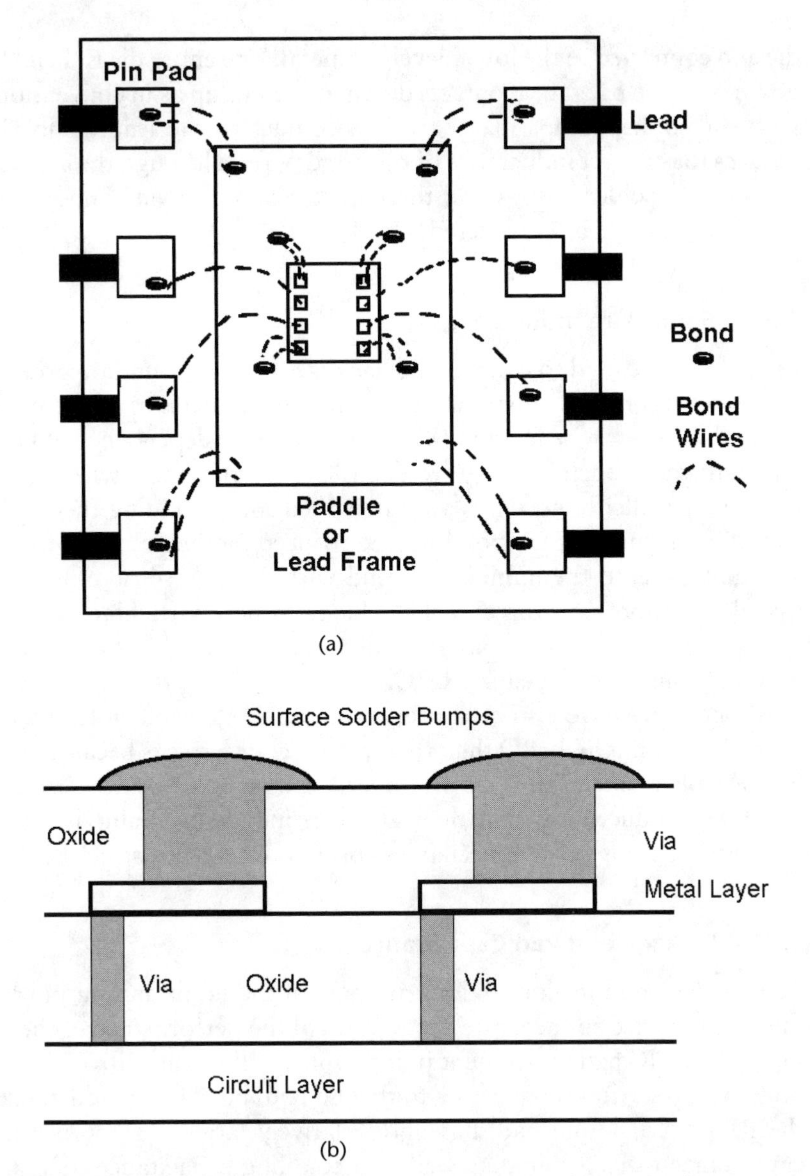

Figure 3.36 Inside look at (a) a generic RF package and (b) flip chip CMOS die.

paddle with an *adhesive* that promotes good thermal conductivity between the silicon die and the paddle. This adhesive may also be electrically conductive to improve grounding. Surrounding the paddle are *package pads* or *leads* that go through the package and provide a connection to the outside world. *Bond wires* are typically used to connect the pads on the RFIC to the package pads for signal or dc input and output. Once the die has been attached to the paddle and the required bond wires connected to the package, the interior of the package is filled with epoxy or other *encapsulant* material, sealing the package and also "freezing" the bond wires in place.

An alternative to this traditional RF packaging technique is the use of *flip chip* technology [Figure 3.36(b)]. Flip chip technology places large (relatively, speaking to the size of the active CMOS elements) solder bumps on the topmost layer of the die and connected to the lower levels of metal. The entire die is then "flipped" upside down onto the package (contacts down instead of up as in conventional packaging), with the solder bumps making direct contact to the leads. Flip chip technology reduces the contact inductance of the bond wires, although the capacitive characteristics of the solder bumps and the inductance of the via connections to the lower metal layers must be considered.

3.5.2 Bond Wire Inductance

Bond wires are used to connect the package pins or grounding pedestal to the RFIC die. A general rule of thumb in computing the inductance of the bond wire is 1 nH per millimeter of wire length [38]. There is a weak dependence of bond wire diameter, with the inductance slightly decreasing as the bond wire diameter increases. Multiple parallel bond wires are often used for grounding the RFIC die since the overall inductance of the bond will be reduced by the number of wires used. There are limitations to the number of bond wires used because only a limited number (usually no more than three) will be able to fit on a particular pad. Multiple ground pads, pins, and bond wires are typically used to make sure there is a low-inductance path to ground for the entire RFIC.

Bond wires have also been used as inductive elements in RFICs [39]. Bond wire inductors exhibit higher Q than their planar counterparts because the looping of the wires tends to make them less susceptible to substrate effects. The drawbacks to bond wire inductors is that their absolute inductance value is difficult to predict accurately because of the mechanical process used in creating the bonds.

3.5.3 Package and Pad Capacitance

The package and its connections are part of the input and output RF path and so must be modeled for accurate description of the performance of the entire package. From the RFIC pad attachment point (which will be capacitive in nature), there is a bond wire (or wires) connecting to the pin pad, putting an inductance in series with the RF path. The pin pad, having a relatively large-area footprint, will exhibit a capacitance to RF ground as well as a coupling capacitance to adjacent pins. The leads (even so-called leadless chip carriers have metallic extensions to the package exterior) will also present an inductance and a small-capacitance component. These effects give the general form for an RF package model, shown in Figure 3.37.

3.5.4 Thermal Properties—Thermal Resistance

Silicon die temperatures as high as 125°C can seriously degrade performance temporarily or catastrophically, so power dissipated by the circuitry must be removed for proper operation of the RFIC. The measure of a package's ability to remove heat from the die and dissipate it into the ambient is termed the *thermal resistance* and is frequently given by the symbol θ (dimensions °C/W). The temperature above the ambient is based on this thermal resistance and the difference between the input powers (dc plus RF) and the output power [40]:

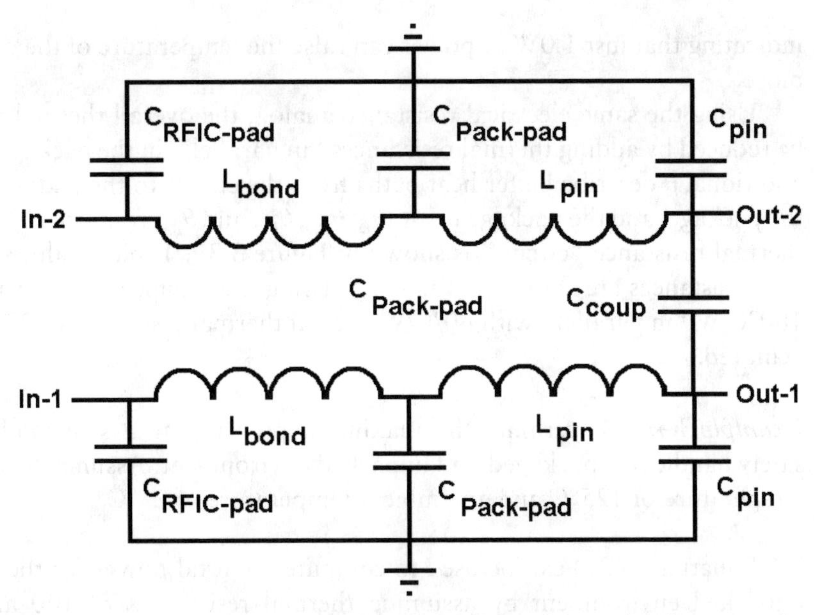

Figure 3.37 RF equivalent circuit for an RF package showing the coupling between adjacent pins as well as the pin and bond wire inductance and pad and package capacitance.

$$T = T_{ambient} + \Theta\left[\left(P_{DC} + P_{RF\text{-in}}\right) - P_{RF\text{-out}}\right];\ \Delta T_{rise} = \Theta\left[\left(P_{DC} + P_{RF\text{-in}}\right) - P_{RF\text{-out}}\right] \quad (3.39)$$

In die format, the thermal resistance of the air-silicon die heat path is relatively high (i.e., it exhibits high resistance to heat flow), $\theta_{AD} = 75°C/W$ or more. The heat path from the device or circuit (which is on the surface of the die) to the silicon is on the order of $\theta_D = 25°C/W$ (Figure 3.38). Thermal resistances add like electrical resistances, so the overall circuit-air thermal resistance can be 70°C/W or more,

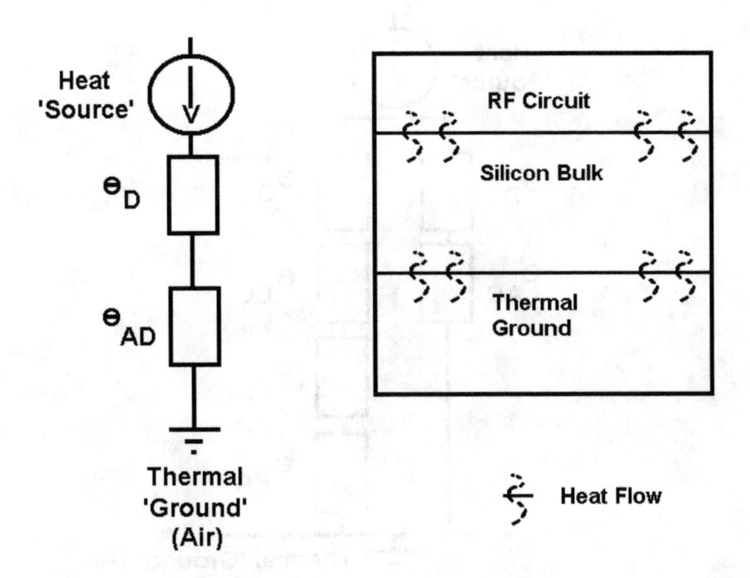

Figure 3.38 Simple thermal model of heat flow from a heat source to a heat sink.

indicating that just 1.0W of power can raise the temperature of the die by 100°C or more.

Using the same electrical resistance analog, the overall thermal resistance may be reduced by adding thermal resistances "in parallel"; in the packaging sense, these additional θ describe better heat paths from the circuit to the paddle, the paddle to the package, and the package to air (θ_D, θ_{CD}, θ_{PP}, and θ_{PA}, respectively). The modified thermal resistance "circuit" is shown in Figure 3.39. Typical values for these thermal resistances are shown in Table 3.5. Using these typical values shows that with 100°C/W "in parallel" with 60°C/W, a total thermal resistance of 37.5°C/W can be achieved.

Example 3.4: Determine the maximum power that a semiconductor die can safely handle in a packaged and unpacked environment. Assume the maximum safe temperature of 125°C and an ambient temperature of 25°C.

Equation (3.39) can be used to compute the total power for the packaged and unpacked environment by assuming thermal resistances of 100 and 37.5°C/W, respectively.

$$T = T_{ambient} + \Theta\left[\left(P_{DC} + P_{RF\text{-in}}\right) - P_{RF\text{-out}}\right] \Rightarrow P_{TOT} = \frac{T - T_{ambient}}{\Theta_{TOT}} = \frac{\Delta T_{rise}}{\Theta_{TOT}}$$

$$P_{TOT} = \frac{T - T_{ambient}}{\Theta_{TOT}} = \frac{125 - 25}{100} = 1\text{W}$$

$$P_{TOT} = \frac{T - T_{ambient}}{\Theta_{TOT}} = \frac{125 - 25}{37.5} = 2.67\text{W}$$

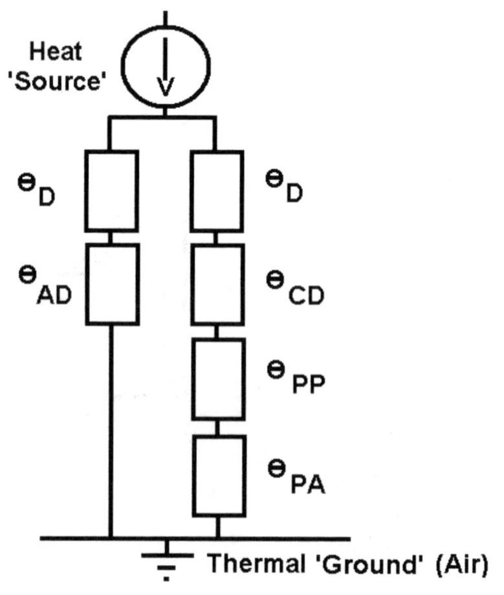

Figure 3.39 Expanded thermal model of heat flow from a heat source to a heat sink through two paths, reducing the overall thermal resistance to heat flow.

Table 3.5 Table of typical thermal resistance values

Θ	*Thermal Resistance (°C/W)*
θ_D	25
θ_{AD}	75
θ_{CD}	5
θ_{PP}	10
θ_{PA}	15

3.6 RFIC Grounding and Signal Isolation

A key component to successful circuit operation is proper grounding, and the RFIC is no exception. Component and device parasitics, undesired circuit coupling or crosstalk, and noise injection can all be controlled or minimized to varying degrees by application of various techniques to ensure proper RF grounding. While GaAs integrated circuits are fabricated on semi-insulating substrates, the CMOS RFIC designer has to work with substrates that are of modest resistivity at best. This section discusses the unique grounding issues CMOS RFICs create as well as introduces several techniques for improving RF grounds and hence circuit performance.

3.6.1 The Grounding Problem

The issues surrounding the attainment of a good RFIC ground involve looking at the problem at both the macro level (packaging) and the micro level (fabrication). At the package level, the RFIC will eventually be placed in some test fixture, and the two grounds have to be connected. The RFIC package ground and the RFIC ground are connected with an inductive bond wire, solder bump, or other inductive connection. The inductance serves to "lift" the RFIC ground from the overall ground as the frequency increases. Multiple ground connections from the die to the package can reduce (but never truly eliminate) the grounding inductance due to the parallel nature of the connections. Digital signals, with their sharp transitions, create "ground bounce" conditions that can inject harmonic-rich signals onto the package ground, where they are easily routed through the silicon die.

3.6.1.1 Substrate Signal Injection

CMOS integrated circuits are typically fabricated on a *p*-type substrate. The doping concentration of the substrate governs the threshold voltage of the *n*MOS device and provides a mechanically stable surface for the IC. The substrate is usually doped relatively high in digital CMOS as a means to minimize the effects of latch-up [41]; however, a highly doped substrate provides plenty of low-impedance paths for RF signals to move throughout the substrate. For passive elements such as the on-chip inductor, the substrate contributes significantly to the losses by the creation of eddy currents. These substrate eddy currents are in direct response to the inductor current and act to reduce the overall inductance. A high-conductivity substrate allows these currents to flow freely, enhancing their influence on the inductor. By using a lower conductivity (higher resistivity) substrate, these losses are reduced.

The effects of substrate signals injected can be observed from the RF equivalent circuit of the MOSFET (Figure 3.40). A highly doped substrate will provide a small R_{bulk}, allowing conductive signal injection into the MOSFET. On the other hand, a lightly doped substrate increases R_{bulk}, reducing this conductive entry path into the MOSFET. Alternatively, R_{bulk} is increased by physical distance, since a physical separation of the sensitive RF circuitry from major sources of substrate noise will provide some increase in isolation. This technique has been used extensively in baseband analog circuitry as a way to improve subsystem isolation and is frequently employed in RFIC designs as well. The signal injection cannot be completely eliminated, however, since capacitive coupling can still occur regardless of doping level.

The impact of the substrate signal injection can be seen by looking at an inverting amplifier with load Z_L and its simple RF equivalent circuit (Figure 3.41). Also shown in the figure is the substrate signal injection V_{SUB}; bulk impedances Z_{bulk} are assumed equal at drain and source to keep the analysis straightforward. Z_{path} is the path impedance from the substrate signal source to the target substrate (the MOSFET). The path impedance increases with increasing path length but decreases with increasing substrate conductivity.

It should be noted that Z_{bulk}, having a resistive component, will have thermal noise characteristics of its own of the form $kTRB$, where $R = \text{RE}\{Z_{bulk}\}$. This thermal noise will be injected from the substrate into the MOSFET, increasing the overall circuit noise, with a subsequent reduction in amplifier noise figure.

The goal of placing analog baseband, digital, and RF circuitry on a single CMOS die is possible, but steps must be taken to significantly reduce the impact of substrate signal injection into sensitive RF circuitry. With transmitter power approaching 30 dBm and receiver power at the antenna of –100 dBm, significant isolation between transmitter and receiver is necessary [42]. The importance of

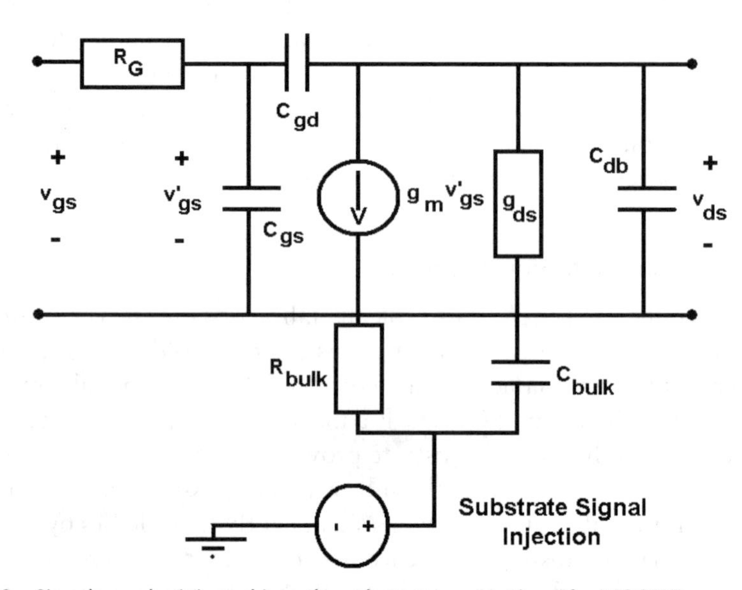

Figure 3.40 Signals can be injected into the substrate connection of a MOSFET.

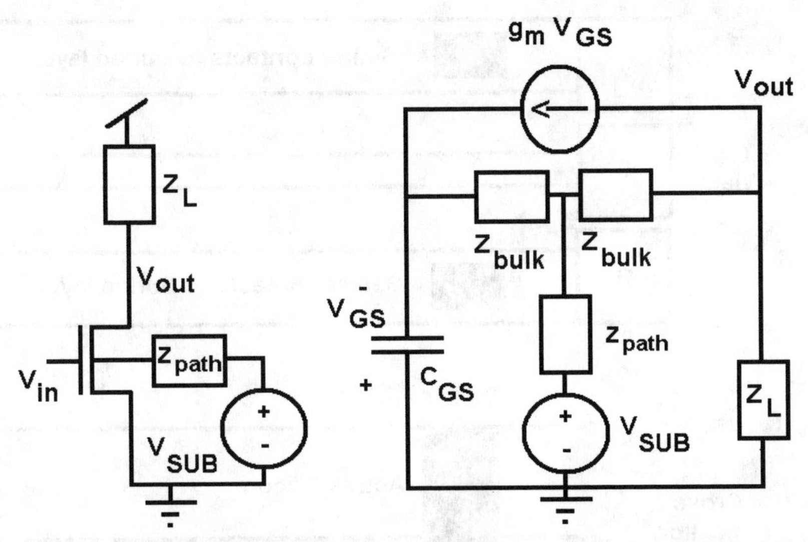

Figure 3.41 Inverting MOS amplifier with substrate signal injection and its simplified RF equivalent circuit. Z_{path} is the impedance of the path from the substrate signal source to the target substrate.

reducing digital switching noise is seen by noting that the sharp edges of digital signals are rich in harmonic content, and with switching signals on the order of volts, even high-order harmonics have enough energy to interfere with RF signals coming into the RFIC in the microvolt range. From the previous example, isolating high-frequency RF signals from digital switching noise requires minimizing digital signal injection into the substrate. Digital system techniques such as gating unused logic and careful sizing of digital clock buffers can be used to reduce the level of digital switching noise injected into the substrate [43]; these designs are outside the scope of this book. One of the first lines of defense against digital switching signal noise is to use differential circuitry wherever possible. Under typical circumstances, the phase of the switching noise is the same (or similar) at the closely located differential pair; when the difference of the two signals is taken, the switching noise (now a common mode signal) is removed. There are a number of other layout and grounding improvements that can be made to reduce the level of digital substrate injection noise between the source of the noise and the sensitive RF circuitry that will be discussed in the next section, on improving grounding and isolation.

3.6.2 Ground and Isolation Improvements

A number of techniques have been used to improve grounding and increase isolation of the RF circuitry from issues involving the substrate. In the section on transmission lines, one technique used for microstrip fabrication was to use the lowest level of metal for the ground plane and the upper levels of metal for the signal lines. A modification to this technique is the use of highly doped *buried layers* under the silicon surface [41]. These highly conductive regions can be grounded and used to shield passive elements from the main substrate (Figure 3.42). A sinker layer is

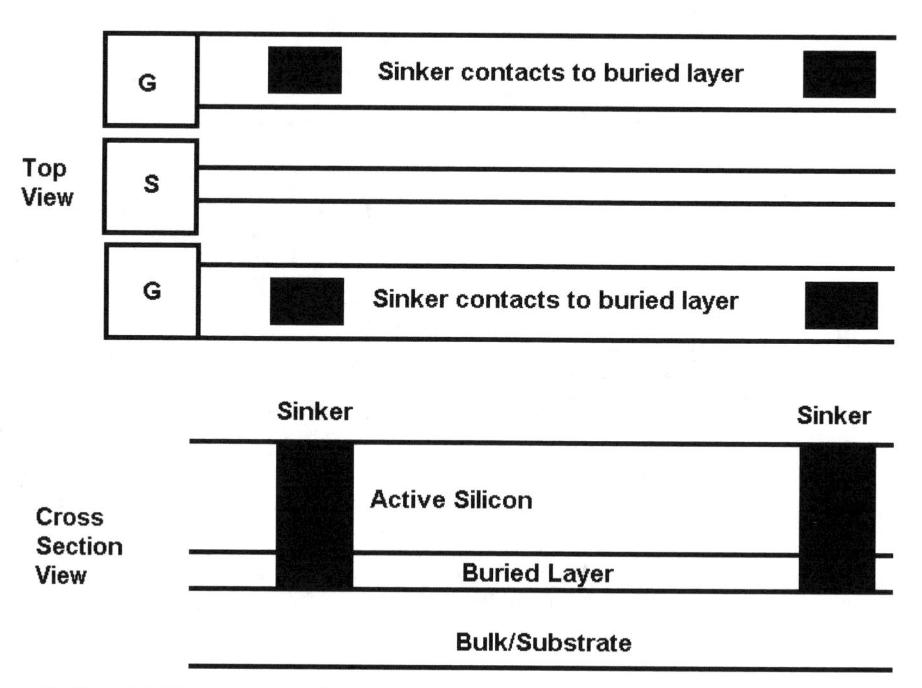

Figure 3.42 A highly doped buried layer connected to top-level ground of a coplanar waveguide can provide a good ground and minimize substrate signal injection. (*After:* [44].)

necessary to connect this buried layer with the die surface, where physical ground connections can be made [44]. Problems can arise with this structure, however, if an undesirable signal gets into the buried layer; there, the signal has a low impedance path throughout the chip. For this reason, the buried layer is often divided into a number of separately grounded sections to minimize the propagation of these signals [45].

Another widely used technique to improve isolation between circuits is through the use of *guard rings*. As the name implies, guard rings surround either the noisy circuit, the sensitive circuit, or both and are fabricated using highly doped regions of material (usually $p+$) and grounded liberally throughout the ring (Figure 3.43). Isolation is a function of the width of the guard ring and the spacing between the ring and the circuitry. Widths as great as 300 μm have been used in some Bluetooth applications with excellent results [46]. Physical separation of the RF circuitry from the remainder of the RFIC is also a widely used technique (Figure 3.44).

Variations on this guard ring theme are the use of deep n wells [42] and triple n wells [47], all used to improve isolation by preventing substrate signals from reaching sensitive RF circuits. A technique that requires postprocessing is the use of micromachining techniques to improve isolation [48]. Another technique that requires micromachining is *deep trench isolation,* where cuts are made in the silicon substrate that are then filled with metal and can create an effective barrier [49]. The use of silicon on insulator (SOI) technology significantly reduces substrate injection because of the extremely high substrate resistance (insulator) coupled with fabrication improvements such as buried ground planes and guard ring isolation [50].

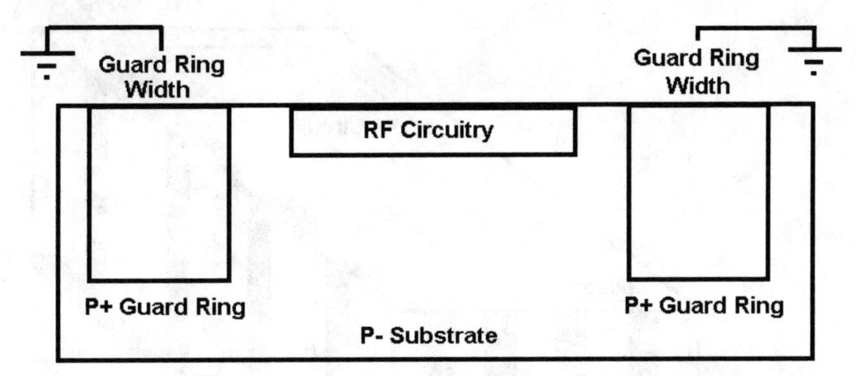

Figure 3.43 A wide grounded guard ring improves isolation by increasing the path length and path impedance for substrate signals. (*After:* [46].)

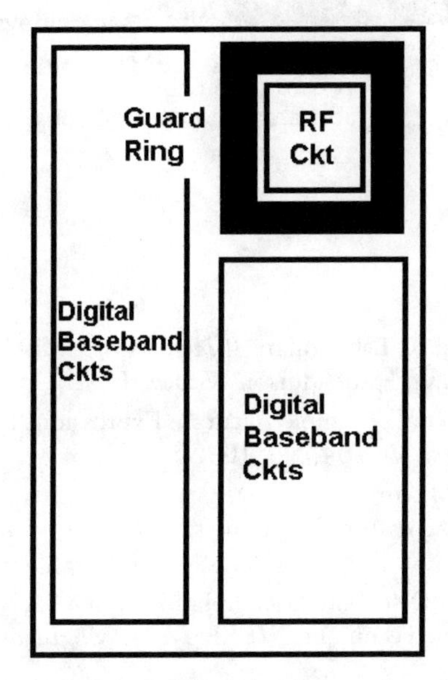

Figure 3.44 Physical separation of RF and digital circuitry can improve isolation.

A unique isolation technique that completely surrounds regions on the IC is the use of a *Faraday cage* (Figure 3.45). Creating this structure requires the availability of through-substrate vias, creating a "fence" structure that can be used to surround noisy circuits, sensitive RF circuits, or both [44]. With both top side and back side ground planes available for shielding, 38 dB of isolation at 5 GHz and 100 μm from the cage was shown.

Circuit design techniques can also be used to improve immunity to substrate coupled signals. As mentioned earlier, differential circuits, with their good common mode characteristics, are used to improve isolation. Another technique is to use different on-chip voltage regulators with large input and output capacitances to dampen switching transients and to provide better digital-to-RF circuitry isolation [52].

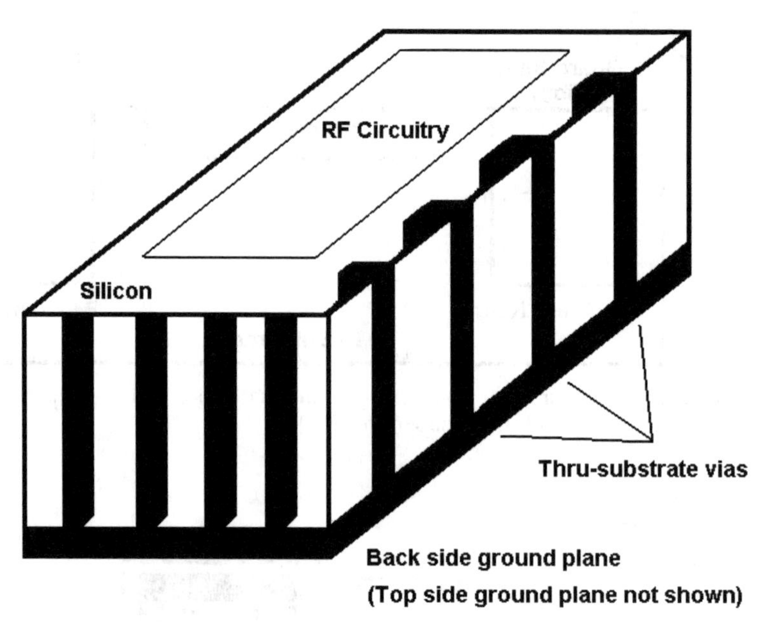

Figure 3.45 Faraday cage fabricated with through-substrate vias to back side ground plane. (*After:* [51].)

References

[1] Weste, N., and K. Eshraghian, *Principles of CMOS VLSI Design: A Systems Perspective*, 2nd ed., Reading, MA: Addison Wesley, 1992.

[2] Mohan, S. S., et al., "Simple Accurate Expressions for Planar Spiral Inductances," *IEEE J. Solid State Circ.*, Vol. 34, No. 10, Oct. 1999.

[3] Niknejad, A. M., and R. G. Meyer, "Analysis, Design and Optimization of Spiral Inductors and Transformers for Si RF IC's," *IEEE J. Solid State Circ.*, Vol. 33, Oct. 1998, pp. 1470–1481.

[4] Aoki, I., et al., "Distributed Active Transformer—A New Power-Combining and Impedance-Transformation Technique," *IEEE Trans. Microwave Theory Techniques,* Vol. 50, No. 1, Jan. 2002.

[5] Chen, Y., et al., "Q-Enhancement of Spiral Inductor with N^+ Diffusion Patterned Ground Shields," *Proc. 2001 IEEE Int. Microwave Symp.*, 2001, Vol. 2, p. 1289.

[6] Yue, Y., and S. S. Wong, "On-chip Spiral Inductors with Patterned Ground Shields," *IEEE J. Solid State Circ.,* Vol. 33, No. 5, pp. 743–751, May 1998.

[7] Weon, D., et al., "High Performance Micro-Machined Inductors on CMOS Substrate," *Proc. 2005 IEEE Int. Microwave Symp.*, 2005, p. 701.

[8] Chua, C. L., et al., "Out-of-Plane High-Q Inductors on Low-Resistance Silicon," *J. Microelectromechanical Syst.*, Vol. 12, No. 6, Dec. 2003, p. 989.

[9] Long, J. R., "Monolithic Transformers for Silicon RF IC Design," *IEEE J. Solid State Circ.*, Vol. 35, No. 9, Sept. 2000.

[10] Zolfaghari, A., A. Chan, and B. Razavi, "Stacked Inductors and Transformers in CMOS Technology," *IEEE J. Solid State Circ.,* Vol. 36, No. 4, April 2001, p. 620.

[11] Das, T., G. Nayak, and P. R. Mukund, "A Generic Macromodel for Coupling Between Inductors and Interconnects for R.F.I.C. Layouts," *Proc. IEEE Int. SOC Conf.*, Sept. 2004, p. 89.

[12] Sevic, J., "Design of Broadband Ununs with Impedance Ratios Less Than 1:4," *High Frequency Electronics*, Nov. 2004.

[13] Ma, K., et al., "800MHz/spl sim/2.5GHz Miniaturized Multi-Layer Symmetrical Stacked Baluns for Silicon Based RF ICs," *Proc. IEEE 2005 Int. Microwave Symp.*, June 2005, p. 283.

[14] Rogers, J., and C. Plett, *Radio Frequency Integrated Circuit Design,* Norwood, MA: Artech House, 2003.

[15] Edwards, T. C., and M. B. Steer, *Foundations of Interconnect and Microstrip Design*, 3rd ed., New York: John Wiley, 2000.

[16] Wadell, B. C., *Transmission Line Design Handbook*, Norwood, MA: Artech House, 1991.

[17] Grabinski, H., B. Konrad, and P. Nordholtz, "Simple Formulas to Calculate the Line Parameters of Interconnects on Conducting Substrates," *Proc. Digest IEEE 7th Topical Meeting on Electrical Performance of Electronic Packaging*, 1998, p. 223.

[18] Sim, S.-P., et al., "A Unified RLC Model for High-Speed On-Chip Interconnects," *IEEE Trans. Electron Devices,* Vol. 50, No. 6, June 2003, p. 1501.

[19] Pozar, D., *Microwave Engineering,* 2nd ed., New York: John Wiley, 1998.

[20] Doan, C. H., et al., "Millimeter-Wave CMOS Design," *IEEE J. Solid State Circ.*, Vol. 40, No. 1, Jan. 2005.

[21] Deutsch, A., et al., "On-Chip Wiring Design Challenges for GHz Operation," *Proc. IEEE,* Vol. 98, April 2001, p. 529.

[22] Weisshaar, A., H. Lan, and A. Luoh, "Accurate Closed-Form Expressions for the Frequency-Dependent Line Parameters of On-Chip Interconnects on Lossy Silicon Substrate," *IEEE Trans. Advanced Packaging*, Vol. 25, No. 2, May 2002, p. 288; see also A. Weisshaar, Workshop Slides, *IEEE Radio Wireless Conf. Workshop*, Jan. 2006.

[23] Rebeiz, G. M., and J. B. Muldavin, "RF MEMS Switches and Switch Circuits," *IEEE Microwave Mag.*, Dec. 2001, p. 59.

[24] De Los Santos, H., et al., "RF MEMS for Ubiquitous Wireless Connectivity," *IEEE Microwave Mag.*, Dec. 2004, Part 1, pp. 36–49, and Part 2, pp. 50–65.

[25] Nguyen, C. T. C., "Vibrating RF MEMS Technology: Fuel for an Integrated Micromechanical Circuit Revolution?" *TRANSDUCERS 2005: Proc. Solid State Sensors, Actuators, and Microsystems Conf.*, June 2005, Vol. 1, p. 243.

[26] Strohm, K. M., et al., "Silicon Micromachined RF MEMS Resonators," *Proc. 2002 IEEE Int. Microwave Symp.* , June 2002, Vol. 2, p. 1209.

[27] Rebeiz, G. M., *RF MEMS Theory, Design, and Technology,* Hoboken, NJ: John Wiley, 2003.

[28] De Los Santos, H., *Introduction to Micromechanical (MEM) Microwave Systems*, Norwood, MA: Artech House, 1999.

[29] Bannon, F. D., III, J. R. Clark, and C. T.-C. Nguyen, "High-Q HF Microelectromechanical Filters," *IEEE J. Solid State Circ.*, Vol. 35, No. 4, April 2000.

[30] Wang, K., A.-C. Wong, and C. T.-C. Nguyen, "VHF Free-Free Beam High-Q Micromechanical Resonators," *IEEE/ASME J. Microelectromechan. Syst.,* Vol. 9, No. 3, Sept. 2000, pp. 347–360.

[31] Wang, J., et al., "1.51-GHz Polydiamond Micromechanical Disk Resonator with Impedance Mismatched Isolating Support," *Proc. 17th Int. IEEE MEMS Conf.*, Jan. 2004, pp. 641–644.

[32] Li, S.-S., et al., "Micromechanical Hollow-Disk Ring Resonators," *Proc. 17th Int. IEEE MEMS Conf.*, Jan. 2004, pp. 821–824.

[33] Nguyen, C., "Frequency-Selective MEMS for Miniaturized Low-Power Communication Devices," *IEEE Trans. Microwave Theory Techniques*, Vol. 47, No. 8, Aug. 1999, p. 1486.

[34] De Wolf, I., "Reliability of MEMS," *Proc. 7th. Int. Conf. on Thermal, Mechanical and Multiphysics Simulation and Experiments in Micro-Electronics and Micro-Systems (EuroSimE)*, 2006, p. 1.

[35] Mellé, S., et al., "Reliability Modeling of Capacitive RF MEMS," *IEEE Trans. Microwave Theory Techniques,* Vol. 53, No. 11, Nov. 2005, p. 3482.

[36] Peroulis, D., S. P. Pacheco, and L. P. B. Katehi, "RF MEMS Switches with Enhanced Power-Handling Capabilities," *IEEE Trans. Microwave Theory Techniques*, Vol. 52, No. 1, Jan. 2004, p. 59.

[37] Lu, Y., L. P. B. Katehi, and D. Peroulis, "High-Power MEMS Varactors and Impedance Tuners for Millimeter-Wave Applications," *IEEE Trans. Microwave Theory Techniques*, Vol. 53, No. 11, Nov. 2005, p. 3672.

[38] Sutono, A., et al., "Experimental Modeling, Repeatability Investigation and Optimization of Microwave Bond Wire Interconnects," *IEEE Trans. Advanced Packaging,* Vol. 24, No. 4, Nov. 2001.

[39] Svelto, F., and R. Castello., "A Bond-Wire Inductor-MOS Varactor VCO Tunable from 1.8 to 2.4 GHz," *IEEE Trans. Microwave Theory Techniques*, Vol. 50, No. 1, Jan. 2002, p. 403.

[40] Sirenza Microdevices, "Determining the Junction Temperature from Device Thermal Resistance for Plastic Semiconductor Devices," Thermal Applications Note SMI-200, 2002.

[41] Blalack, T., Y. Leclercq, and C. P. Yue, "On-Chip RF Isolation Techniques," *Proc. 2002 Bipolar/BiCMOS Circuits and Technology Meeting*, Oct. 2002, p. 205.

[42] Larson, L. E., "Silicon Technology Tradeoffs for Radio-Frequency/Mixed-Signal Systems-on-a-Chip," *IEEE Trans. Electron Devices,* Vol. 50, No. 3, March 2003, p. 683.

[43] Mehta, S. S., et al., "An 802.11g WLAN SoC," *IEEE J. Solid State Circ.*, Vol. 40, No. 12, Dec. 2005, p. 2483.

[44] Hamel, J. S., et al., "Substrate Crosstalk Suppression Capability of Silicon-on-Insulator Substrates with Buried Ground Planes (GPSOI)," *IEEE Microwave Guided Wave Lett.*, Vol. 10, No. 4, April 2000.

[45] Takeshita, T., and T. Nishimura, "A 633 Mb/s Fully Integrated Optical IC with a Wide Tuning Range," *Proc. 2002 IEEE Int. Solid State Circuits Conf.*, Feb. 2002, Vol. 45, p. 258.

[46] van Zeijl, P., et al., "A Bluetooth Radio in 0.18 μm CMOS," *Proc. 2002 IEEE Int. Solid State Circuits Conf.*, Feb. 2002, Vol. 45, pp. 86–87.

[47] Redmond, D., et al., "A GSM/GPRS Mixed Signal Baseband IC," *Proc. 2002 IEEE Int. Solid State Circuits Conf.*, Feb. 2002, Vol. 45, p. 62.

[48] Pham, N., et al., "IC-Compatible Two-Level Bulk Micromachining for RF Silicon Technology," *IEEE Trans. Electron Devices,* Vol. 48, No. 8, Aug. 2001, p. 1756.

[49] Pham, N. P., et al., "A Micromachining Post-Process Module for RF Silicon Technology," *2000 IEDM Tech. Digest*, 2000, p. 481.

[50] Raskin, J., et al., "Accurate SOI MOSFET Characterization at Microwave Frequencies for device Performance Optimization and Analog Modeling," *IEEE Trans. Electron Devices*, Vol. 45, May 1998, p. 1017.

[51] Wu, J. H., et al., "A Faraday Cage Isolation Structure for Substrate Crosstalk Suppression," *IEEE Microwave Wireless Components Lett.*, Vol. 11, No. 10, Oct. 2002.

[52] Connell, L., et al., "A CMOS Broadband Tuner IC," *Proc. 2002 IEEE Int. Solid State Circuits Conf.*, Feb. 2002.

Selected Bibliography

Allen, P., and D. Holberg, *CMOS Analog Circuit Design*, 2nd ed., New York: Oxford University Press, 2002.

Lee, T. H., *The Design of CMOS Radio-Frequency Integrated Circuits*, 2nd ed., Cambridge, UK: Cambridge University Press, 2004.

Razavi, B., *RF Microelectronics*, New York: Prentice Hall, 1997.

CHAPTER 4

Small-Signal MOS Amplifiers for RF

Amplifying stages are needed throughout an RFIC. At the front end stage of a high-performance CMOS RFIC receiver, an amplifier exhibiting low noise and high gain can be a major player in setting the overall noise figure for the entire system. Amplifiers are needed in various stages not only to provide increases in signal amplitude; these amplifiers can also be configured to provide gain in only narrow bands, thereby providing filtering for the system. Active CMOS circuits can also be used to effectively reduce losses in on-chip inductors as well as providing a control function that provides for agile tuning structures. This chapter covers basic amplifier circuits, with a focus on RFIC issues.

4.1 Basic Amplifying Structure

The design of MOSFET RFIC amplifiers requires the designer to toggle between the dc and RF worlds. In the RF world, the designer needs an amplifier that must meet a number of specifications: gain, frequency response, input and output impedance, noise figure, linearity, and power consumption, to name just a few. In turn, these RF characteristics must be set by the dc bias circuits. Competing design parameters often require compromises and trade-offs in the design of amplifiers. The designer should know how the two worlds interact for successful designs to occur. This section covers the basic MOSFET amplifying structure and discusses this interaction.

4.1.1 Single FET with Generalized Load

The most common *n*MOSFET amplifying configuration is the common source configuration shown in Figure 4.1 (its counterpart, the common source *p*MOSFET, is also shown). The figures show the relationship between the complementary circuit schematic diagrams and the complementary power supply and dc bias connections to the physical CMOS circuit. Figure 4.1 also shows the ground (V_{DD}) connections to the *P*-type substrate (*N*-type well) for the *n*MOSFET (*p*MOSFET). The bulk connection on the nMOSFET (well on the *p*MOSFET) helps to ensure that the various PN junctions throughout the IC remain reverse biased under all operating conditions. In addition, since the source and bulk (or well) are at the same potential, $V_{SB} = 0$, yielding a threshold voltage V_T that is not influenced by the body effect (this is not the case for "stacked" transistors). A general impedance $Z_{load} = R_Z + jX_Z$ is connected between the MOSFET drain and one of the rails and is assumed to allow a dc current to flow from the power supply to the MOSFET for proper biasing.

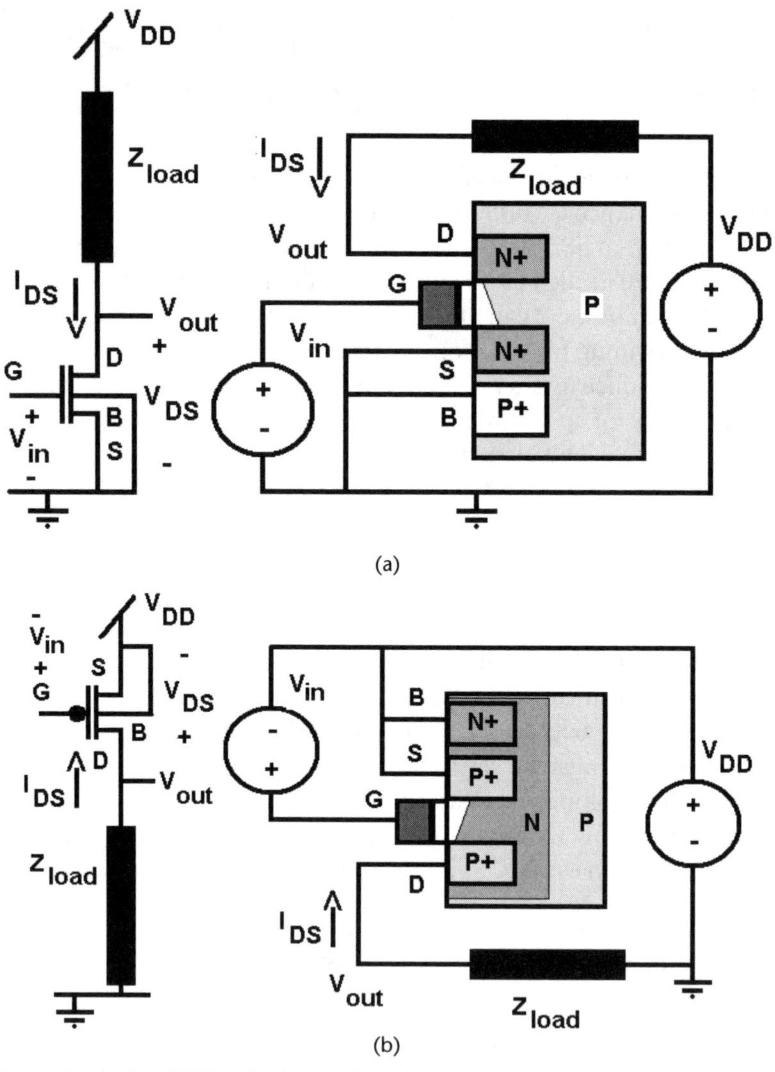

Figure 4.1 Basic (a) *n*MOSFET and (b) *p*MOSFET amplifying structures with a general load, showing the physical connections to the silicon wafer.

At dc, Z_{load} will exhibit some effective resistance R_Z which sets the dc bias voltage at the drain as shown in the dc load line equation:

$$V_{DS} = V_{DD} - I_{DS} R_Z \tag{4.1}$$

I_{DS} is governed by the transistor aspect ratio W/L and the gate bias voltage V_{GS} and so has a direct bearing on the load line characteristics. Load lines can be drawn by looking at (4.1) at the limits of I_{DS} and V_{DS}. Figure 4.2 shows a set of load line curves for values of R_Z = 0.5 kΩ (curve B) and 1.0 kΩ (curve C) based on the MOSFET designed in Example 2.1 and a V_{DD} = 5.0V:

$$V_{DS} = V_{DD}\Big|_{I_{DS}=0} \quad I_{DS} = \frac{V_{DD}}{R_Z}\Big|_{V_{DS}=0} \tag{4.2}$$

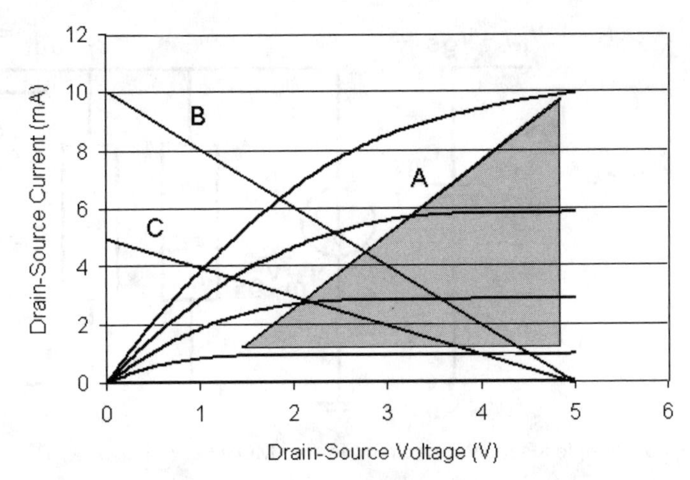

Figure 4.2 Current voltage characteristic for an *n*MOSFET showing load lines and active region of operation.

The choice of operating point depends on the application, but for the most linear small-signal results, the shaded region in Figure 4.2 provides a good trade-off between voltage swing and maintenance of signal fidelity (or linearity). The drain voltage swings from its maxima and minima, swinging the current to its minima and maxima (i.e., a 180° phase shift):

$$V_{DS_{MAX}} = V_{DD} - I_{DS_{MIN}} R_Z \quad V_{DS_{MIN}} = V_{DD} - I_{DS_{MAX}} R_Z \tag{4.3}$$

The voltage minimum should be kept away from the I-V curve knee (the line bordering the shaded area at point A) to keep the signal as linear as possible. The designer sets the dc drain voltage by proper selection of components, keeping in mind that I_{DS} is related to transistor sizing as well as the gate-source voltage (Chapter 2).

Once the dc bias point has been set, the small-signal RF characteristics g_m, C_{GS}, and g_{ds} are then specified. Keeping with the generalized impedance Z_{load}, the RF equivalent circuit can be drawn recalling that at RF (or ac), all independent sources are replaced with their internal impedances (short for dc voltage sources, open for dc current sources).

In Figure 4.3, all capacitances except for C_{GS} have been neglected for the sake of clarity. The input voltage in the common source configuration is v_{gs}, and this voltage is directly impressed across the input gate capacitance. The output voltage at the drain v_{ds} is computed in terms of the input voltage by noting that the total admittance seen by the dependent current source $g_m v_{gs}$ is $g_{ds} + 1/Z_{load}$ so that the voltage gain A_v can be written as

$$A_v = \frac{v_{\text{out}}}{v_{\text{in}}} = \frac{v_{ds}}{v_{gs}} = \frac{-g_m v_{gs} / \left(g_{ds} + Z_{load}^{-1}\right)}{v_{gs}} = -\frac{g_m}{g_{ds} + Z_{load}^{-1}} = -\frac{g_m Z_{load}}{1 + g_{ds} Z_{load}} \tag{4.4}$$

Equation (4.4) shows that both the MOSFET drain conductance and the load govern the overall voltage gain. If the MOSFET is driving a low-impedance load

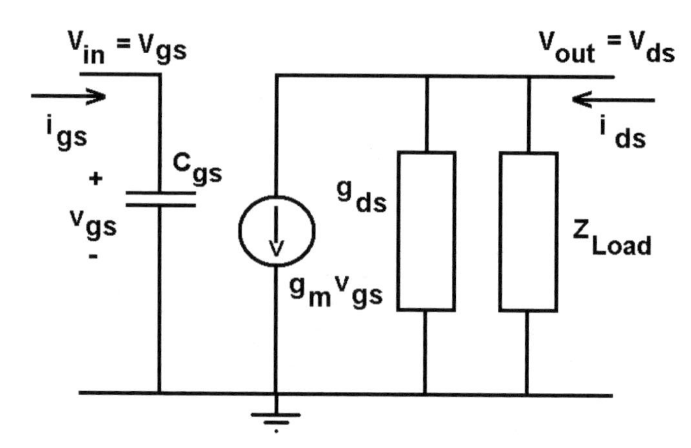

Figure 4.3 Simple RF equivalent circuit for the MOSFET.

(small Z_{load}), then the load characteristics will govern the voltage gain. Conversely, if a high-impedance load is being driven (high Z_{load}), the output characteristics of the MOSFET are an important factor to consider when estimating the amplifier gain. For a high-impedance load where only the g_{ds} term dominates, the voltage gain simplifies to

$$A_v = -\frac{g_m}{g_{ds}} = -g_m r_{ds} \tag{4.5}$$

where $r_{ds} = 1/g_{ds}$, hence the term G_mR or *transconductance amplifier*.

The small-signal output impedance r_{out} of the common source amplifier can be estimated looking back into the drain of the MOSFET (without the load) and with no RF input applied:

$$\frac{i_{ds}}{v_{ds}} = g_{ds} = r_{ds}^{-1} = r_{out}^{-1} \quad r_{out} = \frac{1}{\lambda I_{DS}} \tag{4.6}$$

Equation (4.6) shows that the RF output resistance r_{out} of the MOSFET is actually set by the dc drain source current I_{DS}. Small-signal amplifiers in RFIC applications are often driving the inputs to MOSFETs in the next stage, interstage connections, and device parasitic capacitances, so the loads are primarily capacitive. If all these capacitances are lumped together into a general load capacitance C_{load}, the frequency response of the voltage gain has a single pole at $r_{ds}C_{load}$:

$$\begin{aligned} A_v &= \frac{v_{out}}{v_{in}} = \frac{v_{ds}}{v_{gs}} = -\frac{g_m}{g_{ds} + Z_{load}^{-1}} = -\frac{g_m}{g_{ds} + j\omega C_{load}} \\ &= -\frac{g_m}{g_{ds}}\left(\frac{1}{1 + j\omega r_{ds} C_{load}}\right) = \frac{A_{v0}}{1 + j\omega r_{ds} C_{load}} \end{aligned} \tag{4.7}$$

With capacitive loading, the amplifier gain exhibits a –3-dB corner frequency dependent on the load capacitance and r_{ds}. The resistance r_{ds} is inversely proportional to the dc drain current (Chapter 2), whereas the transconductance is proportional to $I_{DS}^{1/2}$, so the midband gain A_{v0} is proportional to $I_{DS}^{-1/2}$. High gains can be achieved with low dc bias currents (so-called *current starved* operation) and large aspect ratio MOSFETs (W/L):

$$A_{v0} = -\frac{g_m}{g_{ds}} = -\frac{\sqrt{2KP\frac{W}{L}I_{DS}}}{\lambda I_{DS}} = \sqrt{\frac{2KP\frac{W}{L}}{\lambda^2 I_{DS}}} \tag{4.8}$$

A good current source (for the MOSFET amplifier, the $g_m v_{gs}$ dependent current source) should exhibit a high impedance, and low dc bias current MOSFET amplifiers best approximate this source. However, rapid charge or discharge of the load capacitance C_{load}, which translates to rapid voltage swing at high frequencies ($i = C\, dV/dt$), requires large currents, a competing trade-off. In addition, large dc currents require large gate widths W, which increases the device parasitic capacitances and degrades the frequency response of the voltage gain.

Example 4.1: A MOSFET amplifier using 0.5-μm CMOS technology with a 4-kΩ load resistor R_L is to be designed for a midband gain of −10 and have at least a 0.75 V_{peak} voltage swing. Determine the MOSFET sizing and dc bias requirements for this amplifier. Assume that the load resistor dominates the output resistance (ch4-1.txt).

From the voltage swing requirement, the drain current must be able to swing 0.1875 mA_{peak} (0.75V_{peak} across 4 kΩ) and not swing completely to zero. To achieve the voltage gain of −10 requires a transconductance value of

$$A_v = -g_m R_L;\ g_m = -\frac{A_v}{R_L} = -\frac{-10}{4000} = 2.5\,\text{mS}$$

From the current swing requirement, choose a dc bias current of 0.375 mA; the resulting dc bias on the drain is

$$V_{DS} = V_{DD} - I_{DS}R_Z = 3.3 - (0.375)(4) = 1.8\text{V}$$

The required gate width W can be calculated as

$$g_m = \sqrt{2\text{KP}\frac{W}{L}I_{DS}};\ W = L\frac{g_m^2}{2\text{KP}\cdot I_{DS}} = 0.5\frac{(0.0025)^2}{2\cdot 56.3\cdot 10^{-6}\cdot 0.375\cdot 10^{-3}} = 74\ \mu\text{m}$$

The dc gate-source voltage can now be estimated:

$$I_{DS} = \frac{1}{2}\frac{W}{L}\text{KP}(V_{GS} - V_T)^2;\ V_{GS} - V_T = \sqrt{\frac{2I_{DS}}{(W/L)\text{KP}}} = \sqrt{\frac{2(0.000375)}{(74/0.5)56.3\cdot 10^{-6}}} = 0.30\text{V}$$

Figures 4.4 through 4.6 show the results of various circuit responses based on this design using the BSIM parameters in Appendix B. Figure 4.4 shows the dc

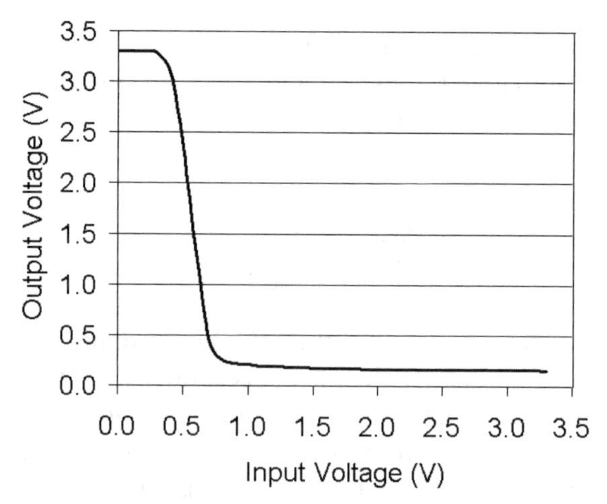

Figure 4.4 The dc transfer characteristic of the amplifier in Example 4.1.

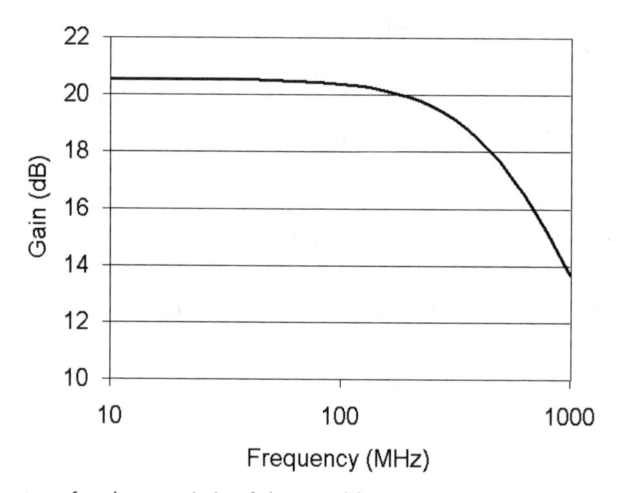

Figure 4.5 The ac transfer characteristic of the amplifier in Example 4.1.

transfer characteristic determined by observing the output voltage as the input voltage V_{GS} is swept from 0 to 3.3V. The amplifying region is located in the vicinity of the largest slope ($0.4V < V_{GS} < 0.8V$) in the dc transfer characteristic; a dc bias of 0.6V biases the input of the amplifier in the high-gain region and provides an output bias on the drain of $V_{DS} = 1.8$V. Figure 4.5 shows the RF frequency response of the amplifier when dc biased in the amplifying region. The initial gain was approximately 17.5 dB, close to the design specification. Iterative adjustments of MOSFET width W brought the gain closer to the 20-dB requirement, as shown in the figure. Figure 4.6 shows the results of a time domain simulation of the circuit, indicating an approximate 1.3V peak swing in the output voltage, which more than meets the problem requirements.

Any change in the circuit specifications will require study of the changes in the MOSFET size and bias conditions. For example, an increase in the dc bias current to 0.5 mA may be necessary if a voltage swing of $1.0V_p$ is needed. The MOSFET will

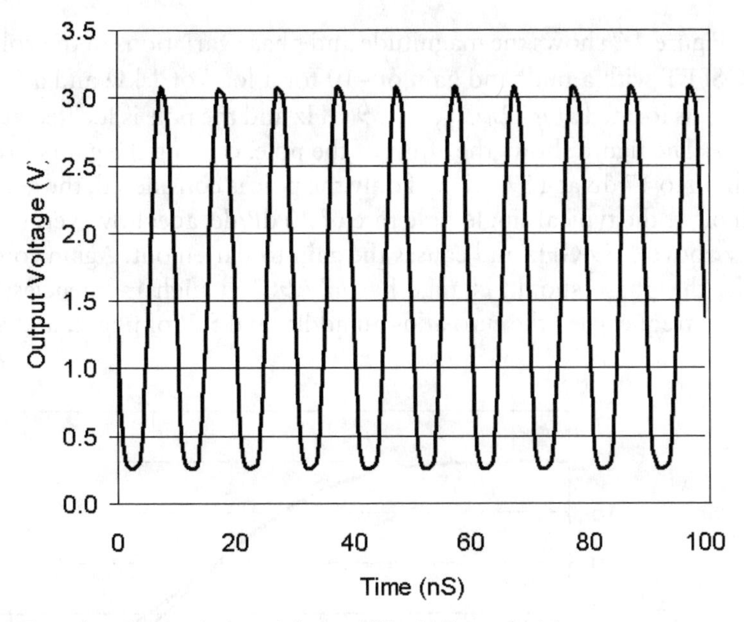

Figure 4.6 Time domain simulation of the amplifier in Example 4.1.

then need to be resized to approximately 55 μm to keep a constant g_m and the gate bias increased about 0.1 to 1.06V to provide the increase in dc current with the smaller MOSFET. The increased drain current will lower V_{DS} to 1.3V.

The common source MOSFET can be also thought of as a *transimpedance amplifier* by noting that the input gate current i_{gs} is related to v_{gs} through the susceptance of C_{GS}:

$$\begin{aligned} i_{gs} &= j\omega C_{GS} v_{gs};\ R_T = \frac{v_{ds}}{i_{gs}} = \frac{v_{ds}}{v_{gs}} \frac{v_{gs}}{i_{gs}} = -\frac{g_m}{g_{ds} + Z_{load}^{-1}} \frac{1}{j\omega C_{GS}} \\ &= j\frac{f_T}{f}\left(\frac{1}{g_{ds} + Z_{load}^{-1}}\right) \end{aligned} \tag{4.9}$$

where R_T is defined as the transimpedance. MOSFETs with high f_T or operating at low frequency provide the capability for effective transimpedance amplifiers.

As the frequency of operation increases, the effects of C_{GD}, the gate-drain feedback capacitance, becomes more evident due to the *Miller effect*. While C_{GD} is considerably smaller than C_{GS} in typical MOSFET structures, Miller feedback causes the input loading by C_{GD} to increase by the gain of the amplifier. From the RF equivalent circuit for the MOSFET common source amplifier including C_{DG}, the gain has a single pole (p_1)and a single zero (z_1), and for reasonable gains, the effective input capacitance is increased dramatically by the g_mR (A_{v0}) product:

$$\begin{aligned} A_v &= -g_m R \frac{1-(j\omega C_{GD}/g_m)}{1+j\omega RC_{GD}} = -A_{v0}\frac{1-(j\omega C_{GD}/g_m)}{1+j\omega RC_{GD}} = -A_{v0}\frac{1-jz_1}{1+jp_1} \\ Z_{in} &= \frac{v_{in}}{i_{gs}} = \frac{1}{j\omega\left[C_{GS}+C_{GD}(1-A_v)\right]} \cong \frac{1}{j\omega\left[C_{GS}+C_{GD}(1+A_{v0})\right]} \end{aligned} \tag{4.10}$$

Figure 4.7 shows the magnitude and phase variations of the voltage gain A_v for a MOSFET with a midband gain of –10 for a load of 1 kΩ and a C_{GD} of 1.0 pF. The zero z_1 is located at $g_m/2\pi C_{GD} = 1.59$ GHz and the pole is located at $1/2\pi RC_{GD} = 159$ MHz. The figure shows the effect of the pole, causing the gain to roll off in a classic fashion to 17 dB at 159 MHz. If only the pole is considered, the gain will continue to roll off at the typical single-pole rate of 20 dB/decade; however, the pole comes into play above 1.59 GHz and causes the gain to flatten out. Again considering only the pole, the phase should change by only 90° at high frequencies; however, at the higher frequencies, the zero adds an additional 90° of phase, and at extremely high

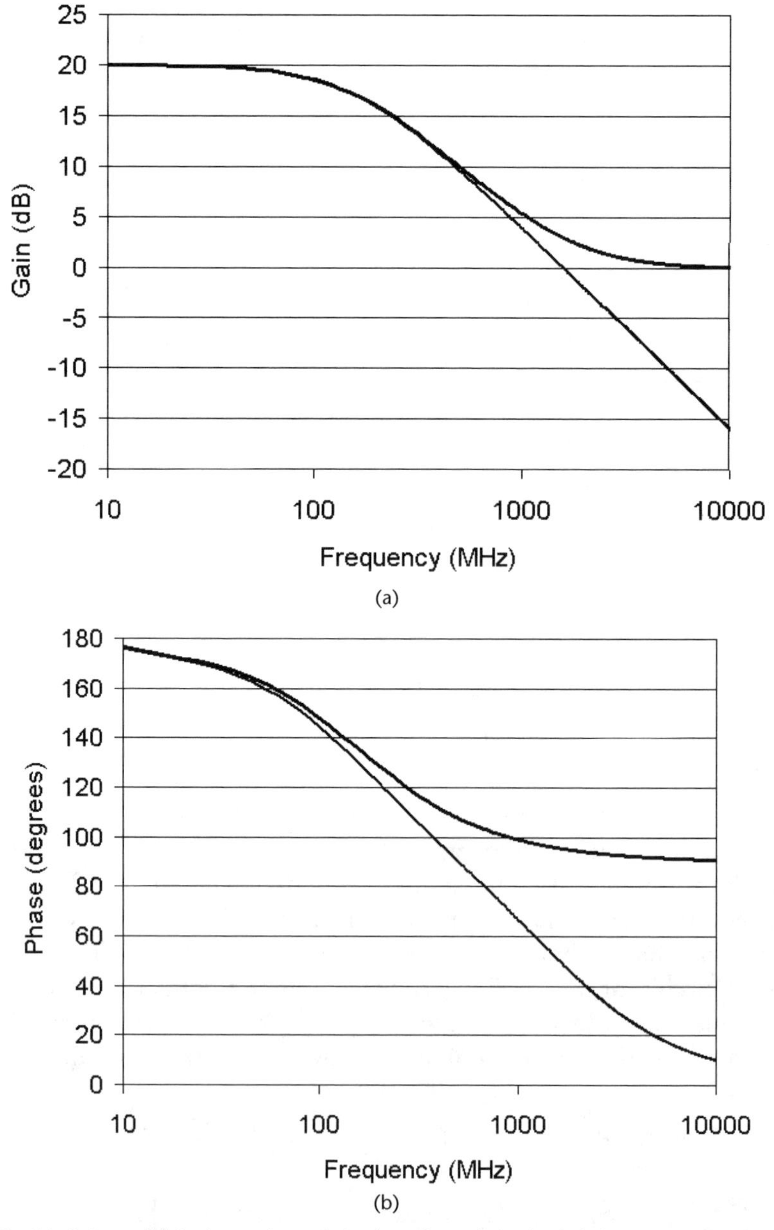

Figure 4.7 (a) Gain and (b) phase plots of single-pole and single-pole/zero transfer characteristic showing flattening of gain and added phase due to the extra zero.

frequencies, the gain will be unity and the phase 0°. At 1.8 GHz the amplifier still has a small level of gain (approximately 2.3 dB) and a phase of 45°. Physically, what is occurring is that above 1.59 GHz, C_{GD} is exhibiting reactance sufficiently low that the input signal is being *fed forward* through the amplifier, flattening the gain.

4.1.2 Amplifier Loading

4.1.2.1 Active Device Loading

While useful for understanding, the purely resistive load is not practical in an RFIC for a number of issues, primarily because of the wafer area needed to achieve a high value of resistance. An alternative to the passive resistor load is to use another MOSFET as an *active load*. Two major types of active loads can be created with either *n*MOS or *p*MOS transistors, the *current source load* and the *MOS diode/active resistor load* [1] (Figure 4.8). For the *p*MOS current source load, the gate is tied directly to a dc voltage source, creating a constant V_{GS} that puts the device in its saturation region. This constant V_{GS} in turn creates a constant I_{DS} (assuming that $|V_{GS}| > |V_T|$):

$$\begin{aligned} I_{DS} &= -\frac{1}{2}\frac{W}{L}\mathrm{KP}(V_{GS} - V_T)^2;\ V_{DS\,\mathrm{MIN}} = V_T - V_{GS} = -\sqrt{\frac{2|I_{DS}|}{\mathrm{KP}\cdot W/L}} \quad p\mathrm{MOS} \\ I_{DS} &= +\frac{1}{2}\frac{W}{L}\mathrm{KP}(V_{GS} - V_T)^2;\ V_{DS\,\mathrm{MIN}} = V_{GS} - V_T = +\sqrt{\frac{2I_{DS}}{\mathrm{KP}\cdot W/L}} \quad n\mathrm{MOS} \end{aligned} \tag{4.11}$$

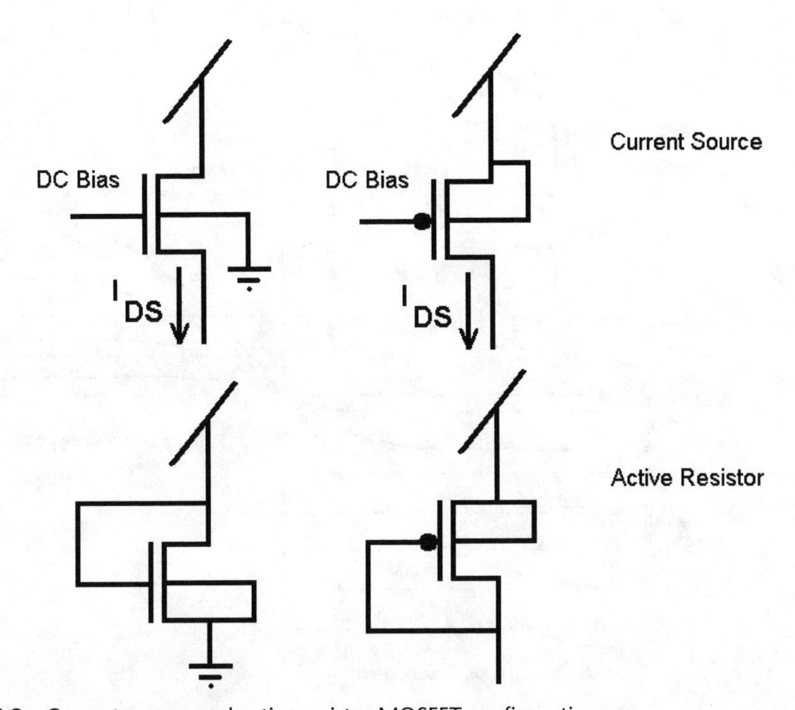

Figure 4.8 Current source and active resistor MOSFET configurations.

Note that in (4.11), the analogous expression for the drain current in the *n*MOS current source has been included.

The current source load current can be varied by the designer through MOSFET sizing, changes in biasing, or both. For a given MOSFET size, the *p*MOS device exhibits a higher resistance over the *n*MOSFET because of its lower channel mobility and hence lower KP. Because the current source load is an active device, a linear load line cannot be drawn; however, a *load curve* analysis can be done that aids in determining the optimum operating point. In designing the current source load, the constant current supplied by the MOSFET by definition is the same current flowing through the amplifying MOSFET, leaving the designer the trade-off between W/L and V_{GS}. A smaller MOSFET can be used to save area, but this would require a larger V_{GS} and corresponding larger V_{DS} drop, possibly limiting the voltage swing at the output of the amplifier.

Figure 4.9 shows an example of an *n*MOSFET amplifier with *p*MOSFET current source load of equal size (74/0.5) with a gate voltage of 2.0V on the *p*MOS device. The two sets of I-V curves can be used to determine the transfer characteristic of the amplifier. Note that $I_{DSn} = -I_{DSp}$ and that $V_{DSp} = V_o - V_{DD}$. The I-V curve for the *p*MOS device can be "flipped" along the V_o axis ($I_{DSn} = -I_{DSp}$) and shifted by V_{DD} for the load curve shown in Figure 4.10(a) as the dotted curve. A dc transfer curve can be drawn from the load curve information presented in Figure 4.10 by noting the input voltage and drain source voltage values at the intersection of the load curve and MOSFET I-V curves [indicated in Figure 4.10(b) as letters *a* through *e*]. A *p*MOSFET with small *W*/*L* will provide a higher gain due to an increase in the transfer characteristic slope but at a lower input dc bias voltage.

The RF load presented by the current source load can be calculated from the RF equivalent circuit of each MOSFET in the circuit (Figure 4.11). Note that since the

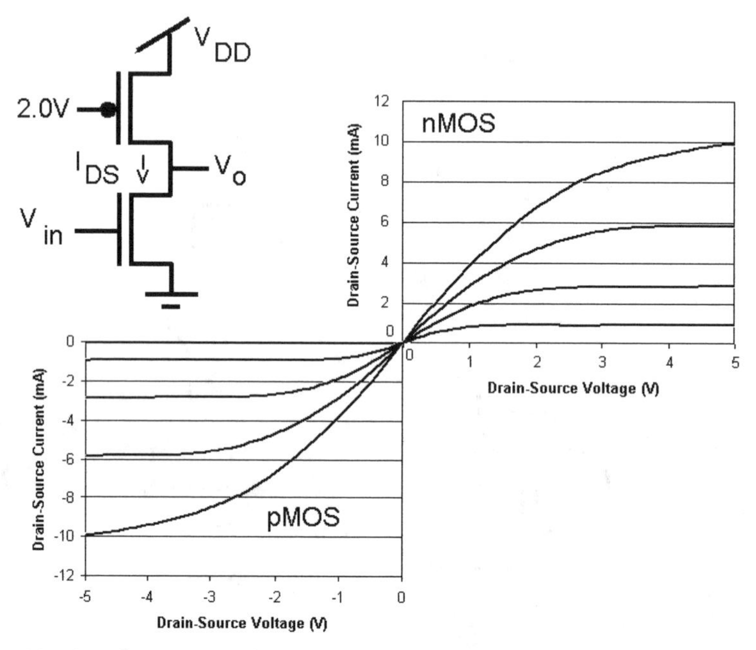

Figure 4.9 Circuit and I-V curves for MOSFETs in a current source loaded amplifier.

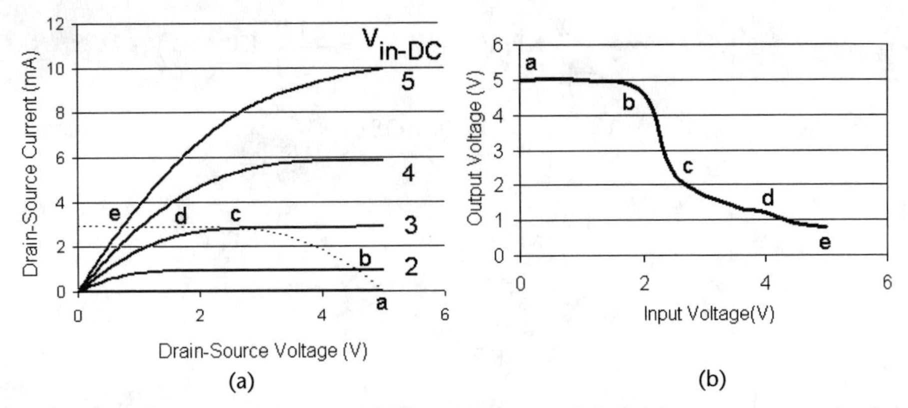

Figure 4.10 (a) The dc load curve and (b) transfer characteristic for the current source loaded amplifier.

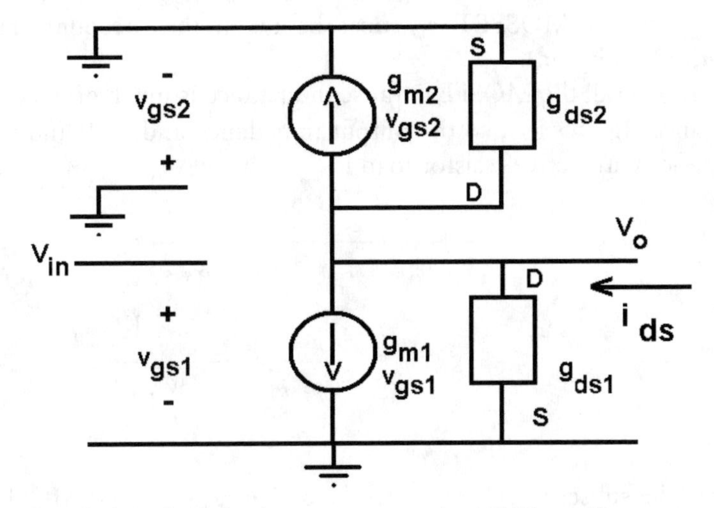

Figure 4.11 RF equivalent circuit for the current source loaded RF amplifier.

gate is connected to a dc source, the gate is effectively RF grounded, making $v_{gs} = 0$ and removing the transconductance current from consideration. The remaining element, $r_{ds\text{-}CS}$, is the output resistance of the current source and loads the drain of the amplifying MOSFET. The output impedance and midband gain of the amplifier are then

$$r_{\text{out}} = r_{ds} \| r_{ds-CS} \quad A_{v0} = -g_m \left(r_{\text{out}} \right); \; r_{ds} = \frac{1}{\lambda I_{DS}} \tag{4.12}$$

where the subscript *CS* refers to the current source load MOSFET. The output impedance of both MOSFETs will be higher for smaller drain currents. The current source load uses an external dc bias to ensure that the MOSFET is in saturation. Another circuit configuration that guarantees that the MOSFET is in saturation is to connect the drain and gate directly: $V_{DS} = V_{GS}$. The drain current and voltage drop for the *p*MOS and *n*MOS active resistor loads are then

$$
\begin{aligned}
I_{DS} &= -\frac{1}{2}\frac{W}{L}\mathrm{KP}(V_{GS}-V_T)^2 = -\frac{1}{2}\frac{W}{L}\mathrm{KP}(V_{DS})^2; \\
V_{DS} &= V_T - \sqrt{\frac{2|I_{DS}|}{\frac{W}{L}\mathrm{KP}}} \quad p\mathrm{MOS} \\
I_{DS} &= +\frac{1}{2}\frac{W}{L}\mathrm{KP}(V_{GS}-V_T)^2 = +\frac{1}{2}\frac{W}{L}\mathrm{KP}(V_{DS})^2; \\
V_{DS} &= V_T + \sqrt{\frac{2|I_{DS}|}{\frac{W}{L}\mathrm{KP}}} \quad n\mathrm{MOS}
\end{aligned}
\tag{4.13}
$$

The RF equivalent circuit for the amplifier with active resistor load looks very similar to that of the current source load with the exception that the gate and drain are tied together. With this configuration $v_{ds} = v_{gs}$, so the transconductance of the active resistor MOSFET g_m must be taken into account (Figure 4.12 for an all-nMOS solution).

In general, the MOSFET transconductance is much greater than the drain conductances ($g_m >> g_{ds}$), so the output impedance and midband voltage gain for the amplifier with active resistor load is

$$
\begin{aligned}
r_{\text{out}} &= \frac{1}{g_{m-AR} + g_{ds-AR} + g_{ds}} \cong \frac{1}{g_{m-CS}}; \\
A_{v0} &= -g_m r_{\text{out}} \cong -\frac{g_m}{g_{m-AR}} = -\frac{\sqrt{\frac{W}{L}\mathrm{KP}}}{\sqrt{\left(\frac{W}{L}\mathrm{KP}\right)_{AR}}}
\end{aligned}
\tag{4.14}
$$

where the subscript AR refers to the active resistor load MOSFET. For the active resistor load, the gain of the amplifier is approximately related to the ratio of the

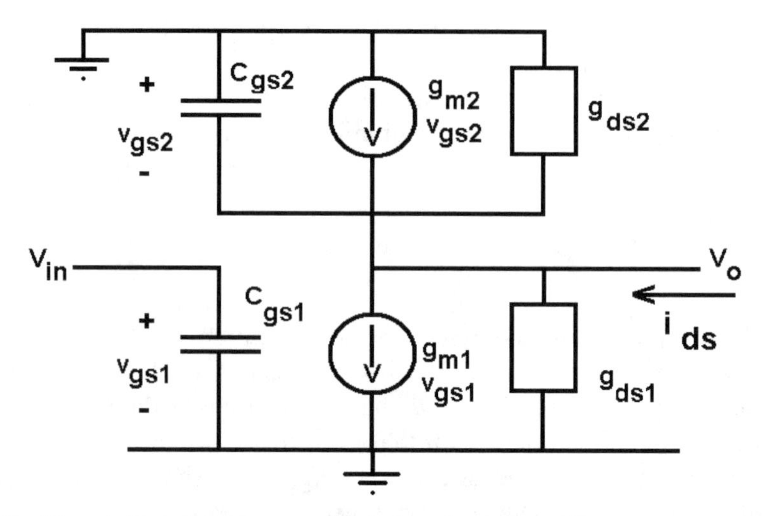

Figure 4.12 RF equivalent circuit for an active resistor–loaded amplifier.

MOSFET aspect ratios W/L. The active resistor load tends to have much lower output resistance than the current source load and so provides relatively small amplifier gain. If the amplifier is driving an external resistance, dc-blocking capacitors are needed to prevent changes in dc operating point.

4.1.2.2 Inductor Loading

Passive resistor and active device loading both create a voltage reduction from the power rail at the drain terminal of the amplifying MOSFET, limiting the voltage swing at the amplifier output. The use of inductive loading provides the required dc bias path to the MOSFET, at the same time providing RF isolation of the amplifier from the power source. Ideally, this inductor exhibits a reactance significantly above that of the load; if so, then the amplifying MOSFET sees only the load and its own internal characteristics, which set the overall circuit gain. Unfortunately, high-value inductors are not feasible in RFIC applications, so there will be inductive loading. Circuit simulation should be performed to ensure that no resonances of the inductor with MOSFET and other circuit capacitances create undesirable performance. As for the other amplifier loads, if the amplifier is driving an external resistive load, dc-blocking capacitors are needed to prevent changes in dc operating point.

4.1.3 Effect of Parasitics

The parasitics of both the MOSFET and the passive elements in the circuit will influence the overall behavior of the amplifier. This section looks at some of these effects.

4.1.3.1 Active Loads

For both types of active loads, the MOSFET heavily doped drain or source region (depending on whether a *p*MOS or an *n*MOS device is used) is connected to the output of the amplifier. These regions exhibit high capacitance that tends to increase the output capacitance (C_{load}) seen by the amplifier, with a corresponding degraded frequency response performance. This capacitance increase is especially evident if an *n*MOS amplifying device and a *p*MOS current source are used. Layout design rules require these two devices to be physically separated, and hence they must have two separate heavily doped drain diffusion regions with their high area and sidewall capacitance contributions. If the same type of MOSFET is used for the amplifier and the active load (usually all *n*MOS), then merging the drain and source connections can significantly reduce this capacitance. However, with this particular circuit topology, the current source or active resistor load has its source at a different potential than the bulk region (ground for *n*MOS, V_{DD} for *p*MOS) and the body effect appears. The body effect creates an additional drain current component to flow in response to the varying source-bulk voltage v_{sb}, effectively increasing the transconductance g_m by a factor η [1]. The bulk effect factor is computed as

$$\eta = \frac{\gamma}{2\sqrt{2|\phi_B| + |V_{SB}|}} \tag{4.15}$$

where the terms γ and Φ_B are defined as the bulk threshold parameter and surface potential, respectively, and are part of the overall SPICE model definition of the MOSFET. The total transconductance in (4.14) for the generalized amplifier including the bulk effect can be written as $(1+\eta)g_m$.

Example 4.2: Replace the 4-kΩ resistor in Example 4.1 with a *p*MOS *current source load* that supplies the same current (0.375 mA) to the circuit and the same voltage drop (1.5V). Determine the midband RF gain of the circuit (ch4-2.txt).

The *p*MOS current source has to provide the same current to the amplifying MOSFET as in the previous example, with the same voltage drop:

$$I_{DS} = -\frac{1}{2}\frac{W}{L}\mathrm{KP}\left(V_{Dsat_{\mathrm{MIN}}}\right)^2; W = L\frac{2I_{DS}}{\mathrm{KP}\left(V_{Dsat_{\mathrm{MIN}}}\right)^2} = 0.5\frac{2\left(0.375\cdot 10^{-3}\right)}{19.1\cdot 10^{-6}\cdot(1.5)^2} = 8.7\mu\mathrm{m}$$

To keep the *n*MOSFET in its amplifying region requires that V_{GS} be set to approximately 0.6V, a reduction from the original design. The gain is approximately 22.5 dB, slightly higher than designed, because of the difference in loading due to the transistor output impedance (Figure 4.13). This design provides a starting point for small adjustments in W or V_{GS} to bring the design closer into specifications.

Example 4.3: Replace the 4-kΩ resistor in Example 4.1 with an *n*MOS *active resistor load* that supplies the same current (0.375 mA) to the circuit and the same voltage drop (1.5V). Determine the midband RF gain of the circuit (ch4-3.txt).

The *n*MOS active resistor has to allow the same current to the amplifying MOSFET as in the previous example with the same voltage drop:

$$I_{DS} = \frac{1}{2}\frac{W}{L}\mathrm{KP}\left(V_{Dsat_{\mathrm{MIN}}}\right)^2; W = L\frac{2I_{DS}}{\mathrm{KP}\left(V_{Dsat_{\mathrm{MIN}}}\right)^2} = 0.5\frac{2\left(0.375\cdot 10^{-3}\right)}{56.3\cdot 10^{-6}\cdot(1.5)^2} = 3\mu\mathrm{m}$$

Note the reduction in gain for the active resistor load (Figure 4.14). The reduction is due to the decreased output resistance of this circuit.

4.1.3.2 Inductor Biasing

The on-chip inductors available in the CMOS RFIC process exhibit relatively low Q. The resistive losses in the inductor will influence the dc biasing of the MOSFET since the current through the inductor will cause a dc voltage drop across the inductor that limits the available voltage swing. This resistive loss can be estimated and should be included in dc bias calculations.

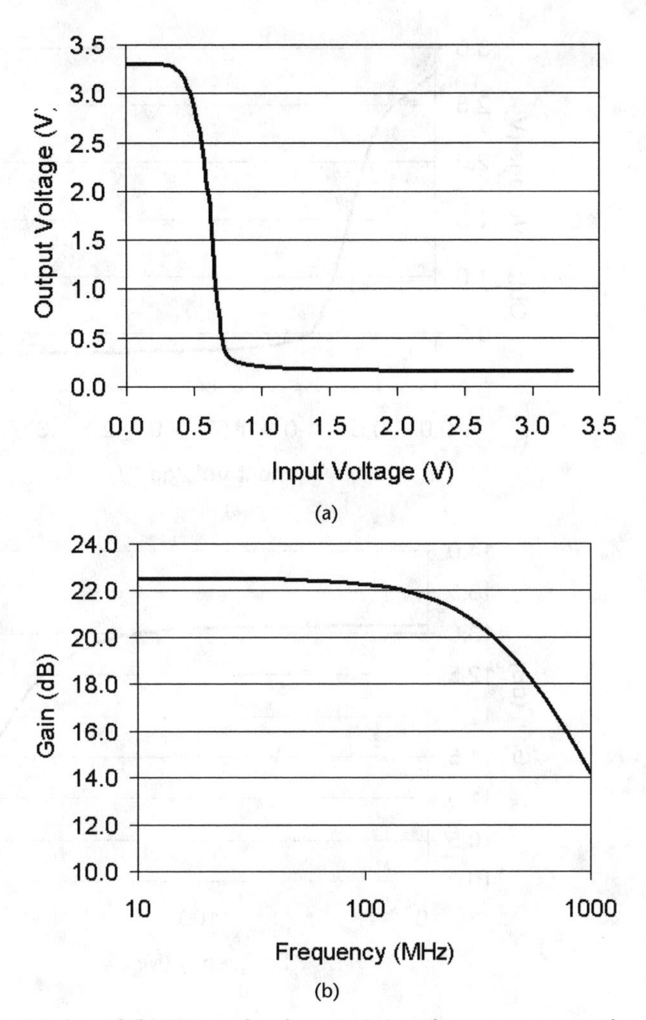

Figure 4.13 The (a) dc and (b) RF transfer characteristics of a current source–load amplifier.

Example 4.4: An on-chip 10-nH inductor for use at 1000 MHz exhibits a Q of 5. Determine the series effective resistance of the inductor.

Using the expression for the series Q of an inductor, the series effective resistance can be calculated:

$$Q = \frac{\omega L}{R}; R = \frac{\omega L}{Q} = \frac{2\pi \cdot 10^9 \cdot 10 \cdot 10^{-9}}{5} = 12.5\Omega$$

For a MOSFET amplifier with a dc bias current of 1.0 mA, the inductor losses create a 0.0125V drop across the inductor.

On-chip inductors adversely affect the RF performance of the amplifier. The total impedance presented by the inductor includes effects due to the oxide capacitance and losses due to bulk effects. Accurate amplifier modeling requires that all parasitic effects be included in any simulation to see if a resonance or other undesired effect takes place. If necessary, the inductor parameters (inductor width,

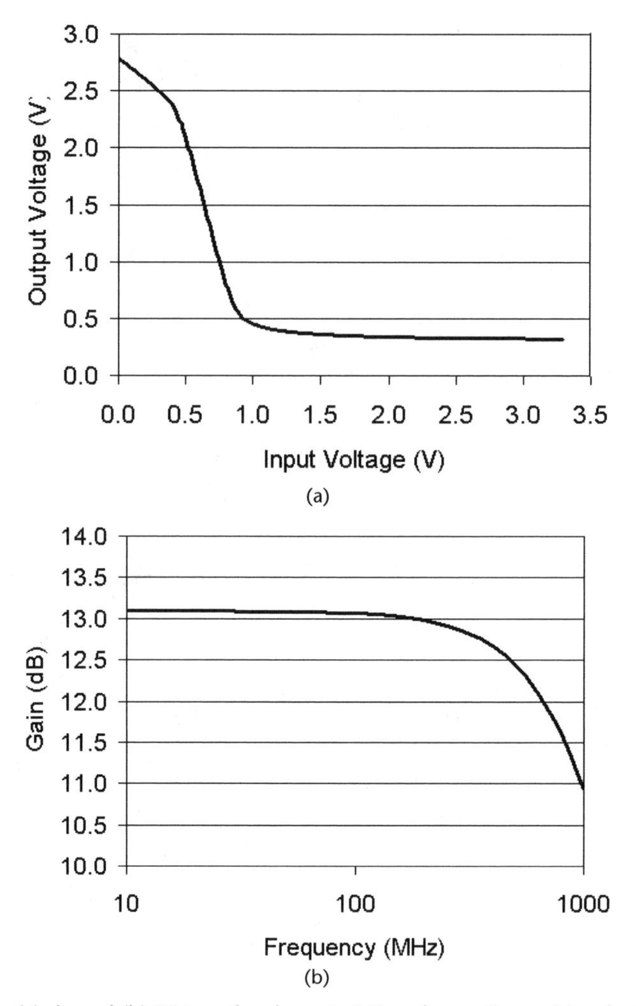

Figure 4.14 The (a) dc and (b) RF transfer characteristics of an active resistor–load amplifier.

number of turns, and turns spacing, for example) may have to be modified to move such effects out of the frequency range of interest.

4.1.3.3 Gate Resistance R_G

In the previous expression for MOSFET amplifier gain, the effect of the gate material resistance R_G was neglected. Gate resistance R_G, however, can be substantial in large aspect ratio MOSFETs and cannot be neglected, especially at high frequencies, where the lowered reactance of the gate input capacitance (C_{GS} and the Miller effect on C_{GD}) makes the two of comparable value. Including R_G in the RF equivalent circuit yields the voltage gain as

$$A_v = -g_m R \frac{1-(j\omega C_{GD}/g_m)}{1+\omega^2 R R_G C_{GD}^2 + j\omega R\left[(1+g_m R)C_{GD} + \frac{R_G}{R}(C_{GD}+C_{GS})\right]} \tag{4.16}$$

In addition to the reduced gain because of the input voltage drop across R_G, there will also be an effect on the power gain due to the gate losses. When R_G is included in the RF equivalent circuit model, the Y-parameters for the circuit can be calculated (see Appendix C for details) and the maximum power gain G_{max} can be computed [2]:

$$G_{max} = \frac{1}{4}\frac{|Y_{21} - Y_{12}|^2}{4\mathrm{RE}(Y_{11})\mathrm{RE}(Y_{22}) - \mathrm{RE}(Y_{21})\mathrm{RE}(Y_{12})} \approx \frac{f_T}{8\pi R_G C_{GD} f^2} \quad (4.17)$$

This expression shows that the power gain is proportional to f_T, rolls off as f^2, and is inversely proportional to R_G. Improvements in G_{max} can be achieved by reducing R_G, which can be reduced through layout with multiple gate fingers. If the gate fingers are connected on only one side, then the gate resistance is reduced by $1/3N$ (where N is the number of gate fingers) from its single-finger value. If the layout permits, connecting the gates on both sides provides another factor of 4 reduction in R_G (Figure 4.15).

In addition to the input gate resistance reduction, merged source-drain connections in the multifingered MOSFET create a reduction in C_{sb} and C_{db}, with a corresponding improvement in frequency response.

4.1.4 Basic Behavioral Model

A behavioral model replaces the detailed circuit model with a mathematical or "black box" representation of the circuit's performance. The behavioral model provides a means to use software to quickly analyze how a complex circuit will react to certain stimuli. A detailed behavioral model consists of a number of so-called *macromodels* that describe each subsystem. Deriving a macromodel for a circuit can be as simple as describing the linear input-output relationship or as complex as deriving mathematical expressions for the amplifier nonlinearities, or thermal effects. For the transconductance amplifiers described in this section, a straightforward linear input-output behavioral model can be described by an ideal transconductor (Figure 4.16). The behavior of the amplifier can be determined by various simulation techniques such as MATLAB or SIMULINK. SPICE can also be employed in behavioral modeling.

4.2 Improvements to the Basic Amplifying Structure

4.2.1 Cascode Circuits

The gain of the simple MOSFET amplifier is directly related to the output impedance as seen by the transistor, which can be either the internal output impedance or other impedance components such as active loads. *Cascode transistors* can be placed between the output and the driven MOSFETs (amplifying or current source loads, for example) and with their gates dc biased. The RF equivalent circuit of the cascode MOSFET on the amplifying MOSFET is shown in Figure 4.17. A small-signal analysis of this structure shows that the output impedance for the cascode amplifier is

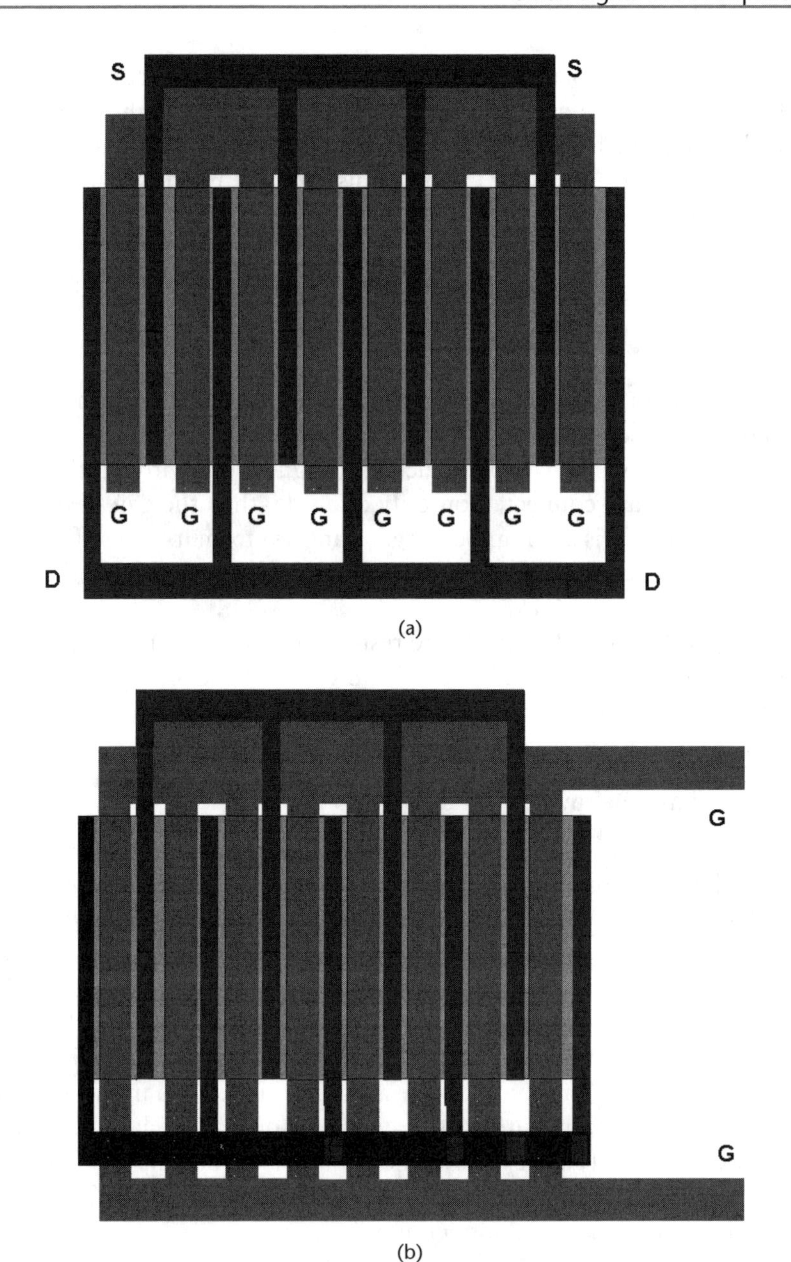

Figure 4.15 Layout of (a) single-sided and (b) double-sided gate connections for multiple gate finger MOSFETs for reduction of R_G.

$$
\begin{aligned}
i_o &= g_{ds1}v_a = g_{m2}v_{gs2} + g_{ds2}(v_o - v_a);\; v_a = -v_{gs2} \rightarrow i_o = -g_{ds1}v_{gs2} \\
r_{\text{out}} &= \frac{v_o}{i_o} = r_{ds2} + r_{ds1}(1 + r_{ds2}g_{m2}) \cong r_{ds1}(g_{m2}r_{ds2})
\end{aligned}
\tag{4.18}
$$

or that the output impedance of the amplifying transistor alone (r_{ds1}) is substantially increased by the gain of the cascode MOSFET ($g_{m2}r_{ds2}$). The high-impedance output provided by the cascode current source load makes the MOSFET output look more like an ideal current source.

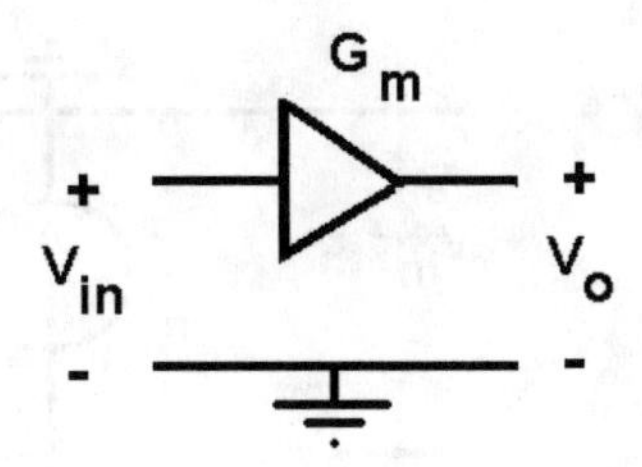

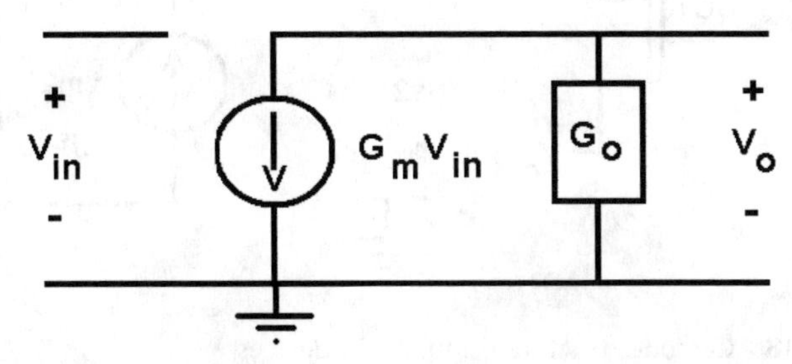

Figure 4.16 Simple behavioral model for a transconductance amplifier.

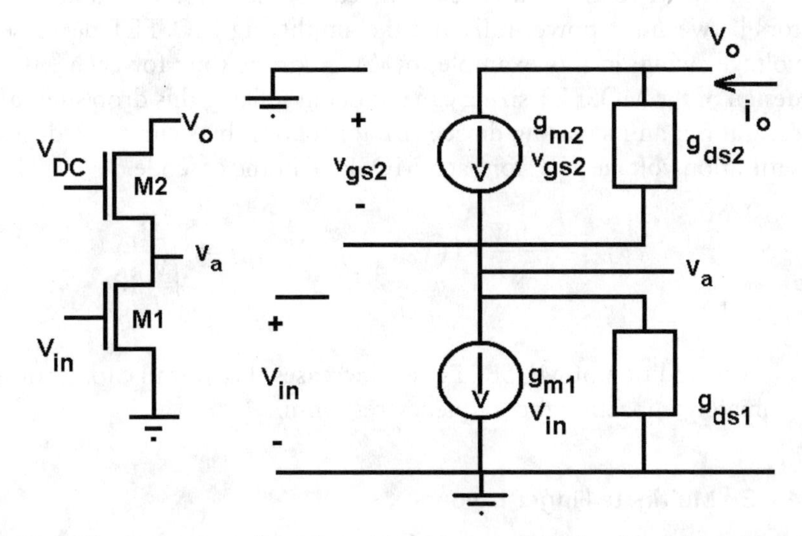

Figure 4.17 Cascode circuit configuration with equivalent circuit for determining the output impedance.

In a similar vein, if a current source load also has a cascode MOSFET added, then its output impedance is increased by the gain of the cascode MOSFET (Figure 4.18).

$$r_{\text{out2-}CS} = r_{ds1}\left(1+ g_{m2} r_{ds2}\right) \cong \left(g_{m2} r_{ds2}\right) r_{ds1} \tag{4.19}$$

The increased output impedance of the circuit dramatically increases the gain:

$$r_{\text{out}} = \left(r_{\text{out}} \| r_{\text{out-}CS}\right);\ A_{v0} = -g_m\left(r_{\text{out}}\right) \cong -g_m^2 r_{ds}^2 \tag{4.20}$$

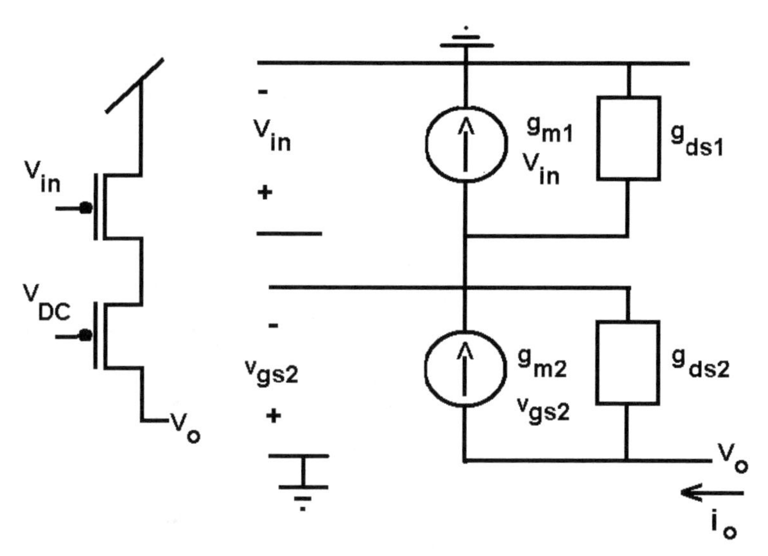

Figure 4.18 Cascode structure using *p*MOS devices.

There are several drawbacks with cascoding. First, the addition of extra transistors between the power rails and the amplifying MOSFET decreases the available voltage swing; in this example, by $2V_{Dsat}$ drops (one for each MOSFET). Careful design of the MOSFET sizing can aid in minimizing this drop since all devices are in saturation; an increasing device size for a given bias current reduces the minimum saturation voltage V_{Dsat} for each MOSFET in the cascode:

$$I_{DS} = -\frac{1}{2}\frac{W}{L}\text{KP}\left(V_{Dsat_{\text{MIN}}}\right)^2;\ V_{Dsat_{\text{MIN}}} = \sqrt{\frac{2I_{DS}}{\frac{W}{L}\text{KP}}} \tag{4.21}$$

The addition of MOSFETs also increases the overall capacitance in the circuit, ultimately impacting the frequency response.

4.2.2 Multigate Finger Layouts

High-current MOSFETs require large gate widths W. Increasing W also increases the areas and perimeters of the drain and source diffusions, thereby increasing C_{db} and C_{sb} in these transistors. Exceedingly large W also increases the distributed effects along the gate, with a reduction in effective transistor action far down the gate. Dividing the MOSFET layout into a number N parallel transistors preserves the dc current handling capability and the transconductance of the MOSFET. This also reduces the parasitic capacitances C_{db} and C_{sb} by sharing the drain and source diffusions between adjacent MOSFETs, reducing the overall area and perimeter. The gate resistance R_G now consists of a number of equal valued resistors in parallel, effectively decreasing R_G by $1/N$. In addition, if both ends of the gate can be tied together, a further reduction by approximately a factor of 4 results. These phenomenon can be seen by way of an example. Consider a MOSFET with an aspect ratio of 240 (120/0.5), as illustrated in Figure 4.19.

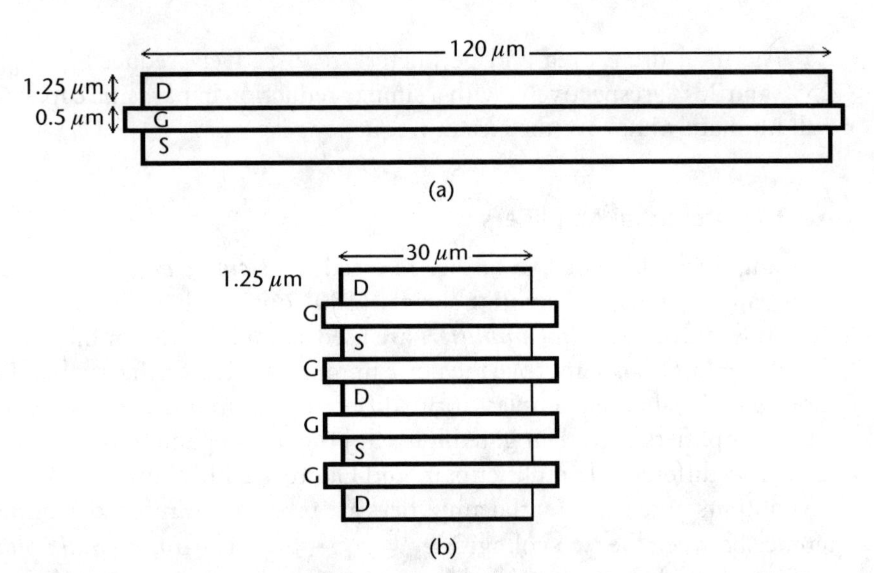

Figure 4.19 (a) Single-gate and (b) multiple-gate finger layout examples.

From the dimensions and conservative layout design rules, the drain/source area (*AD*/*AS*) and perimeter (*PD*/*PS*) can be estimated as

$$AS = AD = 1.25 * W = 150\ \mu\text{m}^2;\ \ PS = PD = 2(1.25 + W) = 242.5\ \mu\text{m}$$

and the gate resistance can be estimated as

$$R_G = \frac{1}{3}\frac{1}{N}\frac{W}{L}R_{sh} = \frac{1}{3}\frac{120}{0.5}R_{sh} = 80R_{sh}$$

Figure 4.19(b) shows an $N = 4$ multifinger layout with the shared drain and source connections for adjacent MOSFETs indicated. Note that there are three drain and two source regions that must all be connected together with metal, hence the need for slightly wider diffusion regions to accommodate the metal-semiconductor contacts. The gate material should be connected on both ends to achieve the most reduction in R_G. The drain (source) is now spread out in three (two) regions of width 30 μm, and the area and perimeter can be computed:

$$AD = (3)(1.25)W = (3)(1.25)30 = 112.5\mu\text{m}^2;\ \ PD = 3[2(1.25 + 30)] = 187.5\mu\text{m}$$

$$AS = (2)(1.25)W = (2)(1.25)30 = 75\mu\text{m}^2;\ \ PS = 2[2(1.25 + 30)] = 125\mu\text{m}$$

where the leading factor of 3 (2) is to take into account the contributions from the drain (source) regions. In a similar fashion, the gate resistance can be estimated as

$$R_G = \frac{1}{3}\frac{1}{N}\frac{W}{L}R_{sh} = \frac{1}{3}\frac{1}{4}\frac{120}{0.5}R_{sh} = 20R_{sh}$$

The total drain area and perimeter are effectively reduced by approximately 25% and 23%, respectively, with a similar reduction in parasitic capacitance, which will ultimately improve the circuit frequency response.

4.2.3 Differential Amplifiers

The amplifiers discussed so far are referred to as *single-ended amplifiers*. In many RFIC applications, differential signals (180° out of phase with one another) are available, and *differential amplifiers* are used as gains stages for these signals. Differential amplifiers also improve the noise immunity is RFICs due to their low common mode gain [1]. In their simplest form, differential amplifiers are two identical single-ended amplifiers with their gate inputs at potentials v_+ and v_- [Figure 4.20(a) shows an *n*MOS differential amplifier pair, and Figure 4.20(b) shows a *p*MOS equivalent]. The outputs of the differential amplifiers are taken at the respective drains as the difference between the two voltages: $v_{out} = v_{out+} - v_{out-}$. The *differential voltage gain* A_{Vd} is defined as the ratio of the differential output voltage to the differential input voltage:

$$A_{Vd} = \frac{v_{out+} - v_{out-}}{v_+ - v_-} \tag{4.22}$$

The RF equivalent circuit of Figure 4.21(a) shows the output voltage at the drain of the differential pair MOSFETs can be written as $v_{out+} = -g_{m1}r_{ds1}v_+$ and $v_{out-} = -g_{m2}r_{ds2}v_-$.

If M1 and M2 are identical sizes and at identical biases, their transconductance and output resistances are also identical (in an ideal world), so the differential voltage gain can be written as

$$A_{Vd} = \frac{v_{out+} - v_{out-}}{v_+ - v_-} = \frac{-g_{m1}r_{ds1}v_+ + g_{m2}r_{ds2}v_-}{v_+ - v_=} = \frac{-g_m r_{ds}(v_+ - v_-)}{v_+ - v_-} = -g_m r_{ds} \tag{4.23}$$

which is identical to the single-ended voltage gain.

Differential amplifiers will typically have some sort of active load on the drain that will set the circuit A_{Vd}. Figure 4.22 shows a differential amplifier with an *n*MOS differential pair and *p*MOS current source loads; A_{Vd} for this topology is

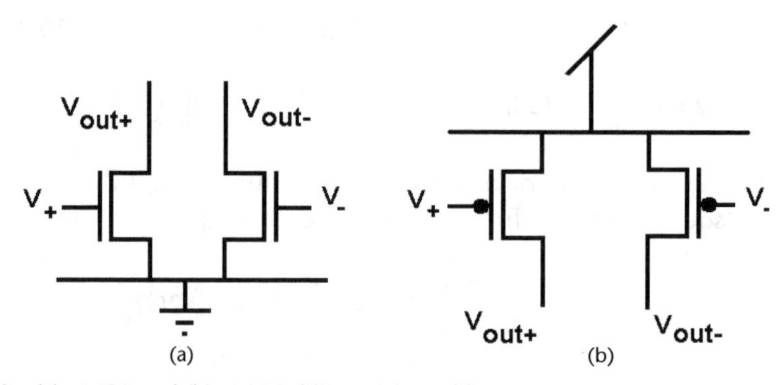

Figure 4.20 (a) *n*MOS and (b) *p*MOS differential amplifier structures.

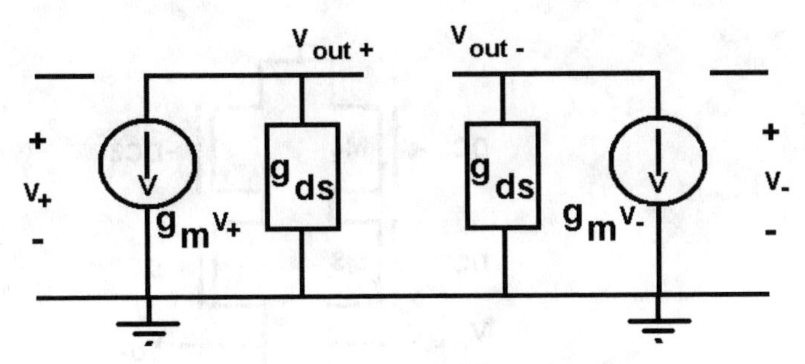

Figure 4.21 RF equivalent circuit of a simple *n*MOS differential pair.

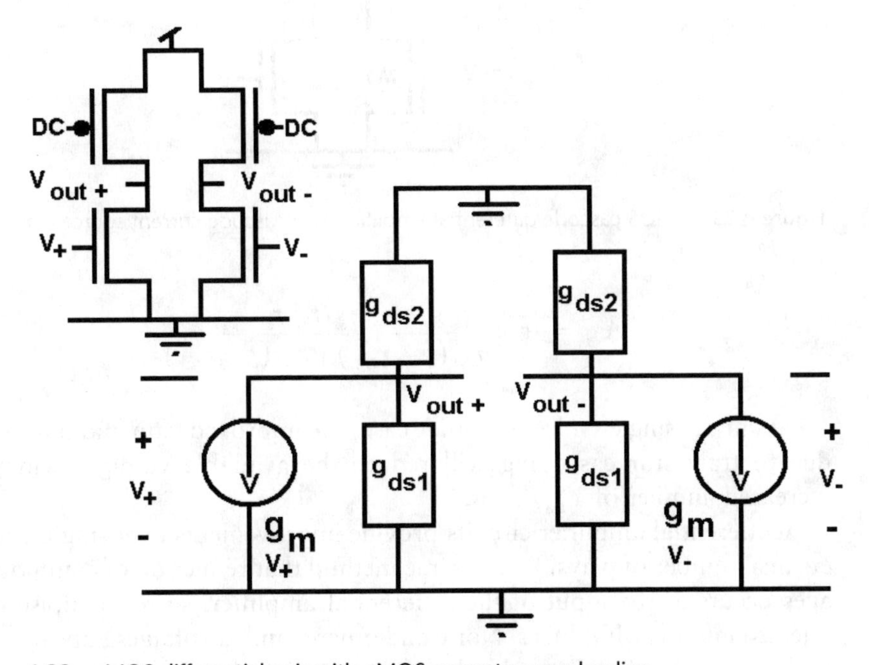

Figure 4.22 *n*MOS differential pair with *p*MOS current source loading.

$$A_{Vd} = -\frac{g_m}{g_{ds1} + g_{ds2}} \tag{4.24}$$

For a differential amplifier with a *p*MOS differential pair (M1) and *n*MOS active resistor loads (M2), the differential gain A_{Vd} can be written as

$$A_{Vd} = -\frac{g_{m1}}{g_{ds1} + g_{m2}} \cong -\frac{g_{m1}}{g_{m2}} \tag{4.25}$$

The differential amplifier's gain and output resistance will benefit from the addition of cascoding transistors as well. Figure 4.23 shows a cascode *n*MOS differential pair with a *p*MOS cascode current source load. The RF equivalent circuit for this cascoded differential amplifier shows that the output impedance is increased by the $g_m r$ of the cascode MOSFETs, increasing the differential voltage gain by the same factor:

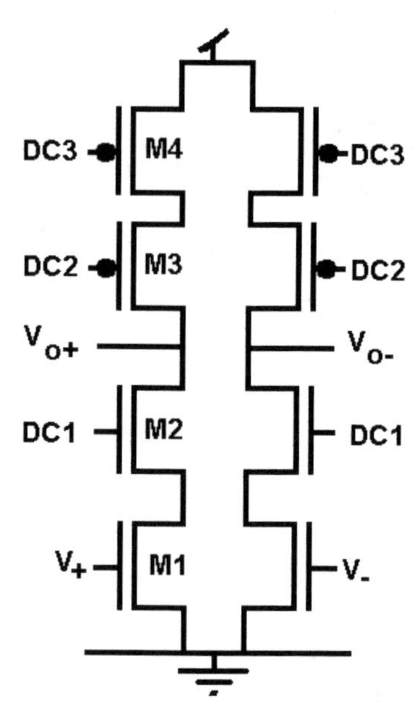

Figure 4.23 *n*MOS cascode differential amplifier with cascode current source loading.

$$A_{Vd} = -g_{m1} \frac{g_{m2} g_{m3} r_{ds1} r_{ds4} r_{ds3} r_{ds2}}{r_{ds1}\left(g_{m2} r_{ds2}\right) + r_{ds4}\left(g_{m3} r_{ds3}\right)} \tag{4.26}$$

As in the single-ended amplifier case, the improved gain and output impedance due to transistor cascoding will reduce the available voltage swing due to the increased number of V_{DS} drops.

Differential amplifier circuits provide improvements over single-ended amplifiers in a number of ways. By the same method that reduction of common mode voltages occurs at the input of the differential amplifier, so too is noise that may be injected into the MOSFETs. Noise and other coupled voltages appearing at the bulk nodes of the MOSFETs tend to be of the same phase since the transistors are physically close together. Since these voltages are of the same phase, they are cancelled at the differential amplifier output as common mode signals. Even-order nonlinearities are also reduced in differential amplifiers. This cancellation can be seen by assuming that the signal v_{in} and $-v_{\text{in}}$ are available at the differential amplifier terminals. If the noninverting differential amplifier output has a nonlinear output of form $v_{o+} = k_1(v_{\text{in}}) + k_2(v_{\text{in}})^2 + k_3(v_{\text{in}})^3 + \ldots$ and the inverting differential amplifier output is of form $v_{o-} = k_1(-v_{\text{in}}) + k_2(-v_{\text{in}})^2 + k_3(-v_{\text{in}})^3 + \ldots$, then the difference between the two output signals removes the second-order (and all even-order) term. The removal of the second-order term tends to also reduce the dc term that arises from the second harmonic of the output signal. This additional dc voltage can be removed by adding a dc blocking capacitor, but at the expense of low-frequency response. This degradation of the low-frequency response will have a negative impact in direct conversion (zero IF) receivers, so differential circuitry is vital for high performance in these circuits.

In general, the input differential pair transistors M1 and M2 are not exactly matched due to variations in W/L or V_T (just to name a few of the possible sources of

MOSFET mismatch). These differences lead to an *input offset voltage* that can skew shift the differential transfer characteristics. Imperfect cancellation of common mode and even-order harmonic terms will also occur in mismatched differential circuits. As a way to mitigate the effect of process variations on the differential amplifier, the MOSFET differential devices are often split into two or more parallel devices with a layout based on a *common centroid* geometry. The process variations are then "spread out" among the transistors, with both sides of the differential pair seeing similar process parameters due to their close physical location.

4.2.4 Current Reuse

Amplifiers consume a considerable current if they are to exhibit good transconductance. For high-gain stages, a number of amplifiers are often employed, each consuming current. A technique known as *current reuse* is often employed to reduce overall dc power consumption by having a single current source bias two circuits. A drawback to this technique is that amplifying transistors are placed in series, thereby causing a reduction in available amplifier voltage swing due to the increased device voltage drops. In developing current reuse circuits, it is important to identify the location of RF grounds (or their equivalents) for each circuit, but not necessarily their dc grounds [3]. The various steps in converting a two-stage amplifier circuit topology to one that allows current reuse are shown in Figures 4.24 through 4.26. Figure 4.24(a) shows this two-stage amplifier with MOSFETs M1 and M2 being dc biased through inductors L1 and L2. The system voltage gain is simply the product of the two amplifier gains:

$$A_v \cong g_{m1} r_{ds1} g_{m2} r_{ds2} \tag{4.27}$$

Closer inspection of the two-stage amplifier shows that the driver stage power supply is an effective RF ground, as is the source of M2 in the driven stage. The driven stage, at least from an RF perspective, can be "stacked" on top of the driver stage [Figure 4.24(b)] with the drain of M1 connected to the gate of M2. Direct application of this technique overlooks, however, the dc biasing of the MOSFETs, primarily M2. Note in Figure 4.24(b) that the source of M2 is at the same dc potential as M1's gate, biasing M2 not in saturation, but actually the device is turned off since $V_{GS} < V_T$. However, if M2 is replaced by a *p*MOS device, then M2's source is at a *higher* potential than the gate, and M2 can be biased into saturation [Figure 4.25(a)]. The output is taken at the source of M2, however, so this circuit behaves as a source follower with reduced gain, thereby limiting the overall system gain. Swapping M2 and L2 positions yields a common source equivalent amplifier [Figure 4.25(b)], where the output is taken at the drain of M2. The final current reuse circuit shown in Figure 4.25(b) has an RF equivalent circuit shown in Figure 4.26.

The equivalent circuit in Figure 4.26 indicates that the system voltage gain is identical to that of the original circuit. In this voltage gain expression, the transconductance terms are for the *n*MOS driver (g_{m1}) and the *p*MOS driven (g_{m2}) FETs. The *p*MOS device must be slightly increased in size by the ratio of the KPs of the two devices to compensate for the reduced intrinsic transconductance of the *p*MOS device. Readjustment of M1's bias voltage will more than likely need to be

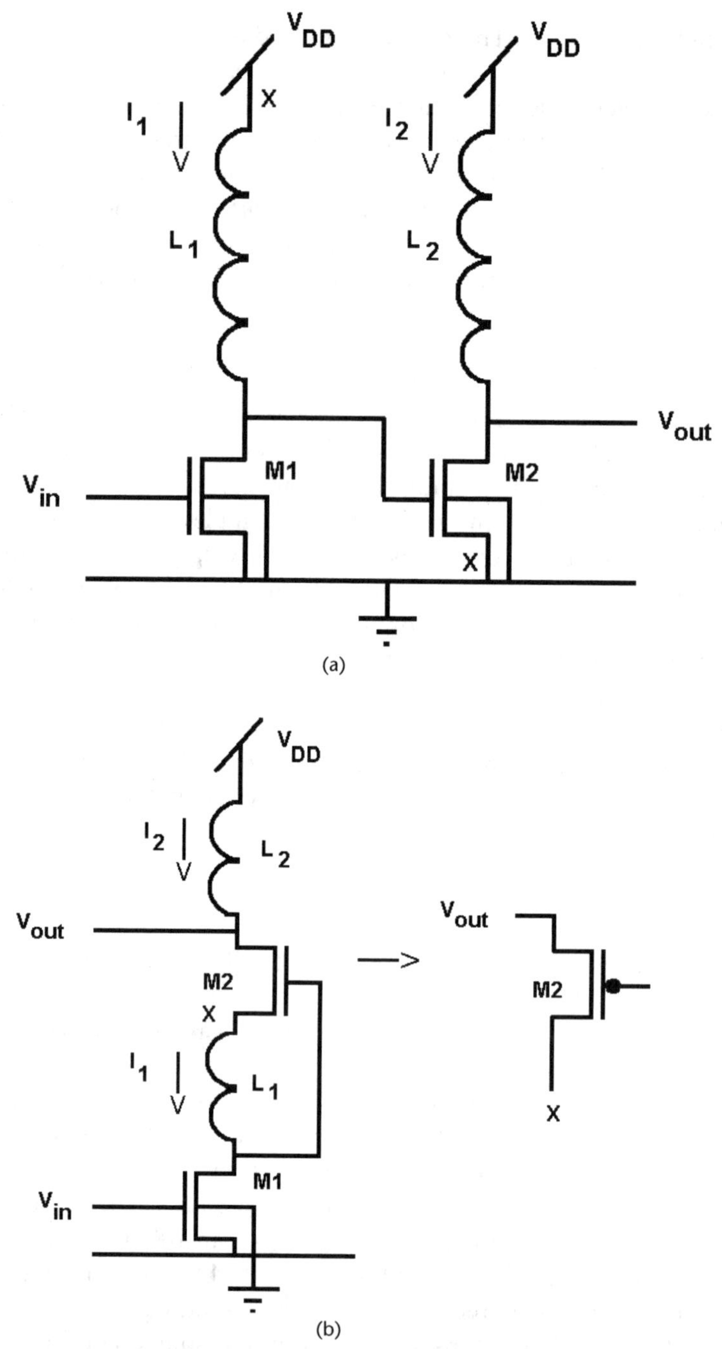

Figure 4.24 Development process for converting a two-stage *n*MOSFET amplifier to a topology suitable for current reuse: (a) two-stage amplifier and (b) conversion of second stage to a *p*MOSFET transconductor.

done. In this circuit, nearly identical gain is achieved with the modified circuit using current reuse (Figure 4.27). Inductive peaking of the amplifier is evident in both simulations. The original circuit shows a total current drain of 20 mA (6 mA for the driver state) and 4-mA current drain for the modified circuit using current reuse. A

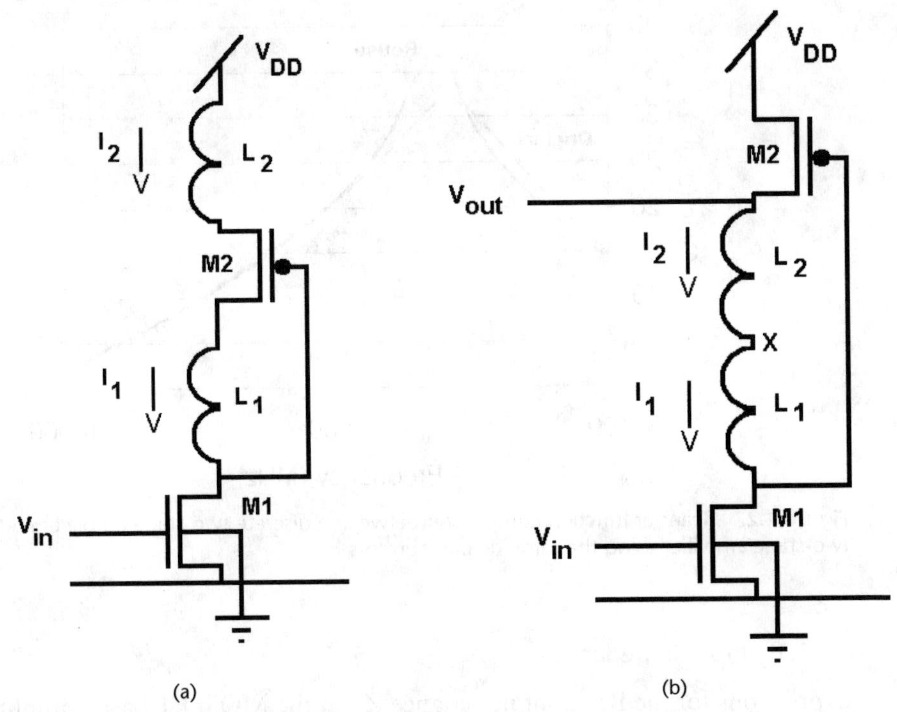

Figure 4.25 (a) Source follower arrangement of two-stage amplifier with *p*MOSFET replacement and (b) common source arrangement of two-stage amplifier with current reuse.

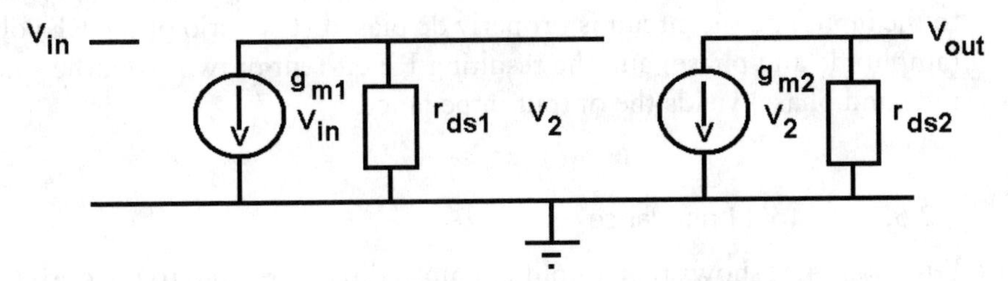

Figure 4.26 RF equivalent circuit of modified amplifier showing gain characteristic identical to that of original circuit.

substantial savings in current for similar gain shows the power of this technique in low-power RFIC designs.

Note that there are now two MOSFETs between the rails; the downside to current reuse is that stacking transistors causes a reduction in output voltage swing. Cascaded stages of differential amplifiers can also use the current reuse technique to improve current budgets.

4.2.5 Input/Output Impedance Modeling Example

This section provides details on using a simulator to determine both the input and output impedance of a MOSFET-based amplifier. The technique can be easily extended to other circuits as well.

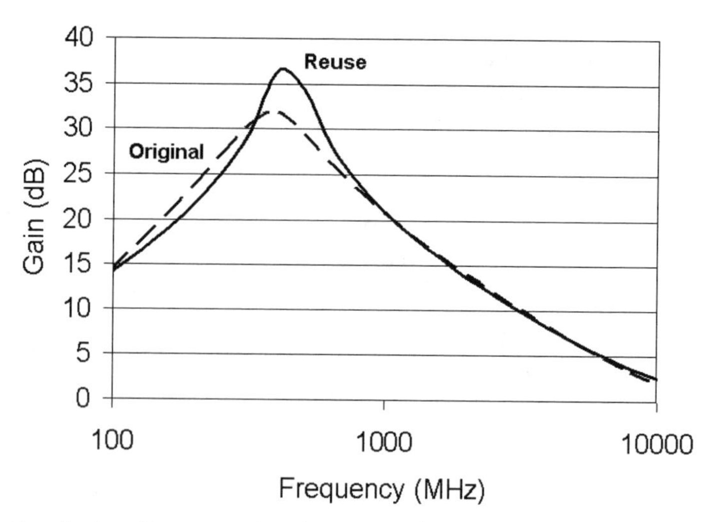

Figure 4.27 Transfer function comparison between a discrete two-stage amplifier and a two-stage amplifier using the same dc bias currents.

4.2.5.1 Input Impedance

Expressions for the RF input impedance Z_{in} of the MOSFET based-amplifier can be determined from the RF equivalent circuit by applying an RF voltage source at the input and calculating the current drawn by this source; the RF input impedance is the ratio of these two terms. The small-signal input impedance can be determined using a circuit simulator in a similar fashion; a small-signal voltage source is applied to the input and the circuit is properly dc biased. The ratio of the RF voltage source (amplitude and phase) and the resulting RF current drawn from the source (amplitude and phase) yields the output impedance.

4.2.5.2 Output Impedance

Equation (4.5) shows that amplifier gain is directly related to the output impedance seen by the transconductance element. Expressions for the RF output impedance Z_{out} can be determined from the RF equivalent circuit by applying an RF current source at the output and calculating the voltage dropped across the output from this source; the output impedance is the ratio of these two terms. The small-signal output impedance can be determined using a circuit simulator in a similar fashion; a small-signal current source is applied to the output and the input is properly dc biased with no RF signal applied. The ratio of the resulting voltage and this current source yields the output impedance. A high-valued resistor should be used in shunt with this current source for convergence purposes and should be at least 10 times the expected output impedance.

Example 4.5: Determine the input and output impedance versus frequency of the amplifier in Example 4.1. Use a 10-GΩ resistor in shunt with the current source (ch4-4.txt).

The circuits for determining the input and output impedance of the amplifier in Example 4.1 are shown in Figures 4.28(a) and (b), respectively. The input impedance is determined by applying a 1-mV signal to the input gate of M1. The current supplied by the source can be used to determine the input impedance Z_{in}. The results of this computation are shown in Figure 4.28(c), where the real and imaginary parts of Z_{in} are plotted versus frequency up to 1000 MHz. The capacitive input nature of M1 is observed in the IM(Z_{in}) graph. For the computation of Z_{out}, the gate is dc biased at the same dc bias voltage as the input impedance calculation and the output is driven with a 1.0-μA source with a 10-GΩ output impedance. Figure 4.28(d) shows the real and imaginary parts of the output impedance.

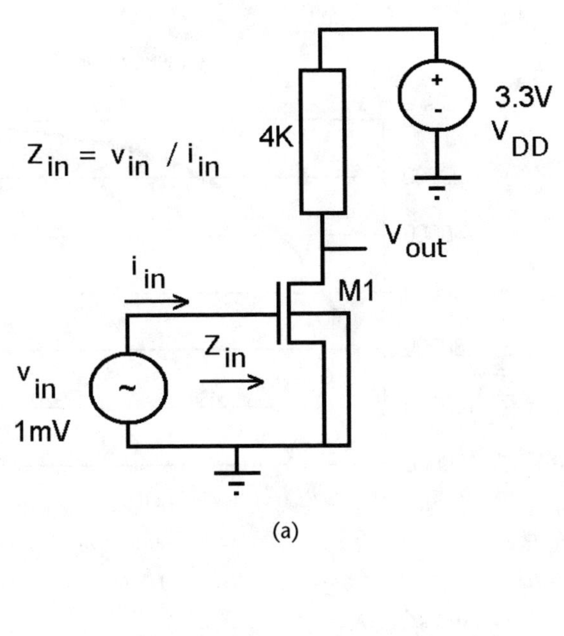

(a)

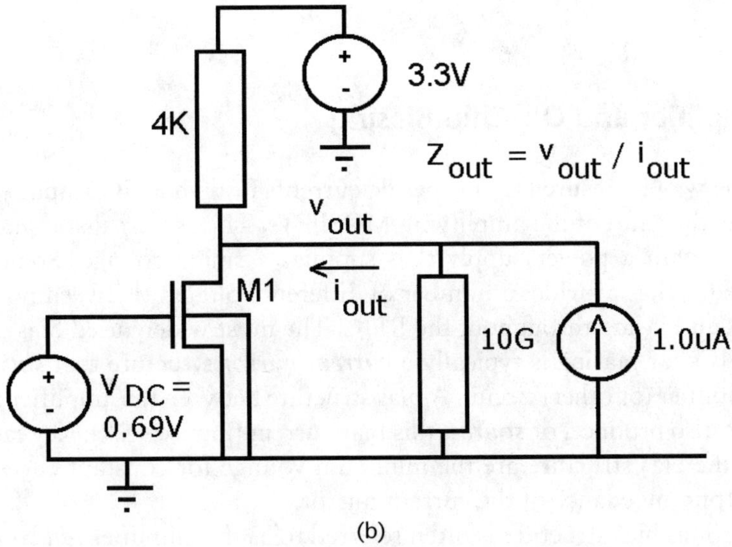

(b)

Figure 4.28 (a) Input and (b) output impedance simulation circuits and real and imaginary parts of the (c) input and (d) output impedance.

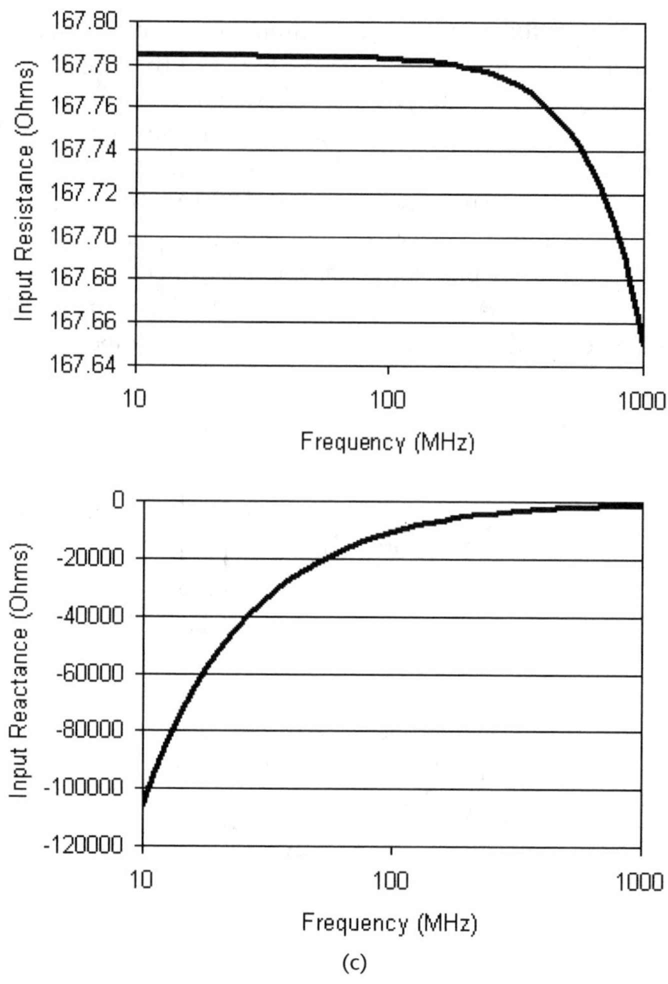

Figure 4.28 (continued)

4.3 Amplifier and On-Chip Biasing

One way to ensure that proper dc currents flow in RFIC amplifiers is to directly dc bias the gates of the amplifying MOSFETs. This is unrealistic since most RFICs are used with a power supply that supplies a single voltage. Some other method is needed that provides a number of different voltages (between both power rails) for circuit biasing throughout the RFIC. The most widely used bias circuit topology in RFICs for biasing is typically a *current mirror* structure that *sinks* I_{bias} through the amplifier (or other) circuit. A bias structure between the amplifier and the power rail can also produce, or *source*, this bias current (Figure 4.29). Key factors in the design of the bias structure are the minimum voltage for constant current at V_{out} and the output impedance of the current mirror.

This bias structure is often referred to as the amplifier *tail current* and is widely used in RFICs to bias most of the circuitry. The tail current (either source or sink) acts to stabilize the dc output voltage and provides *current limiting* to output voltage

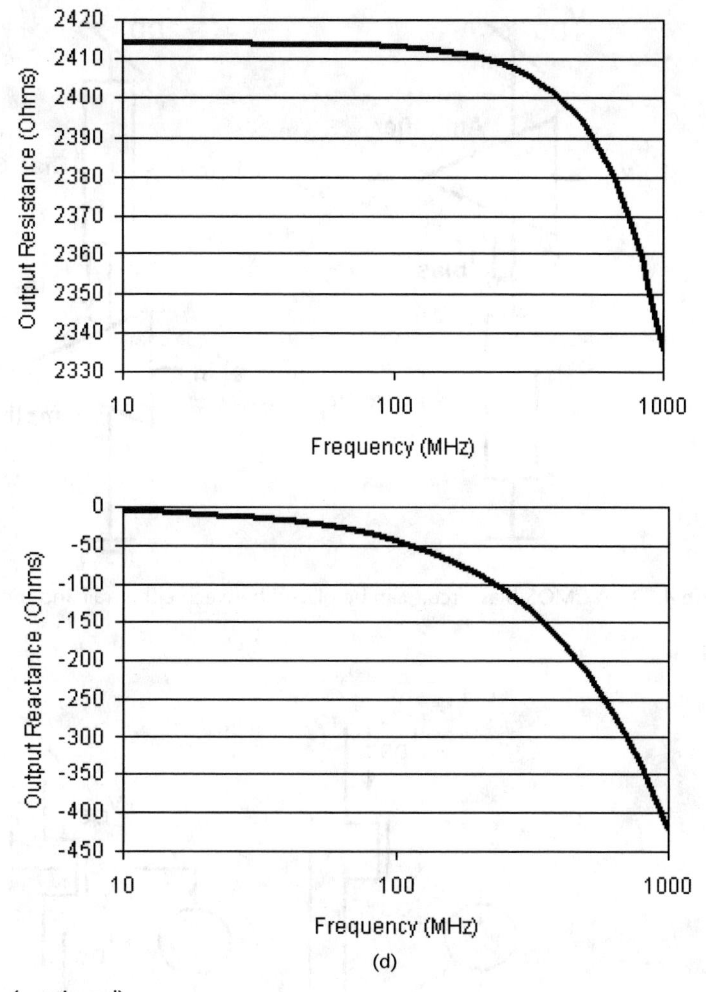

Figure 4.28 (continued)

swings [4]. The tail current source also tends to improve common mode performance in differential amplifiers.

This section explores current sources and current sinks and the design of current mirrors commonly found in RFIC stages.

4.3.1 Current Mirror Structures

An *n*MOS or a *p*MOS transistor can be easily configured as a simple current sink or source, respectively. The current sunk or sourced by the MOSFET is controllable by the designer through the aspect ratio W/L and the applied gate voltage V_{GS} (Figure 4.30):

$$I = \frac{1}{2}\frac{W}{L}\mathrm{KP}\left(V_{GS} - V_T\right)^2 \tag{4.28}$$

The output impedance of these sources/sinks is governed by the source/sink current required and the λ of the MOSFET:

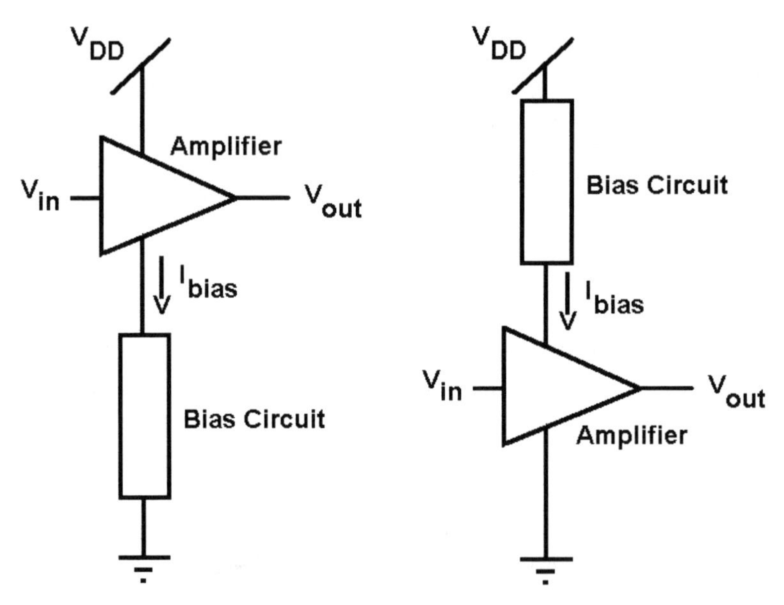

Figure 4.29 A CMOS bias circuit can be placed between either rail and the amplifier.

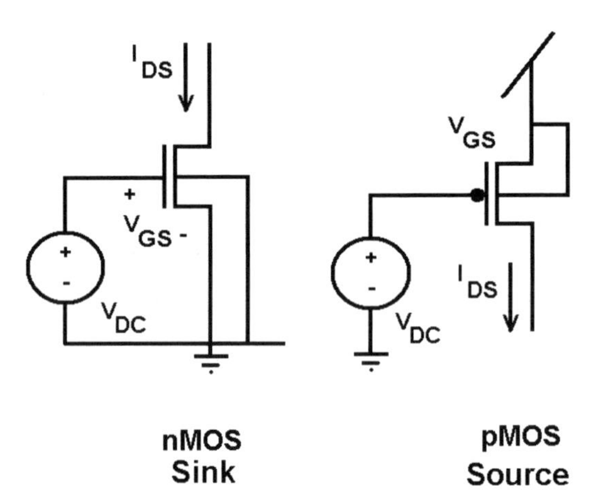

Figure 4.30 Simple *n*MOS current sink and *p*MOS current source.

$$r_{\text{out}} \approx \frac{1}{\lambda I_{DS}} \tag{4.29}$$

Example 4.6: Design a 0.5-μm *n*MOS current sink for 0.5 mA with a gate voltage of 1.0V, as shown in Figure 4.31(a). Simulate the output impedance of the source (ch4-5.txt).

Application of the drain current equation gives the required gate width W at the 1.0V gate bias level:

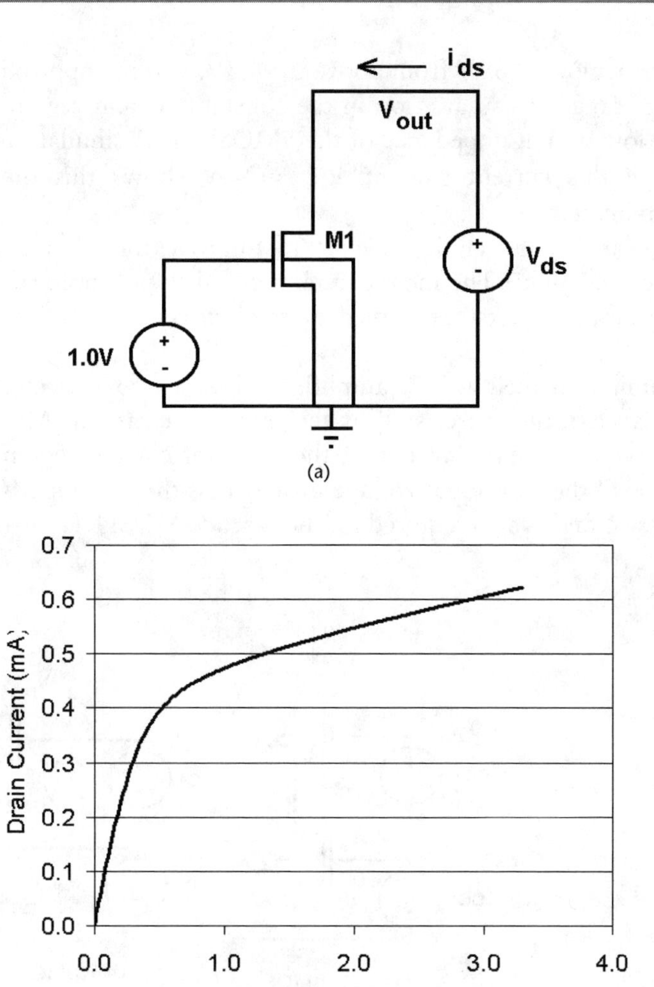

Figure 4.31 (a) *n*MOS current sink current version output voltage. (b) The relatively low output impedance can be seen in the figure for $V_{DS} > 1.0$V.

$$I = \frac{1}{2}\frac{W}{L}\text{KP}(V_{GS} - V_T)^2;\ W = 0.5\frac{2I}{\text{KP}(V_{GS} - V_T)^2}\ \mu\text{m} = 77\mu\text{m}$$

The minimum voltage drop across the MOSFET that keeps the device in the "constant current" region can be computed from (4.13) as

$$V_{DS} = \sqrt{\frac{2I}{\text{KP}\frac{W}{L}}} = 0.34\text{V}$$

Figure 4.31(b) shows the results of the above example. Note that the so-called constant current sunk through the *n*MOSFET is anything but constant over the range of V_{DS} between the minimum voltage computed for the onset of the constant current regime (0.34V from above) to 3.3V, varying approximately 20% over this voltage range. This variation in the constant current region is caused by the relatively low output impedance of the *n*MOSFET. A simulation of the output impedance of this current sink at V_{DS} = 2.0V shows this output resistance to be approximately 16 kΩ.

A variety of techniques exist to improve the output impedance of current sources and sinks. The most common circuit type employs a cascoding transistor; the resulting circuits are termed *cascode current sources* or *cascode current sinks* [Figure 4.32(a)].

Similar to their use in amplifiers, these cascode circuits increase the output impedance of the source/sink by the gain of the cascode MOSFET. A drawback to the cascode current sink is that the constant current region occurs at higher V_{DS} because of the additional voltage drop across the cascode MOSFET. An additional dc bias voltage is also required for the cascode MOSFET. Cascoding the current sink

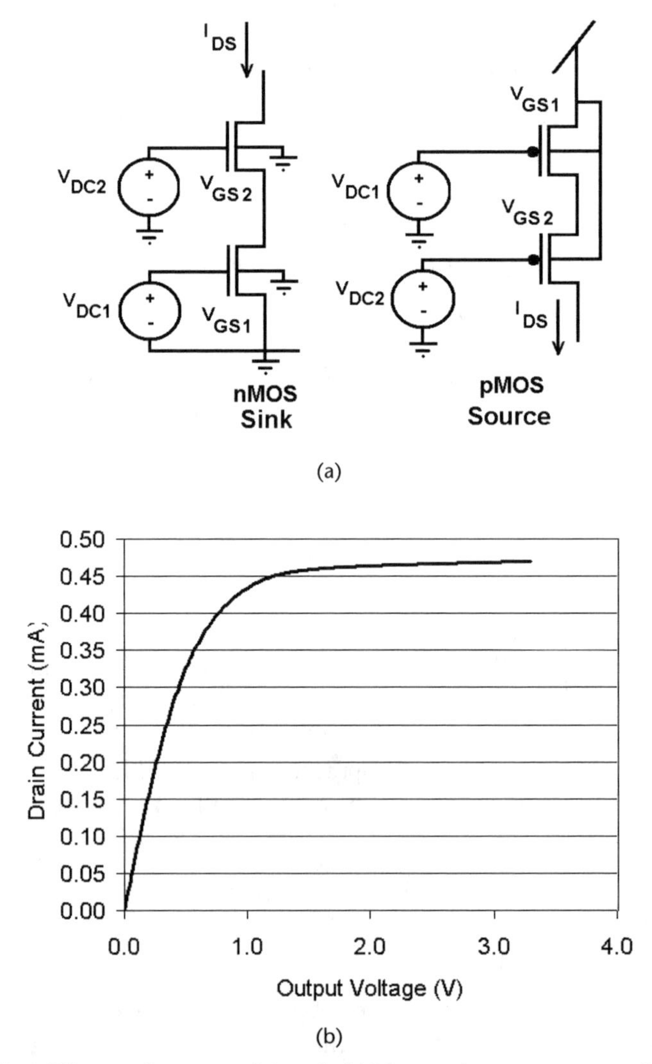

Figure 4.32 (a) *n*MOS cascode current sink and *p*MOS cascode current source; (b) output impedance of *n*MOS cascode current sink, showing the more constant current (ch4-6.txt).

in Example 4.6 with an identical MOSFET with its gate biased at 2.0V shows a marked improvement in the output impedance, as indicated by the nearly constant current above approximately $V_{DS} = 1.2$V [Figure 4.32(b)]. The increased threshold voltage for constant current of approximately $2V_T$ is also seen in the figure.

In many RFIC stages, the currents should track each other for best performance. This current tracking can be performed with a *current mirror* circuit. The concept behind a current mirror is relatively straightforward: from the I_{DS} current equation, if V_{GS} is kept fixed, then the resulting current is controlled solely by the aspect ratio of the MOSFET, *W*/*L*. If two transistors (in this example, *n*MOS current sinks; Figure 4.33) have the same V_{GS}, their currents are "mirrored" by their aspect ratios:

$$I_{DS1} = \frac{1}{2}\left(\frac{W}{L}\right)_1 \mathrm{KP}(V_{GS} - V_T)^2;$$
$$I_{DS2} = \frac{1}{2}\left(\frac{W}{L}\right)_2 \mathrm{KP}(V_{GS} - V_T)^2 = I_{DS1}\frac{\left(\frac{W}{L}\right)_2}{\left(\frac{W}{L}\right)_1} \tag{4.30}$$

A circuit attached to the V_{out} node will then be biased with current I_{DS2}.

Example 4.7: Design an *n*MOS current mirror for currents of 0.5 and 1.0 mA through the two *n*MOS current sinks (ch4-7.txt).

Using the results of Example 4.6 with $W_{M1} = 12\ \mu$m for 0.5 mA, the current mirroring equation yields an aspect ratio for M2:

$$I_{DS2} = I_{DS1}\frac{\left(\frac{W}{L}\right)_2}{\left(\frac{W}{L}\right)_1};\quad \left(\frac{W}{L}\right)_2 = \left(\frac{W}{L}\right)_1 \frac{I_{DS2}}{I_{DS1}} = 24\ \mu\text{m}$$

The resulting current mirror response is shown in Figure 4.34 and verifies that M2 sinks twice the current that M1 does.

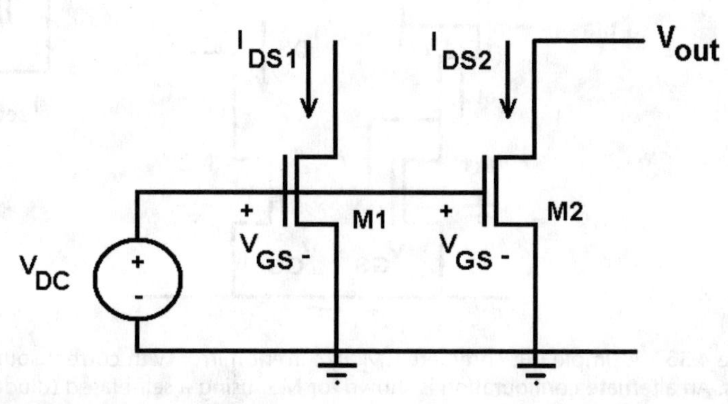

Figure 4.33 Simple *n*MOS current mirror.

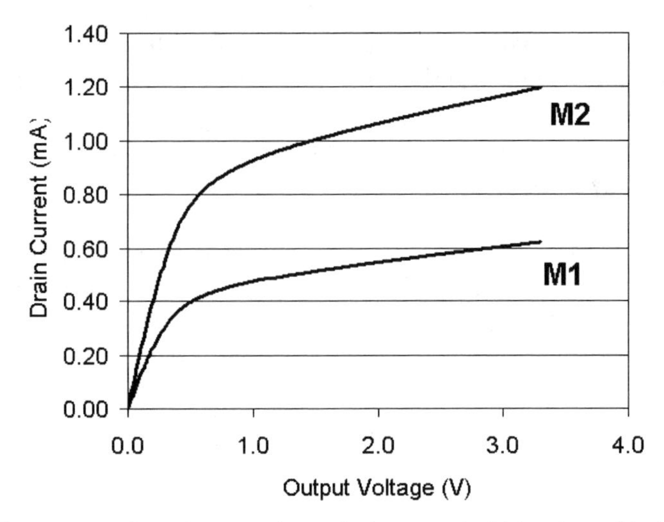

Figure 4.34 The current mirror structure shows that a current of I_0 is mirrored to $2I_0$.

Current mirrors are often configured to act as dc bias references for many circuits in RFICs. This powerful technique takes advantage of the tracking capability of the current mirror to accurately control biasing for a number of circuits. A simple current mirror circuit can be derived by combining the circuit in Figure 4.33 with a *p*MOS current source (M3) and having M1 act as its load (Figure 4.35). The current I_{DS} through M2 can be used to set the bias for additional circuits. The current mirror circuit requires a single dc bias on the *p*MOS current source. The design equations for this simple current mirror are indicated below. Alternatively, M3 can be diode connected to provide a complete self-biasing current mirror.

The source current I_{set} is determined by the dc bias conditions and aspect ratio of the *p*MOSFET, and I_{DS} is mirrored with the M1-M2 pair:

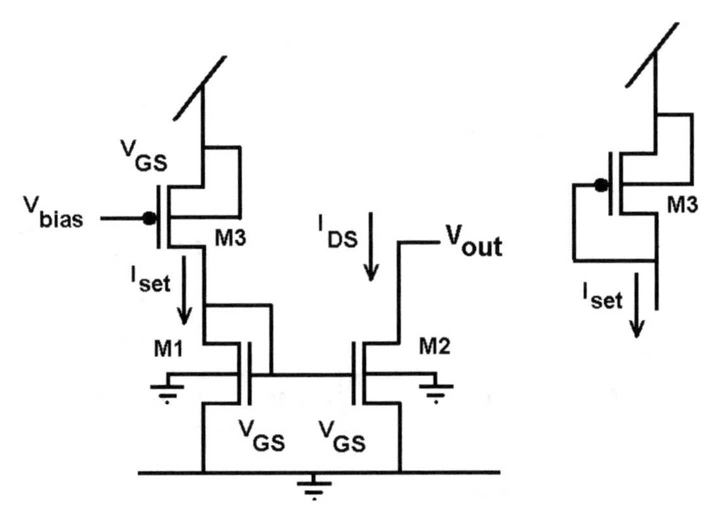

Figure 4.35 A simple but complete CMOS current mirror with current source and mirroring transistors. An alternate configuration is shown for M3, using a self-biased (diode-connected) *p*MOSFET current source.

$$I_{set} = \frac{1}{2}\left(\frac{W}{L}\right)_3 \mathrm{KP}(V_{DD} - V_{bias} - V_T)^2;\ I_{DS} = I_{set}\frac{\left(\frac{W}{L}\right)_2}{\left(\frac{W}{L}\right)_1} \tag{4.31}$$

The active load configuration of M1 guarantees that the device remains in saturation. The current flowing through transistor M1 sets its gate-source voltage:

$$V_{DS1} = V_{GS1} = V_T + \sqrt{\frac{2I_{set}}{\mathrm{KP}(W/L)_1}} \tag{4.32}$$

V_{GS1} should be set for at a few hundred millivolts above V_T to ensure proper mirror action under the usual CMOS fabrication process variations. A higher voltage will reduce the available voltage swing for the bias circuit.

Example 4.8: Design a simple current mirror for 0.5 mA in both the current set and mirrored branches. Assume a 0.5-μm technology with V_{DD} = 3.3V and V_{bias} = 1.5V (ch4-8.txt).

Since the currents in both branches of the current mirror are required to be equal, the aspect ratios of M1 and M2 must also be equal. Current source transistor M3 can be designed by noting that the V_{GS} for this transistor is the difference between the applied bias voltage and V_{DD} (−1.8V), yielding a device width of

$$W = L\frac{2I_{DS}}{\mathrm{KP}(V_{GS} - V_{TP})^2} = 0.5\frac{2(0.5\cdot 10^{-3})}{19.1\cdot 10^{-6}\left[-1.8-(-0.91)\right]^2} = 33\ \mu\mathrm{m}$$

Figure 4.35 shows that the voltage drop across M3 is also its saturation voltage, which can be computed with (4.21) to be 0.9V. The resulting V_{GS} on M1 and M2 is then 2.4V, yielding a width for these MOSFETs of

$$W = L\frac{21_{DS}}{\mathrm{KP}(V_{GS} - V_{TR})^2} = 0.5\frac{2(0.5\cdot 10^{-3})}{53.6\cdot 10^{-6}\left[2.4-0.66\right]^2} = 2.9\ \mu\mathrm{m}$$

A quick simulation shows that this width provides slightly too much current; a final value of 2.6 μm is quickly obtained after only a few iterations using the simulator. The resulting drain currents of MOSFETs M1 and M2 are shown in Figure 4.36. Note that the current in MOSFET M1 is constant; the current in MOSFET M2 varies with the output loading voltage.

There are other configurations of this simple current mirror; a few are shown in Figure 4.37. The design of each type is similar to the previous example.

Figure 4.36 shows that the poor output impedance of the current mirroring transistor M2 does not change simply because a current mirror is employed. Cascode transistors can be employed to improve the output impedance of the current mirroring pair, as shown in Figure 4.38. A drawback to the cascode current

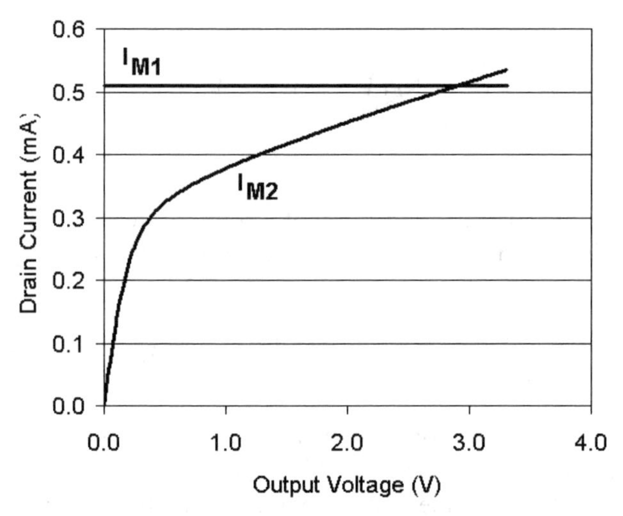

Figure 4.36 CMOS current mirror output currents as a function of output voltage.

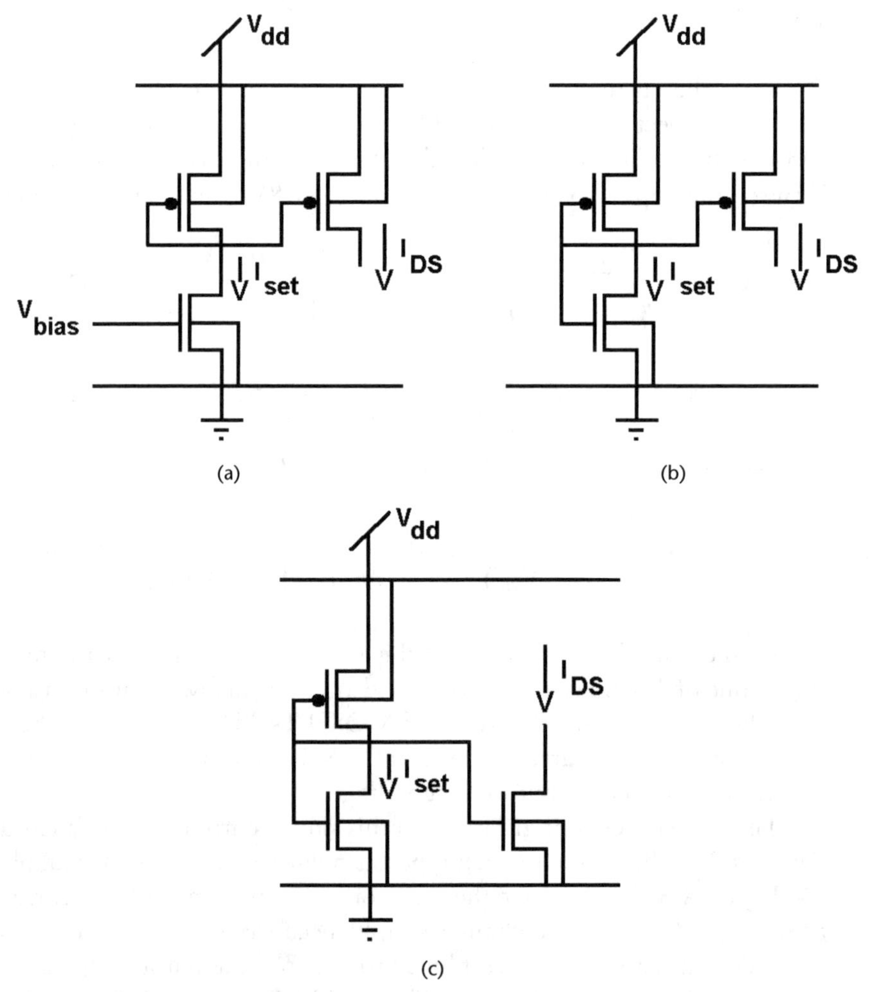

Figure 4.37 Alternative MOS current mirror structures: (a) *p*MOS current mirror, (b) self-biased *p*MOS current mirror, and (c) self-biased *n*MOS mirror.

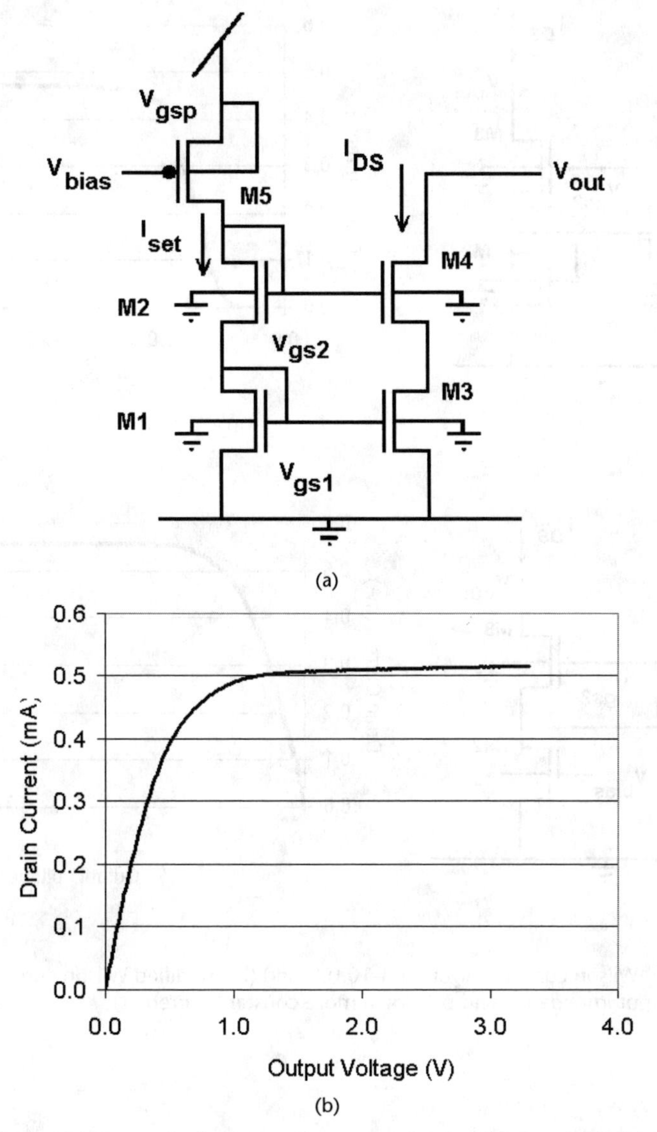

Figure 4.38 (a) Cascode current mirror improves current source/sink and (b) output impedance (ch4-9.txt).

source is evident in the figure; a higher voltage V_{out} is required before the constant current regime is reached. The figure shows that the minimum voltage for constant current is approximately $2V_T$. The improved output impedance is evident from the figure.

Another configuration, the Wilson current mirror (Figure 4.39), uses voltage feedback to improve the output impedance while reducing the number of transistors in the mirror by one [5]. The equivalent circuit for the Wilson current mirror shows that the output impedance is

$$r_{out} = \frac{1}{g_{m2}} g_{m1} r_{ds1} g_{m3} r_{ds3} \tag{4.33}$$

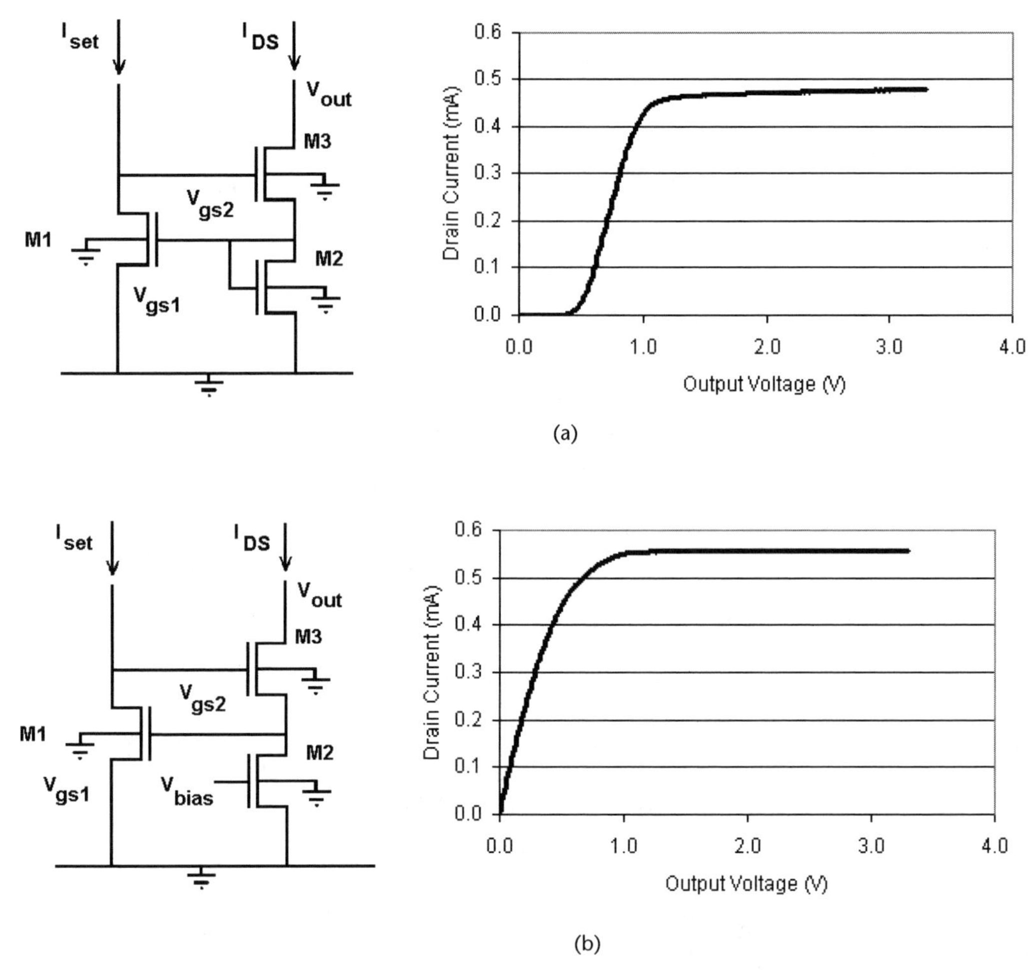

Figure 4.39 (a) Wilson current mirror (ch4-10.txt) and (b) modified Wilson current mirror (ch4-11.txt) improve the output impedance and provide a more constant current.

The factor $1/g_{m2}$ is the output resistance of the active resistor load M2; if the gate of M2 were instead connected to a dc bias, then M2 would become a current sink load with an output impedance of r_{ds2}, improving the output impedance even further:

$$r_{\text{out}} = r_{ds2}\left(g_{m1} r_{ds1} g_{m3} r_{ds3}\right) \tag{4.34}$$

The Wilson current mirror requires both devices to remain in saturation in the constant current region. An estimate of the minimum output voltage for constant current,

$$V_{\text{O-min}} = V_{T2} + \sqrt{\frac{2I}{\text{KP}}}\left[\sqrt{\left(\frac{L}{W}\right)_2} + \sqrt{\left(\frac{L}{W}\right)_3}\right] \tag{4.35}$$

shows that larger aspect ratio MOSFETs can reduce this voltage [1].

4.4 Amplifier Matching

The use of reactive and/or distributed elements in matching or transforming one impedance value to another are widely used in microwave and RF design (Figure 4.40). Most textbooks on microwave engineering devote at least one chapter to impedance matching using both methods [6, 7]. CAD simulators provide either subroutines for matching or matching network synthesis routines to achieve certain user-input parameters and goals. The same techniques can be used by the RFIC designer as well but require that the on-chip reactive element's parasitic resistance and capacitance be taken into account to ensure good simulation results. For frequencies below a few gigahertz, the wavelength is too long for distributed element matching and its attendant real estate requirements. Below approximately 2.0 GHz, lumped element matching is the primary technology even with the parasitic issues.

4.4.1 Classic LC

For ideal narrowband matching (bandwidth of 10% or less), a simple two-element matching circuit topology is often used. There are four possible configurations available, dependent on the relationship between the real parts of the impedances (Figure 4.41). Assume that Z_L can be described as a series resistance and reactance $Z_L = R_L + j\omega L_L$. For $R_L > Z_0$, a reactive element must be placed in shunt with the load and then a series element added to complete the match. The reactances of the matching elements can be found using the following relationships [6]:

$$B = \frac{X_L \pm \left(\sqrt{R_L/Z_0}\right)\sqrt{R_L^2 + X_L^2 - Z_0 R_L}}{R_L^2 + X_L^2}; \; X = \frac{1}{B} + \frac{X_L Z_0}{R_L} - \frac{Z_0}{BR_L} \tag{4.36}$$

Two values for the shunt element B are obtained in this technique. Once the element type is chosen (usually capacitive for dc biasing reasons), the series element X is then specified. For $R_L < Z_0$, a reactive element must be placed in series with the load and then a shunt element added to complete the match:

$$X = \pm\sqrt{(Z_0 - R_L)R_L} - X_L; \; B = \pm\frac{\sqrt{(Z_0 - R_L)/R_L}}{Z_0} \tag{4.37}$$

Example 4.9: A MOSFET's input can be described as a 5.0Ω resistance in series with a gate capacitance of 1.06 pF. Design a matching network for this MOSFET at 1000 MHz that will match it to 50Ω (ch4-12.txt).

Figure 4.40 A matching network is needed to translate a load impedance Z_L to another impedance Z_{in}.

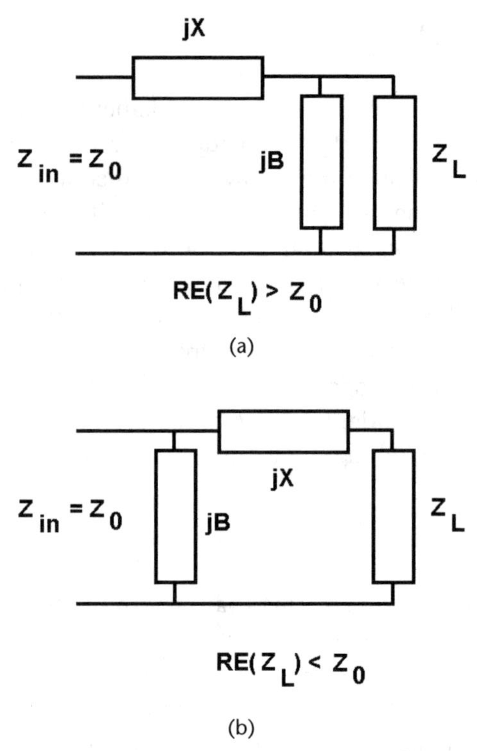

Figure 4.41 Narrowband lumped reactive matching circuit topologies for (a) RE(Z_L) > Z_0 and (b) RE(Z_L) < Z_0.

The MOSFET's input impedance is 5.0 – j150.0Ω at 1000 MHz. Because $R_L <$ 50, the second set of design equations is used. Applying these relationships to the problem's specifications yields the following set of reactances:

$$X = \pm\sqrt{(Z_0 - R_L)R_L} - X_L = \pm\sqrt{(50-5)5} - (-150) = \pm 15 + 150 \Rightarrow j165, j135\Omega$$

$$B = \pm\frac{\sqrt{(Z_0 - R_L)/R_L}}{Z_0} = \pm\frac{\sqrt{(50-5)/5}}{50} = \pm\frac{3}{50} \Rightarrow +j0.06, -0.06\Omega$$

and, solving for the element values,

$$X = j165, j135\Omega \Rightarrow 26.27 \text{ nH}, 21.5 \text{ nH}$$
$$B = +j0.06, -j0.06\Omega \Rightarrow 9.5\text{pF}, 2.65 \text{ nH}$$

The X and B sets are coupled. Using the first set of X and B, simulations show that at zero reactance at the input (1000 MHz), the resistance is 50Ω, verifying the matched condition (Figure 4.42).

4.4.2 Inductive Matching: Source Degeneration

An alternative to the use of classic LC matching for on-chip MOSFETs is to insert inductors in the gate and/or source leads. The MOSFET's C_{GS} is exploited in this

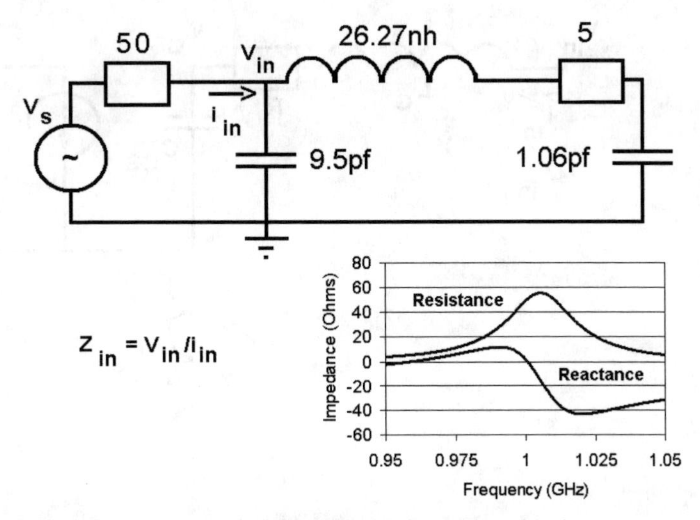

Figure 4.42 Matching network for Example 4.9 with frequency plot showing zero reactance and a match to 50Ω at the design frequency of 1000 MHz.

matching scheme. Using so-called *source degeneration*, these inductors serve two purposes; the first is that the inductors resonate with C_{gs} to remove the inductive component of the input impedance, and the second is that the inductive reactance can be used to move the resistive component to a value more suitable for matching to purely resistive signal sources. A similar technique using only a source inductor can be found in common gate MOSFET circuit topologies. The conditions for source degeneration matching can be seen by looking at the RF equivalent circuit for the common source MOSFET (Figure 4.43).

The input voltage V_{in} is the sum of the voltage drops across L_G, R_G, and C_{GS}, with the addition of the drop across L_s; the current through L_s is enhanced by the transconductance of the circuit:

$$i_{Ls} = i_{\text{in}} + g_m v_{gs} = i_{\text{in}}(1 + g_m / j\omega C_{GS}) \tag{4.38}$$

The ratio of V_{in} to i_{in} gives the input impedance:

$$Z_{\text{in}} = \frac{V_{\text{in}}}{i_{\text{in}}} = \left(R_G + \frac{g_m L_S}{C_{GS}}\right) + j\omega(L_G + L_S) - \frac{j}{\omega C_{GS}} \tag{4.39}$$

The gate resistance R_G is usually a combination of the MOSFET gate resistance and the parasitic resistance of the gate inductor. The two inductors ($L_G + L_S$) are used to resonate C_{GS} and eliminate the reactive component of Z_{in}, leaving only a real component to match to R_s:

$$\omega_0^2 = \frac{1}{(L_G + L_S)C_{GS}} \tag{4.40}$$

If R_G is smaller than the RF source impedance R_S, then the source inductor is specified as $L_S = R_S / \omega_T$ and the gate inductor becomes

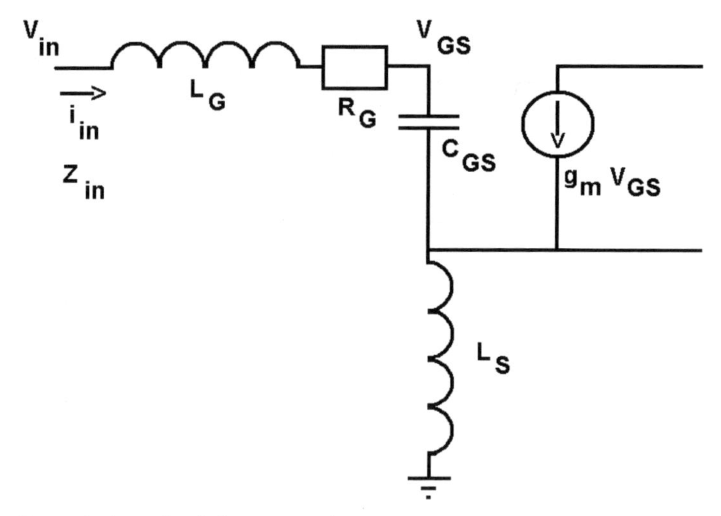

Figure 4.43 RF equivalent circuit for source degeneration matching in a common source MOSFET (the output resistance r_{ds} is assumed to be infinite).

$$L_G = \frac{1}{\omega_0^2 C_{GS}} - \frac{C_{GS} R_S}{g_m} = \frac{1}{\omega_0^2 C_{GS}} - \frac{R_S}{\omega_T} \tag{4.41}$$

Example 4.10: Design a matching network using source degeneration that transforms the input impedance of a common source connected MOSFET of dimensions 500 by 0.5 μm to a purely resistive input impedance at 1000 MHz. Assume ideal reactive elements with $R_g = 0$ and a dc bias of 1.0 mA (ch4-13.txt).

An estimate of the C_{GS} and g_m are necessary to determine the necessary matching components. C_{GS} can be estimated from knowledge of the fabrication parameters and MOSFET size:

$$C_{GS} = C_{ox} WL = \frac{\varepsilon_{ox}}{t_{ox}} WL = \frac{3.9 \cdot 8.85 \cdot 10^{-12}}{135 \cdot 10^{-10}} (500)(0.5) \cdot 10^{-12} = 0.64 \text{ pF}$$

and

$$g_m = \sqrt{2\text{KP} \cdot I_{DS} \cdot W/L} = \sqrt{2 \cdot 56.3 \cdot 10^{-6} \cdot 10^{-3} \cdot (500/0.5)} = 10.6 \text{ mS}$$

To resonate with C_{GS}, the two inductors must sum to

$$(L_g + L_s) = \frac{1}{\omega_0^2 C_{gs}} = 40.9 \text{ nH}$$

The gate and source inductors L_g and L_s can be determined:

$$L_G = \frac{1}{\omega_0^2 C_{GS}} - \frac{C_{GS} R_S}{g_m} = 40.9 \text{ nH} - \frac{(0.62\text{pF})(50)}{0.0106} = 37.9 \text{ nH} \Rightarrow L_s = 3 \text{ nH}$$

The first pass of the design (Figure 4.44) shows the reactive resonance at 1,000 MHz and a real part of approximately 76Ω at that frequency, yielding an input VSWR of 1.5.

In this discussion, the parasitics associated with the lumped elements were neglected (Figure 4.45). On-chip inductors and capacitors exhibit significant parasitics that must be included and optimized if successful matching is to be accomplished (Chapter 3). In some cases, the shunt capacitance of the inductor, for example, can be made part of the matching network itself, eliminating the need for a separate component. In all cases, an iterative process is required to optimize the lumped element layout for optimal matching, all requiring accurate models for the passive elements [8].

A number of techniques have been suggested for this optimization (allstot, for example, using simulated annealing), but all have the general optimization flowchart shown in Figure 4.46 [9]. There are a number of trade-offs involved in the optimization. For example, a high-value inductor can be obtained with many turns of narrow metal; however, this increases the series resistance of the inductor and hence lower its Q. Inductors with wider lines exhibit better Q but at the expense of increased silicon real estate and increased capacitance.

A starting point is necessary in the optimization; a good initialization point is the ideal reactive matching network. Parameters for creating the inductor are input to the model and the related parasitics calculated. CAD is used to determine if a suitable match is obtained; if so, then the process is concluded. If not, the layout is adjusted, the parasitics recomputed, and the match criteria tested again. This process continues until a suitable set of layout parameters for the reactive elements and the desired impedance is obtained.

4.4.3 Example of LC Matching

A 1.2-μm MOSFET was fabricated using a gate width of 2,700 μm (Figure 4.47). Using the flowchart shown in Figure 4.46, parasitics for the inductors and capacitors were computed using standard inductor and capacitor models [10] and were provided as input to a CAD simulator. The resulting CAD layout (including passive element parasitics) is shown in Figure 4.48. Note that C2 has been eliminated in the final layout because the parasitic capacitance of L2 was sufficient.

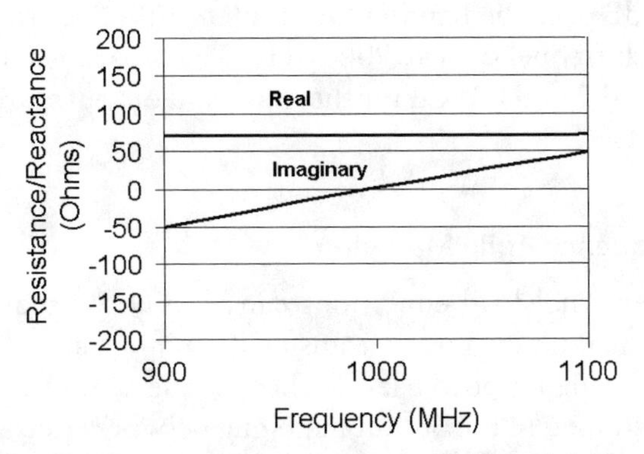

Figure 4.44 Real and imaginary parts (first pass of Example 4.10) of the input impedance of a matched MOSFET using source degeneration.

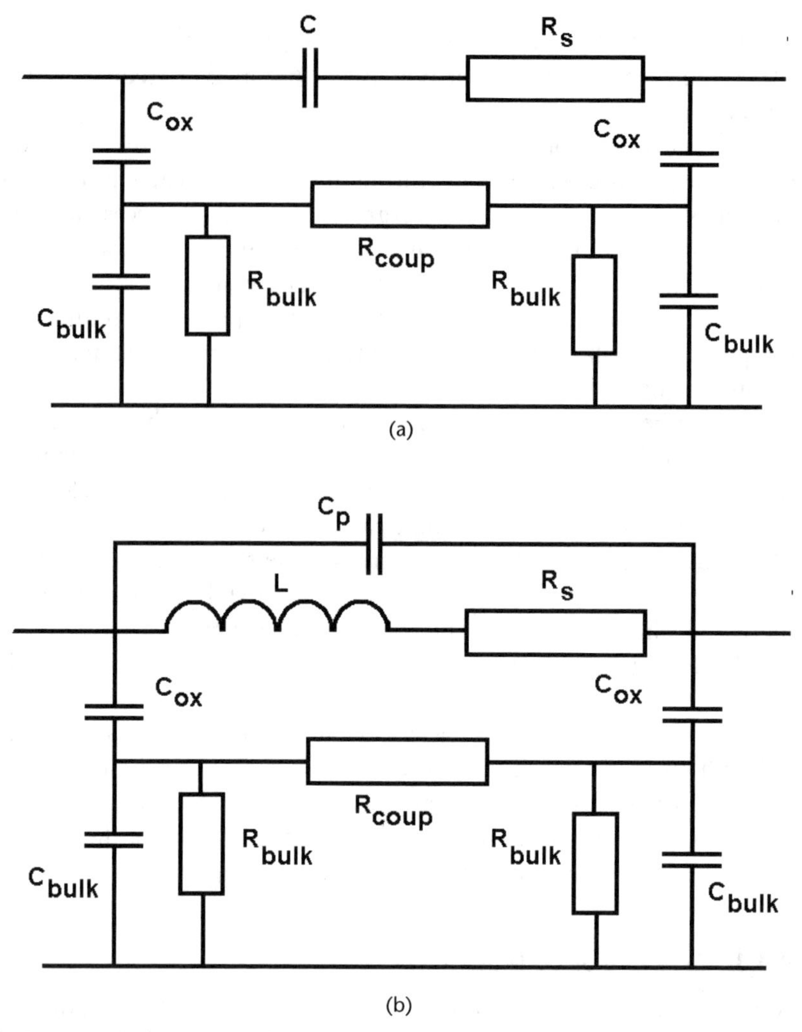

Figure 4.45 RF equivalent circuits for (a) capacitors and (b) inductors, including parasitic elements in the oxide and substrate.

Input and output *S*-parameter measurements showed a return loss of between 16 and 18 dB over the bandwidth of interest (900 ± 100 MHz) (Figure 4.49). The target match frequency was 900 MHz. The *S*-parameter data indicated the initial values of L1, L2 and C1, C2 for the input and output matching structures, as shown in Table 4.1.

4.4.4 Frequency Agile Matching

Changing external load conditions can modify the required matching conditions needed for maximum power transfer or minimization of reflections. For power amplifiers, changing power levels changes the output impedance of the amplifier, requiring retuning of the output matching network. Reconfigurable impedance tuners significantly increase the bandwidth of a matching network by switching in

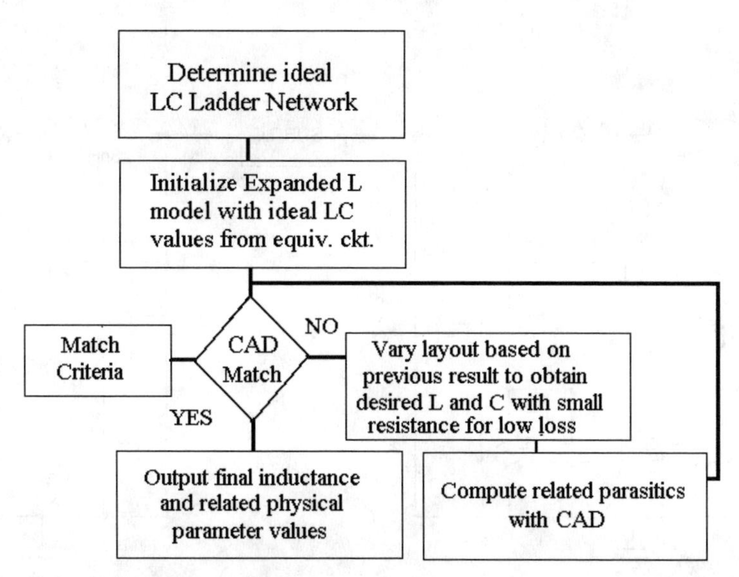

Figure 4.46 Flowchart describing design flow in designing matching networks in the presence of on-chip parasitics in the reactive elements. (*From:* [9]. © 2004 IEEE).

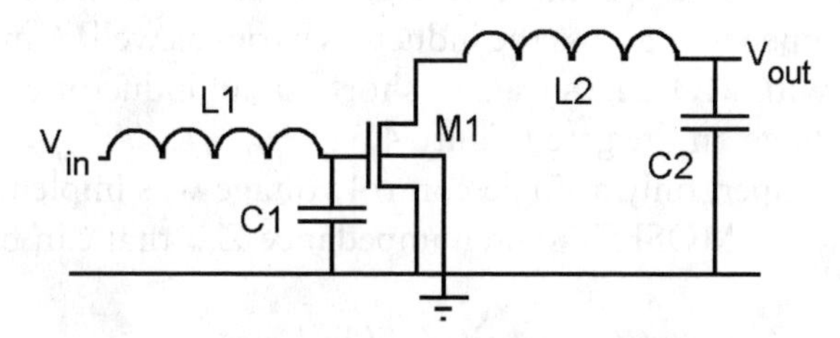

Figure 4.47 Ideal matching network for MOSFET design example.

appropriate combinations of capacitors (and to a lesser extent, inductors) [11–13] or transmission line tuning stubs [14]. Continuous variation of capacitance or inductance can be achieved with a voltage variable capacitor (VVC, or varactor) or voltage variable inductances (VVI) [15]. An important issue for these variable reactive elements is their losses and how they affect the circuit quality factor Q. Stub tuning is more appropriate for high frequency (above approximately 5.0 GHz) due to the physical size of the transmission lines. Coplanar waveguide structures are used in this application [14].

4.4.4.1 FET Switches

MOSFET switches have been used with stacked inductor technologies [15] to implement a voltage variable inductor that can be changed in discrete inductance steps. In Figure 4.50, the total inductance ranges between L and $3L$, depending on the state of the control voltages V_A and V_B.

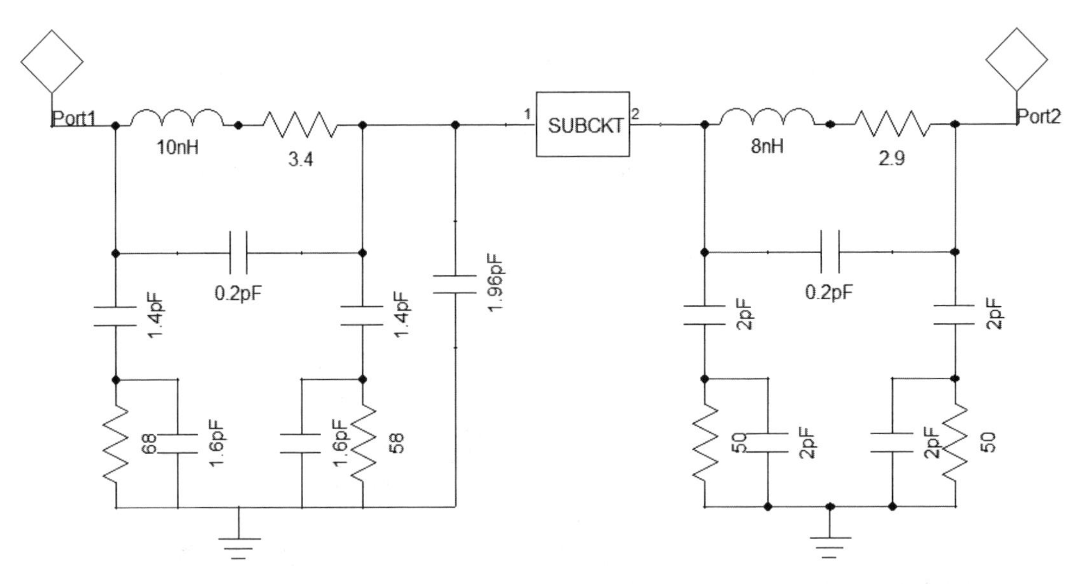

Figure 4.48 LC matching network for simulation, including parasitic effects of the reactive elements.

Alternatively, the control voltage on the MOSFET can be varied so that the shunting impedance across the inductors varies as well. A multilayer set of inductors is created with MOSFETs used to short out an inductor element, depending on the control voltage on the gate (Figure 4.51).

In this paper, only a single control voltage was implemented. The control voltage produces a MOSFET switch impedance Z_{SW} that can be written as

$$Z = \frac{j3\omega Z_{SW}(V)L - \omega^2 L^2}{Z_{SW}(V) + j\omega L} \tag{4.42}$$

When the switch is in its open circuit state, all three inductors are active. As the control voltage goes above V_T, Z_{SW} decreases and, if the switch is ideal, a single inductor impedance is obtained. Varying the control voltage over the range from V_T to the power rail can change the inductance smoothly over a wide range. Figure 4.52 shows a 3:1 variation in inductance over a wide range of voltage. The MOSFET aspect ratio W/L needs to be large enough to provide a low impedance path around the inductor while at the same time keeping the drain/source-to-substrate capacitance small enough to minimize their shunting effect to ground of the RF signal. Equation (4.43) shows the relationship between the two competing elements of the RF MOSFET switch.

$$R_{ON} = \frac{1}{\text{KP}(V_{ctl} - V_T)(W/L)};\; C_{db(sb)} = (WL)\cdot CJ + 2(W + 1.5L)\cdot CJSW \tag{4.43}$$

Q reductions on the order of one-half have been observed with this voltage variable inductor [15]. Simulations should be performed using this technique to determine if the resistive loss due to the reduced Q is prohibitively high.

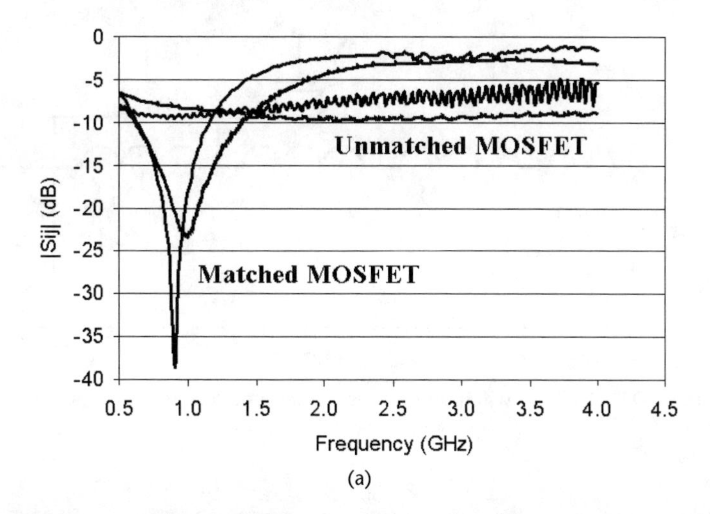

(a)

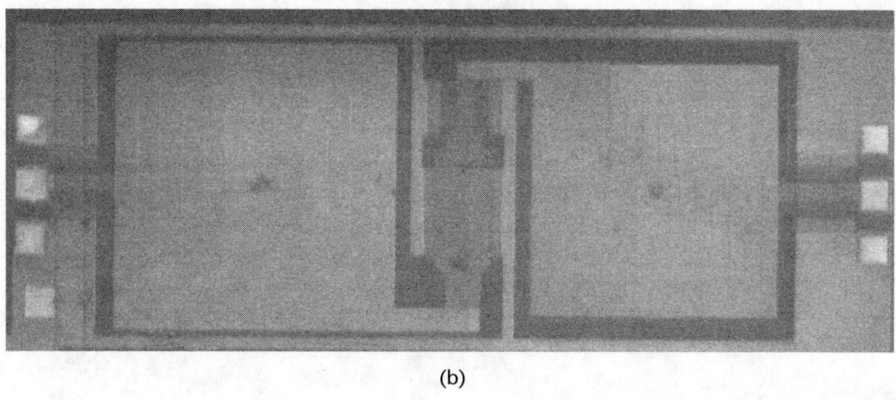

(b)

Figure 4.49 (a) Measured S_{11} and S_{22} for a 1.2-μm MOSFET. (*From:* [9]. © 2004 IEEE.). (b) SEM view of the LC matched MOSFET amplifier (scalable CMOS technology). (Courtesy of The MOSIS Service.)

Table 4.1 Table of Ideal Matching Elements for MOSFET Shown in Figure 4.47

L1	C1	L2	C2
8.65 nH	0.33 pF	7.75 nH	1.99 pF

MOSFETs can be used to switch in banks of capacitors, effectively placing all the elements in parallel (Figure 4.53). For low-loss ground paths, the MOSFETs must exhibit a low impedance in their on state, requiring large aspect ratio devices with a corresponding high parasitic capacitance C_{par}, a major drawback for this technique.

4.4.4.2 MEMS Switches

An alternative to using MOSFET switches with capacitor banks is to instead use MEMS switches. RF MEMS–switched loaded line matching networks have been

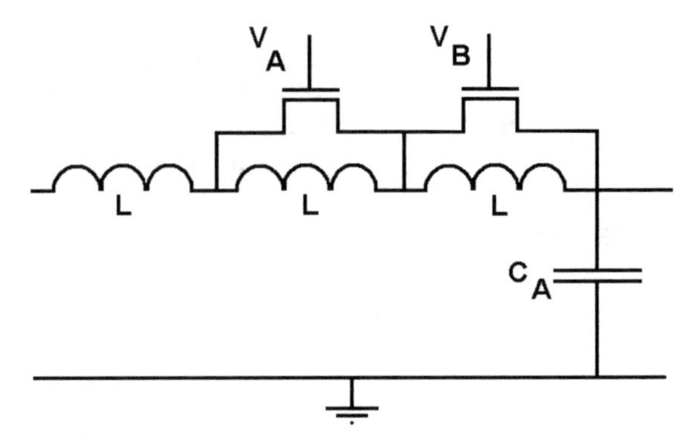

Figure 4.50 MOSFET switching of lumped element components. (*After:* [15].)

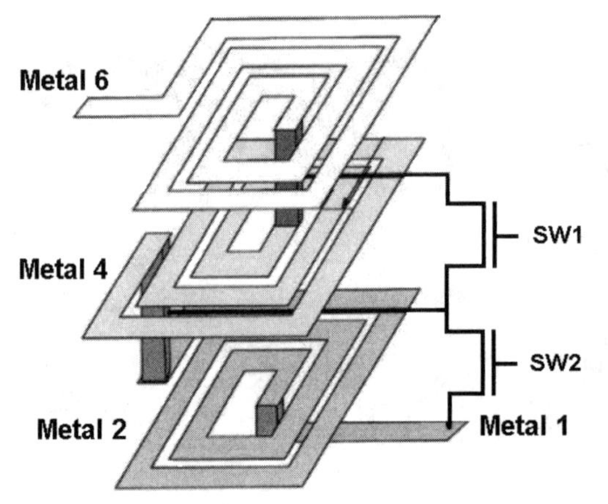

Figure 4.51 Variable inductance using MOSFET switches and on-chip planar inductors. (*After:* [15].)

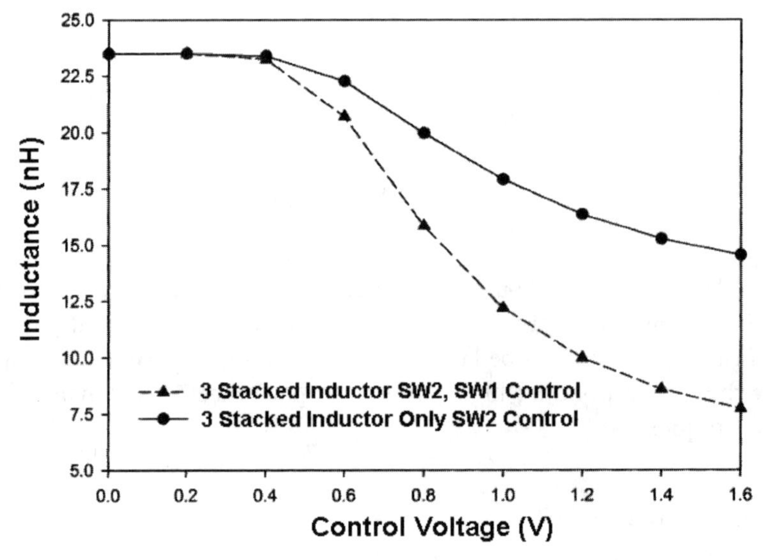

Figure 4.52 Voltage variable inductance is possible using MOSFETs as variable resistance. (*After:* [15].)

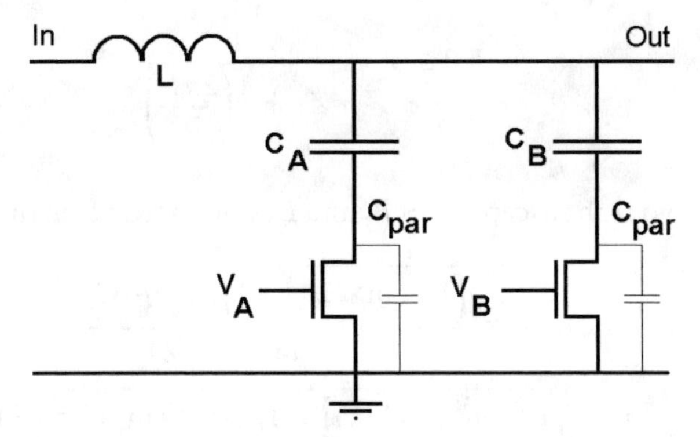

Figure 4.53 MOSFET switches can be used to switch in capacitor banks for filter or tank circuit tuning.

shown to provide wide matching range for amplifier matching at frequencies as high as 60 GHz in CMOS technology. The capacitance switching is accomplished by activating a MEMS switch in series with a fixed capacitance (Figure 4.54) [11–13]. The capacitance of the MEMS switch can vary by as much as 30:1 between the up and down states [11–13], so C_{Fix} is selected to be between the two ranges. The resulting Q of the switched matching network capacitors is approximately

$$Q = \frac{1}{2\pi f C(R_{MEM} + R_{Fix})} \tag{4.44}$$

where the resistances are the parasitic resistances of the relevant capacitance structure.

The switched MEMS capacitor technique has also been used as part of a variable double stub tuner for amplifier matching [14]. The "stubs" in this tuner are replaced by MEMS-switched capacitor banks to provide the variable tuning. The double stub tuning network is useful in RFICs because the transmission line can be of a length *L* consistent with real estate limitations. Matching can then be obtained through the stub design.

In Figure 4.54, the load impedance Z_L is assumed to be a shunt connected circuit: $Y_L = 1/Z_L = G_L + jB_L$. For a given line length *L*, the real part of the admittance G_L can be matched over the range [6, 14]

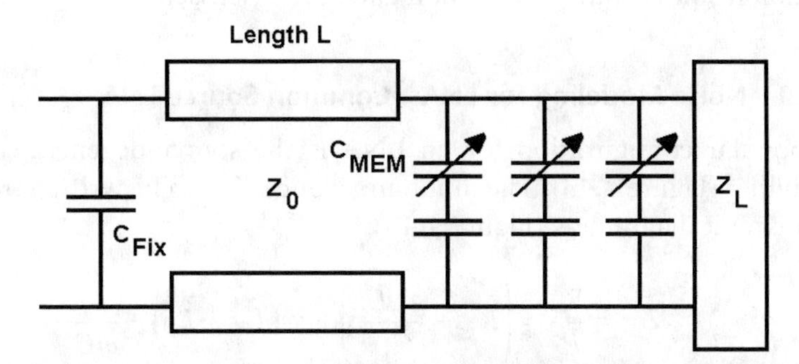

Figure 4.54 MEMs switches can be used to switch in capacitor banks for frequency agile matching.

$$0 \leq G_L \leq \frac{1}{Z_0 \sin^2\left(\frac{2\pi L}{\lambda}\right)} \tag{4.45}$$

with the two stub susceptances B_1 and B_2 calculated using the following equations:

$$\begin{aligned} B_1 &= -B_L \pm \frac{Y_0 + \sqrt{\left[1+\tan^2(2\pi L/\lambda)\right]G_L Y_0 - G_L^2 \tan^2(2\pi L/\lambda)}}{\tan(2\pi L/\lambda)} \\ B_2 &= \frac{\pm Y_0 \sqrt{\left[1+\tan^2(2\pi L/\lambda)\right]G_L Y_0 - G_L^2 \tan^2(2\pi L/\lambda)} + G_L Y_0}{G_L \tan(2\pi L/\lambda)} \end{aligned} \tag{4.46}$$

The required capacitance can then be calculated using

$$C_{Match} = \frac{B}{2\pi f} \tag{4.47}$$

4.5 Low-Noise Amplifiers

Low-noise amplifiers (LNAs) are a crucial system in an RFIC receiver since the LNA is the first stage in the receiver chain and therefore helps set the entire system noise figure. As discussed in Chapter 2, the two primary sources of MOSFET noise are the drain channel noise $\bar{i}_{dN}^2$ and the gate resistance induced noise V_{gN}^2. Noise analyses of common LNAs often assume that only the input-referred drain channel noise and the gate noise need to be considered [16–19]:

$$\bar{i}_{dN}^2 = 4kT\gamma g_{d0}B;\ \bar{i}_{dN-in}^2 = 4kT\gamma g_{d0}B\left(\frac{f}{f_T}\right)^2\ \ \overline{V}_{gN}^2 = 4kTR_G B \cong 4kT\left(\frac{1}{kN}\frac{W}{L}R_S\right)B \tag{4.48}$$

Another source of noise in real MOSFETs is the inductors in the gate and source leads used for input matching (for example, using source degeneration). The inductors have relatively low Q, and the corresponding resistive losses will contribute to the overall LNA noise performance. Since these are resistive losses, their noise voltage contributions are of form $4kTRB$. Two common LNA circuit topologies, the common source and the common gate, are discussed.

4.5.1 Noise Modeling for LNAs: Common Source LNA

A popular configuration for an LNA is the source-degenerated common source amplifier (Figure 4.43). The input impedance Z_{in} can be written from the RF equivalent circuit diagram as (Figure 4.43):

$$Z_{\text{in}} = \frac{V_{\text{in}}}{i_{\text{in}}} = \left(R_g + \frac{g_m L_s}{C_{gs}}\right) + j\omega\left(L_g + L_s\right) - \frac{j}{\omega C_{gs}} \tag{4.49}$$

The input matching configuration and conditions for matching where L_g and L_s resonate with C_{gs} were discussed previously.

Transconductance g_m, C_{gs}, and R_G are functions of the MOSFET W and L. This dependence on gate width W has important ramifications when looking at the noise characteristics of the common source (CS) LNA (Figure 4.55). The effective transconductance of the circuit at resonance can be written as the ratio of the output current from the MOSFET with respect to the source voltage V_S:

$$G_m = \frac{i_o}{V_s} = \frac{f_T}{f}\frac{1}{R_s}\frac{1}{\left(1+\dfrac{\omega_T L_s + R_g}{R_s}\right)} \tag{4.50}$$

The contributions from the various noise mechanisms can be seen by looking at the noise equivalent circuit, which includes the noise impedances and contributions from the source, the gate inductor, R_G, and the MOSFET drain current noise.

The noise figure F of the LNA can be computed from the definition

$$F = \frac{SNR_{\text{in}}}{SNR_{\text{out}}} = \frac{P_{sig}/P_{N\text{-in}}}{P_{sig}/P_{N-\text{out}}} = \frac{P_{N-\text{out}}}{P_{N\text{-in}}} = \frac{\bar{i}^2_{N-\text{out}}}{\bar{i}^2_{N\text{-in}}} \tag{4.51}$$

where $\bar{i}^2_{N-\text{out}}$ and $\bar{i}^2_{N\text{-in}}$ are the total input and output noise currents. The relationship between the output noise current and the source noise voltage in the absence of any other noise contributions is given by the effective transconductance G_m:

$$i_{N-in} = G_m V_{s-N} \tag{4.52}$$

The mean square output noise current can be written in terms of the effective transconductance G_M and the source resistance R_S:

$$\bar{i}^2_{N-in} = G_m^2 \overline{V}^2_{sN} = G_m^2 (4kTR_s) = \frac{\omega_T^2}{\omega^2}\frac{1}{R_s}\frac{4kT}{\left[1+\dfrac{\omega_T L_s + R_g}{R_s}\right]^2} \tag{4.53}$$

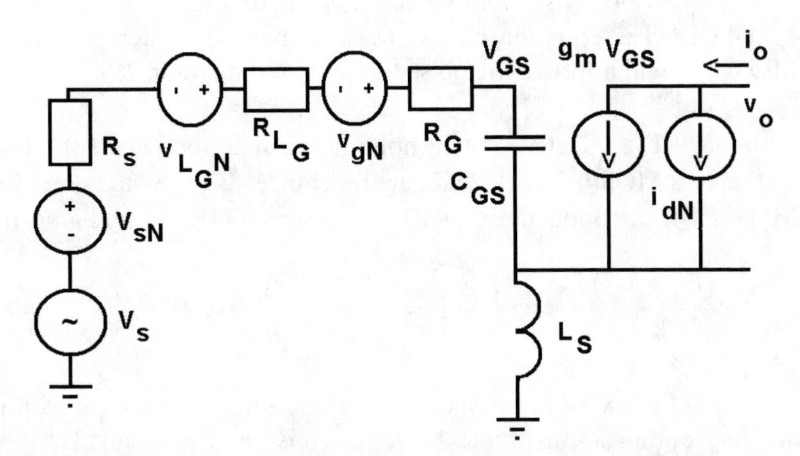

Figure 4.55 Input and output noise contributions for the source-degenerated MOSFET LNA.

The contributions from the other noise sources in the circuit are made under the assumption that all the noise sources are uncorrelated; this allows simple addition of their noise contributions. The total input noise contributions come from two sources: the input noise source V_{sN} and the resistive noise contribution from R_g. This value of R_g is the total resistance in the gate lead; this can be both the inherent MOSFET gate resistance as well as any parasitic resistance in the gate inductor. The V_{sN} contribution was derived previously in (4.53); the gate resistance noise source V_{gN} appears as v_{gs} and so also appears as a noise current at the output ($i_{\text{out}-gN}$):

$$v_{gs} = \frac{1}{j\omega R_s G_{gs}} \frac{V_{gN}}{\left[1+\frac{\omega_T L_s + R_g}{R_s}\right]} = \frac{1}{j\omega R_s C_{gs}} \frac{\sqrt{4kTR_g}}{\left[1+\frac{\omega_T L_s + R_g}{R_s}\right]}$$
$$i_{\text{out}-gN} = g_m v_{gs} = g_m \frac{1}{j\omega R_s C_{gs}} \frac{\sqrt{4kTR_g}}{\left[1+\frac{\omega_T L_s + R_g}{R_s}\right]} \tag{4.54}$$

The total output noise current is the sum of that due to the source noise, the gate resistance R_g noise, and the drain channel noise:

$$\bar{i}^2_{N-\text{out}} = \bar{i}^2_{N-\text{in}} + \bar{i}^2_{dN} + \bar{i}^2_{\text{out}-gN} = \bar{i}^2_{\text{out}-sN} + 4k\gamma T g_{d0} + \bar{i}^2_{\text{out}-gN}$$
$$\bar{i}^2_{N-\text{out}} = \bar{i}^2_{N-\text{in}} + 4k\frac{\gamma}{\alpha} T g_m + \frac{g_m^2 4kTR_g}{\omega^2 R_s^2 C_{gs}^2} \frac{1}{\left[1+\frac{\omega_T L_s + R_g}{R_s}\right]^2} \tag{4.55}$$

where $g_m = \alpha g_{d0}$. Substituting (4.55) into the noise figure expression (4.51), the noise figure F of the LNA can be written as

$$F = \frac{\bar{i}^2_{N-\text{out}}}{\bar{i}^2_{N-in}} = 1 + \frac{R_g}{R_s} + \frac{\gamma}{\alpha} g_m R_s \left(\frac{f}{f_T}\right)^2 \left[1+\frac{\omega_T L_s + R_g}{R_s}\right]^2 \tag{4.56}$$

Equation (4.56) shows that the noise figure improves with the use of high-f_T MOSFETs. The equation also shows that high-Q inductors and multigate finger MOSFETs will improve the noise figure by reduction in R_g.

Example 4.11: Determine the noise figure in decibels in a 1.0-GHz LNA in a 50Ω system of a MOSFET with 25Ω gate resistance, 8Ω inductor series resistance, and a MOSFET transconductance of 10 mS. Assume $\alpha = \gamma = 1$ and an f_T of 25 GHz.

$$F = 1 + \frac{(R_g + R_{lg})}{R_s} + R_s g_m \left(\frac{f}{f_T}\right)^2 = 1 + \frac{(25+8)}{50} + (0.01)(50)\left(\frac{1}{25}\right)^2 = 1.66 \rightarrow 2.2 \text{ dB}$$

The influence of noise figure with transconductance g_m is a bit more complex, and the optimization of the LNA for noise and gain involves trade-offs; what improves the gain or frequency response may degrade the noise performance, or vice versa. This trade-off between optimum noise performance and gain is widely known in discrete device technologies (GaAs MESFETs, for example), where the noise properties of the transistor are used in the LNA design process. The optimum noise figure follows the equation [6]

$$F = F_{\min} + \frac{4G_N}{Z_0}\left|Z_S - Z_{opt}\right|^2 \tag{4.57}$$

where G_N is the equivalent noise conductance and Z_S is the source impedance presented to the MOSFET. In general, the optimum source impedance for best noise figure, Z_{opt}, is *not* the optimum source impedance for maximum gain. Since the input is not well matched under conditions of lowest noise, input reflection coefficients can be high. Introducing a slightly nonoptimal value of Z_S can improve the gain, but the overall noise figure will be greater than $F_{\min}$. Higher values of G_N result in higher noise figures when nonoptimal noise matching occurs.

If the term in brackets in (4.56) is assumed to be unity, the minimum noise figure can be found by taking the derivative of this equation and setting it equal to zero. By substituting this resulting value of into the equation $F - F_{\min}$, (4.57) can be written in terms of the LNA circuit and MOSFET parameters:

$$F = F_{\min} + \frac{g_m}{R_S}\left(\frac{f}{f_T}\right)^2 \left|R_S - R_m\right|^2 \tag{4.58}$$

Equation (4.57) shows that the term G_N can be minimized by using high-f_T MOSFETs. A similar analysis looking at the impact of multiple gate finger MOSFETs shows that the minimum noise figure $F_{\min}$ is independent of N, the number of gate fingers.

The optimum device parameters for integrated LNAs are not well known because the devices themselves have not even been fabricated yet. Therefore, an in-depth look at (4.56) is instructive to see the trade-offs in LNA gain (through dc biasing and MOSFET sizing) and noise figure. For a given f_T process, smaller device aspect ratios W/L and bias currents I_{DS} improve the noise figure through lower g_m, indicating that high power dissipation is not necessary for good noise performance [20]. However, as g_m is lowered to obtain better noise performance, the gate inductance L_g has to increase to maintain matching requirements. The increase in L_g has a corresponding increase in R_g, with a corresponding increase in noise from this source. Increasing g_m reduces the inductance needed for matching (and hence the losses) but increases the noise contribution from the drain channel. The optimal LNA design requires investigation of the effect of gate width, dc bias, and gain. Models of the circuit elements (both passive and active) and bias conditions are combined to determine the best W for noise performance. Especially important is modeling R_g/R_s; as g_m changes, L_g and hence R_g must change to maintain the matching condition.

Writing R_g in terms of the inductor Q_g,

$$Q_g = \frac{\omega L_g}{R_g};\ \frac{R_g}{R_s} = \frac{1}{Q_g}\left(\frac{1}{\omega C_{gs}} - \frac{f}{f_T}\right) = \frac{1}{Q_g}\left(\frac{1}{\omega C_{ox} WL} - \frac{f}{f_T}\right) \tag{4.59}$$

yields an expression for the noise figure F that shows its dependence on MOSFET parameters:

$$F = 1 + \frac{1}{Q_g}\left(\frac{1}{\omega R_S C_{ox} WL} - \frac{f}{f_T}\right) + \frac{\gamma}{\alpha}\sqrt{2\text{KP} \cdot I_{DS} \cdot W/LR_s}\left(\frac{f}{f_T}\right)^2 \left[1 + \frac{1}{Q_g}\left(\frac{1}{\omega R_S C_{ox} WL} - \frac{f}{f_T}\right) + \frac{\omega_T L_s}{R_s}\right]^2 \tag{4.60}$$

Cascode circuits are also employed in LNAs to improve the gain characteristics of the stage, to improve isolation, and to reduce the Miller effect capacitance [16].

4.5.2 Noise Modeling for LNAs: Common Gate LNA

Another MOSFET configuration that provides good low-noise characteristics at high frequencies is the common gate (CG) configuration. A simplified circuit for the CG LNA is shown in Figure 4.56. This circuit is not used in exactly this form; the gate is at a lower potential than the source and therefore below the threshold voltage. Several biasing configurations are used to ensure proper gate bias while creating the effective grounded gate; one such example is shown in Figure 4.57 [21]. Here, M1 is the CG LNA circuit, biased using a current mirror structure with M2. Capacitor C is used to ensure that the gate of M1 is RF grounded.

The basic CG LNA still uses the source degeneration inductor L_s as used in the CS LNA, but the RF input is placed at this node. This inductor provides a parallel resonance to the gate capacitance and matches the input of the LNA. The relatively low Q of on-chip inductors is a benefit for the CG LNA, making this structure more tolerant of process variations [16]. The RF equivalent circuit for the CG LNA is shown in Figure 4.58. The input impedance looking into the source can be written (neglecting the gate resistance R_G and drain capacitance C_{DS} for the moment) as

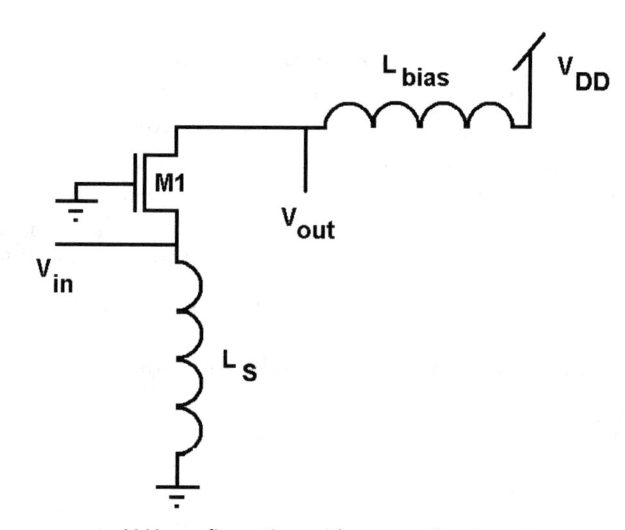

Figure 4.56 Common gate LNA configuration with source degeneration.

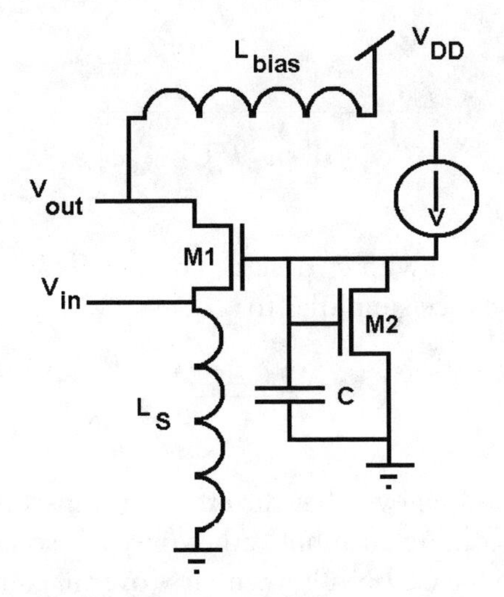

Figure 4.57 Biasing scheme for the common gate LNA.

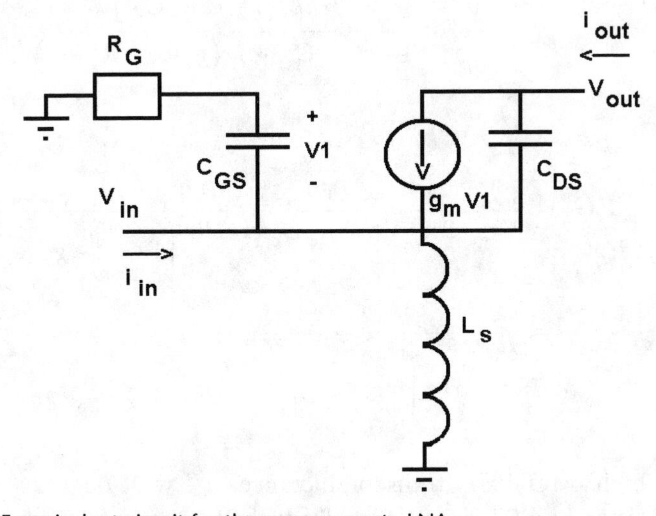

Figure 4.58 RF equivalent circuit for the common gate LNA.

$$Z_{\text{in}} = \frac{V_{\text{in}}}{i_{\text{in}}} = \frac{j\omega L_s}{1 + j\omega g_m L_s - \omega^2 L_s C_{gs}} \tag{4.61}$$

If L_s is chosen to resonate C_{gs} (neglecting C_{ds}), the input impedance becomes simply $1/g_m$. For input matching, the requirement for the design of the MOSFET is

$$Z_{in} = R_s = g_m^{-1} = \left(2\text{KP} \cdot I_{DS} \cdot W/L\right)^{-1/2} \tag{4.62}$$

The effective transconductance for the CG LNA can be written as

$$G_m = \frac{i_{\text{out}}}{V_s} = \frac{\dfrac{g_m}{R_s} j\omega L_s}{1 - \omega^2 L_s C_{gs} + j\omega\left(g_m L_s + \dfrac{L_s}{R_s}\right)} \tag{4.63}$$

At resonance and with g_m designed to match the source impedance R_S, the effective transconductance simplifies to

$$G_m = \frac{g_m}{1 + g_m R_s} = \frac{1}{2R_s} \tag{4.64}$$

Equation (4.64) shows that the effective transconductance is not dependent on the device transconductance but rather only the source impedance when matched. This is similar to the CS LNA but with less overall gain [16]. Including the gate resistance (Figure 4.58) modifies the input impedance and effective transconductance when matched to

$$Z_{in} = \frac{V_{in}}{i_{in}} = \frac{j\omega L_s\left(1 + j\omega R_g C_{gs}\right)}{1 - \omega^2 L_s C_{gs} + j\omega\left(g_m L_s + R_g C_{gs}\right)} \tag{4.65}$$

and

$$\begin{aligned} G_m &= \frac{i_{\text{out}}}{V_s} = \frac{1}{2R_s} \frac{1}{1 + j\omega R_g C_{gs}} = \frac{1}{2R_s} \frac{1}{\left(1 + j\dfrac{\omega}{\omega_T} g_m R_g\right)} \\ \left|G_m\right| &= \frac{1}{2R_s} \frac{1}{\sqrt{1 + \left(\dfrac{\omega}{\omega_T} g_m R_g\right)^2}} \end{aligned} \tag{4.66}$$

A high MOSFET transconductance g_m will degrade the effective transconductance, and G_m degrades as $1/f$ as the frequency increases. The gate resistance affects the linear characteristics of the CG LNA and also contributes to the overall circuit noise. Figure 4.59 shows a noise equivalent circuit for the CG LNA, including the noise source for the gate resistance. The noise figure F for the CG LNA can be derived in a manner similar to that for the CS LNA. The input noise source transferred to the output is a component of the total output noise current $i_{N\text{-out}}$:

$$\bar{i}_{N\text{-in}}^2 = \left|G_m\right|^2 \overline{V}_{sN}^2 = 4kT\left|G_m\right|^2 R_s \tag{4.67}$$

The drain current noise contribution is the standard drain channel noise given in (4.55).

The gate resistance noise transfers to the output through v_{gs} and becomes the contribution i_{gN}^2. Neglecting for the moment the noise contribution from R_g, the noise figure for the CG LNA is defined in the same manner as in the CS case and can be written as

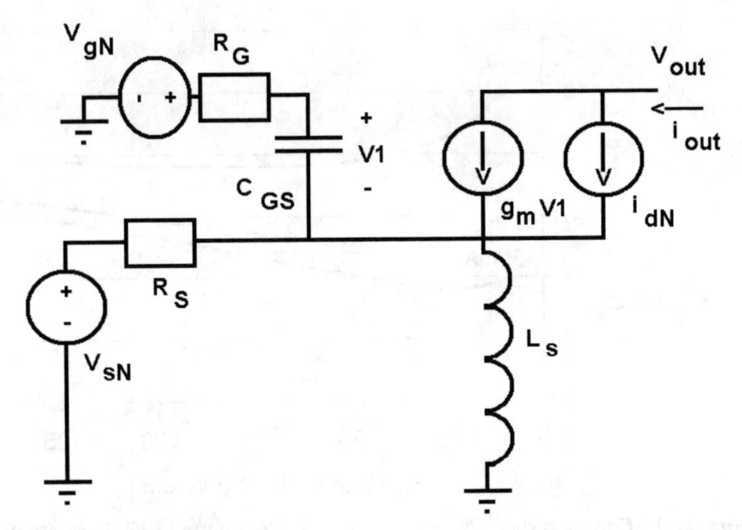

Figure 4.59 Noise equivalent circuit for the CG LNA showing contributions from the source, gate resistance, and drain channel noise.

$$F = 1 + \frac{\bar{i}_{dN}^2}{\bar{i}_{sN}^2} = 1 + \frac{\gamma}{\alpha} \frac{1}{g_m R_s} = 1 + \frac{\gamma}{\alpha} \frac{1}{\sqrt{2\text{KP} \cdot I_{DS} \cdot W/L}} R_s \tag{4.68}$$

The expression for the noise figure shows that large MOSFET transconductance will minimize the noise figure. However, the matching requirement shows that the product $g_m R_s$ is nearly unity, and the noise figure is then relatively independent of MOSFET construction [16]. When the gate resistance is included, however, unity $g_m R_s$ is only an approximation for resistive matching, and MOSFET parameters do influence F, including the construction parameters through R_g. The gate noise source V_{gN} gets transferred to the output as an additional noise current. The third term in (4.69) is the additional term due to the gate resistance. Note that this expression reduces to the previous expression for F if the gate resistance R_g is small:

$$\begin{aligned} F &= 1 + \frac{\bar{i}_{gN}^2 + \bar{i}_{dN}^2}{\bar{i}_{sN}^2} = 1 + \frac{\gamma}{\alpha} \frac{1}{g_m R_s} + \frac{R_g}{R_s} \\ &= 1 + \frac{\gamma}{\alpha} \frac{1}{\sqrt{2\text{KP} \cdot I_{DS} \cdot W/L}} R_s + \frac{1}{R_s} \frac{1}{kN} \frac{W}{L} R_{SH} \end{aligned} \tag{4.69}$$

Equation (4.69) shows that there is an optimal gate width W that will provide minimum noise figure. In addition, higher biases and multigate finger MOSFETs will also improve noise figure (Figure 4.60). The optimum noise figure, however, does not necessarily meet the matching requirement of $1/R_s$, as was noted in the common source case [6].

Alternatively, the minimum noise figure can be found by taking the derivative of F with respect to W and setting this derivative equal to zero:

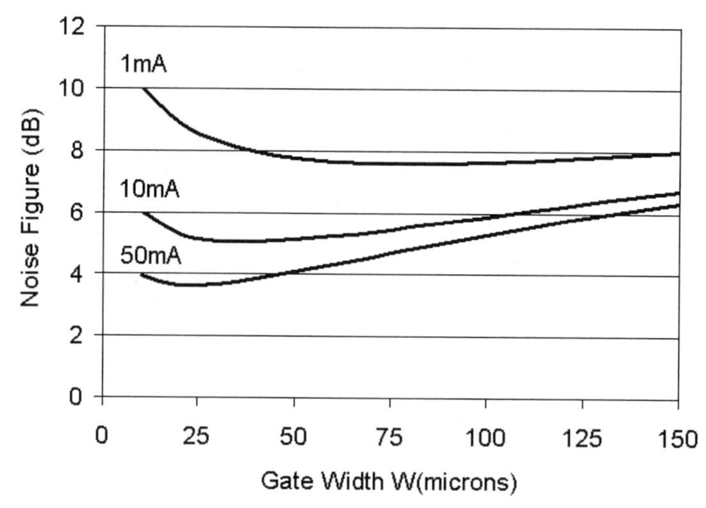

Figure 4.60 Common gate LNA noise figure as a function of MOSFET gate width for 1.0-, 10-, and 50-mA dc bias.

$$W_{opt} = L\left[\frac{1}{2}\frac{\gamma}{\alpha}\frac{R_s kN}{\sqrt{2\text{KP}\cdot I_{DS}}\cdot R_{SH}^{\;2}}\right]^{2/3} \tag{4.70}$$

Modifications to the basic CG circuit have been shown to improve the CG noise figure even further. A feedback resistor across the drain to source of the MOSFET creates a CG LNA with resistive feed-through [Figure 4.61(a)]. The input impedance is more controllable with this external resistor (instead of relying on the MOSFET r_{ds}) [22]. The G-boosted CG LNA [Figure 4.61(b)] uses a low-gain inverting feedback stage between the input and the gate [16]. The G_m boost increases the effective transconductance by the amount $1 + A_v$, with the noise figure contribution from the drain being reduced by the same $1 + A_v$ factor. A simple unity gain stage can provide a 1.9-dB improvement in noise figure [16].

4.5.3 Modeling Example

Determine the optimal gate width W for a MOSFET used in a common source LNA in 0.5-μm technology at 1.0-, 3.0-, and 10.0-mA dc bias currents at 1.0 GHz. Assume a gate inductor Q of 7, $\gamma/a = 1$, $\text{KP} = 56.3 \times 10^{-6}\ A/V^2$, $C_{ox} = 2.6 \times 10^{-3}\ \text{F/m}^2$, and $R_s = 50\Omega$.

The equation for noise figure F in (4.60) shows that the optimal gate width for noise figure depends on bias current. Using these parameters as input, the noise figure F can be calculated as a function of gate width and dc bias current. The results of these computations, Figure 4.62, show that there is a minimum noise figure F for various combinations of dc bias current and gate width.

For 1.0 mA, $W = 500\,\mu$m; for 3.0 mA, $W = 750\,\mu$m; and for 10 mA, $W = 950\,\mu$m. These gate widths correspond to device g_m of 10.6, 22.4, and 46.2 mS, respectively. Note that better noise figures occur for higher amplifier power dissipation. The two inductances L_g and L_s can be determined from application of (4.39) and (4.41). For

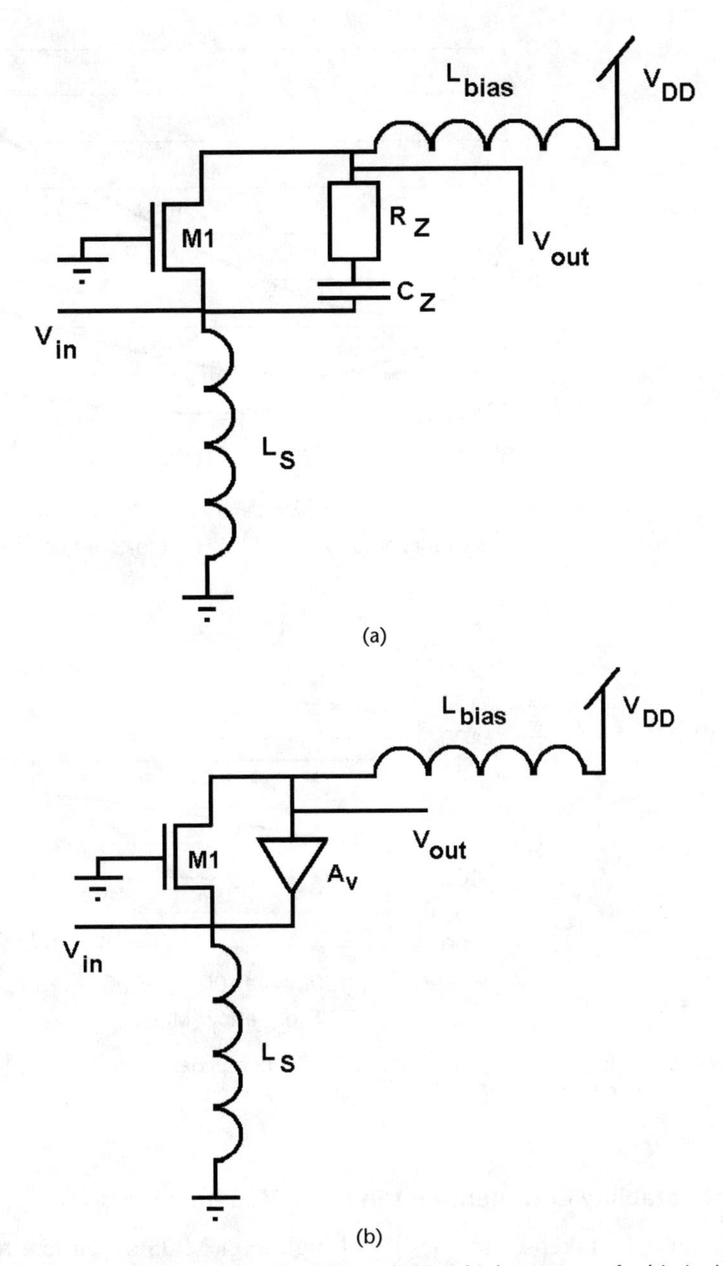

Figure 4.61 CG LNA with (a) resistive feed-through (C_z is high reactance for biasing) and (b) G_m boosting amplifier A_v.

the 1.0-mA case, the inductances L_g and L_s are 35 and 3 nH, respectively. These inductances reduce to 19.3 and 1.3 nH, respectively, for the 10-mA case. CAD simulation for this circuit (Figure 4.63) shows that at 1.1 GHz, the input impedance is approximately 72Ω and is purely real (imaginary part is zero at the design frequency). This amplifier shows 2 dB of gain at the design frequency. The effective transconductance G_m is 8 mS at 1.0 GHz, close to the theoretical value of 10 mS. The frequency can be reduced to the 1.0-GHz design frequency with just a few minutes spent optimizing the circuit.

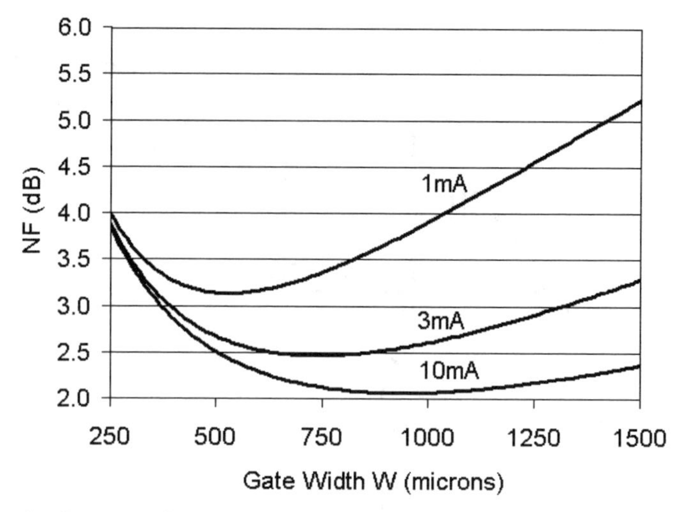

Figure 4.62 Noise figure as a function of gate width W for the common source LNA.

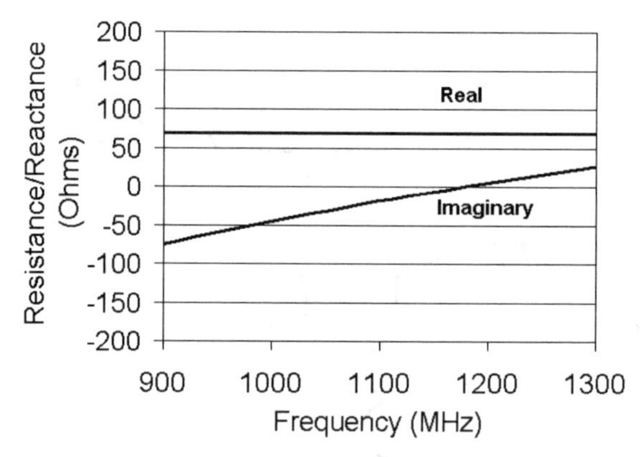

Figure 4.63 Real and imaginary parts of the input impedance for the common source LNA with source degeneration (ch4-14.txt).

4.5.4 Stability Considerations in MOS RFIC Amplifiers

Care must be taken in the design of high-gain MOS amplifiers that the amplifiers remain stable under all input and output loading conditions, such as when the amplifier looks into a filter out of band. A widely used metric for study of the stability characteristics based on S-parameters involves the use of the *stability factor K*, given by the expression

$$K = \frac{1 - |S_{11}|^2 - |S_{22}|^2 + |\Delta|^2}{2|S_{21}S_{12}|}$$
$$\Delta = S_{11}S_{22} - S_{12}S_{21} \tag{4.71}$$

Unconditionally stable amplifiers exhibit a combination of K greater than unity and Δ less than one. From the expression for K, MOSFETs with high S_{21} (which yield high gains) require that the feedback parameter S_{12} be small. Inspection of the *S*- and *Y*-parameter expressions in Appendix C show that the main feedback element governing amplifier instability is the feedback capacitance C_{gd}. Unwanted feedback can be reduced in MOS amplifiers in a number of ways. The most obvious is to reduce C_{gd} by using the minimum gate width *W* transistor needed for a specific gain characteristic. Equation (C.1) also indicates that a large gate resistance R_G at the input (either external or part of the polysilicon gate resistance) will reduce the effects of C_{gd} feedback as the frequency increases. A large shunt resistance at the amplifier output (either externally applied or through g_{ds}) or with resistive feedback provides alternative means for improving amplifier stability. A drawback to the use of these resistive approaches to feedback reduction, however, is that they also produce a significant reduction in amplifier gain.

A number of circuit topologies and modifications have been studied that can improve the stability of MOS amplifiers. The MOS cascode amplifier circuit improves overall stability by reducing the gain at the output of the driver transistor (v_d/v_{in} on the drain of M1 in Figure 4.17), which makes the device more unilateral. Further improvements using the cascode topology involve the use of reactive gate lead terminations on the cascoding MOSFET (M2 in Figure 4.17) [23, 24]. Another circuit type that shows improved stability over the common source amplifier is the common gate configuration as shown in Figure 4.58. Here, the feedback capacitance C_{ds} is usually much smaller than C_{gd} in the common source configuration, improving stability and isolation [20]. Noise figure for the common gate amplifier is somewhat higher than that for the common source amplifier, however.

References

[1] Allen, P., and D. Holbert, *CMOS Analog Circuit Design*, 2nd ed., Oxford, UK: Oxford University Press, 2002.

[2] Manku, T., "Microwave CMOS—Device Physics and Design," *IEEE. J. Solid State Circ.*, Vol. 34, No. 3, March 1999, p. 277.

[3] Zhou, J. J., and D. J. Allstot, "Monolithic Transformers and Their Application in a Differential CMOS RF Low-Noise Amplifier," *IEEE J. Solid State Circ.*, Vol. 33, No. 12, Dec. 1998.

[4] Hajimiri, A., and T. H. Lee, "Design Issues in CMOS Differential LC Oscillators," *IEEE J. Solid State Circ.*, Vol. 34, No. 5, May 1999.

[5] Wilson, G. R., "A Monolithic JFET-npn Operational Amplifier," *IEEE J. Solid State Circ.*, Vol. SC-3, No. 5, Dec. 1968, p. 341.

[6] Pozar, D., *Microwave Engineering*, 2nd ed., New York: John Wiley and Sons, 1998.

[7] Maatthaei, G., L. Young, and E. M. T. Jones, *Microwave Filters, Impedance Matching Networks and Coupling Structures*, Dedham, MA: Artech House, 1980.

[8] Melendy, D., and A. Weisshaar, "A New Scalable Model for Spiral Inductors on Lossy Silicon Substrate," Proc. *2003 IEEE Int. Microwave Symp.*, 2003, pp. 1007.

[9] Caverly, R., et al., "Modeling and Characterization of Integrated Inductors and Transformers for Amplifier Applications in Silicon High Frequency Systems," *Proc. 5th Topical Meeting on Silicon Monolithic Integrated Circuits for RF Systems (SiRF04)*, Sept. 2004.

[10] Mohan, S. S., et al., "Simple Accurate Expressions for Planar Spiral Inductances," *IEEE J. Solid State Circ.*, Vol. 34, No. 10, Oct. 1999.

[11] Vaha-Heikkila, T., and G. M. Rebeiz, "A 20–50 GHz Reconfigurable Matching Network for Power Amplifier Applications," *Proc. 2004 IEEE Int. Microwave Symp.*, 2004, p. 717.

[12] Lu, Y., et al., "A MEMS Reconfigurable Matching Network for a Class AB Amplifier," *IEEE Microwave Wireless Components Lett.*, Vol. 13, No. 10, Oct. 2003.

[13] Brown, E. R., "RF-MEMS Switches for Reconfigurable Integrated Circuits," *IEEE Trans. Microwave Theory Techniques*, Vol. 46, No. 11, Nov. 1998, p. 1868.

[14] Papapolymerou, J., et al., "Reconfigurable Double-Stub Tuners Using MEMS Switches for Intelligent RF Front-Ends," *IEEE Trans. Microwave Theory Techniques*, Vol. 51, No. 1, Jan. 2003.

[15] Park, P., et al., "Variable Inductance Multilayer Inductor with MOSFET Switch Control," *IEEE Electron Device Lett.,* Vol. 25, No. 3, March 2004.

[16] Allstot, D., X. Li, and S. Shekhar, "Design Considerations for CMOS Low-Noise Amplifiers," *Proc. 2004 IEEE Radio Frequency Integrated Circuits Symp.*, 2004, p. 97.

[17] Svelto, F., et al., "Implementation of a CMOS LNA Plus Mixer for GPS Applications with No External Components," *IEEE Trans. VLSI Syst.,* Vol. 9, No. 1, Feb. 2001, p. 100.

[18] Wang, B., J. Hellums, and C. Sodini, "MOSFET Thermal Noise Modeling for Analog Integrated Circuits," *IEEE J. Solid State Circ.*, Vol. 29, No. 7, July 1994, pp. 833–835.

[19] Andreani, P., and H. Sjöland, "Noise Optimization of an Inductively Degenerated CMOS Low Noise Amplifier," *IEEE Trans. Circuits Sys.. II*, Vol. 48, No. 9, Sept. 2001.

[20] Shaeffer, D., and T. H. Lee, "A 1.5V, 1.5 GHz CMOS Low Noise Amplifier," *Proc. 1996 Symp. on VLSI Circuits*, 1996, p. 32.

[21] Razavi, B., "A 60-GHz CMOS Receiver Front-End," *IEEE J. Solid State Circ.*, Vol. 41, No. 1, Jan. 2006, p. 17.

[22] Guan, X., and A. Hajimiri, "A 24 GHz CMOS Front End," *IEEE J. Solid State Circ.*, Vol. 39, No. 2, Feb. 2004, p. 368.

[23] Komijani, A., A. Natarajan, and A. Hajimiri, "A 24-GHz, +14.5-dBm Fully Integrated Power Amplifier in 0.18-μm CMOS," *IEEE J. Solid State Circ.*, Vol. 40, No. 9, 2005, p. 1901.

[24] Hossein, S., M. Lavasani, and S. Kiaei, "A New Method to Stabilize High Frequency High Gain CMOS LNA," *Proc. 2003 ICECS*, 2003, p. 982.

Selected Bibliography

Lee, T. H., *The Design of CMOS Radio-Frequency Integrated Circuits*, 2nd ed., Cambridge, UK: Cambridge University Press, 2004.

Rogers, J., and C. Plett, *Radio Frequency Integrated Circuit Design*, Norwood, MA: Artech House, 2003.

CHAPTER 5

Ancillary CMOS Circuits and Measurements

The material in the previous chapter focused on a single type of RFIC circuit, the RF amplifier. Although amplifiers are extremely important for increasing signal amplitude while at the same time introducing minimal noise, a number of other on-chip functions or applications require a different set of solutions. This chapter addresses a few of these functions with examples of circuits (and accompanying analysis) used to improve device performance. Q-enhancement, stage buffering, and gain control circuits are covered in this chapter. Filters are briefly covered to show how they may be integrated into gain blocks for conservation of silicon real estate. A section on the basics of power harvesting (dc power generation from an externally applied RF signal) is also covered. This circuit type is finding increasing use in self-powered RFICs such as are found in radio frequency identification (RFID) applications.

The design principles described in this and previous chapters are just one part of the overall RFIC development process. Closing the design-fabrication-test loop often requires that measurements be performed on circuit prototypes to ensure that the design and simulation phase has considered important parasitic and other circuit nonidealities. The focus of the final section in this chapter is the influence of on-chip measurement structures on the design RFIC circuit element. Techniques are also covered on how to remove or deembed the desired RFIC circuit measurement parameters from those of the overall measurement data.

5.1 Ancillary CMOS RFIC Circuits

5.1.1 Negative g_m Circuits (Q-Enhancement)

The low Q of RFIC passive elements, primarily inductors, is a major issue in the design of these circuits. While layout techniques such as using upper layer metals can improve the inductor Q, values greater than 10 are difficult to obtain. From the Q expression of an inductor, it is easy to see that reducing the series resistance of the inductor will improve the Q:

$$Q = \frac{\omega L}{R} \tag{5.1}$$

Active circuits termed Q-enhancement circuits are often used with inductors to improve their Q. These circuits operate on the principle of creating a negative resistance or negative g_m that will partially cancel the inductor or other resistance, thereby improving Q. Care must be taken, however, that not all the resistance is removed because of stability issues (i.e., an oscillator may be created; see Chapter 6).

A simple negative resistance circuit can be constructed of a MOSFET with source-connected capacitance, as shown in Figure 5.1 with its associated RF equivalent circuit.

From the equivalent circuit diagram, the input impedance can be written as

$$
\begin{aligned}
Z_{\text{in}} &= \frac{V_{\text{in}}}{I_{\text{in}}} = \frac{I_{\text{in}}/j\omega C_{gs} + \left(I_{\text{in}} + g_m v_{gs}\right)/j\omega C}{I_{\text{in}}} = 1/j\omega C_{gs} + \left(1 + g_m/j\omega C_{gs}\right)/j\omega C \\
&= -\frac{g_m}{\omega^2 C C_{gs}} - j\omega\left(\frac{1}{C_{gs}} + \frac{1}{C}\right) = -\frac{f}{f_T}\frac{1}{\omega C} - j\omega\left(\frac{1}{C_{gs}} + \frac{1}{C}\right)
\end{aligned} \tag{5.2}
$$

Note that Z_{in} has a negative real part that can be used to partially cancel resistive loss in the circuit. An inductor is needed in shunt with the capacitor to provide the dc path; the inductor reactance should be much larger than the capacitance (Figure 5.2). A high-impedance current source can also be used in place of the inductor; the bias source can be adjusted for the desired g_m that provides cancellation of the resistive losses at the frequency of interest. A more practical negative g_m circuit uses cross-coupled MOSFETs (Figure 5.3); the negative resistance occurs between the two drain connections.

V_{in} is the difference between the two drain voltages. Assuming $g_{ds} = 0$ for each MOSFET, the RF equivalent circuit allows the relationships between V_{in} and I_{in} and the RF equivalent circuit gate source voltages and output currents to be written as

$$
\begin{aligned}
V_{\text{in}} &= V_{ds1} - V_{ds2} = V_{gs2} - V_{gs1} \\
I_{\text{in}} &= g_{m1} \cdot V_{gs1} + sC_{gs2} \cdot V_{gs2} = -\left(g_{m2} \cdot V_{gs2} + sC_{gs1} \cdot V_{gs1}\right)
\end{aligned} \tag{5.3}
$$

Subsequent mathematics yields the following relationship (assuming matched devices) for the impedance across the two drain terminals, $V_{\text{in}}/I_{\text{in}}$:

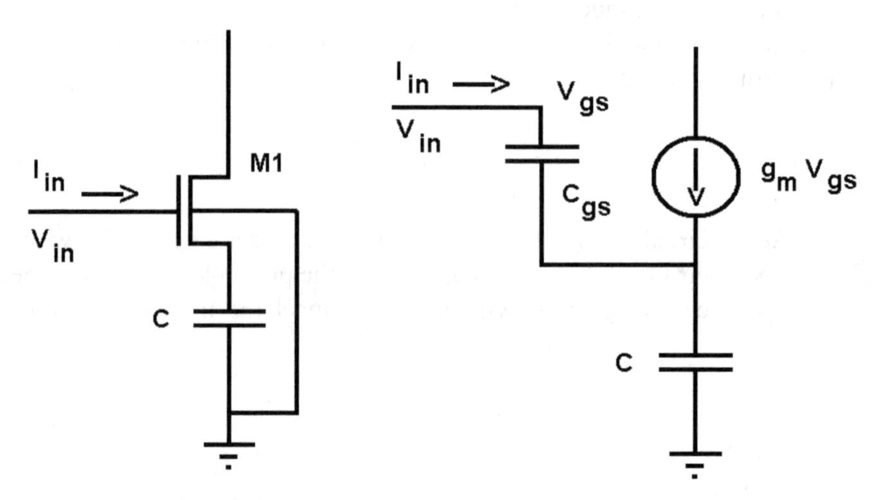

Figure 5.1 Conceptual negative resistance circuit and its RF equivalent circuit.

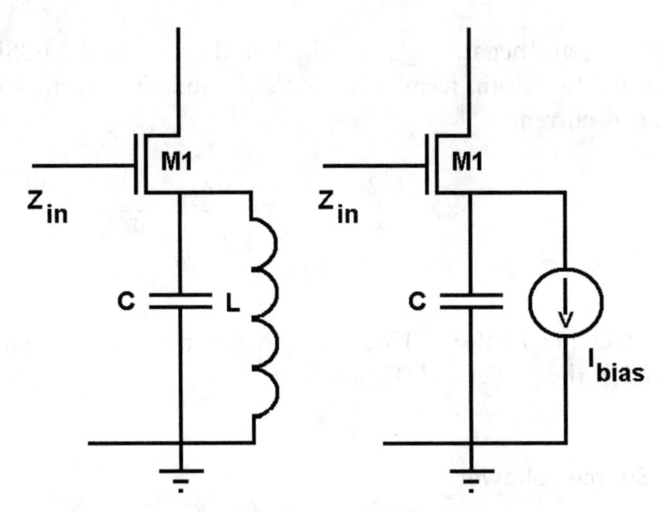

Figure 5.2 Possible biasing arrangements for the negative g_m circuit.

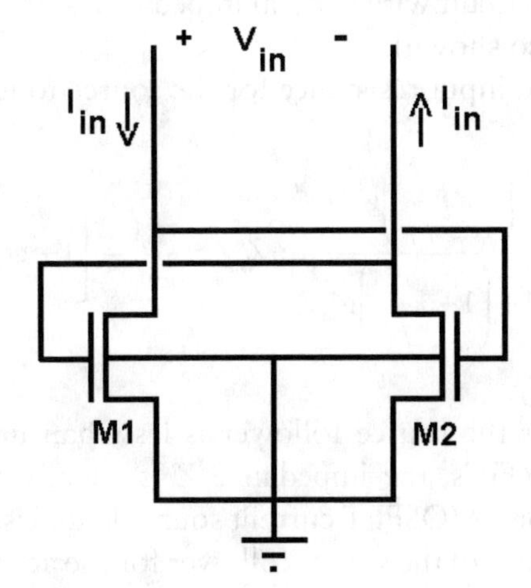

Figure 5.3 Cross coupled MOSFETs create a practical negative g_m circuit. The negative resistance is between the MOSFET drains.

$$Z_{\text{in}} = \frac{V_{\text{in}}}{I_{\text{in}}} = \frac{2}{sC_{gs} - g_m} = -2\frac{g_m + sC_{gs}}{g_m^2\left(1+\frac{\omega^2}{\omega_T^2}\right)} = \frac{-2}{\left(1+\frac{\omega^2}{\omega_T^2}\right)}\left(\frac{1}{g_m} + j\frac{\omega}{\omega_T}\right) \tag{5.4}$$

If the frequency of operation is significantly lower than the unity current gain frequency f_T, then the resistance seen between the two drain terminals can be written as

$$R_{\text{in}} = \text{RE}(Z_{\text{in}}) = \text{RE}\left(\frac{-2}{\left(1+\frac{\omega^2}{\omega_T^2}\right)}\left(\frac{1}{g_m} + j\frac{\omega}{\omega_T}\right)\right) \Rightarrow -\frac{2}{g_m} \tag{5.5}$$

hence, the term, negative g_m circuit. For the matched MOSFETs, the resistance between the two drain terminals is $-2/g_m$, a function of the MOSFET aspect ratio and dc bias current:

$$R_{in} = -\frac{2}{g_m} = -\frac{2}{\sqrt{2 \cdot KP \cdot I_{DS} \cdot \frac{W}{L}}} \tag{5.6}$$

The p-channel MOSFET devices can also be used as negative g_m circuits at the other end of the power rail (Figure 5.4).

5.1.2 Source Follower

A widely used buffering stage is the source follower. This circuit exhibits high input resistance and hence light loading of the driver circuit. Figure 5.5 shows an nMOS source follower circuit with general impedance Z in the source lead (the RF equivalent circuit is also shown).

The gain and input resistance for the source follower are

$$A_v = \frac{\left(1 + j\frac{f}{f_T}\right)g_m Z}{1 + \left(1 + j\frac{f}{f_T}\right)g_m Z};\ Z_{in} = \frac{1}{j\omega C_{gs}}\left[1 + g_m Z\left(1 + j\frac{f}{f_T}\right)\right] \tag{5.7}$$

The gain for the source follower is less than unity and exhibits a high input impedance. In RFICs, the impedance Z is usually resistive and can be an active MOSFET load or a MOSFET current source load. Using the subscript Z for the load parameter, the gain of the source follower for the active resistor load and the current source load are as follows:

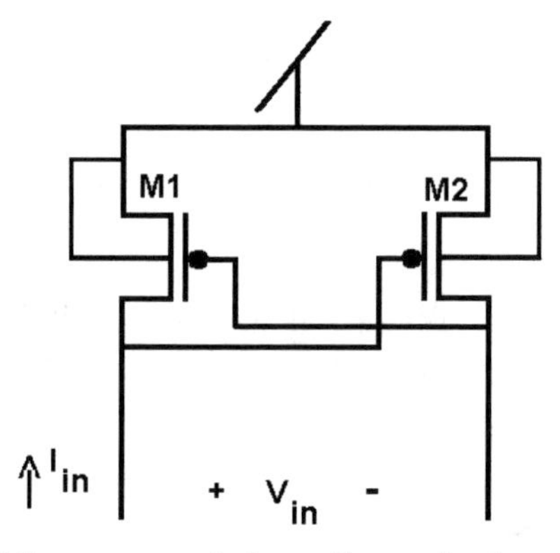

Figure 5.4 pMOS version of the cross-coupled negative g_m circuit.

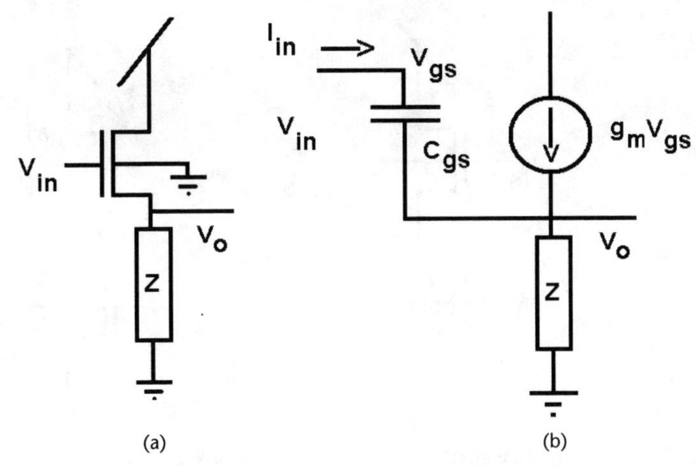

Figure 5.5 (a) *n*MOSFET source follower circuit and (b) its RF equivalent circuit.

$$\text{Active resistor: } A_v = \frac{\left(1+j\dfrac{f}{f_T}\right)\dfrac{g_m}{g_{mZ}}}{1+\left(1+j\dfrac{f}{f_T}\right)\dfrac{g_m}{g_{mZ}}} \approx \frac{\dfrac{g_m}{g_{mZ}}}{1+\dfrac{g_m}{g_{mZ}}} = \frac{\sqrt{\dfrac{(W/L)}{(W/L)_Z}}}{1+\sqrt{\dfrac{(W/L)}{(W/L)_Z}}} \tag{5.8}$$

$$\text{Current source: } A_v = \frac{\left(1+j\dfrac{f}{f_T}\right)\dfrac{g_m}{g_{dsZ}}}{1+\left(1+j\dfrac{f}{f_T}\right)\dfrac{g_m}{g_{dsZ}}} \approx \frac{\dfrac{g_m}{g_{dsZ}}}{1+\dfrac{g_m}{g_{dsZ}}} = \frac{\sqrt{\dfrac{2KP(W/L)}{\lambda^2 I_{DS}}}}{1+\sqrt{\dfrac{2KP(W/L)}{\lambda^2 I_{DS}}}} \tag{5.9}$$

For the active resistor load, the gain can be designed simply through the aspect ratios of the two MOSFETs. For the current source load, the gain expression is more complex; however, g_{ds} is usually much smaller than g_m in MOSFETs, so the ratio g_m/g_{ds} is quite large, providing a gain for the current source load close to unity. Note that in Figure 5.6 MOSFET M1 will suffer from the body effect if the bulk is tied directly to ground (bulk tied to V_{DD} for *p*MOS devices), requiring consideration of the additional transconductance current term $g_{mb}v_{sb}$. The overall effect of this additional term is to modify the g_m in the design expressions above to $g_m(1+\eta)$ where $g_{mb} = \eta_{gm}$.

5.1.3 Simple Automatic Gain Control Circuits

Communication protocols using RFICs can require dynamic ranges of 80 dB or more. Under certain conditions, large-amplitude signals may saturate amplifiers and other subsystems; a lengthy relaxation time may be required for these circuits to come out of saturation, and information transmitted during that circuit relaxation time may be lost. Automatic gain control (AGC) circuits are used to dynamically vary the gain of amplifier stages to minimize the saturation possibilities in these circuits. AGC circuits contain three major parts: a variable gain amplifier (VGA), a signal

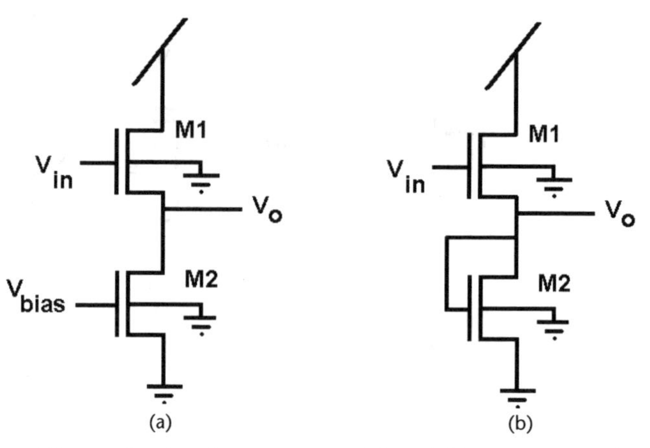

Figure 5.6 Source follower circuits with (a) current sink and (b) active resistor loads.

sampler, and a control circuit that controls the VGA (Figure 5.7). The sampler looks at the amplitude of the output signal V_{out} and continuously feeds that peak value to the control circuit. Since the sampler looks at the peak value, a peak detector is often used for this stage. The control block looks at the peak value and compares this value with some reference and then feeds this information to the VGA in the form of a control voltage or current. The VGA, therefore, should have some means of varying its gain in response to the control-in signal from the control block.

While the VGA and sampler are usually analog/RF in nature, the control block is often implemented in digital technology and is part of the overall digital control for the system. Digital control is preferred since the AGC circuit can be programmed for any AGC characteristic. To maintain tight control of the gain, two or more VGAs are often used, one for "coarse" and the other for "fine" gain adjustment [1]; a digital control block can be programmed to handle the gain apportionment tasks. For amplifiers, the voltage gain is proportional to the input MOSFET transconductance g_m. The transconductance, in turn, is proportional to the aspect ratio W/L and the dc bias current I_{DC}. Since the size is set at fabrication, a simple VGA can be created by adjusting the dc bias through the transconductor [2],

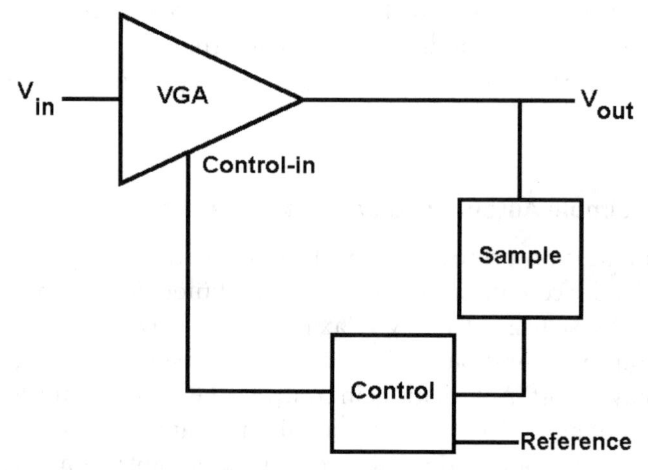

Figure 5.7 Block diagram of an AGC circuit.

although since g_m is only proportional to $I_{DC}^{1/2}$, this dependence is weak. One alternative to this technique is to use a resistively degenerated differential amplifier (RDDA); this technique has good large-signal characteristics [2, 3]. Figure 5.8 shows a simple RDDA, where MOSFETs M1 and M2 and the current sources are designed as the typical differential amplifier. MOSFET MR provides the source degeneration, where the control line on the gate is a voltage that controls the amount of degeneration in an active resistor mode. Using the abbreviation R_{MR} for the active resistance of MOSFET MR, the differential gain for this circuit is

$$A_v = \frac{A_{v0}}{1+g_m R_{MR}} \approx \frac{A_{v0}}{1+g_m \dfrac{1}{2\text{KP}\cdot(W/L)(V_{control}-V_T)}}$$

$$\approx \frac{A_{v0}}{1+\sqrt{2\text{KP}(W/L)_{MI} I_{DC}}\ \dfrac{1}{2\text{KP}\cdot(W/L)_{MR}(V_{control}-V_T)}} \tag{5.10}$$

$$\approx \frac{A_{v0}}{1+\sqrt{\dfrac{(W/L)_{MI} I_{DC}}{2\text{KP}\cdot(W/L)_{MR}^2 (V_{control}-V_T)^2}}}$$

For control voltages near threshold, the gain is very low, but it increases as the control voltage moves above V_T.

Example 5.1: Design a VGA with 1.0 mA of bias current with a differential gain of 5 (14 dB) at 1000 MHz. Use 1-kΩ resistors for the active load and tail (Figures 5.8 and 5.9). Plot the output gain versus control voltage on M3 (ch5-1.txt).

The design equations for this circuit show that $W/L = 800$ for M1 and M2 and that a dc bias of 200 mV above threshold will provide the required gain. A 400-μm

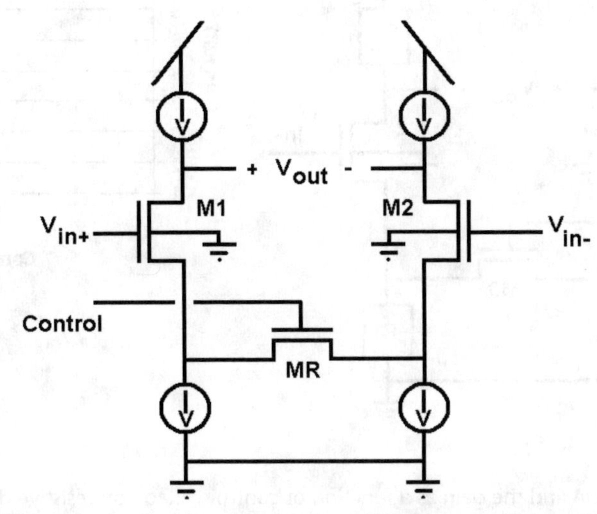

Figure 5.8 Variable gain differential amplifier using resistive degeneration via MOSFET MR to control the gain.

gate width MOSFET (M3) provides the greatest gain variation in the range of 1.25 to 1.55V. Figure 5.9 shows that the gain is linear in dB and is related to the control voltage over a small range (0.25V), typical for this type of VGA [4].

An alternative VGA implementation suitable for digital control uses a binary-weighted transistor array [1]. In this concept, current source sections are selectively turned on and off, providing discrete steps in current that can be used to change the gain of an amplifier. This technique removes the need for a DAC to control the VGA. This technique also provides a means to approximate an exponential current gain characteristic, providing another linear-in-dB gain relationship:

$$I_{out} = e^{x} \approx \frac{1+x}{1-x} \qquad x = b_1 2^1 + b_0 2^0 \tag{5.11}$$

For discrete values of I_{in}, twos-complement arithmetic provides a means of dealing with the negative sign needed in the denominator:

$$-I_{in} = \bar{I}_{in} + I_{DC} = \bar{b}_1 2^1 I_{DC} + \bar{b}_0 2^0 I_{DC} + I_{DC} = \bar{b}_1 2^1 I_{DC} + b_0 2^0 I_{DC} \tag{5.12}$$

so the output current in terms of the input current becomes

$$e^{x} \approx \frac{1 + b_1 2^1 + b_0 2^0}{1 + \bar{b}_1 2^1 + b_0 2^0} \tag{5.13}$$

To illustrate the exponential current amplifier, a simple 2-bit version is shown in Figure 5.10. The digital control bit ($b_1 b_0$) controls the gate voltage to pass to MOSFETs M3 or M4 (MC or MD), selectively applying V_{bias} to their gates. A high output impedance is possible with this configuration since M3 and M4 (MC or MD) also act as cascode MOSFETs for their respective current sources I_{DC} and $2I_{DC}$ [1].

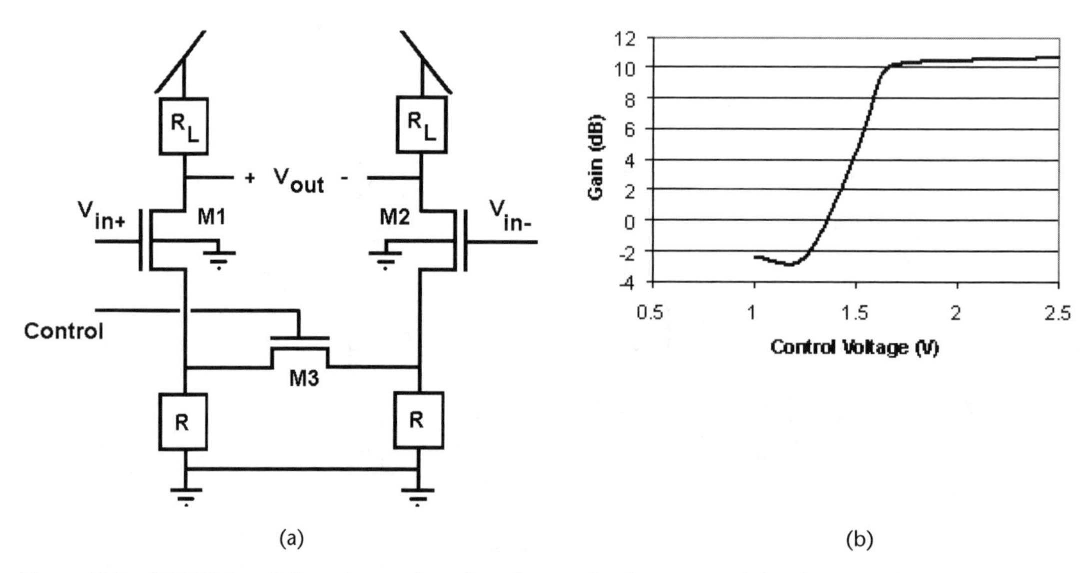

Figure 5.9 (a) RDDA and the gain as a function of control voltage on resistive degeneration MOSFET M3. (b) Gain versus control voltage response. (After: [4].)

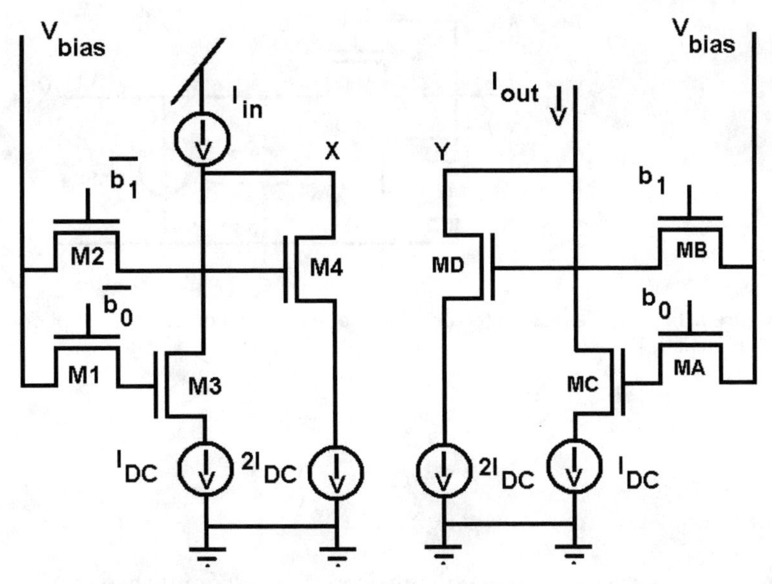

Figure 5.10 Two-bit exponential current amplifier. (After: [1].)

M3 and M4 (MC or MD) are sized in proportion to their current sources, but switching MOSFETs M1 and M2 (MA and MB) can be minimum size. Table 5.1 shows the current I_{in} as a function of digital control word b_1b_0.

Six or more bits of digital control are possible with this technique [1]. However, as the number of bits increases, the MOSFET sizing increases as well, severely capacitively loading nodes X and Y. This capacitive loading limits the high-frequency response for even a modest number of bits. However, this type of current amplifier is excellent in the lower frequency IF stage of an RF receiver.

Diode peak or envelope detectors have been used for decades in discrete systems. For RFICs, diode-connected MOSFETs can replace the diodes on a nearly one-for-one basis to implement a peak detector [2, 3, 5]. A simple peak detector is shown in Figure 5.11. Transistor M1 is a diode-connected MOSFET acting as a rectifier, C is used to "store" the peak value, M2 is used to provide a dc return path or for matching for the circuit, and I_{time} sets the decay time of the capacitor.

For rapid charging, M1 should have a high W/L to reduce its effective on-state resistance. For discharging of C for rapid tracking, the current source I_{time} sets the discharge rate so that the peak detector is ready for the next AGC sample. The current needed can be estimated using the expression

Table 5.1 Input and Output Current as a Function of Digital Input Word

b_1b_0	I_{in}	I_{out}
0 0	$3I_{DC}$	0
0 1	$2I_{DC}$	I_{DC}
1 0	I_{DC}	$2I_{DC}$
0 0	0	$3I_{DC}$

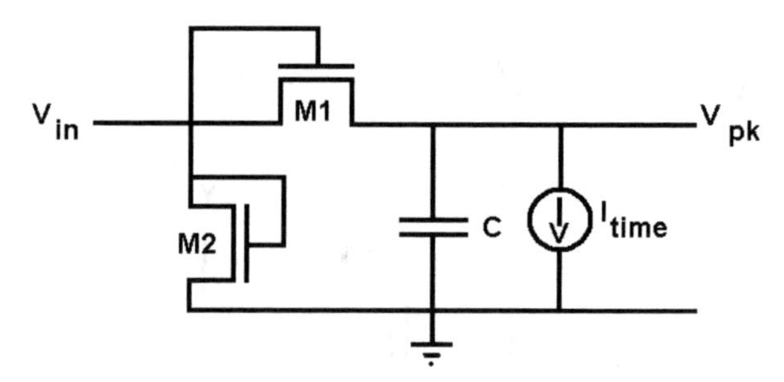

Figure 5.11 Simple peak detector using diode-connected M1. (M2 is optional and may be used for matching.)

$$I_{time} = C\frac{\Delta V}{\Delta T} \tag{5.14}$$

where ΔV is the amount of voltage drop to be accomplished in ΔT time. A simple single MOSFET configured as current source is often used for this circuit; one of the higher impedance current sources described earlier can also be used if required. There are trade-offs on the response time of this detector. If a high I_{time} is selected for rapid discharge of C, the current source will "steal" rectified current from C and lower the charging rate. Figure 5.12 shows a $2.0V_p$, 1000-MHz RF signal input to this simple peak detector. A limitation of this simple peak detector is evident from the figure: the peak output is limited to $V_{\text{in-}peak} - V_T$. For this reason, a preamplifier is often added to the circuit to compensate for this reduced output. The figure also shows that the output voltage discharges to 1/e of its value in approximately 60 ns. For C = 1.0 pF, the required sinking current can be approximated as

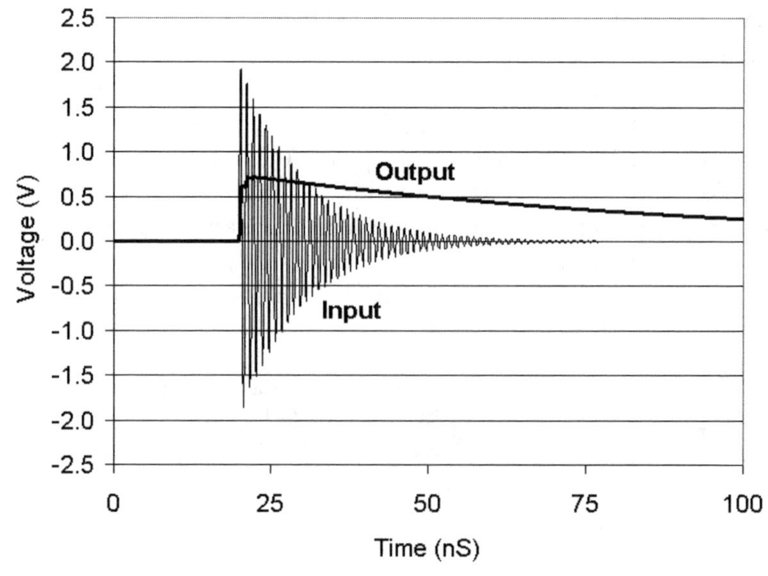

Figure 5.12 Transient simulation of response time of peak detector and decay due to current source (ch5-2.txt).

$$I_{time} = C\frac{\Delta V}{\Delta T} = 10^{-12}\frac{0.5}{60 \cdot 10^{-9}} = 8.3\ \mu\text{A} \tag{5.15}$$

This type of peak detector increases in output voltage as the input voltage increases, the opposite to that needed for the VGA. A voltage-voltage converter or low-gain inverting amplifier are needed to change the slope of the peak detector so that it can be fed directly to the VGA. This slope inversion is not needed in digital AGC since control can be programmed to compensate for the slope differences. More complex AGC circuits can be found in the references listed in the Selected Bibliography at the end of the chapter.

5.2 Ancillary Passive CMOS RFIC Circuits

5.2.1 Generation of dc from Applied RF Power

Some means must be made available to power the active components (transponders, amplifiers, and so forth) in so-called passive RFID or implanted systems. There are two main types of power generators for these applications: antenna fed and inductive coupling. Rectification and smoothing of the resulting dc voltage is required for both to provide power to CMOS electronics in these applications. There are many types of diode rectifier circuits and voltage multipliers; Figure 5.13 shows a few of these architectures [6].

In MOS processes, junction diodes tend to exhibit high leakage currents and poor high-frequency performance. However, by tying the gate and drain together, the MOSFET active resistor's I-V characteristic shows diodelike behavior (Figure 5.14), with no current conduction occurring until $|V_{GS}| > |V_T|$ (so-called diode-connected MOSFET).

Antenna-fed systems rely on resonant antennas to provide the RF energy directly to the voltage converter circuits [7]. These antennas are typically quarter-wave dipoles, either linear or folded [8]. Inductive coupling uses two coils in proximity to each other to transfer energy through the mutual inductance between the two coils [9]. An applied RF signal in one of the coils (termed the primary) induces a current in the secondary coil that is then rectified and fed to the RFID circuit. For many RFID applications, the antennas and coils are physically large and are often located off chip. Figure 5.15 shows an example of a power harvesting circuit based on a full-wave rectifier circuit [9]. The capacitor C is on chip and should exhibit a low reactance at the magnetic induction frequency to smooth the rectified voltage for use by the remainder of the RFIC. The secondary coil may be on chip or off chip.

Figure 5.16 shows a simulation of the inductively coupled power harvester based on this principle. Two 100-nH coils are used with a 1.0-GHz external signal applied. The coupling between the two coils is 0.1, and each *p*MOS device has an aspect ratio of 175/0.5. An on-chip 10-pF capacitor is used to smooth out the rectified signal. The simulation indicates that a 7.0V peak signal applied to the primary can induce up to 3.3V rectified and smoothed output voltage. The voltage reaches approximately 3.8V 500 ns after application of the activating signal. Load calculations show that this simple circuit can supply 50 μA with little voltage sag. Lower

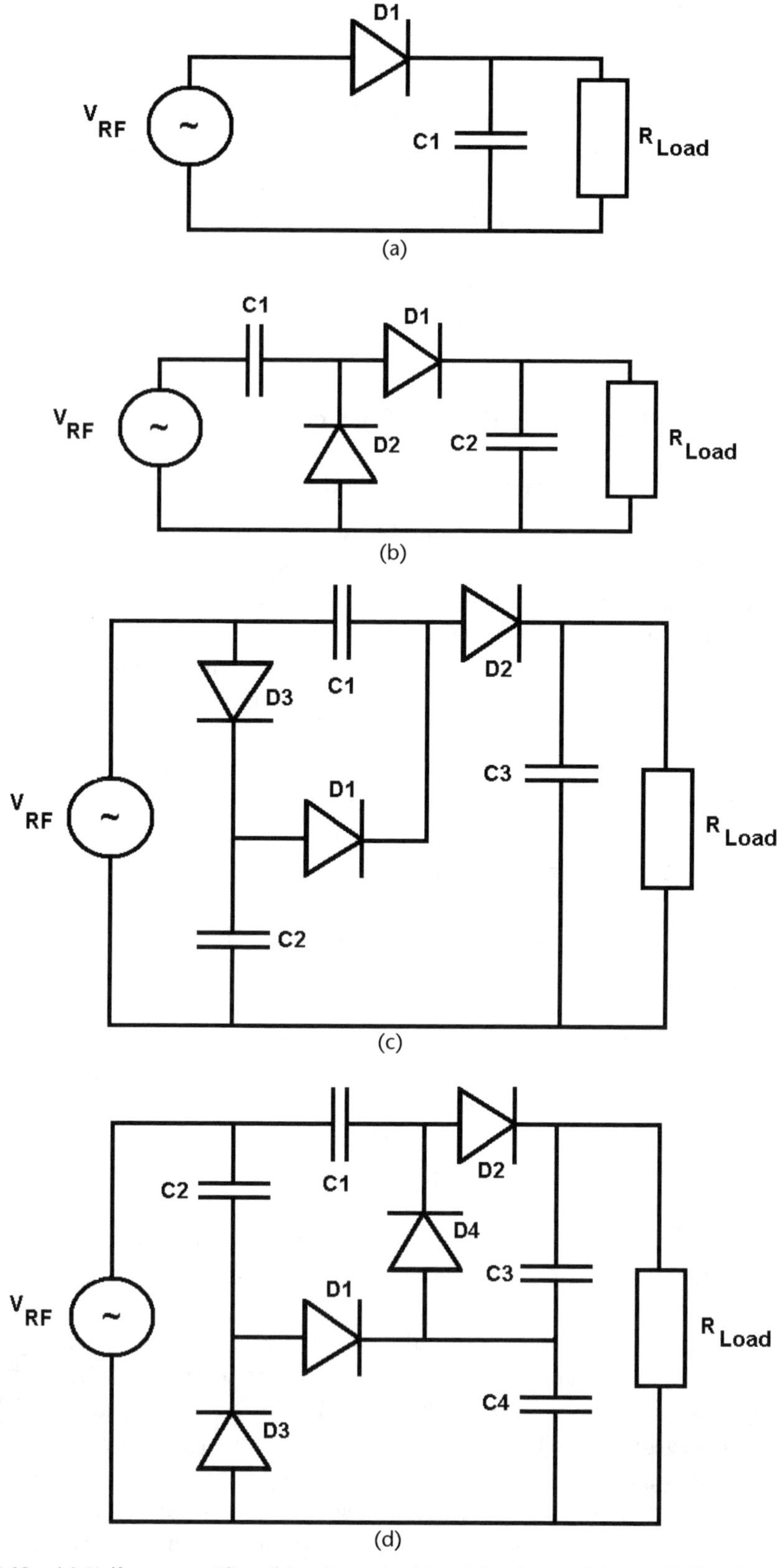

Figure 5.13 (a) Half-wave rectifier, (b) voltage doubler, (c) voltage tripler, and (d) voltage quadrupler.

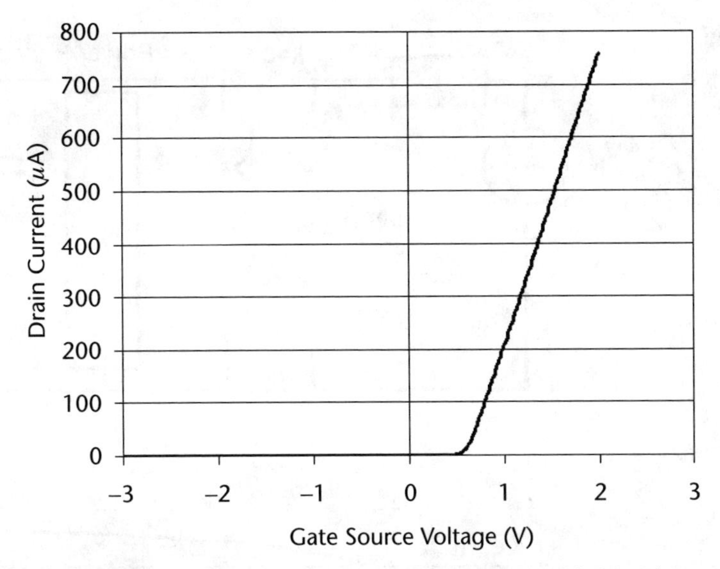

Figure 5.14 The *n*MOSFET active resistor shows rectifying behavior (ch5-3.txt).

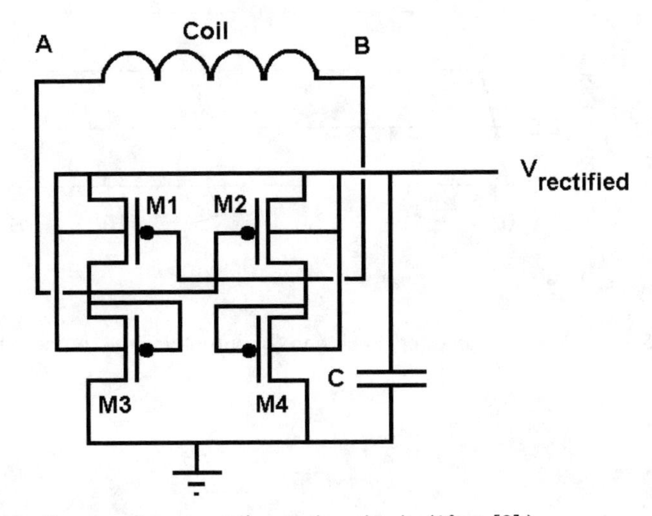

Figure 5.15 Inductive coupling power harvesting circuit. (After: [9].)

coupling will reduce the voltage available to the rectifier. Increasing the RF input to the primary coil will increase the available dc current. Limiter circuits are often used to ensure that high primary energizing powers do not exceed the voltage breakdown of the devices [8]. Low-power voltage regulation is often added to provide a more constant voltage in the presence of varying RF power levels. The regulators work as long as the RF-generated voltage is greater than the dropout voltage of the regulator.

For antenna-fed systems, studies have shown that if the input capacitance of the diode-connected MOSFET is tuned out (such as a bond wire connecting the RFIC to the external antenna), the input voltage amplitude is equal to [7]

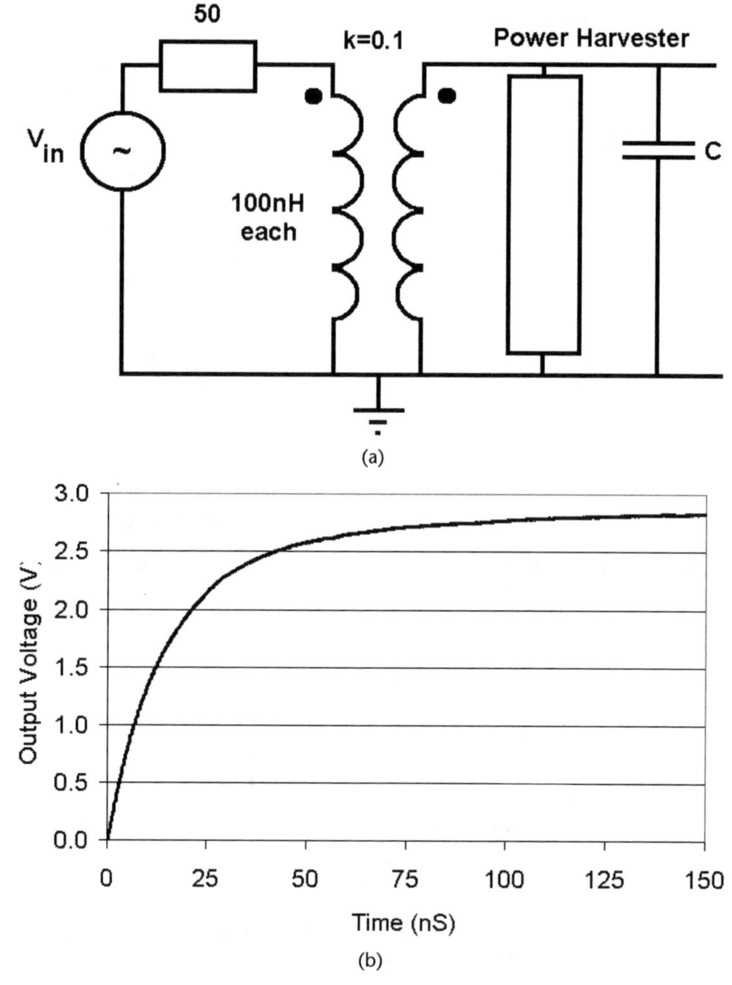

Figure 5.16 (a) Power harvester circuit and (b) output voltage response to a 7.0V_{peak} input RF signal (ch5-4.txt).

$$v_{in} = 2\sqrt{2P_A R_{ant}} \frac{R_{in}}{R_{in} + R_{ant}} \tag{5.16}$$

where P_A is the available power from the antenna, R_{in} is the real part of the input MOSFET impedance, and R_{ant} is the antenna impedance real part. Equation (5.16) shows that to increase the input voltage amplitude, the antenna should match the input resistance of the MOSFET rectifier. For a 50Ω antenna matched to the rectifier, +20 dBm will provide a voltage of 3.1V.

5.2.2 Active Inductor Circuits

Passive inductors in RFICs are lossy, exhibit low Q, and consume large amounts of IC real estate for even modest values of inductance. The positive inductive reactance of an inductor can be created over a relatively narrow bandwidth, however, with the use of an active inductor or gyrator circuit. An active inductor can be created using the simple circuit shown in Figure 5.17 (also shown is a simple RF equivalent circuit). An external resistor is placed between the MOSFET gate and a dc gate bias.

The impedance looking into the source of M1 Z_{in} can be written as

$$Z_{in} = \frac{1 + j\omega R_G C_{gs}}{g_m + j\omega C_{gs}} = \frac{1}{g_m} \frac{1 + j\omega R_G C_{gs}}{1 + j\frac{f}{f_T}} \tag{5.17}$$

If the frequency of operation is well below f_T, the impedance Z_{in} becomes

$$Z_{in} \approx \frac{1}{g_m} + j\frac{\omega R_G C_{gs}}{g_m} = \frac{1}{g_m} + j\frac{\omega R_G}{\omega_T} = \frac{1}{g_m} + j\frac{f}{f_T} R_G \tag{5.18}$$

Because an assumption of $f/f_T << 1$ was used to obtain the form for Z_{in}, R_G needs to be large to obtain reasonable values of inductive reactance. In addition, the device g_m should be large for reasonable active inductor Q:

$$Q = \frac{X_L}{R} = \frac{\frac{f}{f_T} R_G}{\frac{1}{g_m}} = \frac{f}{f_T} g_m R_G \tag{5.19}$$

Larger g_m is created with larger device size, which increases the parasitic capacitance in the device; of most importance is C_{sb}, which is a function of the source area and perimeter and serves as a shunt capacitance to ground at the input Z_{in}. Since C_{sb} is in shunt with the inductor, minimizing this parasitic capacitance improves the active inductor self-resonant frequency:

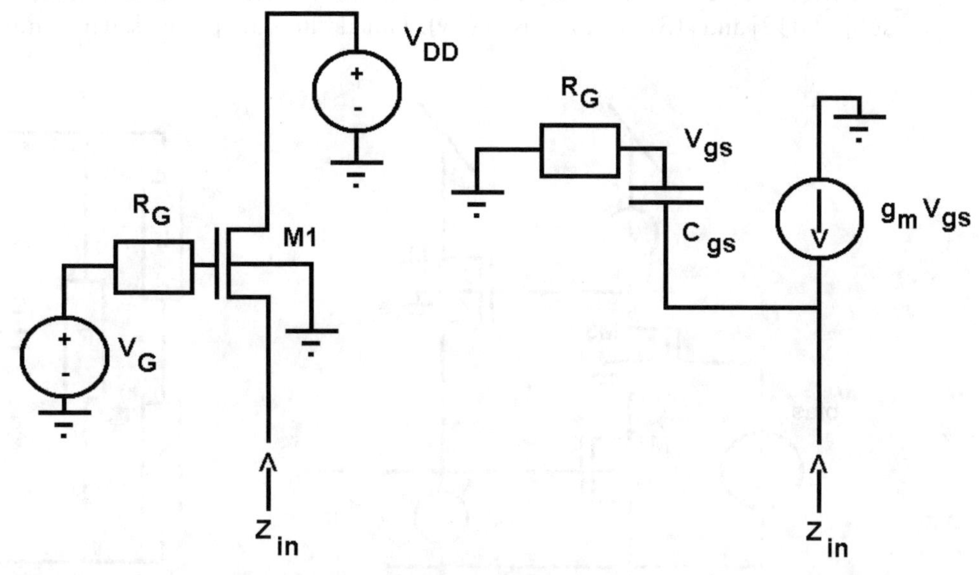

Figure 5.17 Active inductor circuit and its RF equivalent circuit.

$$\omega_0 = \frac{1}{\sqrt{LC_{sb}}} = \frac{1}{\sqrt{\frac{R_G}{\omega_T} C_{sb}}} = \sqrt{\frac{\omega_T}{R_G C_{sb}}} \tag{5.20}$$

An active inductor circuit that actually exploits the device parasitics has been developed [10]. A cascode active inductor circuit has been proposed (Figure 5.18 and its inductor equivalent circuit) and contains three MOSFETs and two current sources (which at their minimum will be another two MOSFETs). The active inductor parameters for this circuit can be found from the equivalent circuit to be

$$L = \frac{C_{gs2}}{g_{m1} g_{m2}} = \frac{1}{g_{m1} \omega_{T2}}; \ R_L = \frac{1}{g_{m1} g_{m2}} \frac{g_{ds1} g_{ds3}}{g_{m3}} jC = C_{gs1}; \ R = \frac{1}{g_{m2}}$$
$$Q = \frac{\omega L}{R} = \frac{f}{f_T} \frac{g_{m2} g_{m3}}{g_{ds1} g_{ds3}} \tag{5.21}$$

The active inductor self-resonant frequency is

$$\omega_0 = \sqrt{\frac{g_{m1} g_{m2}}{C_{gs1} C_{gs2}}} = \sqrt{\omega_{T1} \omega_{T2}} \tag{5.22}$$

The Q of the active inductor may be further enhanced by combining the previous two techniques with the addition of resistive feedback on MOSFET M3, as shown in Figure 5.19 [11].

5.3 Tuned Amplifiers

The field of filter design is very broad and numerous texts have been written on the subject ([12] and [13] to name just a few). Tuned filters are often used in combination

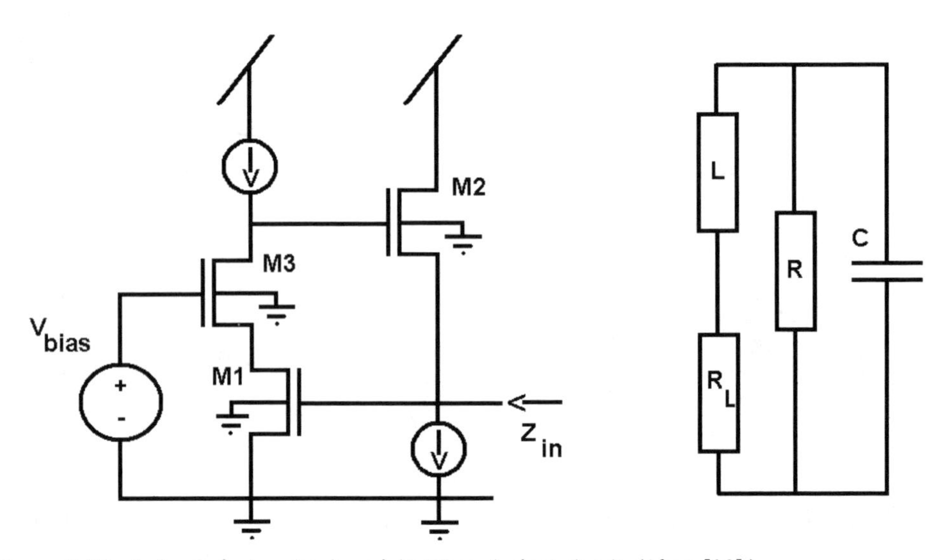

Figure 5.18 Active inductor circuit and its RF equivalent circuit. (After: [10].)

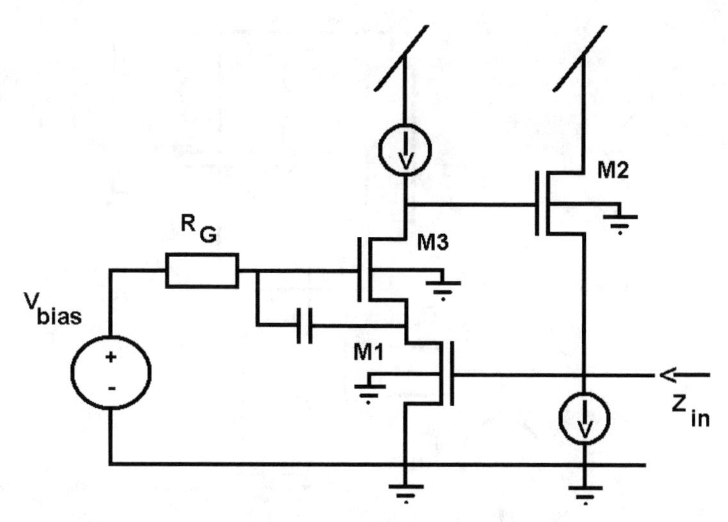

Figure 5.19 Circuit for enhancing the Q of the active inductor. (After: [11].)

with amplifiers to combine functions and save silicon real estate. The filter portion of these amplifiers can be designed using techniques similar to those used in designing classical lumped element LC prototype filters (Butterworth, Chebyshev, elliptical, and so forth) [14]. Higher order filters can be created by cascading several filter sections. The area of high-frequency filter design is quite broad, with new techniques and circuit improvements continually being advanced. This section covers only a few basic filtering circuits to provide the reader the foundation for better understanding of this concept. The reader is encouraged to investigate the rich variety of literature expanding on this basic theme.

5.3.1 LC Tuned Filters

The basic LC tuned amplifier consists of the basic MOSFET transconductance amplifier with an inductor and capacitor at the output [Figure 5.20(a)].

The Q of the filter is governed by the combined Q of the inductor and the capacitor that make up the loading; usually, the inductor is the passive element limiting the overall filter Q. The impedance Z for the RLC load can be written in terms of L, C, and parallel Q as

$$Z = R\frac{j\frac{1}{Q}\frac{f}{f_0}}{\left(1-\omega^2 LC\right)+j\frac{1}{Q}\frac{f}{f_0}} = R\frac{j\frac{1}{Q}\frac{f}{f_0}}{\left(1-\frac{f^2}{f_0^2}\right)+j\frac{1}{Q}\frac{f}{f_0}};\ Q=\frac{R}{\omega_0 L};\ \omega_0^2=\frac{1}{LC} \quad (5.23)$$

At resonance ($f = f_0$), the impedance Z reduces to the parallel equivalent resistance R.

The transfer characteristic of the circuit is calculated the same way as the gain:

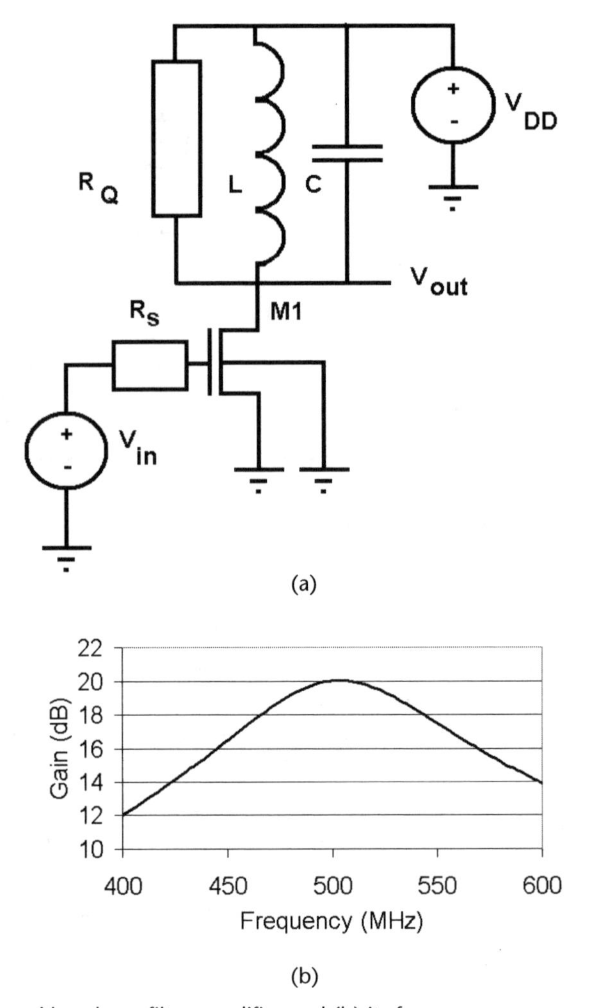

Figure 5.20 (a) Tuned bandpass filter amplifier and (b) its frequency response.

$$H(\omega) = A_v = -\frac{g_m}{g_{ds} + Z^{-1}} \cong -g_m Z = -g_m R \frac{j\frac{1}{Q}\frac{f}{f_0}}{\left(1-\omega^2 LC\right) + j\frac{1}{Q}\frac{f}{f_0}} \tag{5.24}$$

The transfer function peaks at the resonance frequency $(LC)^{-1/2}$, with the resistive portion of the tuned circuit ($g_m R$) determining the overall gain of the circuit, creating a bandpass configuration. The shunt capacitance can be a combination of the parasitic capacitance of the inductor and an external element. Higher Q bandpass filters can be obtained by using one of the Q-enhancement techniques described earlier.

Example 5.2: Design a 500-MHz LC tuned bandpass filter that exhibits a gain of –10 at the resonant frequency at the output. Assume that the inductor Q is 5 and an ideal capacitor (ch5-5.txt).

Several ways can be used to approach the solution to this problem. One way to start is to determine the inductance and capacitance at a resonance frequency of 500 MHz. The relatively low inductor Q means that to achieve a high parallel resistance, a small inductance should be used. This is not too much of a problem since on-chip inductors are of small value anyway; choosing a 10-nH inductance yields a parallel resistance of 158Ω by application of (5.24). The resonating capacitor for this circuit is 10 pF. To achieve the required gain specification of –10 at the resonant frequency, the *n*MOSFET transconductance g_m needs to be calculated from the expression for the RF gain. At the resonance frequency, the reactance of L and C cancel, leaving only the parallel resistance R_Q to set the gain. Using the standard gain equation for a simple resistively loaded amplifier ((4.5), where $Z_{load} = R_Q$) yields the required transconductance:

$$|A_v| = 10 = g_m R_Q \rightarrow g_m = \frac{|A_v|}{R_Q} = 63 \text{ mS}$$

Assuming a dc gate voltage of 1.0V yields an *n*MOSFET aspect ratio of $W/L =$ 3,290:

$$g_m = \left[\text{KP}\left(\frac{W}{L}\right)(V_{GS} - V_T)\right] \rightarrow \left(\frac{W}{L}\right) = \frac{g_m}{\left[\text{KP}(V_{GS} - V_T)\right]} = 3{,}290$$

The resulting bandpass characteristic for this tuned amplifier is shown in Figure 5.20(b).

Using similar circuit reasoning, an amplifier exhibiting band notch behavior can be created with a series LC network at the output (Figure 5.21). Inductive biasing is used in this example to ensure the proper dc bias on the MOSFET transconductor.

Actual LC tuned filter design is complicated by the complex nature of the inductors and capacitors that make up the tuned circuit. An iterative process is required to optimize the IC layout parameters for these elements so that the desired bandpass or band notch frequency is obtained. The relatively low inductor Q values, however, will ultimately limit these filter types to relatively low-Q filters and hence relatively wide bandwidths. Undesired capacitance such as interconnect capacitance or interwinding capacitance can degrade the filter performance [14].

Standard LC ladder filter synthesis of bandpass and band notch filters exhibit a very large spread in calculated component values that can exceed two orders of magnitude or more. Besides the design challenge of fabricating the required inductors and capacitors to the resolutions needed for a good filter response, the compensation of parasitics becomes difficult since large L and small C values (or vice versa) are often encountered in LC ladder filters [10]. The parasitics of one reactive element can often swamp the desired value of its reactive counterpart (for example, the shunt parasitic capacitance of a large-valued inductor may far exceed that of the shunt capacitance needed for branch resonance). The use of coupled resonator filters provides an alternative to stand-alone LC ladder filters. The ability to replace the second-order ladder bandpass structure with a second-order coupled resonator filter can be seen by deriving the voltage transfer function for the two

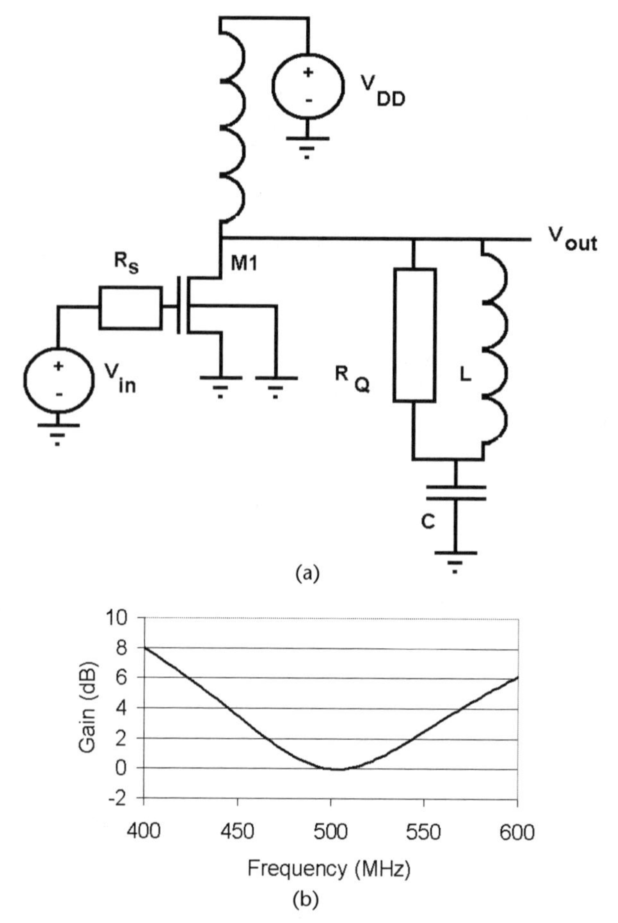

Figure 5.21 (a) Tuned band notch amplifier and (b) its frequency response.

circuits. Figure 5.22 shows a second-order LC ladder network for a bandpass structure.

For simplicity, the inductors, capacitors, and resistances are assumed to be equal. The voltage transfer function for the ladder circuit, $H(f) = V_{out}/V_{in}$, can be written as:

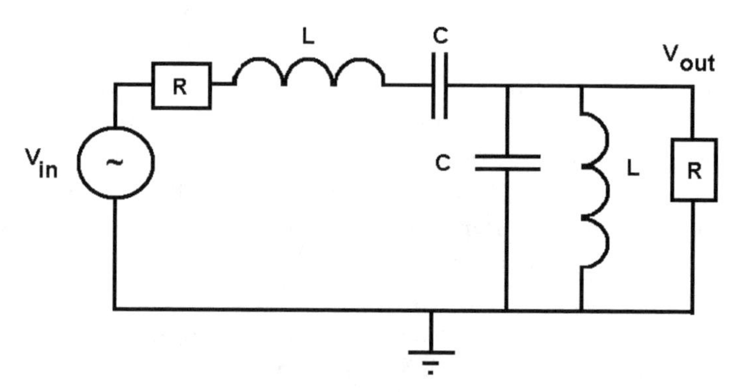

Figure 5.22 Second-order bandpass ladder filter structure.

$$H_{ladder}(f) = \frac{V_{out}}{V_{in}} = \frac{\frac{1}{Q^2}}{\left[j\left(\frac{f}{f_0} - \frac{f_0}{f}\right)\right]^2 + \frac{2}{Q}\left[j\left(\frac{f}{f_0} - \frac{f_0}{f}\right)\right] + \frac{2}{Q^2}} \tag{5.25}$$

where

$$Q = \frac{R}{\omega_0 L}$$

At resonance $(f = f_0)$, the transfer function is 0.5. For the coupled resonator filter, coupling between filter sections takes place via the mutual inductance exhibited between two on-chip transformers. An ideal two-pole coupled resonator filter is shown in Figure 5.23, where M is the mutual coupling between the two inductors L_1 and L_2. The coupling coefficient k can be written as

$$k = \frac{M}{\sqrt{L_1 L_2}} \tag{5.26}$$

For identical inductances, capacitances, and resistances, the transfer function V_{out}/V_{in} can be written as [14]

$$H_{coupled}(f) = \frac{V_{out}}{V_{in}} = \frac{-j\frac{k}{Q}\left(\frac{f_0}{f}\right)}{\left[j\left(\frac{f}{f_0} - \frac{f_0}{f}\right)\right]^2 + \frac{2}{Q}\left[j\left(\frac{f}{f_0} - \frac{f_0}{f}\right)\right] + \frac{1}{Q^2} + \left(k^2\right)\left(\frac{f_0}{f}\right)^2} \tag{5.27}$$

where

$$k = \frac{M}{L};\ Q = \frac{R}{\omega_0 L\left(1 - k^2\right)}$$

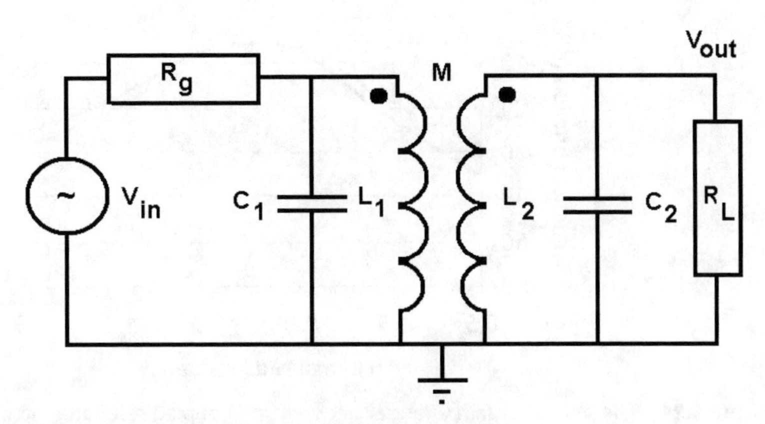

Figure 5.23 RF equivalent circuit for a coupled resonator filter.

At resonance ($f/f_0 \sim 1$), $|H(f)|$ reduces to

$$|H(f)| = \frac{kQ}{1+(kQ)^2} \tag{5.28}$$

and for small kQ (loose inductor coupling, low inductor Q), the response goes to kQ. The ladder circuit and the coupled resonator circuit can be made to approach each other if the product of the coupling and inductor Q is unity: $kQ = 1$. The similarity of the two transfer functions is shown in Figure 5.24, where (5.25) and (5.27) are plotted versus normalized frequency. An inductor Q of 5 and coupling coefficient k of 0.2 were assumed.

The Q of on-chip inductors is usually less than 5, requiring transformer coupling coefficients of at least 0.2 for direct replacement of the ladder network with a coupled resonator. However, this transformer coupling requirement can be relaxed somewhat if the inductor Q were higher; Q-enhancement circuits are usually used in this approach. The cross-coupled negative g_m circuit can be employed to increase the Q of the circuit and is required on both sides of the coupled resonator filter. The bias for the negative g_m circuit can be applied through the center of the resonator (since this is a differential circuit, this point is an effective RF ground) (Figure 5.25).

A shunt loss element was assumed in the model for the LC resonator. While it is mathematically more tractable to use a shunt resistor to describe the resonant circuit loss, the primary loss element is actually the series resistance of the inductor. Translating the series resistance of the inductor to a shunt resistance in the model makes this shunting resistance frequency dependent (in addition to the frequency dependence of the series resistance of the inductor itself) [15]:

$$R_{shunt} = R_{series}\left(1+\frac{\omega^2 L^2}{R_{series}^2}\right) \tag{5.29}$$

This in turn makes the Q of the coupled resonator filter frequency dependent:

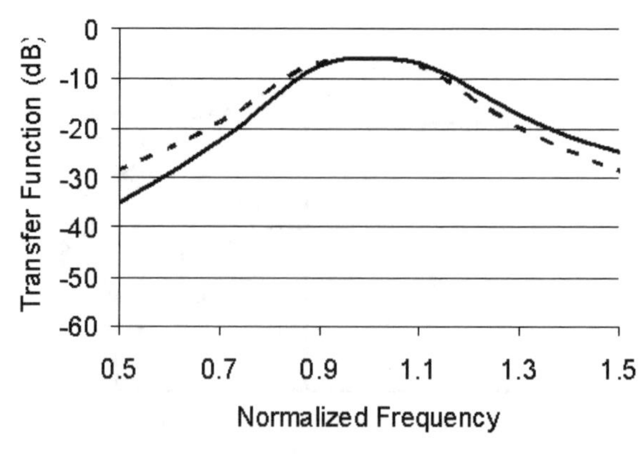

Figure 5.24 The second-order LC ladder network and coupled resonator filter have similar frequency responses near resonance.

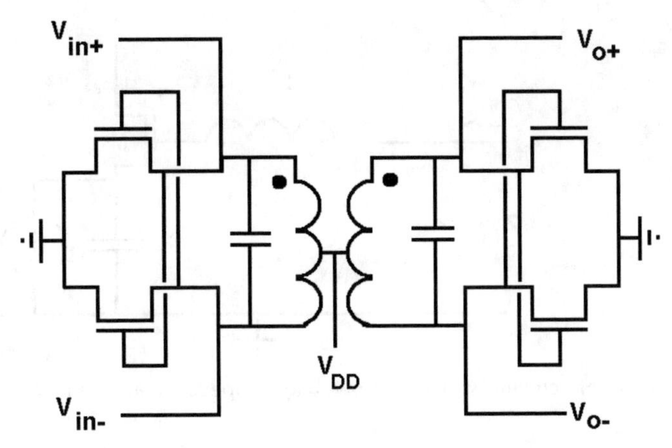

Figure 5.25 Negative g_m circuits can enhance the Q of coupled resonator filters.

$$Q(\omega) = \frac{R}{\omega_0 L\left(1-k^2\right)} = R_{series}\left(1+\frac{\omega^2 L^2}{R_{series}^2}\right)\frac{1}{\omega_0 L\left(1-k^2\right)} \tag{5.30}$$

Equations (5.29) and (5.30) show that the Q will increase with frequency; the Q-enhancement circuit will then have to track the changing resonator loss to compensate so that the filter response will stay equivalent to its LC ladder counterpart. The Q-enhanced resonator circuit gets its improved Q from the negative resistance of the cross-coupled MOSFETs. One way for this negative resistance to change in the as-fabricated circuit is by changing I_{DS}:

$$R_{\text{in}} = \text{RE}(Z_{\text{in}}) = -\frac{2}{g_m} = -\frac{2}{\sqrt{2\cdot \text{KP}\cdot I_{DS}(W/L)}} \tag{5.31}$$

As the resonator Q changes with frequency, a feedback circuit can be employed to vary the gate voltage on the negative g_m tail current source, thereby changing the bias current [14, 16]. In the vicinity of resonance, the requirement $k\text{Q} = 1$ makes the resonator filter equivalent to its LC ladder counterpart. As R_{shunt} increases with frequency, $-1/g_m$ should increase as well, implying a reduction in I_{DS}. An AGC-type circuit topology can be employed to detect the resonance peak at a specified I_{DS} and then track this peak using a peak detector.

The simple band notch filter discussed earlier in its simple form does not have adequate Q for most applications. Q-enhancement circuits, however, can also be used to improve the performance of the simple LC structure. One such notch filter uses a single MOSFET in a negative g_m circuit (Figure 5.26) to achieve a notch depth of more than 20 dB [17, 18].

The input impedance Z_{in} for the circuit illustrated in Figure 5.26 can be written from knowledge of the RF equivalent circuit as

$$Z_{\text{in}} = \frac{V_{\text{in}}}{i_{\text{in}}} = \frac{1}{j\omega C_f} + j\omega L + R_L - \frac{g_m}{\omega^2 C_f C_{gs}} + \frac{1}{j\omega C_{gs}} + R_g \tag{5.32}$$

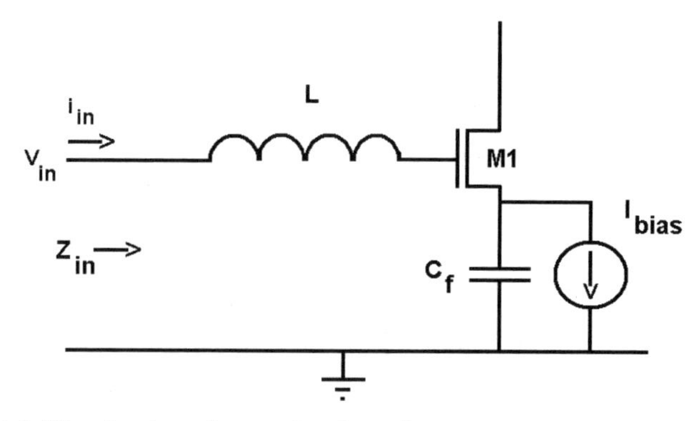

Figure 5.26 Notch filter structure shows a low impedance at resonance.

where R_L and R_g are the inductor and M1 gate parasitic resistances, respectively. The current source I_{bias} is used to set M1 transconductance g_m to cancel the resistive portion of the input impedance, thereby adding a zero to the expression for Z_{in}:

$$Z_{\text{in}} = \frac{1}{j\omega C_f} + j\omega L + \frac{1}{j\omega C_{gs}} = \frac{1-\omega^2 LC_{eq}}{j\omega C_{eq}} = \frac{1-\dfrac{f^2}{f_0^2}}{j\omega C_{eq}}; \tag{5.33}$$

$$C_{eq} = \frac{C_f C_{gs}}{C_f + C_{gs}}; \; f_0 = \frac{1}{2\pi\sqrt{LC_{eq}}}$$

A coupled resonator version of the bandstop filter has shown a 40-dB notch depth [16]. The coupled resonator version of this circuit (Figure 5.27) creates an input impedance term dependent on the mutual inductance of the coupled inductors, M. As in the filter circuit described above, the current source I_{bias} is used to set the proper g_m that provides a cancellation of the resistive losses that then increases the Q:

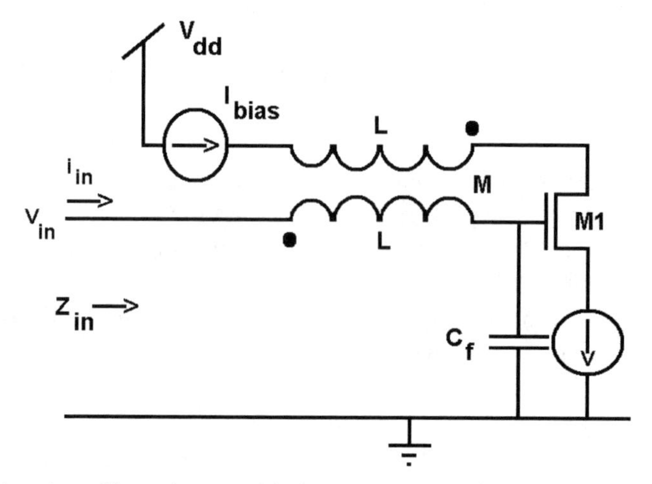

Figure 5.27 A bandstop filter using a coupled resonator. Note the winding polarity.

$$Z_{\text{in}} = \frac{1}{j\omega}\left(\frac{1}{C_f} + \frac{1}{C_{eq}}\right) + j\omega L + R_L + R_g - \frac{Mg_m}{C_{eq}};\ C_{eq} = C_{eq} = \frac{C_f C_{gs}}{C_f + C_{gs}} \tag{5.34}$$

Note the polarity of the dot convention in the coupled inductors; this connection places the transconductance current source of M1 out of phase with the input, providing the phase difference needed for resistive cancellation.

5.3.2 SAW Filtering

In modern communication circuits, surface acoustic wave (SAW) filters are used to provide excellent bandpass filter selectivity with relatively low insertion loss (several decibels). While these filters are widely used, they are almost exclusively off-chip components and hence are only mentioned here. SAW filters can find their way into single packaged circuits, however, as part of a multichip or of multipart modules.

5.3.3 Polyphase Filters

Polyphase filters are widely used in RFIC systems for providing the proper phase differences, usually in multiples of 90°, for use in circuits such as image reject mixers. While not "true" filters since their primary use is generating quadrature phases, these circuits can typically be implemented using circuit topologies similar to those of traditional filter circuits. Simple RC or RL circuits can provide 90° phase shifts, but these circuits are sensitive to parasitics as well as process mismatches. The 90° phase difference between two signals can be seen with two RC filter structures, one configured as a low-pass and the other as a high-pass filter (Figure 5.28).

For the low-pass side, the phase shift can be written as

$$\phi_{LP} = -\tan^{-1}(\omega RC) \tag{5.35}$$

and for the high-pass side, the phase shift is

$$\phi_{HP} = \tan^{-1}\left(\frac{1}{\omega RC}\right) \tag{5.36}$$

The phase difference between the two outputs for $\omega RC = 1$ becomes

$$\Delta\phi = \phi_{HP} - \phi_{LP} = \tan^{-1}(1) - \left[-\tan^{-1}(1)\right] = \frac{\pi}{2} \Rightarrow 90° \tag{5.37}$$

A drawback to this technique is that the signals are of equal amplitude only at ωRC = 1 and that both signals suffer 3-dB attenuation at that frequency (Figure 5.28). In addition, if R or C of one filter section is not matched with its counterpart in the other section, phase and amplitude errors will occur that can have a major impact on the circuit's performance. For example, a 5% variation in component difference between the two filter sections can provide up to 2° of phase difference and 0.2 dB of amplitude difference. Increasing the number of elements in the filter will

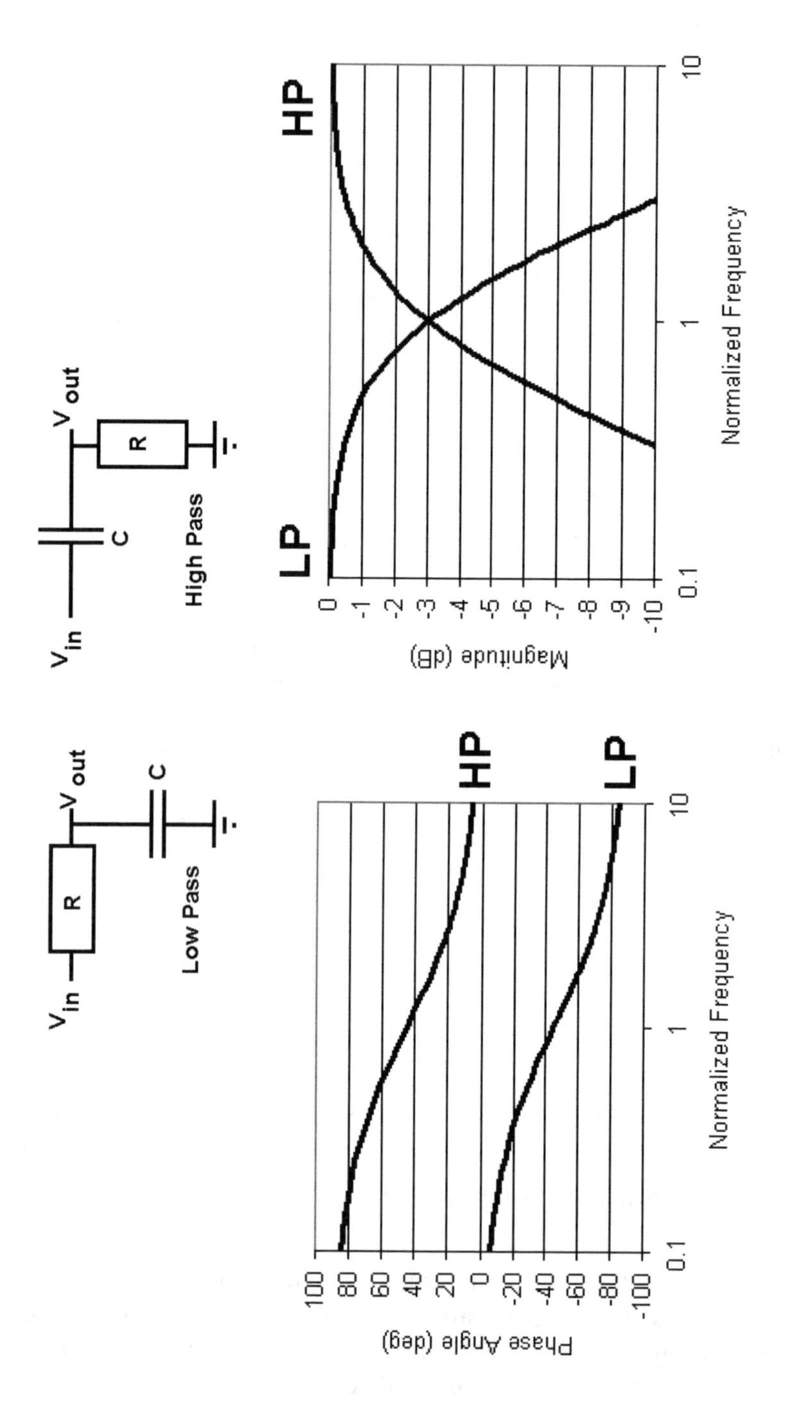

Figure 5.28 Low-pass and high-pass RC filters used to create a 90° phase difference between the two outputs.

"spread out" possible component variations, providing better phase and amplitude matching. A four-section polyphase filter (two low-pass and two high-pass sections) provides not only a 90° phase shift at a single port but also four quadrants of phase shift (Figure 5.29).

Choosing an RC product for unity frequency ($F = 1$ Hz), the ideal polyphase filter shows a full 90° phase shift between each port (Figure 5.30).

Varying one resistor in the polyphase network by 10% (indicated by the symbol R in Figure 5.29) exhibits a worst case phase error of 1.7° (Figure 5.31) over the entire frequency range, showing that the process variations' effects on phase shift in this circuit have been significantly reduced from the simple RF phase shift network of Figure 5.28.

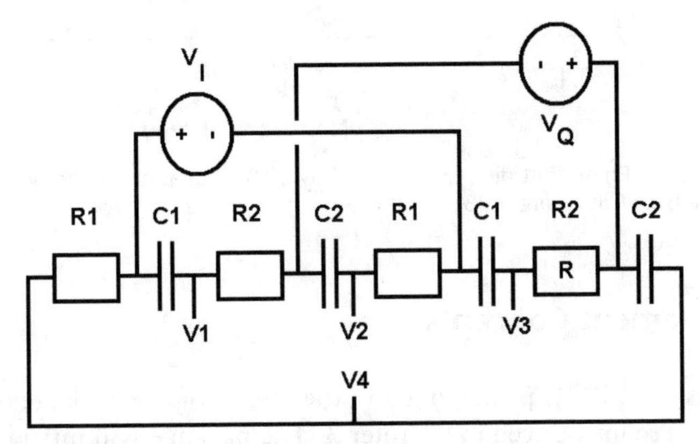

Figure 5.29 Four-pole polyphase filter suitable for use in an RFIC application. V_I and V_Q are the in-phase and quadrature (90° phase shift) input signals.

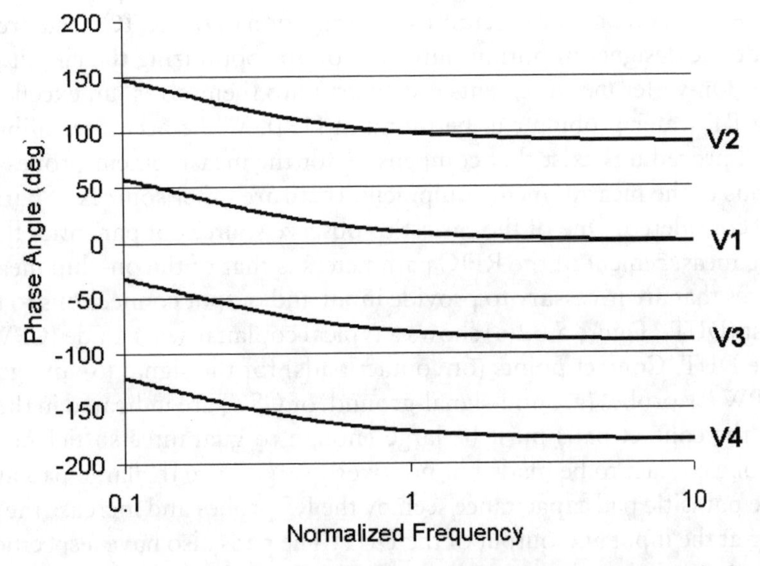

Figure 5.30 All four quadrants of 90° phase shift are available using a passive polyphase filter.

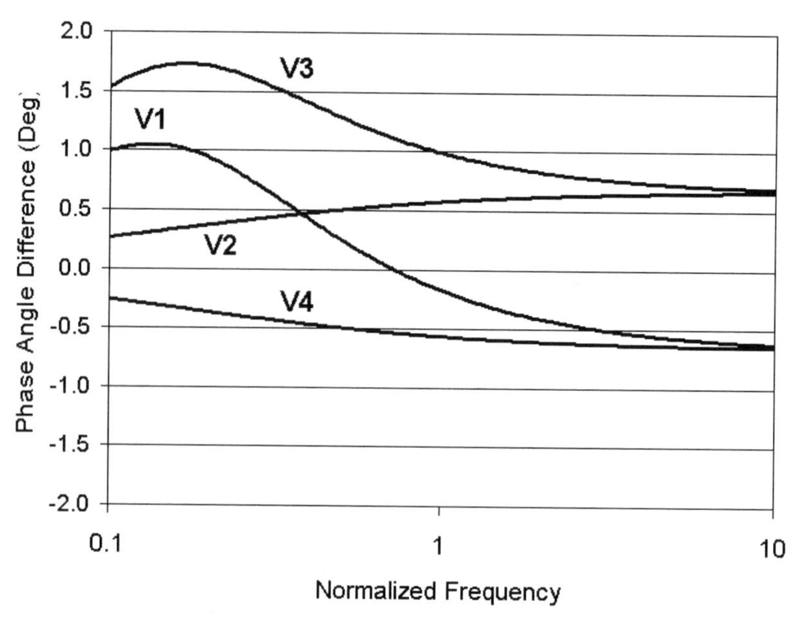

Figure 5.31 Phase shift difference caused by a 10% variation in one polyphase filter component value (resistor *R* in Figure 5.28).

5.4 Measurement Concepts

The basics of RFIC packaging and the effects of the package on overall RF performance were introduced in Chapter 3. The package will introduce a variety of reactive elements (package capacitance and bond wire inductance to name just a few) into the system that will degrade the performance of the RFIC from its unpackaged version. In the prototyping phase of RFIC development, these unwanted parasitics that are introduced by the package have the potential to mask the true performance of the circuit, so RF or on-wafer measurements are often performed. On-wafer measurements give a more accurate representation of the RFIC circuit response and provide the designer important information for optimizing the circuit performance.

On-wafer measurements are an art unto themselves (an excellent review article on RF wafer probing can be found in [19]). While accurate calibration standards and procedures exist that compensate for the measurement probes and the connections to the measurement equipment, there are other sources of parasitics that must be considered. One of the easiest to observe sources of parasitics that can influence the measurement of the RFIC parameters is that of the on-chip measurement structures that are necessary to provide input and output connections to the device under test (DUT). Figure 5.32(a) shows a typical coplanar waveguide (CPW) connection to the DUT. Contact points (or contact pads) for the signal (S) and ground (G) of the CPW RF probes (ground-signal-ground, or GSG) are indicated in the figure. The size of the contact pads must be large enough to guarantee sufficient area for a good probe contact to be made but not overly large since the large pad area will increase the parasitic pad capacitance seen by the RF probes and increase the capacitive loading at the input and output of the DUT. The pads also have a specific separation distance, or pitch, as indicated in the figure. Typical values of contact pad and pitch are

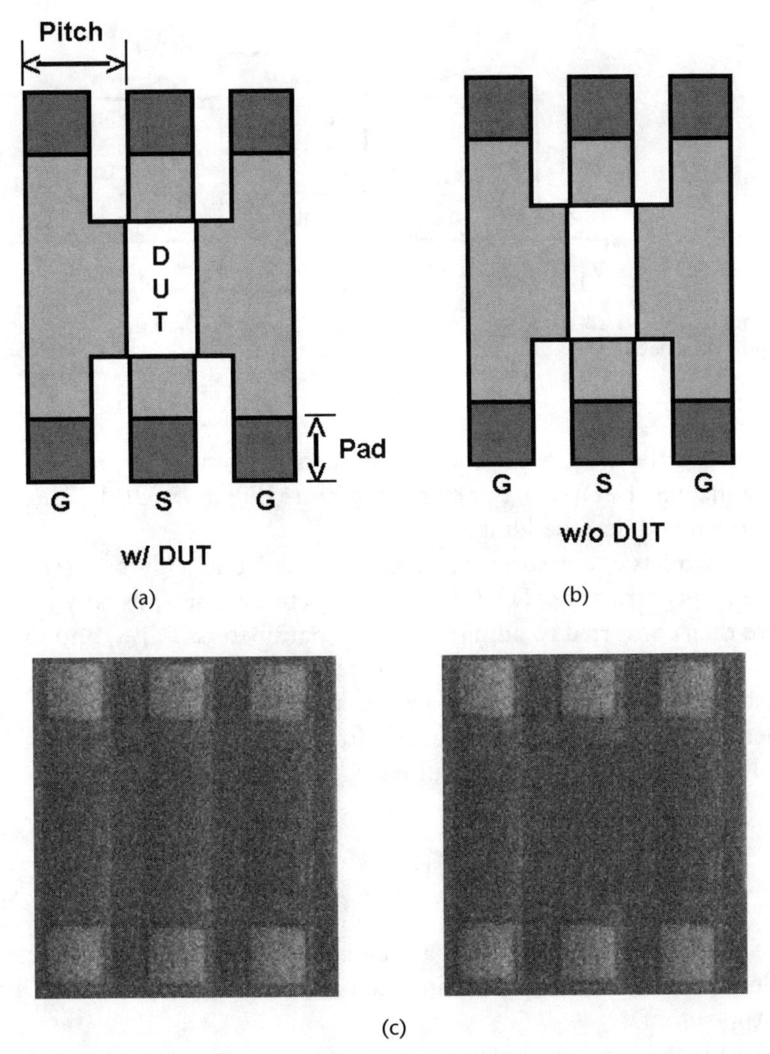

Figure 5.32 Coplanar waveguide test structure (a) with DUT in place and (b) without DUT in place but used for deembedding the DUT from the test structure. The probe configuration for this measurement structure is GSG. (c) SEM view of short and open circuit test structures (scalable CMOS technology). (Courtesy of The MOSIS Service.)

50 and 100 μm, respectively. The pad size and pitch can vary considerably from these typical values, however, with smaller values for higher frequency measurements and larger values for lower frequency or high-power measurements. One of the parameters required when ordering from the manufacturer of the RF wafer probes is this pitch distance.

The measurement structure shown in Figure 5.32(a) shows that the signal line (S) has a certain length that will introduce phase shift and possibly a small amount of loss into the measurement of the DUT. In addition, there will also be effects due to substrate coupling and losses introduced because of the measurement structure. Various techniques are employed to "back out," or deembed, the desired DUT characteristics from the overall measurement data taken at the CPW contact pads (which includes all the above mentioned parasitics). The simplest technique used to

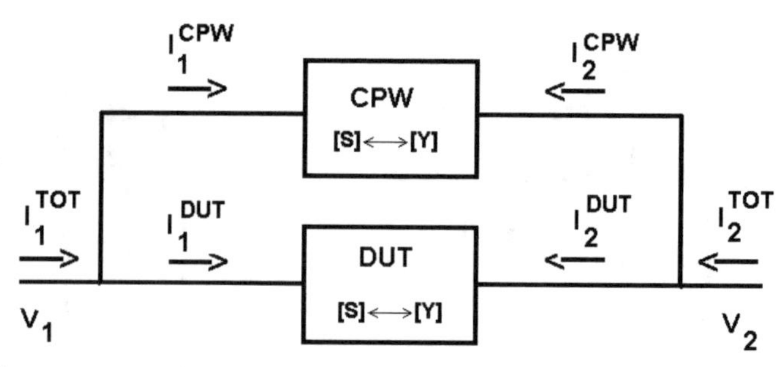

Figure 5.33 The *Y*-parameters of a parallel set of networks are the sum of the *Y*-parameters of the individual networks.

deembed the DUT is to fabricate an identical test structure on the RFIC wafer but not include the DUT in this empty structure [Figure 5.32(b)]. This is often referred to as open pad deembedding.

Two sets of measurements are then taken, one with the DUT in place and one on the empty structure. The two measurement data sets, usually in *S*-parameter form, are then converted to admittance or *Y*-parameters ($[Y]_{TOT}$ and $[Y]_{CPW}$) (see the conversion formulas for *S*- to *Y*-parameters and *Y*- to *S*-parameters in Appendix D). A useful property of admittance network parameters is that for any number of parallel networks expressed in *Y*-parameter form, the total *Y*-parameter matrix for the overall circuit is simply the sum of each set of individual network parameters (Figure 5.33):

$$[Y]_{TOT} = [Y]_{DUT} + [Y]_{CPW} \quad [Y]_{DUT} = [Y]_{TOT} - [Y]_{CPW} \tag{5.38}$$

Deembedding the DUT parameters is then a simple subtraction of the matrix elements; the *S*-parameters are then obtained by a [Y] to [S] matrix conversion (Appendix D).

A more sophisticated technique not only uses the open circuit test structure [Figure 5.42(b)] but also includes additional test structures, with the DUT replaced by a variety of open and short circuit layout topologies [20]. Measurements of the test structures are performed (both with and without the DUT), the data sets are then converted to *Z*- or *Y*-parameters, and the DUT parameters are deembedded by subtracting selected *Z*- or *Y*-parameter entries (details in [20, 21]). More advanced deembedding techniques using other "dummy" test structures (such as specific through lines) can be used to increase deembedding accuracy and ultimately improve knowledge of the DUT performance [22].

References

[1] Elwan, H. O., and M. Ismail, "Digitally Programmable Decibel-Linear CMOS VGA for Low-Power Mixed-Signal Applications," *IEEE Trans. Circuits Systems–II,* Vol. 47, No. 5, May 2000, p. 388. (See also Elwan, H. W., T. B. Tarim, and M. Ismail, "Digitally Programmable dB-Linear CMOS AGC for Mixed-Signal Applications," *IEEE Circuits Devices Mag.,* July 1998, p. 8.)

[2] Tadjour, S., F. Behbahani, and A. Abidi, "A CMOS Variable Gain Amplifier for a Wideband Wireless Receiver," *Proc. 1996 Symp. on VLSI Circuits,* 1996, p. 86.

[3] Harjani, R., O. Birkenes, and J. Kim, "An IF Stage Design for an ASK-based Wireless Telemetry System," *Proc. 2000 Int. Symp. Circuits and Systems,* Vol. 1, 2000, p. 52.

[4] Wei Lin, C., Y.-Zen Liu, and K. Y.J. Hsu, "A Low Distortion and Fast Settling Automatic Gain Control Amplifier in CMOS Technology," *Proc. 2004 Int. Symp. Circuits Sys.,* Vol. 1, 2004, p. 541.

[5] Razavi, B., *RF Microelectronics,* Upper Saddle River, NJ: Prentice Hall, 1997.

[6] *ARRL Handbook,* Newington, CT: American Radio Relay League, 1987.

[7] Karthaus, U., and M. Fischer, "Fully Integrated Passive UHF RFID Transponder IC with 16.7-μW Minimum RF Input Power," *IEEE J. Solid-State Circuits,* Vol. 38, No. 10, Oct. 2003, p. 1602.

[8] Curty, J.-P., et al., "Remotely Powered Addressable UHF RFID Integrated System" IEEE J. Solid-State Circuits, Vol. 40, No. 11, Nov. 2005, p. 2193.

[9] Sauer, C., et al., "Power Harvesting and Telemetry in CMOS for Implanted Devices" *IEEE Trans. Circuits Systems*—I, Vol. 52, No. 12, Dec. 2005, p. 2605.

[10] Zhuo, W., J. Pineda de Gyvez, and E. Srinchez-Sinencio, "Programmable Low Noise Amplifier with Active-Inductor Load," *Proc. ISCAS '98,* Vol. 4, May 1998, p. 365.

[11] Yang, J.-N., Yi-C. Cheng, and C.-Yi Lee, "A Design of CMOS Broadband Amplifier with High-Q Active Inductor," Proc. *3rd IEEE Int. Workshop System-on-Chip for Real-Time Applications,* June 2003, p. 86.

[12] Maatthaei, G., L. Young, and E. M. T. Jones, *Microwave Filters, Impedance Matching Networks, and Coupling Structures,* Dedham, MA: Artech House, 1980.

[13] Temes, G. C., and J. W. LaPatra, *Introduction to Circuit Synthesis and Design,* New York: McGraw-Hill, 1977.

[14] Li, D., and Y. Tsividis, "Design Techniques for Automatically Tuned Integrated Gigahertz-Range Active LC Filters," *IEEE J. Solid-State Circuits,* Vol. 27, No. 8, Aug. 2002, p. 967.

[15] Pozar, D., *Microwave Engineering,* 2nd ed., New York: John Wiley and Sons, 1998.

[16] Kuhn, W. B., "Design of Integrated RF Bandpass Filters and Oscillators for Low-Power Radio Receivers," *Proc. 9th IEEE Int. ASIC Conf.,* Sept. 1996, p. 87.

[17] Soorapanth, T., and S. S. Wong, "A 0-dB IL 2140 30 MHz Bandpass Filter Utilizing Q-Enhanced Spiral Inductors in Standard CMOS," *IEEE J. Solid-State Circuits,* Vol. 37, No. 5, May 2002, p. 579.

[18] Nguyen, T.-K., and S.-G. Lee, "A 5.2 GHz Image Rejection CMOS Low Noise Amplifier Using Notch Filter," *Proc. 46th IEEE Int. Midwest Symp. Circuits Systems Symp.,* Vol. 3, Dec. 2003, p. 1231.

[19] Wartenberg, S. A. "Selected Topics in RF Coplanar Probing," *IEEE Trans. Microwave Theory Tech.,* Vol. 51, No. 4, April 2003, p. 1413.

[20] Vandamme, E., D. Schreurs, and C. van Dinther, "Improved Three-Step De-Embedding Method to Accurately Account for the Influence of Pad Parasitics in Silicon On-Wafer RF Test-Structures," *IEEE Trans. Electron Devices,* Vol. 48, No. 4, April 2001, p. 737.

[21] Kolding, T., "A Four-Step Method for De-Embedding Gigahertz On-Wafer CMOS Measurements," *IEEE Trans. Electron Devices,* Vol. 47, No. 4, April 2000, p. 734. (See also Kolding, T., "On Wafer Calibration Techniques for Giga-Hertz CMOS Measurements," *Proc. IEEE 1999 Int. Conf. Microelectronic Test Structures,* Vol. 12, March 1999, p. 5.)

[22] Cho, M., et al., "A Shield-Based Three-Port De-Embedding Method for Microwave On-Wafer Characterization of Deep-Submicrometer Silicon MOSFETs," *IEEE Trans. Microwave Theory Tech.,* Vol. 53, No. 9, Sept. 2005, p. 2926.

Selected Bibliography

Lee., T. H., *The Design of CMOS Radio-Frequency Integrated Circuits,* 2nd ed., Cambridge, UK: Cambridge University Press, 2004.

Rogers, J., and C. Plett, *Radio Frequency Integrated Circuit Design,* Norwood, MA: Artech House, 2003.

CHAPTER 6

CMOS Oscillator Circuits

Oscillators are used in a many of the blocks in an RFIC. In the receive section, oscillators are used in conjunction with mixers for frequency shifting (primarily by downconversion) of the received RF signal to a lower frequency so that signal processing can be more easily performed. On the transmit side, the oscillator generates a signal that is encoded with the information to transmit and then frequency shifted (primarily upconverted), amplified, and sent to the antenna for transmission.

"Why can a high school kid running a PA system make an oscillator, but a professor with Lyapanov[1] functions and ADS cannot?" [1]. While a humorous quote to be sure, the quote really gets to the crux of the oscillator design problem. Although an amplifying structure is indeed required to create an oscillator (the PA system), a host of other issues arise as part of the oscillator design:

- The oscillator frequency should be controllable.
- The oscillator signal amplitude should be controllable.
- The oscillator should have reasonable power consumption.
- The oscillator should be stable and controllable.
- The oscillator signal should not cause undue interference with adjacent signals.
- We need to know how on-chip components influence the design and operation of oscillators.

These and other oscillator issues are discussed in this chapter. The chapter begins with a review of general feedback principles (no Lyapunov functions seen here, but we do discuss the Barkhausen[2] criteria). With these general feedback principles, the design of fixed-frequency oscillators is covered from several complementary perspectives and applied to both lumped element as well as mechanical and digital based oscillators. Frequency control of oscillators is reviewed and how the various resonators in the oscillator ultimately influence the noise level in these circuits. A number of design examples are provided to help in understanding of these concepts.

1. In the late 1800s, A. M. Lyapunov (a former student of P. L. Chebyshev) investigated the use of various functional forms to test the stability of control and nonlinear systems [2]. He also proved the central limit theorem, a fundamental tenet in probability.
2. H. G. Barkhausen was a German electrical engineering professor in the early 1900s. He first observed discrete jumps in magnetic field phenomena that were subsequently named the Barkhausen effect. His study of transmitters and oscillators led to oscillation criteria that also bear his name.

6.1 Review of General Feedback Principles

6.1.1 General Feedback Systems

Two blocks are required to describe the general feedback system: a gain block and a feedback block. Although we can easily write a block diagram, the main design issues are in the details of how these two circuits interact and how this interaction causes oscillation to occur. Figure 6.1 shows a block diagram of a general feedback system with an amplifier having an *open loop gain* A and *feedback network of gain* B. Input and outputs X_{in} and X_{out} can be voltages, currents, a combination of the two, or other parameters.

Assuming that the feedback is additive with input, the output parameter X_{out} can be written as

$$X_{out} = A(X_{in} + BX_{out}); \; X_{out}(1 - AB) = AX_{in} \tag{6.1}$$

with the ratio of the output parameter to the input parameter given by

$$\frac{X_{out}}{X_{in}} = \frac{A}{1 - AB} \tag{6.2}$$

The denominator of (6.2) goes to zero when $AB = 1$, implying that there is an output without a driving input signal, just what is needed for an oscillator. In other words, the loop gain (AB) has to be equal to unity (one with a phase angle of zero) in order for an oscillation to occur. This is the *Barkhausen criteria* and they allow the circuit designer to look at the amplifier and feedback blocks to determine the conditions for oscillation. Note that some feedback analyses have the term ABX_{out} in (6.1) as $-ABX_{out}$ because they use a difference rather than a summing node in the input branch. In this case, the unity condition still applies, but the phase angle is 180°; the analysis is still the same. For general feedback systems, both A and B are functions of frequency, which makes the study of feedback systems so interesting. For most feedback systems, such as control systems, stability is of utmost concern because the system must be well behaved for all conditions. If the amplifier gain is extremely high (i.e., A is large), the voltage gain is simply related to the feedback structure itself:

$$\frac{X_{out}(\omega)}{X_{in}(\omega)} = \frac{A(\omega)}{1 - A(\omega)B(\omega)} \Rightarrow -\frac{1}{B(\omega)} \tag{6.3}$$

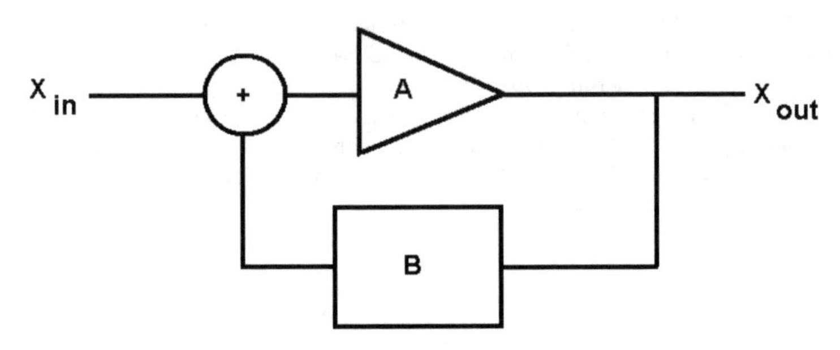

Figure 6.1 Block diagram of a general feedback system.

For this discussion, however, the focus will be on how to determine gain and feedback circuits needed to create an oscillator using MOSFETs as the source of open loop gain. For the case of an oscillator, we want the system to oscillate as opposed to simply amplify (as the axiom states, "if you want to build an oscillator, build an amplifier ... it'll oscillate"). In any case, the loop gain AB goes to unity.

6.1.2 Gain/Phase Margins

The key factor in feedback system stability is the loop gain, $A(\omega)B(\omega)$. Although the so-called *Nyquist criteria* are used to mathematically determine feedback system stability [3], the resulting mathematics are difficult to apply in practice. Knowledge of the amplitude and phase responses of the loop gain AB, however, provide good stability rules of thumb for use by the CMOS RFIC designer. For reasonably stable systems, the design goal is to make sure the magnitude of the loop gain $|AB|$ is less than unity at zero phase. The smaller the magnitude, the greater the stability. This *gain margin* is often chosen to be anywhere from 10 to 20 dB. Conversely, if the magnitude of AB is unity, the system is still stable if the phase is far from 0°. The *phase margin* is often chosen to be 30° to 60° to ensure system stability. Figure 6.2 shows Bode plots of the loop gain amplitude and phase showing these two margins. For the loop gain to satisfy the oscillation criterion, however, $|AB|$ must be 1 and the phase of AB must be equal to 0°, implying that the gain and phase margins are zero. The same Bode plots and techniques can be used to determine the frequency (or frequencies) where an oscillation will occur. These plots of loop gain will be visited in

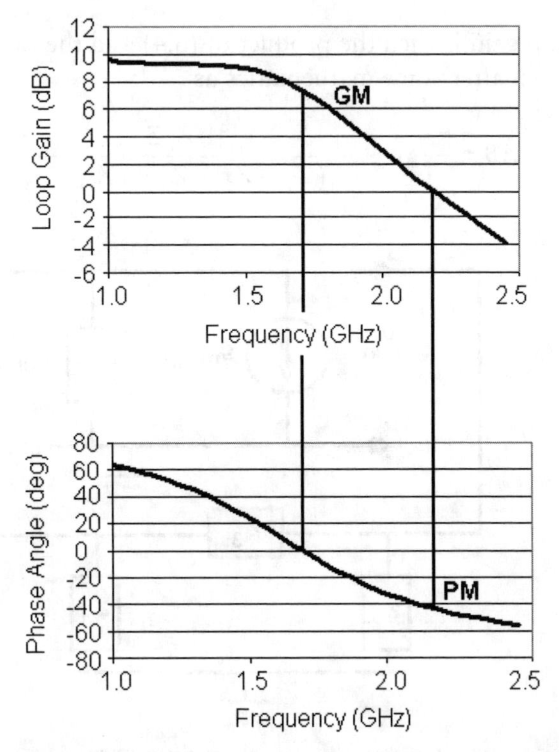

Figure 6.2 Gain margin (GM) and phase margin (PM) of the loop gain. In this case, the gain margin is 7 dB and the phase margin is 42°.

the next section and used to define oscillation criteria for general oscillating feedback networks.

6.1.3 Reactance Oscillators

The fundamental oscillator type is often called the *reactance oscillator* since the feedback network is composed of only reactive elements. The loop gain for the reactance oscillator can be derived by looking at an example of a common source *n*MOSFET amplifier (general transconductance g_m and output resistance r_{out}) with a pi-network of general reactances Z_i surrounding the structure as shown in Figure 6.3. In the figure, the biasing networks are not shown (an issue of clarity) but must be included to bias the *n*MOSFET into its amplifying region (open loop gain greater than 1). The feedback term B is easily determined as the ratio of V_{in} to V_{out}:

$$B = \frac{V_{in}}{V_{out}} = \frac{Z_2}{Z_2 + Z_3} \tag{6.4}$$

The gain, not including feedback (termed the *open loop gain*) is a function not only of the MOSFET RF equivalent circuit parameters but also the loading by the feedback network itself. The effective load seen by the MOSFET is the output resistance r_{out} in parallel with the reactance of Z_1 in parallel with the series term $Z_2 + Z_3$:

$$V_{out} = -g_m V_{in}\left(r_{out} \| Z_B\right);\ Z_B = \frac{Z_1(Z_2 + Z_3)}{Z_1 + Z_2 + Z_3} \tag{6.5}$$

The loop gain is then the product of (6.4) and the ratio V_{out}/V_{in} in (6.5), which can be written after some mathematics as

$$AB = -g_m r_{out} \frac{Z_1 Z_2}{r_{out}(Z_1 + Z_2 + Z_3) + Z_1(Z_2 + Z_3)} \tag{6.6}$$

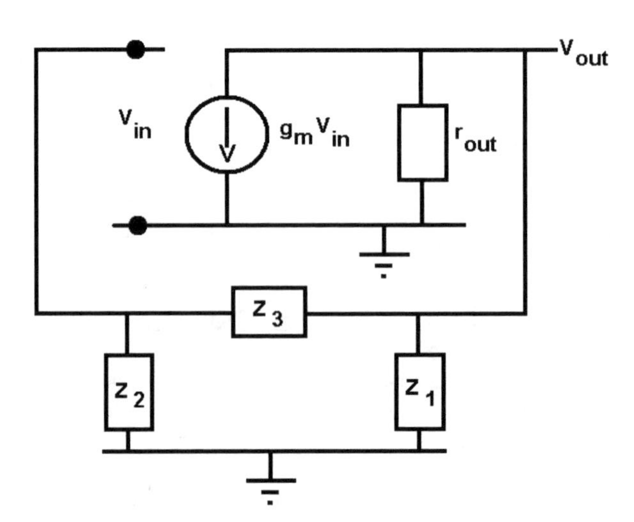

Figure 6.3 Generalized oscillator structure with MOSFET RF equivalent circuit and three reactive element feedback pi-networks.

For pure reactances ($Z_i = jX_i$), (6.6) can be written as

$$AB = g_m r_{out} \frac{X_1 X_2}{-X_1(X_2 + X_3) + jr_{out}(X_1 + X_2 + X_3)} \tag{6.7}$$

Recall from the oscillation criteria that AB had to be unity with zero phase for oscillation to occur. From (6.7), $AB = 1$ only occurs if the imaginary term in the denominator of (6.7) is equal to zero, which requires one of the reactances to have the *opposite* sign from the others:

$$X_1 + X_2 + X_3 = 0 \tag{6.8}$$

This result implies that the feedback network will contain both positive and negative reactive elements, inductors and capacitors, respectively. Using (6.8) yields a simplified expression for the loop gain AB:

$$AB = g_m r_{out} \frac{X_1 X_2}{-X_1(X_2 + X_3)} = g_m r_{out} \frac{X_2}{X_1} \tag{6.9}$$

For AB to be unity with zero phase, X_2 and X_1 must be of the *same* reactance sign, and the ratio should reduce AB to unity. For X_2 and X_1 of the same reactance, reactance X_3 must be

$$X_3 = -(X_1 + X_2) \tag{6.10}$$

Assuming X_2 and X_1 are capacitors and X_3 an inductance, (6.10) gives the resonant frequency or frequency of oscillation of the circuit (provided (6.9) is satisfied):

$$X_3 = \omega L = -(X_1 + X_2) = \frac{1}{\omega}\left(\frac{1}{C_1} + \frac{1}{C_2}\right)$$

$$\omega_{tank}^2 = \frac{\left(\frac{1}{C_1} + \frac{1}{C_2}\right)}{L} = \frac{1}{L\frac{C_1 C_2}{C_1 + C_2}} \tag{6.11}$$

This type of oscillator is generally referred to as the basic *reactance oscillator* and contains two capacitors and one inductor that make up the so-called *tank circuit* (Figure 6.4).

When power is initially applied to the feedback amplifier circuit, the only input to the amplifier is wideband random thermal noise. This small thermal noise voltage is amplified and band limited by the resonator in the feedback loop where the loop gain will be larger than unity. As the oscillation proceeds, amplifier gain A will decrease due to decreasing transconductance g_m until the loop gain is unity at zero phase angle at the oscillation frequency.[3]

3. Recall in Chapter 2 that MOSFETs operating in the subthreshold have very high transconductances that gradually decrease as the MOSFET reaches the strong inversion operating region.

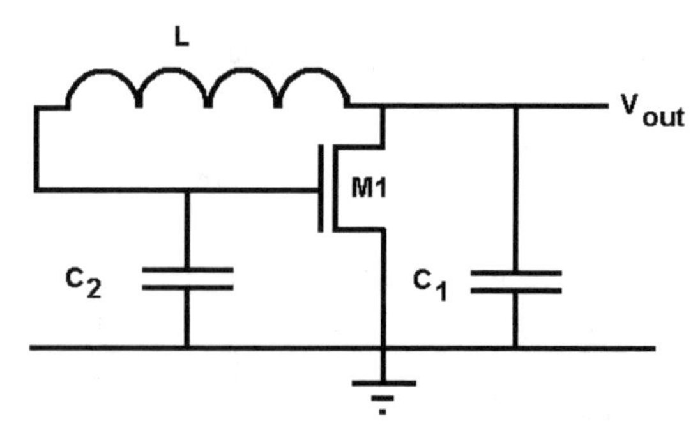

Figure 6.4 Basic reactance oscillator (bias network not shown but required for biasing M1 into its amplifying regime).

The simple circuit shown in Figure 6.4 brings up a number of issues common to all CMOS RFIC oscillator designs. The first issue is the inductance. Equation (6.6) assumes that the impedances in the feedback network have no resistance. From the discussion of passive elements in Chapter 3, on-chip inductors in CMOS RFIC technology usually exhibit relatively poor Q's and therefore introduce additional loss into the oscillator circuit. These losses must be considered because they have an impact on the loop gain and therefore on the circuit conditions required for oscillation. The second issue is the frequency-determining capacitances C_1 and C_2. The RF equivalent circuit in Figure 6.3 neglected to show the input MOSFET capacitance C_{in}. Fortunately, this parameter can be made part of feedback network capacitance C_2, indicating that an external capacitance required for setting the frequency will be less than that calculated in (6.9) and (6.11) because of input capacitive loading by the MOSFET. The capacitive loading effects are not just limited to the input. The output capacitance of the MOSFET can be substantial because of the diffusion region making up this MOSFET terminal. This output capacitance C_{out} also reduces the external capacitance C_1 needed for oscillation. The oscillator is not operating all by itself and so capacitive loading of the output (C_{load}) is also a part of C_1 and will affect the operating frequency if significant. The parasitics in the feedback loop also impact the loop gain response. Recall from the discussion of on-chip inductors in Chapter 3 that not only does the inductor have a finite resistance that limits its Q, but the inductor also exhibits a significant shunt capacitance by virtue of the long trace length. Figure 3.11 shows that this shunt capacitance can be modeled as a capacitance C_{ox} equally distributed at both terminals of the inductor, thereby adding to feedback network capacitances C_1 and C_2. Assuming ideal capacitors but a lossy inductor ($Z_3 = R_3 + jX_3$), the loop gain equation can be modified as follows:

$$AB = g_m r_{out} \frac{C_1}{C_2} \frac{1}{1 + \frac{1}{Q} \frac{\omega \omega_c}{\omega_{tank}} \left(1 + \frac{C_1}{C_2}\right) \left[1 + \left(\frac{\omega}{\omega_c}\right)^2\right]} \tag{6.12a}$$

$$R_3 X_1 + r_{out}(X_1 + X_2 + X_3) = 0 \tag{6.12b}$$

where $\omega_c = 1/r_{out}C_1$ and $Q = X_3/R_3$. The resonance condition has to overcome these additional losses (6.12b), with the resonance condition becoming

$$\begin{aligned}\omega_{res}^2 &= \frac{\left(\frac{1}{C_1}+\frac{1}{C_2}\right)}{L}\left(1+\frac{R_3}{r_{\text{out}}}\frac{C_2}{C_1+C_2}\right) \\ &= \omega_{tank}^2\left(1+\frac{1}{Q}\frac{\omega_{tank}L}{r_{\text{out}}}\frac{C_2}{C_1+C_2}\right) = \omega_{tank}^2\left(1+\frac{1}{Q}\frac{\omega\omega_c}{\omega_{tank}}\right)\end{aligned} \tag{6.13}$$

Equation (6.13) shows that for finite inductor Q's, the resonant frequency is slightly increased. The feedback circuit corner frequency ω_c is typically much less than the frequency of oscillation. The inductor Q impacts this feedback corner frequency by increasing the impact of r_{out} by reducing its value [(Qr_{out}) gets smaller as the Q reduces with high inductor series resistance]. Putting these nonidealities together shows that the loop gain and resonant frequency expression can be written as

$$AB = g_m r_{\text{out}}\frac{C_{1,ext}+C_{\text{out}}+C_{load}+C_{ind}}{C_{2,ext}+C_{\text{in}}+C_{ind}}\times \frac{1}{1+\frac{\omega L}{Qr_{\text{out}}}\left\{1+\left[\omega\left(C_{1,ext}+C_{\text{out}}+C_{load}+C_{ind}\right)r_{\text{out}}{}^2\right]\right\}} \tag{6.14a}$$

$$\omega^2 = \omega_{tank}^2\left(1+\frac{1}{Q}\frac{1}{\omega_{tank}}\frac{1}{r_{out}\left(C_{1,ext}+C_{out}+C_{load}+C_{ind}\right)}\right);$$

$$\omega_{tank} = \sqrt{\frac{1}{L\frac{\left(C_{1,ext}+C_{out}+C_{load}+C_{ind}\right)\left(C_{2,ext}+C_{in}+C_{ind}\right)}{\left(C_{1,ext}+C_{out}+C_{load}+C_{ind}\right)+\left(C_{2,ext}+C_{in}+C_{ind}\right)}}} \tag{6.14b}$$

Equations (6.14a and b) show that careful and accurate modeling of both active devices and passive elements in the reactance oscillator is necessary for successfully predicting both the loop gain magnitude AB as well as the oscillator freqency ω. The reader is referred to Chapters 2 and 3 to review the design and analysis equations for these critical circuit parameters.

Example 6.1: Determine the loop gain magnitude and phase for a simple MOSFET reactance oscillator feedback network. Assume that $g_m = 0.01$S, $r_{\text{out}} = 10$ kΩ, L = 10 nH, C_1 and C_2 are 1.0 and 10.0 pF, respectively, and the inductor series resistance varies from 0 to 12Ω in 3Ω steps.

The magnitude and phase of (6.6) is computed and shown in Figure 6.5. The resonant frequency is easily seen to be approximately 1.7 GHz, indicating that this is the probable frequency of oscillation. Note that the phase goes to zero at the same point for all values of inductor resistance [Figure 6.5(b)], but the loop gain [Figure 6.5(a)] tells a more interesting story. For a lossless inductor ($R_3 = 0$), there is plenty of gain available for the circuit to oscillate ($|AB| > 0$ dB). However, as the inductor losses increase, the loop gain at 0° phase shift decreases, and for an inductor resistance greater than about 6Ω, the loop gain drops below 0 dB ($|AB| < 1$), indicating

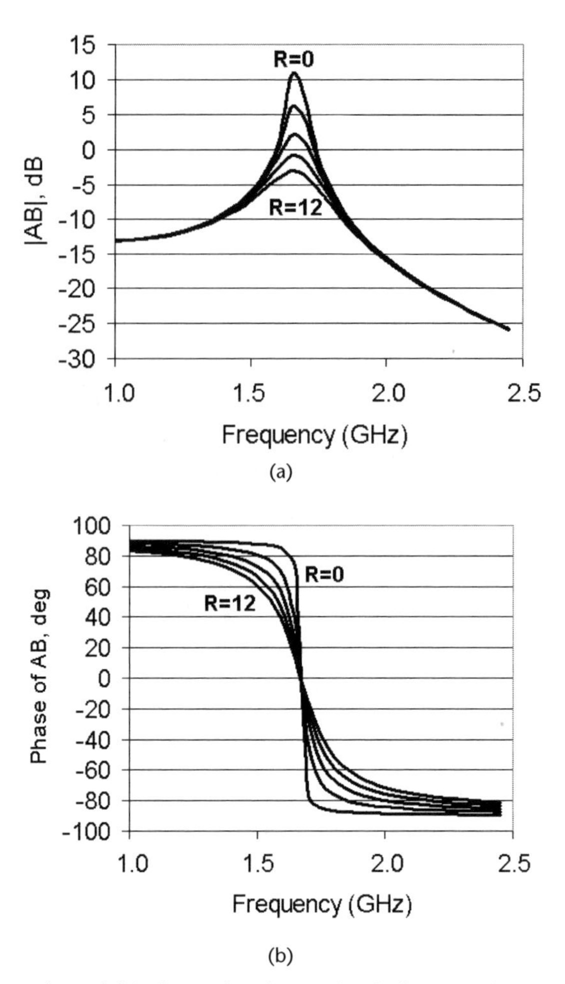

Figure 6.5 (a) Magnitude and (b) phase plots for varying inductor resistance for a simple MOSFET reactance style oscillator.

that any oscillation that does begin will be dampened by the additional circuit losses. The 6Ω inductor resistance corresponds to an inductor Q of 18, which is quite high for on-chip inductors. However, note that the phase shift is not dependent on gain (at least to first order), so increases in MOSFET gain (through increases in g_m) can be used to compensate for low-Q inductors. This "game" is not without limit, however, because both C_{in} and C_{out} are also functions of MOSFET sizing and will increase as g_m increases with increasing MOSFET gate width W. The increasing capacitance will also lower the oscillation frequency.

Example 6.2: A MOSFET's width cannot be extended without bound to improve g_m to compensate for losses and improve loop gain without the added capacitance influencing the resonant frequency. Investigate the effect of increasing gate width and MOSFET bias (V_{gs}–V_T) on the loop gain and resonant frequency. Assume a gate length of 0.5 μm and the values in Table 6.1.

The parameters listed in Table 6.1 were programmed into (6.14a) to investigate how the loop gain AB varies with W and $V_{gs} - V_T$. Figure 6.6(a) shows the results of

Table 6.1 Simulation Parameters for Example 6.2

Parameter	*Value*
L	10 nH
Q	10
$C_{1,ext}$, $C_{2,ext}$	1.0 pF
KP	56×10^{-6} A/V^2
λ	0.1
C_{ox}	2.5 fF/μm^2
C_{area}	0.43 aF/μm^2
$C_{sidewall}$	0.3 aF/μm

these calculations and indicates that there is a broad peak in loop gain *AB*'s dependence on gate width *W* beyond a certain value of *W*. Increasing W beyond this point does not improve the loop gain significantly, but raises the equivalent capacitance seen by the inductor and hence decreases the frequency of oscillation [Figure 6.6(c)]. Figure 6.6(b) selects particular bias points to show the loop gain's dependence on gate width and bias. Note that for higher bias levels, the loop gain begins to saturate ($V_{gs} - V_T = 2.0$) or even peak and start of roll off ($V_{gs} - V_T = 3.0$), indicating that MOSFET sizing should be thoroughly investigated when designing a MOSFET-based reactance oscillator.

6.1.4 Classic Reactance Oscillator Circuits

The previous discussion focused on the common source nMOSFET amplifier as the gain block in a general reactance oscillator. MOSFETs in common gate (CG) and common drain (CD) amplifying configurations, however, may provide easier biasing and better oscillator performance and are often used instead of the common source configuration. A grounded gate oscillator with the general feedback network reactances Z_i can be defined as shown in Figure 6.7(a). With an RF bypassed gate connection to a dc supply providing the gate RF ground, this makes the CG oscillator type particularly easy to gate bias.

The common drain (or source follower) oscillator as shown in Figure 6.7(b) can also be created using similar element definitions for the feedback network reactances Z_i. The RF ground in Figure 6.7(b) can be a direct connection to the dc power supply (a shunt capacitor is also often used), making this oscillator circuit a particularly good, easy choice to provide drain bias current.

If Z_1 and Z_3 are capacitors and Z_2 is an inductor [Figure 6.8(a)], the CG oscillator is then referred to as a *Colpitts* oscillator as originally envisioned by Colpitts. Equation (6.10) indicated that using two inductors (X_1 and X_3) and one capacitor (X_2) will also satisfy the resonance criteria of the feedback oscillator system. This configuration is referred to as the classic *Hartley* oscillator [Figure 6.8(b)] but is rarely used in CMOS RFIC design because of poor inductor Q and the associated increase in losses the amplifier must overcome. In addition, inductors are inefficient users of silicon real estate so why double that inefficiency with two inductors (although tapped inductors are often used).

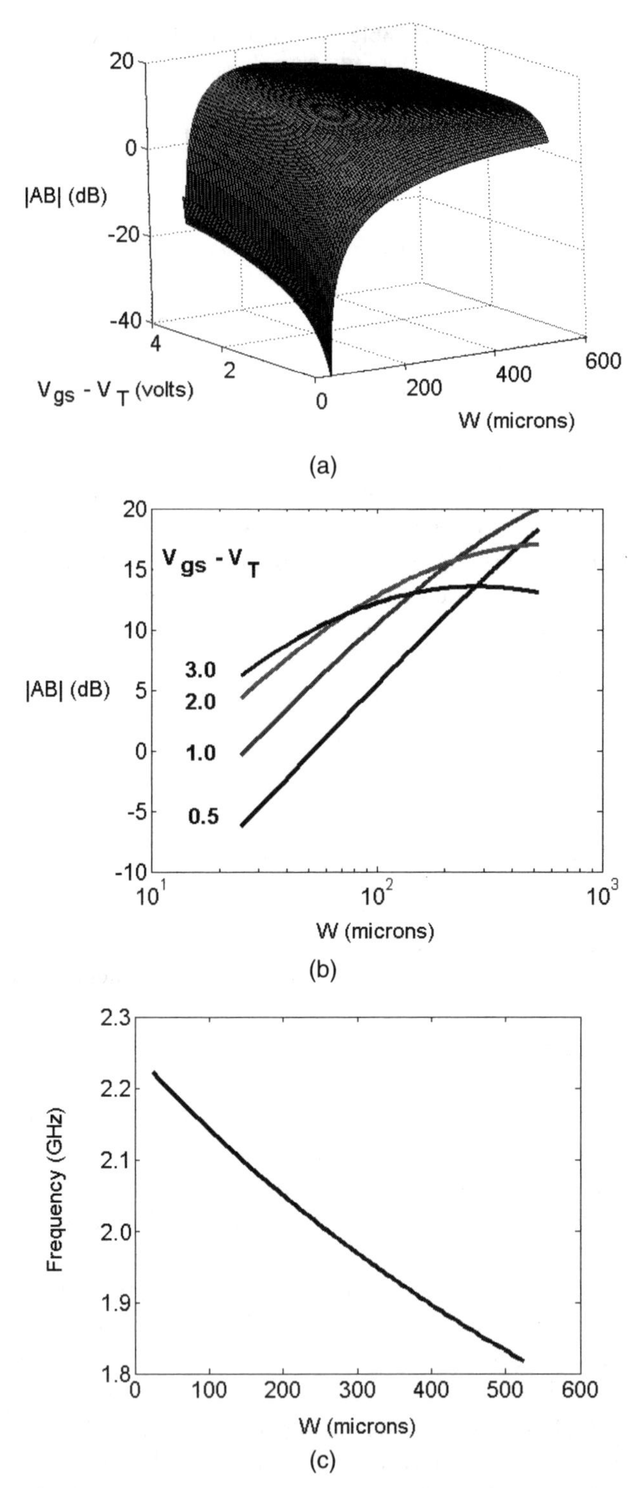

Figure 6.6 (a) Feedback network loop gain versus gate width and MOSFET bias. (b) Loop gain versus gate width for selected MOSFET bias levels. (c) Resonant frequency versus gate width for a MOSFET reactive oscillator.

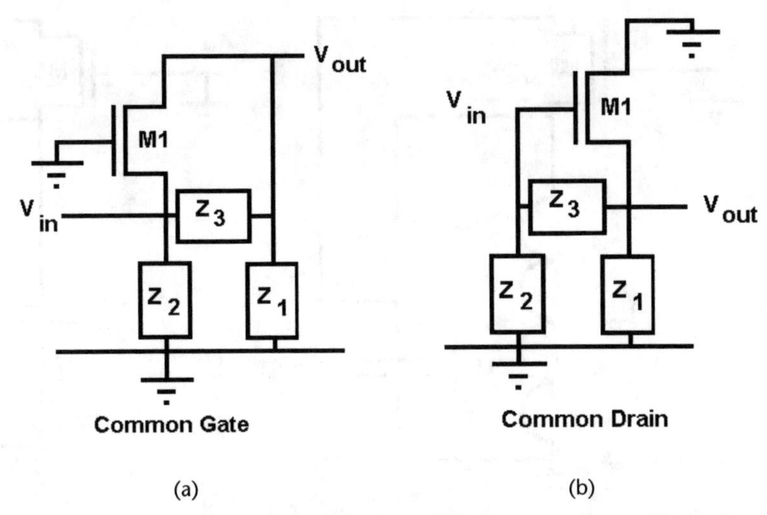

(a) (b)

Figure 6.7 General (a) common gate and (b) common drain reactance oscillator circuits. (Bias networks not shown but needed to bias M1 as an amplifier.)

The feedback element X_3 does not necessarily have to be a single component but can be a series LC network [Figure 6.8(c)] with total reactance given by

$$X_3 = \omega L_3 \left[1 - \left(\frac{\omega_3}{\omega} \right)^2 \right]; \omega_3 = \frac{1}{\sqrt{L_3 C_3}} \tag{6.15}$$

This configuration can also be thought of as a Colpitts oscillator, but is more often referred to as a *Clapp*-style oscillator. The addition of the series capacitor C_3 allows the use of a larger inductor value while still maintaining the required inductive reactance that resonates with the tank capacitors. The Clapp oscillator can theoretically store more energy, which raises the Q of the tank circuit (which also aids in the reduction of oscillator phase noise) [4]. However, the larger inductors come with increased resistive losses and capacitive parasitics which must be considered. Other oscillator configurations such as the *Seiler* and *Vackar* are variations on the basic Colpitts and Hartley themes [4].

6.2 Fixed-Frequency Oscillators

From the amplifier discussion in Chapter 4, the use of differential circuits is widespread in typical CMOS RFICs because of the inherent reduction of common mode noise sources. Differential oscillator circuits are widely used to complement such circuitry and this section investigates some of these oscillator circuits.

6.2.1 Single-Stage Amplifier with LC Tank Load

Figure 6.5 shows that the reactance oscillator undergoes an 180° phase shift over frequency, from +90° to −90°, with the zero phase shift frequency indicating the frequency of oscillation. A high-Q parallel RLC circuit [Figure 6.9(a)] undergoes a

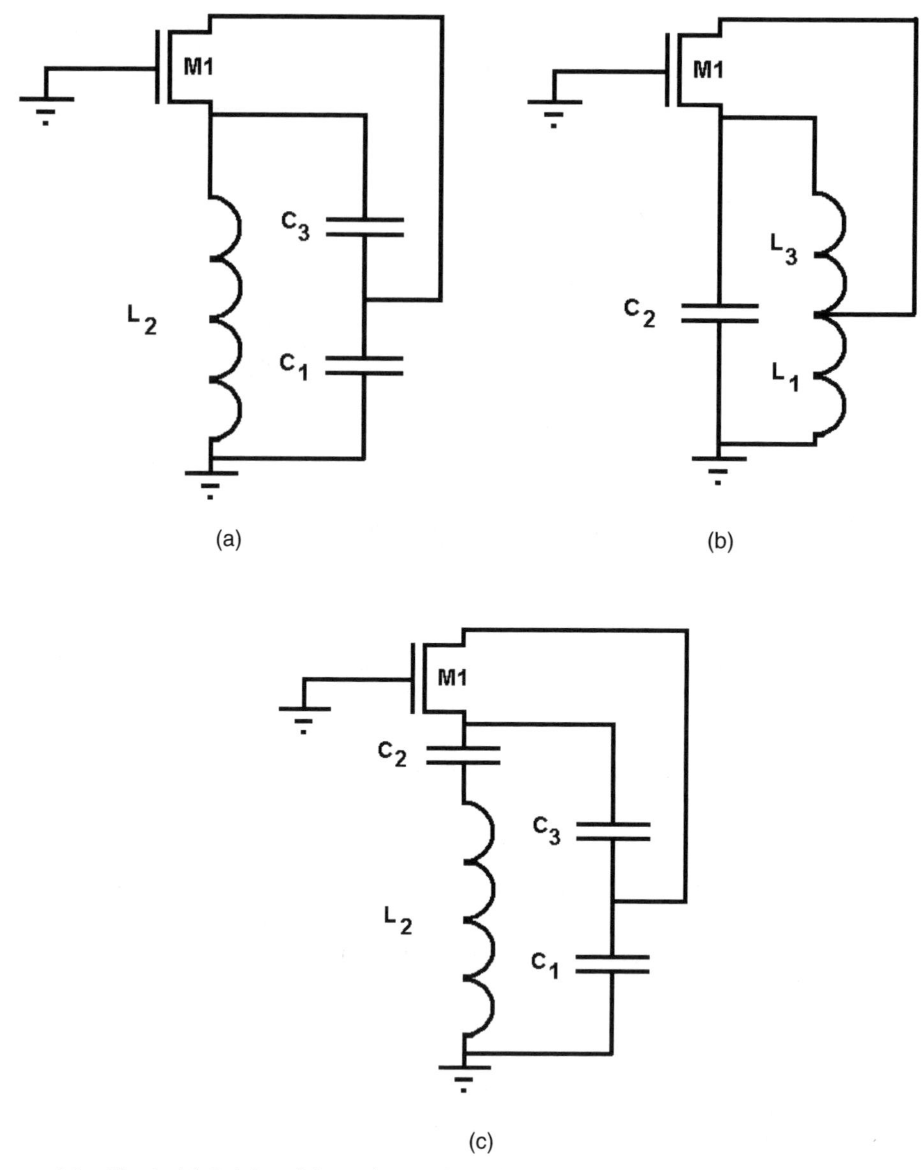

Figure 6.8 Classic (a) Colpitts, (b) Hartley, and (c) Clapp-style oscillator structures based on a common gate *n*MOSFET amplifier (bias networks not shown).

phase change in the vicinity of resonance as indicated by its equivalent impedance expression:

$$Z(\omega) = \frac{j\omega L_p}{\left(1-\omega^2 L_p C\right) + j\dfrac{\omega L}{R_p}} = \frac{j\omega L_p}{\left(1-\dfrac{\omega^2}{\omega_0^2}\right) + j\dfrac{1}{Q_p}\dfrac{\omega}{\omega_0}};\ \omega_0^2 = \frac{1}{L_p C};\ Q_p\ \frac{R_p}{\omega_0 L} \quad (6.16)$$

The inductor's series resistance is replaced by a standard parallel equivalent resistance and inductance, R_p and L_p, respectively, in (6.16):

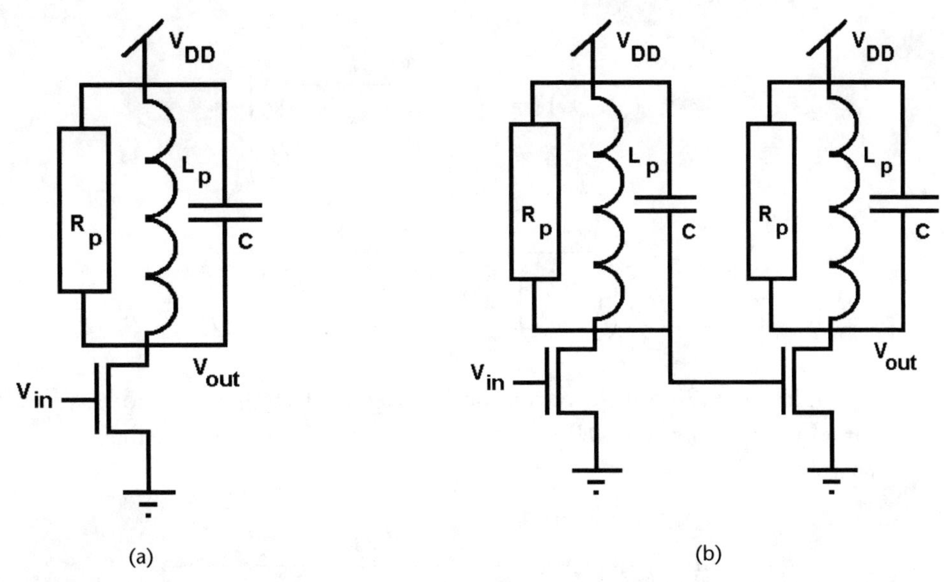

Figure 6.9 (a) Single-stage (b) and two cascade stages of an *n*MOSFET amplifier with RLC resonator load.

$$R_p = R_s\left(1+Q_s^2\right);\ L_p = L_s\left(\frac{1+Q_s^2}{Q_s^2}\right);\ Q_s = \frac{\omega L_s}{R_s} \Rightarrow Q_s = Q_p \tag{6.17}$$

A passive circuit has no gain associated with it, but if the parallel RLC circuit is configured as the load of a MOSFET-based amplifier [Figure 6.9(a)], the output voltage gain of the amplifier with this impedance Z is given by

$$A(\omega) = -g_m Z(\omega) = -\frac{j\omega g_m L_p}{\left(1-\frac{\omega^2}{\omega_0^2}\right)+j\frac{\omega}{\omega_0}\frac{1}{Q_p}} \tag{6.18}$$

Equation (6.18) assumes that the MOSFET output resistance r_{ds} is much greater than R_p and that R_p dominates the output resistance. For a high-Q circuit, this circuit also undergoes a similar phase change, but the circuit can exhibit gain at frequencies near resonance. In terms of building an oscillator, however, unless a feedback network as previously described is used, no particular advantage is achieved. However, this changes dramatically if two of these amplifiers are cascaded as discussed in the next section.

6.2.2 Feedback Cascade of Two Amplifiers

If two of the previous section's RLC tuned amplifiers are cascaded [Figure 6.9(b)], the cascade gain is simply the result in (6.18) squared:

$$A_{TOT} = (-g_m Z)^2 = \left(\frac{j\omega g_m L_p}{\left(1 - \frac{\omega^2}{\omega_0^2}\right) + j\frac{\omega}{\omega_0}\frac{1}{Q_p}} \right)^2$$

$$= -\frac{(\omega g_m L_p)^2}{\left(1 - \frac{\omega^2}{\omega_0^2}\right)^2 + 2j\frac{1}{Q_p}\frac{\omega}{\omega_0}\left(1 - \frac{\omega^2}{\omega_0^2}\right) - \frac{1}{Q_p^2}\frac{\omega^2}{\omega_0^2}} \tag{6.19}$$

In Figure 6.9(b), if the output node is tied back to the input node, a feedback factor of unity is obtained ($B = 1$) and the loop gain for this feedback system can be written as

$$AB = -\frac{(\omega g_m L_p)^2}{\left(1 - \frac{\omega^2}{\omega_0^2}\right)^2 + 2j\frac{1}{Q_p}\frac{\omega}{\omega_0}\left(1 - \frac{\omega^2}{\omega_0^2}\right) - \frac{\omega^2}{\omega_0^2}\frac{1}{Q_p^2}} \tag{6.20}$$

At resonance $(\omega = \omega_0)$, the loop gain reduces to

$$AB = (\omega_0 g_m L_q Q_p)^2 = (g_m R_p)^2 \tag{6.21}$$

which shows that if the MOSFET transconductance is sufficient for (6.21) to be unity (or greater for gain margin purposes), the circuit will act as an oscillator, with output frequency given by

$$\omega^2 L_p C = 1;\ \omega_{osc} = \frac{1}{\sqrt{L_p C}} \tag{6.22}$$

Redrawing Figure 6.9(b) to show the feedback connection provides the simplest form of the widely used CMOS RFIC *cross-coupled LC oscillator* (Figure 6.10). This type of oscillator provides a differential output voltage at V_{out} and is ideal for driving mixers and other circuits with differential inputs. Equation (6.22) indicates that high-Q resonator circuits (large R_p) have relaxed gain requirements for the amplifier. However, on-chip inductors are known for their relatively low Q's, and so the MOSFET amplifying transistors must have sufficient gain (through W/L sizing) to overcome these losses. A loop gain margin of 10 to 15 dB is typically sufficient to ensure reliable oscillator startup [3].

Figure 6.10 shows an oscillator whose negative-going portion of the waveform is limited by gate-source voltage V_{gs} on the transistors (for saturation, V_{gs} must be greater than V_T) yielding *voltage-limited* operation. The addition of a current source or current sink (transistor Mb) changes the output voltage swing to be more *current limited* (Figure 6.11). The current source or sink bias voltage (V_b) may be derived using a current mirror structure either specific to the oscillator circuit or from a system-wide bias circuit. Current limiting of oscillators is the preferred method of

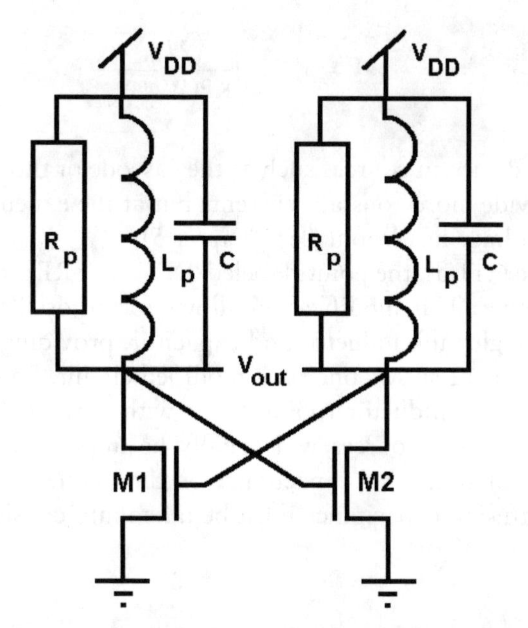

Figure 6.10 Simplest form of the CMOS RFIC cross-coupled LC oscillator.

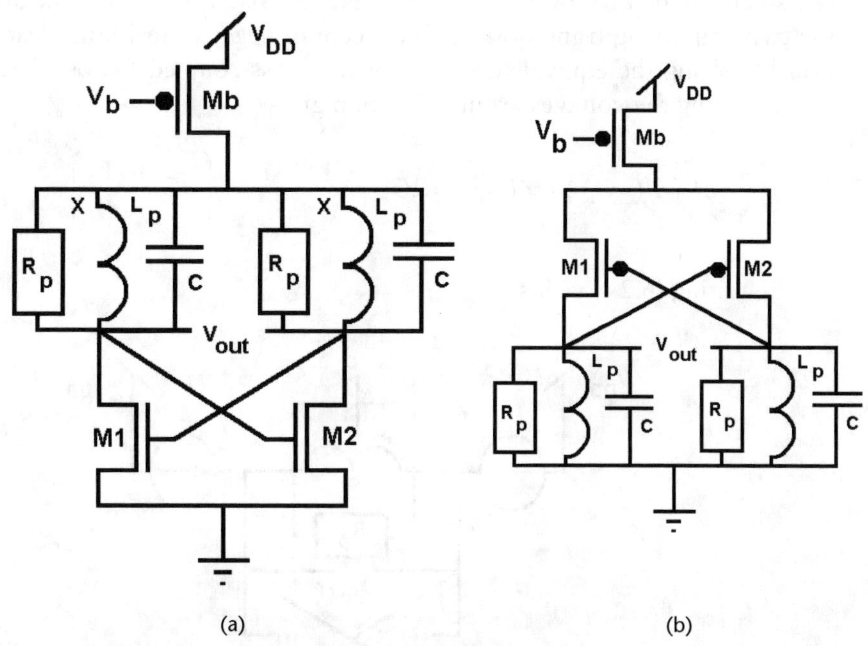

Figure 6.11 CMOS cross-coupled LC oscillator with (a) *n*MOSFETs with tail current sink and (b) *p*MOSFETs with tail current source.

operation but suffers from a somewhat reduced voltage swing due to the need to keep the current source/sink in its constant current region. For the simple current sink shown in Figure 6.11(a) or (b), the voltage drop across Mb is given by

$$V = V_T + \sqrt{\frac{2I_{DS}}{\mathrm{KP}(W/L)_{Mb}}} \tag{6.23}$$

Improved current sources such as the cascode or those based on Wilson current mirrors provide more constant currents but at the expense of higher voltage drops and lower voltage headroom (see Chapter 5).

In Figure 6.11(a), the points labeled "X" are at RF ground potential, so they can be tied together. This modification allows for an oscillator structure that is composed of a single tank inductor and capacitor, providing significant savings in real estate. Figure 6.12 shows one such modified circuit. Power can be applied using a center tap on the inductor (point Y) or with current sources tied at the output. Because the current sources will usually be implemented using pMOSFETs, the introduction of additional capacitance on the frequency-determining nodes due to the drain diffusion capacitance must be taken into consideration during the design process.

6.2.3 Negative G_m Perspective

The cross-coupled LC oscillator shown in Figure 6.10 or 6.11 can also be termed a *negative* G_m or $-G_m$ *oscillator*. This term comes from the observation that the cross-coupled pair M1/M2 provides a negative resistance between the drain terms of the two amplifying transistors. This concept of negative resistance leads to oscillation by noting the equivalent circuit of the cross-coupled LC oscillator in Figure 6.13. Writing the voltages around the loop gives

$$ZI + I/(-G_m) + ZI = 0 \Rightarrow 2G_m = \frac{1}{Z} = \frac{1}{R_p} + j\left(\omega C - \frac{1}{\omega L_p}\right) \tag{6.24}$$

Rewriting (6.24) yields

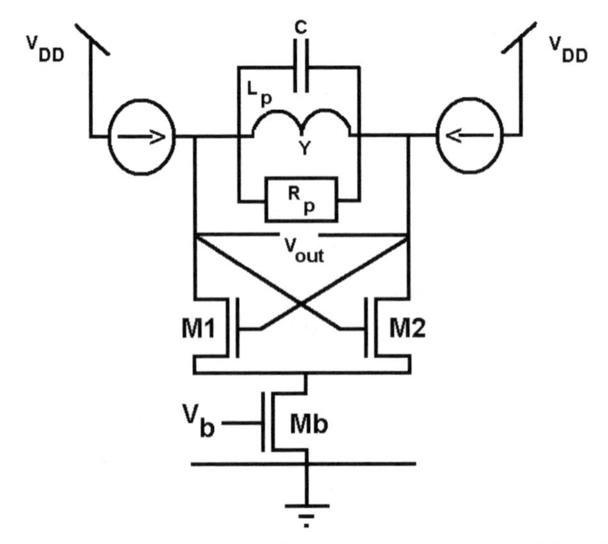

Figure 6.12 LC oscillator with a single LC tank. Biasing can be applied either through the current sources or by an inductor tap (point *Y*).

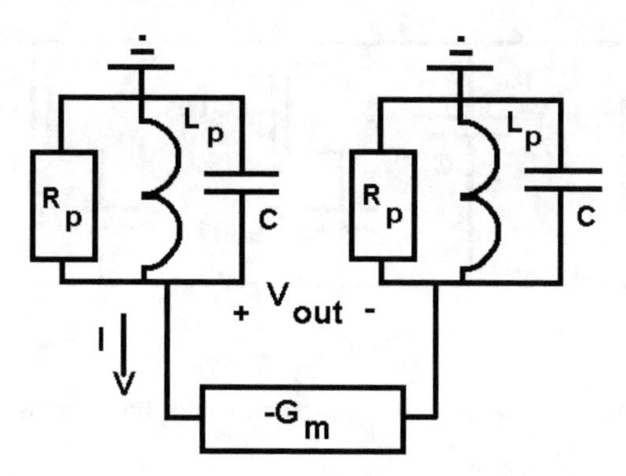

Figure 6.13 RF equivalent circuit of the negative resistance oscillator.

$$-2G_m + \frac{1}{R_p} = 0 = j\left(\omega C - \frac{1}{\omega L_p}\right) \tag{6.25}$$

implying that an oscillating current will flow only under the conditions in (6.24) (i.e., $G_m = 1/2R_P$). The frequency of oscillation is determined by the reactive portion of (6.25), and the negative resistance of the cross-coupled amplifiers cancels the loss portions of the LC tank circuit [5].

6.2.4 Coarse Frequency Control

The resonant frequency of the oscillator can be changed by changing the tank circuit inductance, the capacitance, or both. *Band-switching* oscillators have been designed that provide coarse tuning by switching one or more capacitors [6–8] or inductors [6] into the tank circuit (capacitive band switching is shown in Figures 6.14 and 6.15). Because of large real estate issues associated with inductors, inductive band switching is used less often than capacitive band switching. MOSFET RF switch design is an important consideration in this type of frequency control. When the MOSFET-based RF band switch is activated, the on-state resistance R_s is in series

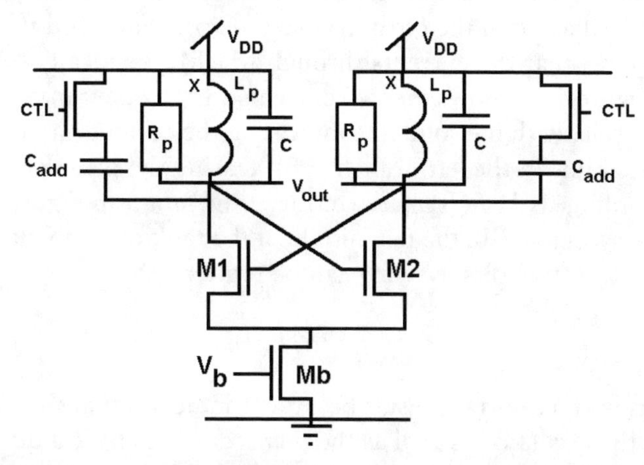

Figure 6.14 Additional capacitance can be switched into the tank circuit (C_{add}) to provide coarse oscillator tuning or band switching.

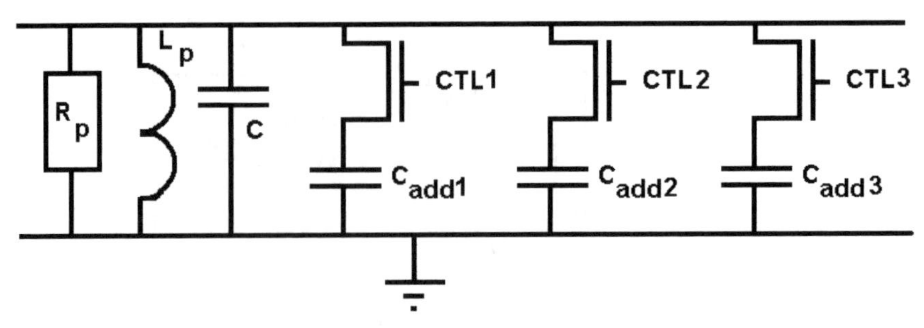

Figure 6.15 Multiple bands can be switched using the same oscillator circuit by capacitor banks.

with the added capacitor. If the MOSFET switch resistance R_s is too large, this degrades the capacitor Q and hence the Q of the entire tank circuit [6]. Improving the switch on resistance requires increasing the switching MOSFET gate width W, which also increases the drain region capacitance contribution, increasing the effective capacitance seen by the inductor and thereby lowering the oscillator frequency [6]. Multiple bands may be switched with capacitor banks but the additional capacitance introduced from each switch must be considered. A frequency span of 500 MHz with a center frequency of 2.1 GHz has been achieved with this technique [8]. Alternatively, MEMS switches can be used to switch in either inductance sections or used in capacitor bank switching [9]. Fine-tuning of the oscillator within each band is discussed in Section 6.4.

6.2.5 Oscillator Design Specifications: Voltage Swing and Q

An important parameter for an oscillator is its oscillating frequency. However, eventually this oscillator will have to drive another stage (at least) on the RFIC, and so another important specification for the oscillator is the level of available *voltage swing* the circuit can achieve. At resonance, Figure 6.11 shows that the parallel equivalent resistance R_p is the load that M1/M2 sees and that the voltage generated across this resistance governs the output voltage, V_{out}. In this circuit, the bias current I_{bias} is set by MOSFET Mb with one-half of the I_{bias} flowing through each branch of the cross-coupled LC oscillator. Once the oscillator has started up and stabilized, the differential nature of the cross-coupled LC oscillator and the constant I_{bias} through Mb requires that the currents through M1/M2 be ideally 180° out of phase at all times. Figure 6.16 shows the two currents varying their maximum amounts through one RF cycle, with the solid line indicating the current on the right branch (RB) and the dashed line in the left branch (LB). During the peak RF swing on RB, the peak output voltage is ½ R_pI_{bias} (at arbitrary time unit 5 in Figure 6.16), whereas on the peak RF swing on LB, the magnitude of the tank output voltage is the same, implying that the total voltage swing can be written as

$$\Delta V = R_p I_{bias} \tag{6.26}$$

This is an important result because it indicates that the current sink (or source) biasing the oscillator as well as the *inductor losses* have a direct bearing on the output voltage swing.

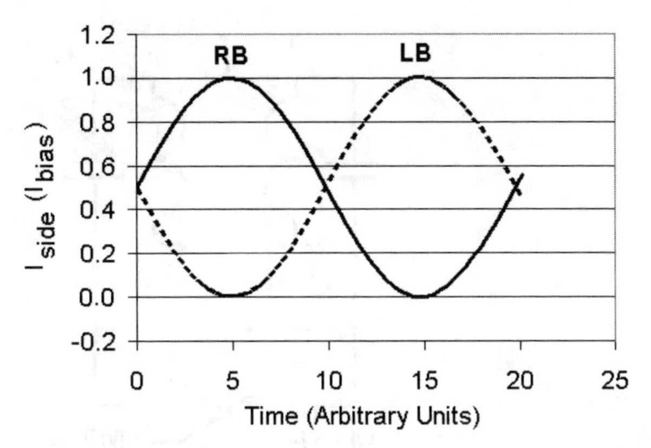

Figure 6.16 RF current swing on each branch (RB, right branch; LB, left branch) of the cross-coupled LC oscillator. The average value of current flowing through both branches is $\frac{1}{2}I_{bias}$.

Equation (6.26) can be recast into a more physically meaningful expression by noting the relationship between the parallel equivalent tank resistance R_p and the inductor series resistance R_s (6.17):

$$\Delta V = R_s I_{bias}\left(1+Q^2\right) \tag{6.27}$$

Equation (6.27) gives one reason (others are given later in the chapter) why oscillator designers are always worried about tank circuit Q: The output voltage swing of the tank circuit increases as the Q of the circuit increases. This is not just a linear relationship either; doubling the inductor Q creates a fourfold increase in tank voltage swing. Equation (6.27) assumes that the Q is only determined by the inductor; if the tank capacitors have relatively low Q values or the MOSFET has a relatively low output resistance, these must be taken into consideration as part of the equivalent parallel tank resistance R_p. An alternative viewpoint of this result is to note that for a given bias current I_{bias} (such as that set by the system power requirements), the output voltage swing can be improved by enhancing the Q of the inductor.

Example 6.3: Determine the bias current needed for a 3.0V peak tank voltage with an inductor exhibiting series resistance of 5Ω and a Q of 10.

From (6.17), the bias current required to provide a 3.0V oscillator tank voltage with the effective parallel resistance R_P is found to be

$$I_{bias} = \frac{\Delta V}{R_p} = \frac{\Delta V}{R_s\left(1+Q^2\right)} = \frac{3}{5\left(1+10^2\right)} = 5.94 \text{ mA}$$

6.2.6 Modeling/Design Example

This section (ch6-1.txt) details the design of a 1.0-GHz oscillator using a cross-coupled LC oscillator with tail current source (Figure 6.17). A few assumptions used in this example are that inductors with Q of 10 are available for the design and the dc

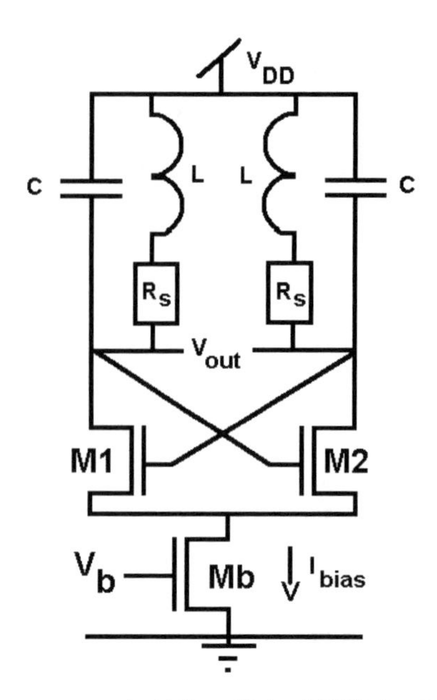

Figure 6.17 Circuit used for cross-coupled LC oscillator SPICE simulation.

drawn from the 3.3V power supply is limited to 4.0 mA. Note that 0.5-μm gate length MOSFETs with KP = 56×10^{-6} A/V^2 and $\lambda = 0.1$ are available for the design.

The current sink is one of the first circuit elements of the LC oscillator to look at since this circuit limits the overall voltage swing. The current sink must support the full 4.0-mA circuit current as well as have a relatively low saturation voltage. Assuming a current sink control voltage of 1.66V, the width of MOSFET Mb can be computed as follows:

$$\frac{W}{L} = \frac{2I_{DS}}{\text{KP}\left(V_{gs} - V_T\right)^2_{Mb}} = 142 \Rightarrow W = 71\ \mu\text{m}$$

The resulting minimum voltage to keep Mb in saturation is $V_{gs} - V_T = 1.0$V, thereby reducing the total available swing for the tank circuit to 2.3V. For design purposes, assume the peak tank voltage $\Delta V = 2.0$V. With this 2.0V peak voltage, the maximum parasitic series resistance of the tank inductor L_S can be found:

$$\Delta V = R_s I_{bias}\left(1+Q^2\right) \Rightarrow R_s \le \frac{\Delta V}{I_{bias}\left(1+Q^2\right)} = \frac{2}{0.002\left(1+10^2\right)} = 10\Omega$$

For a conservative design, keep this parasitic series resistance to 5Ω or less. The tank inductance L_s is related to the Q and series resistance R_s by (6.17), indicating that inductor's value must be at least

$$Q = \frac{\omega L_s}{R_s} \Rightarrow L_s = \frac{QR_s}{\omega} = \frac{(10)(5)}{2\pi \cdot 10^9} = 7.96\ \text{nH}$$

and, using this value of inductance, the resonating capacitor C to provide the 1000-MHz tank circuit resonance should be

$$\omega_{osc} = \frac{1}{\sqrt{LC}} \Rightarrow C = \frac{1}{\omega_{osc}^2 L} = \frac{1}{\left(2\pi \cdot 10^9\right)^2 7.96 \cdot 10^{-9}} = 3.18 \text{ pF}$$

To ensure oscillator startup, (6.21) indicates that the loop gain *AB* must be unity. However, to overcome other losses in the oscillator circuit, a higher gain is often needed. Assuming a gain of 10 for each amplifying MOSFET implies that the transconductance of M1/M2 must be at least

$$g_m R_p = 10 \Rightarrow g_m = \frac{10}{R_p} = \frac{10}{R_s\left(1+Q^2\right)} = \frac{10}{5\left(1+10^2\right)} = 20 \text{ mS}$$

The aspect ratio for M1/M2 can be determined from knowledge of the required transconductance and bias current through the devices:

$$g_m \sqrt{2I_{DS} \cdot \text{KP}\frac{W}{L}} \Rightarrow \frac{W}{L} = \frac{g_m^2}{2I_{DS} \cdot \text{KP}} = \frac{4 \cdot 10^{-4}}{2(0.002)56 \cdot 10^{-6}} = 1{,}785 \Rightarrow 1{,}800$$

At this stage, all parameters for the initial simulation run have been specified. Note that a larger R_P coming from a higher inductor Q will tend to reduce the required transconductance. However, as the Q increases, the impact of the output resistance of the MOSFET r_{ds} begins to be seen and must be considered for an accurate design under these conditions. The first-pass SPICE simulation based on these derived oscillator parameters showed MOSFET Mb sinking a current of approximately 5.0 mA. The aspect ratio of Mb was reduced slightly until the required 4.0-mA current flow was achieved. A transient simulation was then performed with the results shown in Figure 6.18(a) where the differential output voltage is plotted versus time. The results show the classical oscillator startup behavior, with the output increasing with time until the oscillator stabilizes. Detailed investigation of the startup region shows the small voltage variations gradually amplified until the oscillator self limits. The peak output voltage of 2.1V is quite close to the design value of 2.0V. Taking an FFT of the time waveform gives the frequency spectrum of the differential output. The first design pass showed that the oscillation frequency was approximately 800 MHz, significantly different than the 1.0-GHz design frequency. The major reason for the difference in the design oscillation frequency is that the loading associated with the large-area MOSFETs M1/M2 was not considered. Reducing the external capacitance C on each branch of the oscillator to 1.59 pF moved the resonant frequency to 1.0 GHz [Figure 6.18(b)]. A reduction in aspect ratio of M1/M2 from 1,800 to 600 showed little impact on the overall oscillator voltage swing although the resonant frequency shifted upward slightly due to the decreased drain capacitance of the smaller MOSFET current sink. This frequency shift would more than likely not be noticed in the actual fabricated circuit because the effects of parasitics on the oscillator frequency are larger. An alternative to using transient analysis is to use a harmonic balance (HB) simulator to verify the

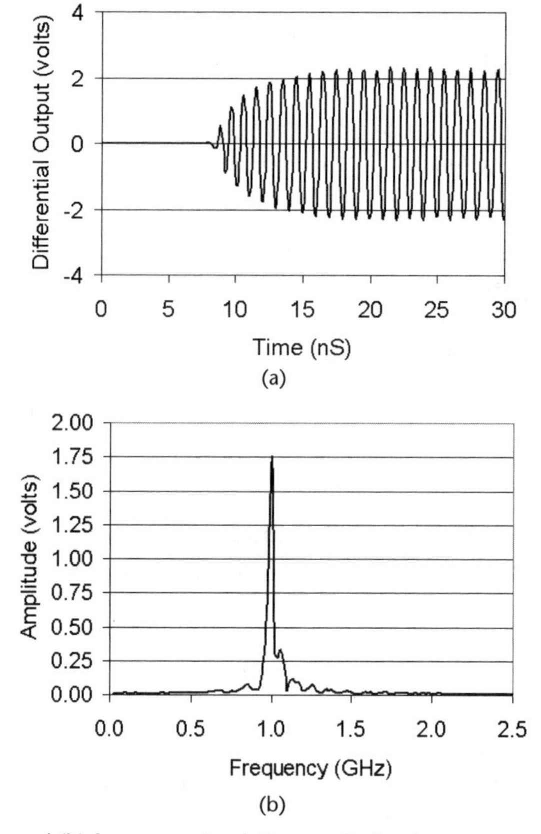

Figure 6.18 (a) Time and (b) frequency simulation results for the cross-coupled LC oscillator.

oscillator operation. Using such an HB simulator shows a similar peak output voltage at 1000 MHz, verifying the SPICE transient solution.

Many CMOS RFIC layout tools have a library of fixed components available to the designer. These *standard cell libraries* contain well-defined and previously modeled components that can be dropped into a design with the knowledge that, at least for the standard cell component, there can be some degree of confidence that the RF characteristics will remain constant between fabrication process runs. With this in mind, a second design example (ch6-2.txt) uses fixed 10-nH inductors ($Q = 10$) that could be from one such standard cell library. The other design parameters from the first example will remain the same. Since the current requirement has not changed, the MOSFET Mb's aspect ratio W/L will initially remain the same at 142. The series resistance of the inductor can be computed as

$$R_s = \frac{\omega L}{Q} = \frac{2\pi \cdot 10^9 \cdot 10 \cdot 10^{-9}}{10} = 6.28\Omega$$

and the resonating capacitor as

$$\omega_{osc} = \frac{1}{\sqrt{LC}} \Rightarrow C = \frac{1}{\omega_{osc}^2 L} = \frac{1}{\left(2\pi \cdot 10^9\right)^2 10 \cdot 10^{-9}} = 2.53 \text{ pF}$$

The loop gain expression provides an indication of the transconductance required for MOSFETS M1 and M2:

$$g_m R_p = 10 \Rightarrow g_m = \frac{10}{R_p} = \frac{10}{R_s\left(1+Q^2\right)} = \frac{10}{6.28\left(1+10^2\right)} = 15.9 \text{ mS}$$

which is slightly less than that needed for the previous set of design conditions, with the resulting aspect ratio of 1,200 required:

$$g_m = \sqrt{2I_{DS} \cdot \text{KP}\frac{W}{L}} \Rightarrow \frac{W}{L} = \frac{g_m^2}{2I_{DS} \cdot \text{KP}} = \frac{2.5 \cdot 10^{-4}}{2(0.002)56 \cdot 10^{-6}} = 1{,}116 \Rightarrow 1{,}200$$

The resulting simulation showed good oscillator startup, a 2.0V peak tank voltage swing, and an oscillation frequency of 850 MHz. Reducing the external capacitance C on each branch to compensate for the loading of the MOSFETs moved the resonant frequency to 1.0 GHz.

6.2.7 Mechanical-Based Oscillators

Highly stable oscillators traditionally used piezoelectric material (crystals) to set the oscillator frequency. Recently, MEMS have been used to create highly stable frequency sources in silicon-on-chip systems. This section covers crystals from a perspective that introduces terminology that is used when describing MEMS oscillators as well.

6.2.7.1 Crystal-Based Oscillators

The most common *piezoelectric material* used in making frequency standard *crystals* is quartz. In piezoelectric material, a mechanical strain creates a small electrical force. Conversely, an electrical stimulus translates into mechanical force that causes the crystal to vibrate at a specific frequency. The mechanical vibration frequency of the crystal is directly related to its dimensions. The relationship between the mechanical vibration and electrical resonance of the crystal can be modeled using standard RLC equivalent circuit models. In the case of a crystal, however, the RF equivalent circuit values are not true electrical parameters (that is, the crystal is not a lumped element circuit), but are described in terms of *motional parameters* to distinguish the crystal resonance parameters from their true electrical counterparts. The distinct advantage in using crystals in oscillators is that they exhibit extremely high Q values (in the thousands to hundreds of thousands). The effect of this high Q on the motional parameters can be seen by simply looking at a series equivalent circuit for the crystal with its corresponding Q equation:

$$Q = \frac{\omega_0 L_m}{R_m} = \frac{1}{\omega_0 R_m C_m};\ \omega_0 = \frac{1}{\sqrt{L_m C_m}} \tag{6.28}$$

where the subscript m is used to denote a motional parameter. To achieve an extremely high Q for a set motional resistance, the motional inductance L_m must be extremely large with a correspondingly motional capacitance that is extremely small. Being crystalline in nature, quartz crystals have different properties depending on how the crystals are cut with respect to their various crystal planes. The resonant frequency of quartz crystals is relatively constant over the temperature range 0° to + 50°C for so-called AT-cut crystals. AT-cut crystals, however, are quite sensitive to thermal stresses, which can be a major issue in applications where frequency stability is critical. SC-cut (SC-stress-compensated) crystals, on the other hand, are cut at such a crystalline angle as to minimize these stresses. SC-cut crystals have a relatively narrow range of near zero temperature sensitivity around 85°C. SC-cut crystals are often used in extremely stable oscillators where special thermal control units (often termed crystal ovens) are employed to maintain this temperature. Crystal-based oscillators also do not necessarily have to operate at the fundamental resonant frequency of the crystal. Various odd-order harmonics of the fundamental (third overtone, fifth overtone, etc.) can also be used for applications at frequencies significantly above the "cut" frequency.

Crystal resonators can replace one or more reactance elements in the general oscillator structure. For example, Figure 6.19 shows a Colpitts oscillator where a crystal resonator (XTAL) replaces the feedback inductor. Note that two large inductors (RFC1 and RFC2) are added in the circuit for biasing purposes; their reactance is assumed to be very high at the frequency of oscillation. The frequency of oscillation of the circuit can be slightly varied by the addition of a series LC network at point X, creating what is often referred to as a *variable crystal oscillator* (VXO) [4]. The frequency variation with this circuit is quite small, often less than a fraction of a percent. Large values of L and C can be used but at the risk of losing oscillator stability, which is what the crystal was used for in the first place.

Example 6.4: Determine the motional inductance and capacitance of a quartz crystal that exhibits a Q of 1,000, a motional resistance of 1,000Ω and a resonance frequency of 159 MHz.

Using (6.28) for the Q and crystal resonance frequency, L_m and C_m can be computed as follows:

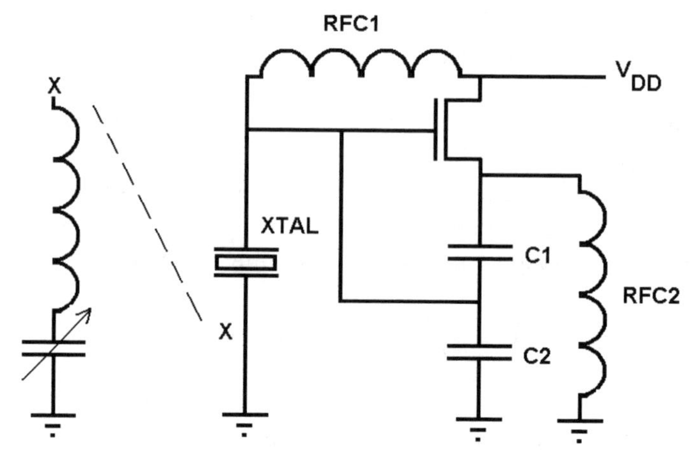

Figure 6.19 Colpitts crystal oscillator with VXO circuit shown as inset.

$$L_m = \frac{QR_m}{\omega_0} = \frac{1{,}000 \cdot 1{,}000}{10^8} = 10^{-2}\ \text{H}$$

$$Q = \frac{1}{\omega_0 R_m C_m}; \Rightarrow C_m = \frac{1}{\omega_0 R_m Q} = \frac{1}{10^8 \cdot 1{,}000 \cdot 1{,}000} = 10\ \text{fF}$$

From this example, the motional parameters for the crystal are indeed quite different than one would expect in a lumped element RLC circuit. However, the resonant frequency is calculated the same as in (6.28).

6.2.7.2 MEMS-Based Oscillators

As previously discussed in Chapter 3, MEMS structures have natural vibrational resonance frequencies not unlike the traditional quartz crystals, with the exception being the fact that MEMS can be integrated into silicon-based systems with various processing steps. Like its crystal counterpart, MEMS resonators are also often described by their motional parameters and behave, at least from an analytic sense, the same way as crystal resonators. For MEMS resonators, the output that is fed back to the amplifying structure is a current [10] and, since the MEMS is voltage driven, a *transresistance* amplifier should be used (Figure 6.20). For a MEMS membrane of width W_r, electrode width W_e, and gap g_0, the current output and the motional resistance R_m of the MEMS membrane can be written as [10, 11]

$$I = V_P \frac{W_r W_e}{g_0^2} \omega_0 X; \; R_m = \frac{1}{Q} \frac{k_r}{\omega_0 V_P^2} \frac{g_0^4}{(\varepsilon_0 W_r W)^2} \tag{6.29}$$

where X is the amplitude of the vibration, and k_r is the beam stiffness parameter. The resonant frequency can be approximated from (3.38). Equation (6.29) shows that a narrow gap or wide beams increase the current and reduce the motional resistance. There are limits to reducing the gap or widening the beam, however [10].

A transresistance amplifier can be created using one of the single or differential amplifier topologies described earlier, but with the addition of a MOSFET at the input to the amplifying transistors as shown in Figure 6.21(a) (differential amplifier topology shown) with the RF equivalent circuit of the left branch shown in Figure 6.21(b) [10, 12]. Analysis of the RF equivalent circuit in Figure 6.21(b) shows that the transresistance R_T can be written as

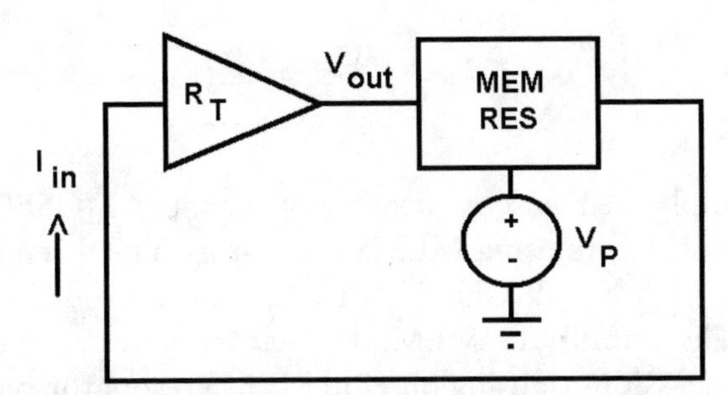

Figure 6.20 Block diagram of a MEMS resonator oscillator using a transresistance (R_T) amplifier. (*After:* [10].)

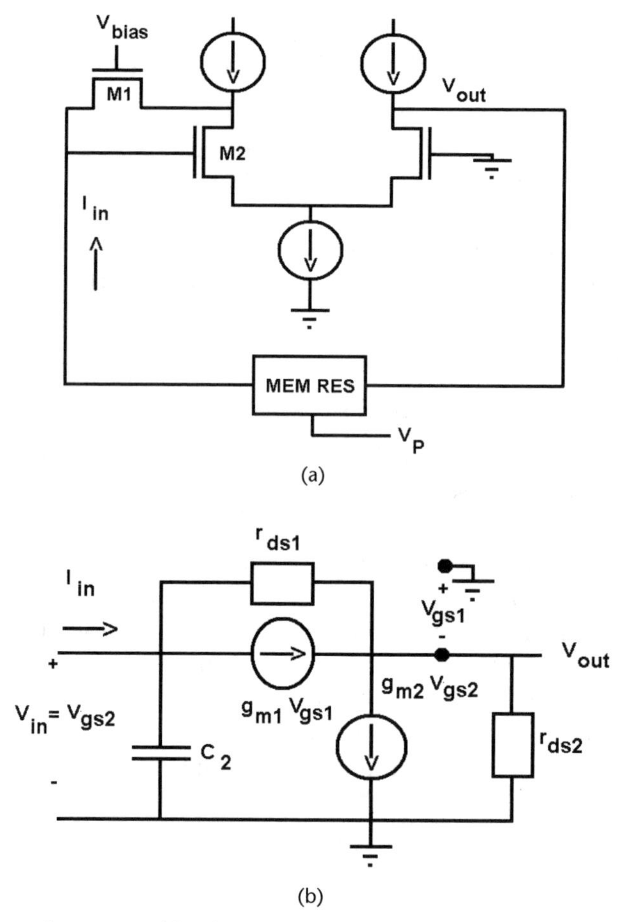

Figure 6.21 Transresistance amplifier for use in MEMS resonator oscillator: (a) simplified circuit and (b) RF equivalent circuit of left branch.

$$
\begin{aligned}
R_T &= \frac{V_{out}}{I_{in}} \\
&= -\frac{g_{m2} r_{ds1}^2 r_{ds2}}{(1+g_{m1} r_{ds1}) g_{m2} r_{ds1} r_{ds2} + (1+j\omega r_{ds1} C_2)(r_{ds1} + r_{ds2} + g_{m1} r_{ds1} r_{ds2})}
\end{aligned}
\tag{6.30}
$$

If it is assumed that the transconductances g_{m1} and g_{m2} of the MOSFETs are large and the cutoff frequency term $j\omega r_{ds1} C_2$ is small, then (6.30) can be written as

$$R_T = \frac{V_{\text{out}}}{I_{\text{in}}} = -\frac{1}{g_{m1}} \tag{6.31}$$

which is simply the transconductance of the input MOSFET M1.

Stable oscillators using MEMS resonators have been developed for use at frequencies as high as 1.9 GHz [12] using thin-film bulk acoustic wave resonators (FBARs). The stability of MEMS oscillators is improving to such an extent that there are discussions of using on-chip MEMS resonator-based oscillators as atomic clocks [13].

6.3 Ring Oscillator

The *ring oscillator* is a popular circuit for generating on-chip signals and is often used by IC fabrication houses as a test structure for evaluating the quality of the fabrication process. As the term *ring* implies, the ring oscillator is a ring or cascade of an odd number of CMOS inverters with the final inverter output tied directly back to the input inverter (feedback term $B = 1$). Being based on digital inverters, the ring oscillator ideally provides a square-wave output with the usual sharp edge transitions. In practice, however, the waveform is not truly square but shows rounding at the transition edges. Nevertheless, the output still exhibits much sharper transitions than the LC-based oscillators described earlier. Each inverter has a certain digital time delay τ between stages with each binary transition; this delay, termed the *inverter pair delay,* is the sum of the rise and fall times of the individual (and identical) inverters, τ_R and τ_F, respectively. With N-odd inverters, the total propagation delay governs the oscillation frequency:

$$f_{osc} = \frac{1}{2N\tau} = \frac{1}{N(\tau_R + \tau_F)} \tag{6.32}$$

The following discussion covers the basic of the inverters making up the ring oscillator as well as the conditions for oscillation and various alternative ring oscillator circuits.

6.3.1 Basic CMOS Inverter and Ring Oscillator

The inverter delay time is related to the length of time it takes to charge/discharge the capacitance seen by each node of the inverter cascade (Figure 6.22). In its simplest form, the CMOS inverter consists of a single *n*MOSFET and *p*MOSFET device with the gates of each tied to the input and the drains of each MOSFET tied to the output. The capacitive loading seen by each inverter is the sum of the input

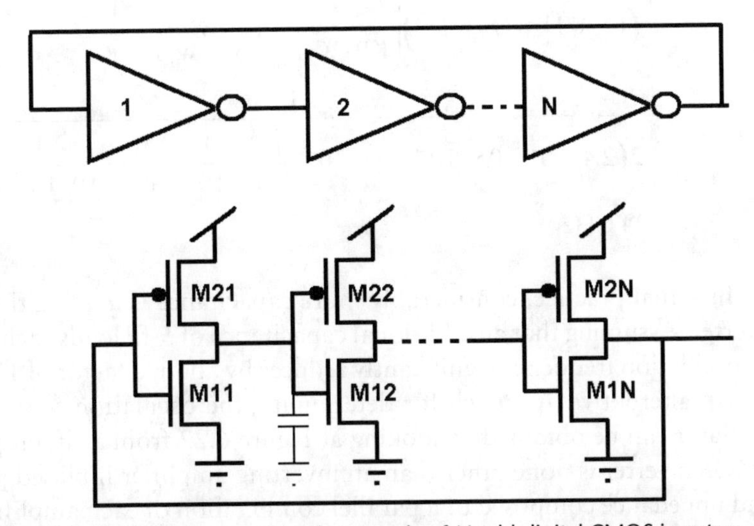

Figure 6.22 The ring oscillator consists of a cascade of *N*-odd digital CMOS inverters.

capacitance of the next inverter in the cascade, the parasitic drain capacitances of the MOSFETs, and any interconnect capacitance between the two stages (which will be summarily neglected in the following discussion). From a circuit perspective, the delay time is computed as the length of time it takes for *p*MOSFET M2*i* to charge the inverter load capacitance C_{L-i+1} (τ_r, the rise time) and to discharge this capacitance through *n*MOSFET M1*i* (τ_f, the fall time). Taking the average delay as one-half the sum of these two delay times [14], the ring oscillator oscillation frequency can be approximated as

$$f_{osc} = \frac{1}{2N\tau} = \frac{1}{N(\tau_R + \tau_F)} = \frac{1}{N}\frac{1}{C_L\left(\left.\frac{1}{\mathrm{KP}\cdot W/L}\right|_n + \left.\frac{1}{\mathrm{KP}\cdot W/L}\right|_p\right)} \tag{6.33}$$

Assuming the load capacitance is simply the input gate capacitance of each MOSFET, the ring oscillator frequency can be written as

$$f_{osc} = \frac{1}{N}\frac{1}{\left(C_{ox}WL|_n + C_{ox}WL|_p\right)\left(\left.\frac{1}{\mathrm{KP}\cdot W/L}\right|_n + \left.\frac{1}{\mathrm{KP}\cdot W/L}\right|_p\right)} \tag{6.34}$$

Equation (6.34) indicates that f_{osc} is proportional to L^{-2} and implies that technology improvements through reduced gate lengths increase the ring oscillator frequency for a given N.

Example 6.5: For a three-stage ring oscillator, determine the frequency of oscillation using the parameters in Table 3.1. Assume $W = L = 0.5\ \mu$m.

Substituting values from the table yields an oscillation frequency of

$$\begin{aligned} f_{osc} &= \frac{1}{3}\frac{1}{\left(C_{ox}WL|_n + C_{ox}WL|_p\right)\left(\left.\frac{1}{\mathrm{KP}\cdot W/L}\right|_n + \left.\frac{1}{\mathrm{KP}\cdot W/L}\right|_p\right)} \\ &= \frac{1}{3}\frac{1}{2\left(2.4\cdot 10^{-3}\cdot 0.5\cdot 10^{-6}\cdot 0.5\cdot 10^{-6}\right)\left(\left.\frac{1}{56\cdot 10^{-6}\cdot 1}\right|_n + \left.\frac{1}{19\cdot 10^{-6}\cdot 1}\right|_p\right)} \\ &= 3.94\ \text{GHz} \end{aligned}$$

In actual practice, considerably more capacitance is loading the output of each inverter. Assuming that an additional capacitance of 5-fF loads each inverter output, the oscillation frequency significantly reduces by about a factor of 10 to 430 MHz.

An alternative approach for determining the oscillation frequency of the ring oscillator can be obtained by looking at Figure 6.22 from a circuit perspective. The CMOS inverter is none other than an inverting amplifier if biased properly. With a load impedance composed of a parallel combination of each amplifier's r_{ds} (r_{out}) and input C_L, the cascade gain (N stages) can be written as

$$A = (-1)^N \left(\frac{g_m r_{ds}}{1 + \omega r_{ds} C_L} \right)^N = (-1)^N \left(\frac{g_m r_{ds}}{\sqrt{1 + (\omega r_{ds} C_L)^2}} \right)^N e^{-jN \tan^{-1} \omega r_{ds} C_L} = AB \qquad (6.35)$$

which is also the loop gain since $B = 1$. For oscillation to occur, the loop gain phase must go to zero and for N-odd stages. This criterion yields

$$\omega_{osc} = \frac{1}{r_{ds} C_L} \tan \frac{\pi}{N} \qquad (6.36)$$

which also indicates that the oscillation frequency decreases as the number of stages increases. The other half of the Barkhausen criteria requires that the magnitude of the loop gain be unity:

$$AB = 1 = \left(\frac{g_m r_{ds}}{\sqrt{1 + \left(\tan \frac{\pi}{N} \right)^2}} \right)^N \Rightarrow g_m r_{ds} = 1 + \left(\tan \frac{\pi}{N} \right)^2 \qquad (6.37)$$

Increasing the number of inverters then reduces the gain requirements of each inverter but also decreases the oscillation frequency. Equation (6.36) indicates that the oscillation frequency is related to r_{ds}, which is also inversely proportional to the dc bias current through the MOSFETs ($1/\lambda I_{DS}$), making the oscillation frequency a function of dc bias current. This feature is revisited in a later section.

6.3.2 Single/Differential Ring Oscillators

In CMOS RFICs, many circuits are differential in nature because of the improvements in noise and common mode suppression. In keeping with this differential circuit plan, the ring oscillator can also be configured in differential form in a similar manner as its circuit of origin, the differential amplifier. Two inverter stages are required and configured for the inverted and noninverted inputs as shown in Figure 6.23 with the corresponding schematic symbol [14].

The differential ring oscillator can be created by cascading N differential inverters with the outputs tied to input complements and the feedback connection going to the corresponding input ($B = 1$) as shown in Figure 6.24.

Various improvements to the basic differential CMOS ring oscillator have been proposed. A more linear I-V characteristic can be achieved by the use of so-called Maneatis loads instead of the single *p*MOSFET loads [Figure 6.25(a)] [15, 16]. Latching the outputs of the differential inverters using the circuit configuration of Figure 6.25(b) can keep sharp digital edges even if the overall delay time is large if a low frequency is required [17]. These modified inverters can replace each inverter in the ring oscillator cascade shown in Figure 6.23.

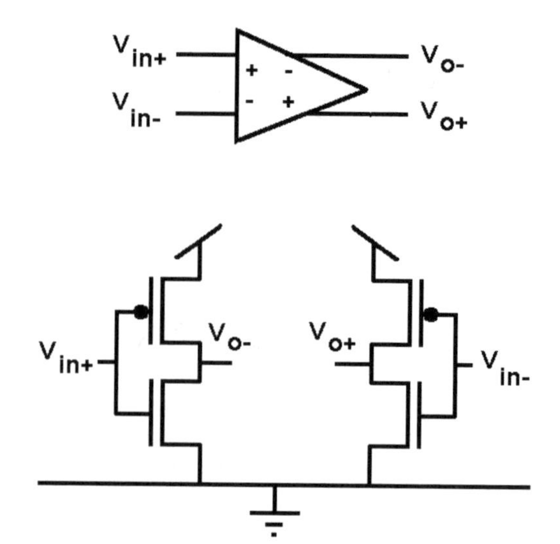

Figure 6.23 Simple differential CMOS inverter based on inverting amplifier.

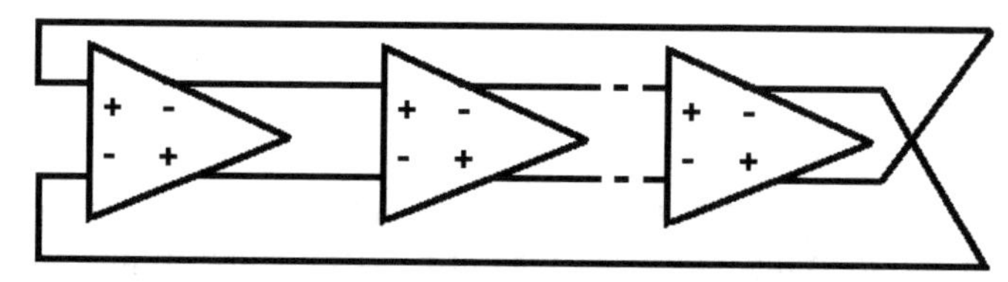

Figure 6.24 Basic differential CMOS ring oscillator.

6.4 Voltage Control of Oscillators

A fixed-frequency oscillator is a great circuit but is limited in use for communications systems with multiple channels unless its frequency of oscillation can somehow be made to change with the application of some external stimulus. For CMOS RFIC oscillators (and indeed most discrete and integrated oscillators), the external signal is most often a voltage that can be used to change a reactive element in the LC tank circuit, usually the overall tank capacitance. The combination of a variable-voltage capacitor (or varactor) in an oscillator tank circuit gives rise to the *voltage-controlled oscillator* (VCO). In discrete RF circuits, the varactor is usually a diode but in CMOS RFICs, MOSFETs and MEMS of various configurations and structures are used instead of diode varactors (although similar circuit schematic symbols are often used). This section looks at voltage control of both LC and ring oscillators using different types of varactors.

6.4.1 Location in the Tank Circuit

Because the frequency of oscillation is controlled by the tank circuit, the tank is a natural place to locate the tuning varactor. The varactor $C_v(V)$is usually placed in shunt with a fixed capacitance C_{fixed} for tuning. Figure 6.26 shows two oscillator types with possible sites for the tuning varactor location.

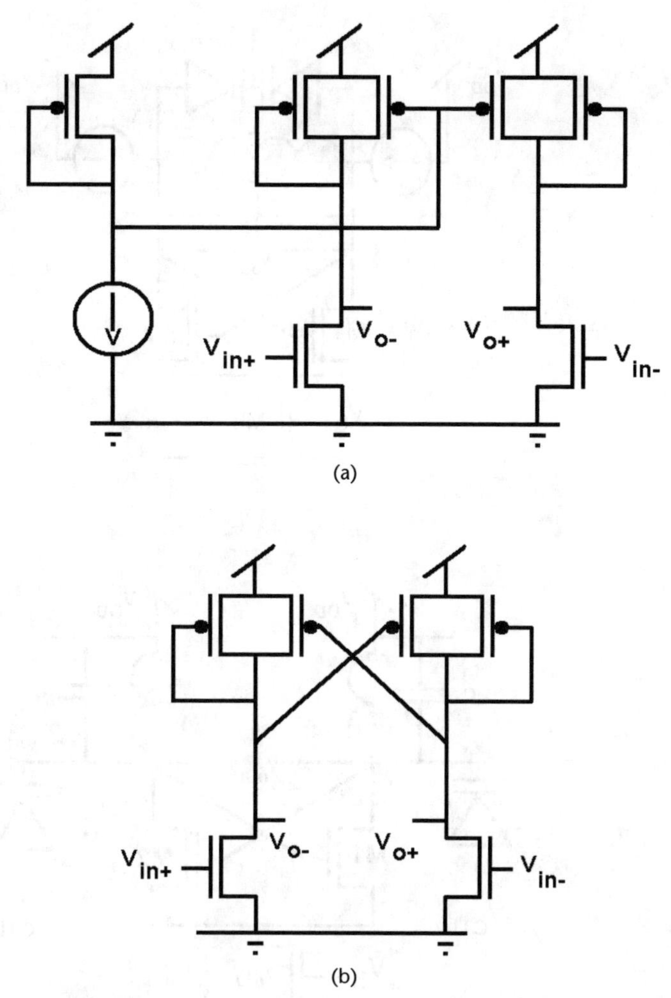

Figure 6.25 Modifications to the basic differential ring oscillator: (a) Maneatis loading and (b) latching loads. (*After:* [15, 17].)

The oscillation frequency can then be written as

$$f_{osc} = \frac{1}{2\pi}\frac{1}{\sqrt{LC_{TOT}}} = \frac{1}{2\pi}\frac{1}{\sqrt{LC_{fixed}}}\frac{1}{\sqrt{1+\dfrac{C_v(V)}{C_{fixed}}}} \tag{6.38}$$

The varactor is not a perfect voltage variable capacitance, however. It will always exhibit some losses that can be described by a lumped series resistance and a corresponding value for varactor Q. Because the varactor is in the LC tank circuit, the varactor should exhibit a high Q so that the overall tank circuit Q will not be degraded. This is typically the case for MOSFET varactors because of the relatively low on-chip planar inductor Q.

Equation (6.38) shows that the dependence of f_{osc} on applied voltage can be relatively complex. For the sake of discussion, however, suppose the oscillation frequency varies linearly with applied VCO control voltage V_{CTL}:

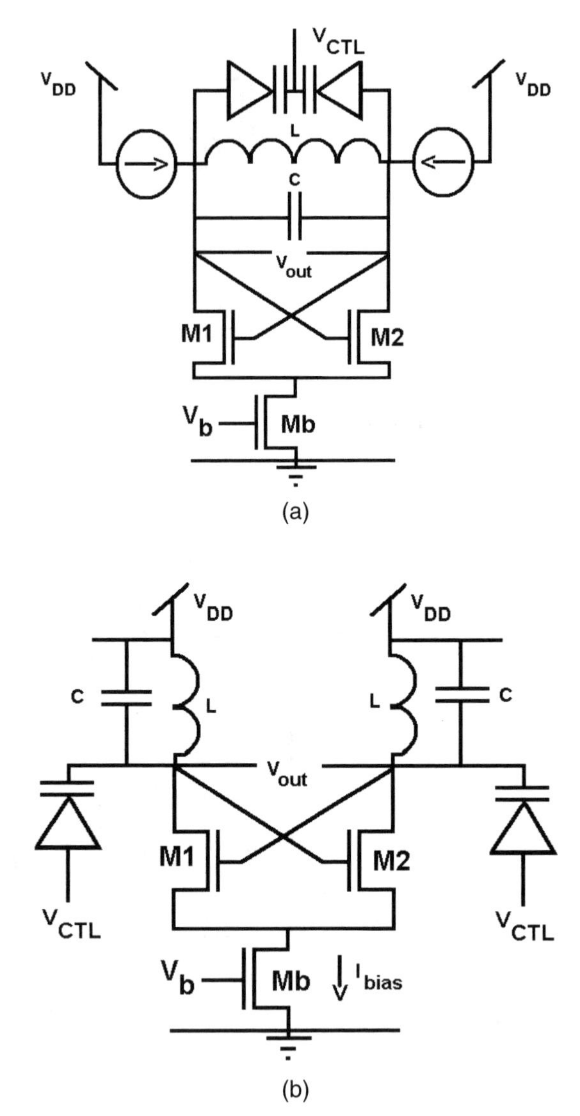

Figure 6.26 Possible locations for LC oscillator tuning varactors: (a) single inductor tank and (b) inductor tank dual.

$$f_{osc} = f_0 + K_{VCO} V_{CTL} \tag{6.39}$$

The change in f_{osc} with control voltage, $df_{osc}/dV_{CTL} = K_{VCO}$, is often referred to as the *VCO sensitivity* or *VCO gain* and describes how much the oscillator frequency varies per volt of applied control voltage. Using this definition for K_{VCO} and applying it to (6.38), the VCO gain can be written in terms of the CV characteristic of the varactor:

$$\frac{df_{osc}}{dV_{CTL}} = K_{VCO} = -\frac{1}{2}\frac{f_{osc}}{C}\frac{dC}{dV_{CTL}} \tag{6.40}$$

indicating that a varactor with a high control voltage sensitivity will provide a high tuning range. The dependence on capacitance variation with voltage implies that smaller inductors should be chosen so that larger capacitors, and hence larger dC/dV, can be used to maximize the tuning range. The lower valued inductors also have less series resistance.

6.4.2 Variable Capacitance Devices

The two main varactor types used in controlling f_{osc} in CMOS RFIC VCOs are MOSFETs operating in accumulation, depletion, or inversion mode and MEMS tunable capacitors. PN junction type varactors are not often used because their tuning ranges are usually less than 10%.

6.4.2.1 MOSFET Varactors

The most often used MOSFET varactor in CMOS applications is the accumulation mode varactor since it has the highest capacitance per unit area, better Q, and wider voltage tuning range than inversion/depletion MOSFET or PN junction varactors [18]. The gate terminal of the MOSFET varactor is connected to the tank to minimize capacitive loading by the parasitic capacitances so as to not overload the tank circuit. In this configuration, any gate resistance R_g associated with the MOSFET varactor will degrade the varactor Q and hence could impact the overall tank Q. Multiple gate fingered MOSFETs can be employed here to reduce the effective series resistance and raise the MOSFET varactor Q. The control voltage is applied to the drain source connection (Figure 6.27). The control voltage is at RF ground and so the varactor is essentially in shunt with the tank capacitor.

One difficulty when simulating an LC VCO is that accumulation mode varactors are more difficult to model that inversion or depletion mode MOSFET varactors. MOSFET varactors can have Q's of 15 or more, which are somewhat

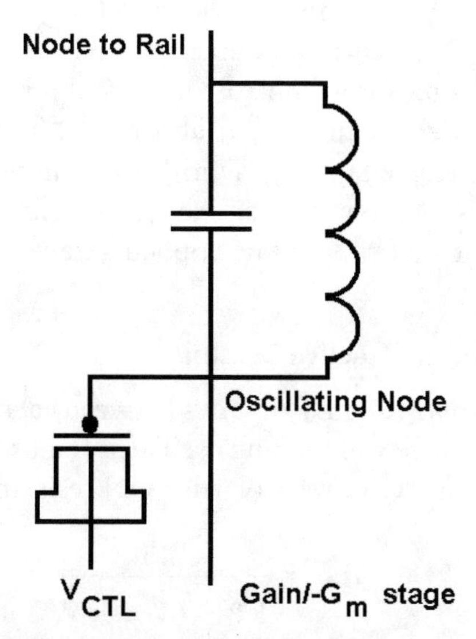

Figure 6.27 The gate of the MOSFET varactor is tied to the tank circuit with the drain source connection tied to the control voltage. A *p*MOSFET varactor is shown in this figure.

higher than those of on-chip planar spiral inductors and do not dramatically degrade the overall tank Q. Guard rings and other isolation structures are required to reduce substrate couplings of oscillator signals (or substrate noise to the tank) through the varactor (and are a good idea for oscillators no matter what type of varactor is used). A detailed discussion of the CV characteristic of the various MOSFET varactor types can be found in Chapter 2, Section 2.3.3.

6.4.2.2 MEMS Varactors

An alternative to the MOSFET varactor is to exploit the dependence of the gap distance g_0 on applied voltage of MEMS beams and membranes [19, 20]. Using the parallel plate capacitance approximation $C = \varepsilon_0 A/g_0$, any applied stimulus that decreases g_0 will increase the capacitance, providing the required voltage variable capacitance. MEMS-based varactors have much higher Q's and wider capacitance swings with voltage than their MOSFET counterparts. Specialized fabrication and postprocessing techniques can create varactors with capacitance tuning ranges of 300% or more [21], which corresponds to very wide VCO tuning ranges. While conceptually simple, major research efforts continue into the creation and fabrication of novel varactor structures to improve MEMS CV tuning ranges.

6.4.3 Voltage Control of Ring Oscillators

The ring oscillator frequency is controlled by the cascade inverter time delay, which in turn is governed by the charge/discharge rate of the capacitors loading each inverter. Two methods are widely used to turn the ring oscillator into a VCO: (1) by adjusting the current to the load capacitors or (2) by varying the charging resistance seen by the load capacitors.

6.4.3.1 Tail Current Adjustment

The charge/discharge rate can be varied by changing the capacitor charge/discharge current. Adding a tail current source to the basic CMOS inverter circuit [Figure 6.28(a)] provides a means to control the current through this source with the applied gate voltage. During the time V_{in} is at a logic "0" (0V), the *p*MOSFET is on and charges each load capacitance C_L. During the complementary logic state ($V_{in} = V_{DD}$), the *n*MOSFET is on, but the discharge current flow will be governed by the pull-down current sink M3 and its applied gate voltage.

6.4.3.2 Series MOSFET Active Resistor

The addition of an *n*MOSFET in series between each inverter can be used to control the oscillation frequency of the ring oscillator [Figure 6.28(b)]. In this case, the series MOSFET acts as an active resistor with ON resistance given by

$$R_{\mathrm{ON}} = \frac{1}{\mathrm{KP}\dfrac{W}{L}(V_{CTL} - V_T)} \tag{6.41}$$

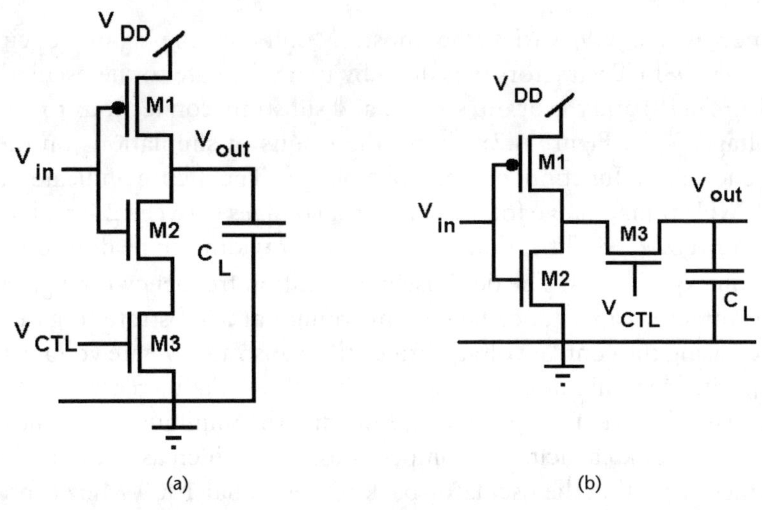

Figure 6.28 (a) Tail current adjustment and (b) series-connected MOSFETs can be used to vary the ring oscillator frequency with voltage control.

Large values of R_{ON} can be achieved with control voltages close to the threshold. Small ON resistance requires large W/L ratios, which tend to increase the parasitic capacitances at the drain and source, increasing the loading seen by each inverter and thereby lowering the top frequency that can be achieved with the ring oscillator. This type of structure is useful up to a few tens of megahertz.

6.4.4 VCO Design Example

Extend the example of Section 6.2.6 to create a VCO (ch6-3.txt). We would like this VCO to exhibit a VCO gain of 50 MHz/V using a pMOSFET inversion mode varactor. (The inversion mode was chosen for simulation simplicity.) Assuming 0.5-μm CMOS technology, determine the device width W of the varactor and verify the tuning characteristic.

The fixed LC oscillator in the example of Section 6.2.6 used a 0.95-pF tank capacitor to set the nominal 1.0-GHz oscillation frequency. The necessary dC/dV characteristic of the MOSFET varactor can be estimated from the VCO gain equation:

$$K_{VCO} = -\frac{1}{2}\frac{f_{osc}}{C}\frac{dC}{dV_{CTL}} \Rightarrow \frac{dC}{dV_{CTL}} = -2K_{VCO}\frac{C}{f_{osc}} = -2\left(5\cdot 10^{7}\right)\frac{0.95\cdot 10^{-12}}{1\cdot 10^{9}} = 0.095\text{ pF/V}$$

The first pass at a MOSFET varactor gate capacitance estimate can be done by using the gate oxide capacitance equation and assuming a 1V tuning range:

$$C_g = WLC_{ox} \Rightarrow WL = \frac{C_g}{C_{ox}} = \frac{0.095\cdot 10^{-12}}{2.6\cdot 10^{-3}} = 36\ \mu\text{m}^2 \Rightarrow W = L = 6\ \mu\text{m}$$

This calculation just provides a starting point for simulation purposes. Detailed knowledge of a representative varactor CV relationship will show the exact voltage range where CV varies the most. Modifying the example circuit to include the *p*MOSFET varactors was done by tying the gate to the oscillation node (V_{out} of Figure 6.18) and the drain, source, and substrate connections to an external control voltage, V_{CTL}. Figure 6.29 shows the results of simulations on the oscillation frequency as a function of control voltage. The figure indicates an approximate 34-MHz tuning range for VCO control voltages between 2.0 and 3.3V for a K_{VCO} of 26.5 MHz/V. Modifications to the varactor width can be done during further simulation experiments. The decrease in oscillation frequency for $V_{CTL} = 0$ occurs due to the varactor capacitance being a maximum at a substrate to gate voltage of 3.3V. Increasing the control voltage brings this substrate-to-gate voltage toward 0V and a capacitance minimum, creating a positive K_{VCO}. An increase in MOSFET width will increase the overall capacitance, lowering the minimum frequency. A reduction in the fixed tank capacitance compensates for the increased capacitive loading of the varactor, putting the oscillator back to a nominal 1000-MHz center.

6.5 Oscillator Phase Noise and Estimation

The introduction of noise into oscillator circuits creates a different set of problems than noise injected into amplifiers. Oscillator noise tends to broaden the spectrum of the ideally pure single-frequency tone that the perfect oscillator generates (Figure 6.30). Oscillator noise is highest near the carrier (*close to carrier noise*) and rolls off with frequency, eventually blending in with the overall circuit thermal noise. This broadened oscillator spectrum has an impact on both transmitters and receivers.

A transmitter with a broadened spectrum may introduce additional noise into adjacent channels, decreasing the adjacent channel SNR (Figure 6.31). On the receive end, the spectral broadening due to oscillator noise is transferred to both the undesired signal (or the signal in an adjacent channel) as well as the desired signal in the desired channel. The noise of the undesired signal is spread into the bandwidth of the desired signal. This phenomenon is termed *reciprocal mixing* (Figure 6.32). This can be a major contributor to increased noise in communication systems with closely spaced channels.

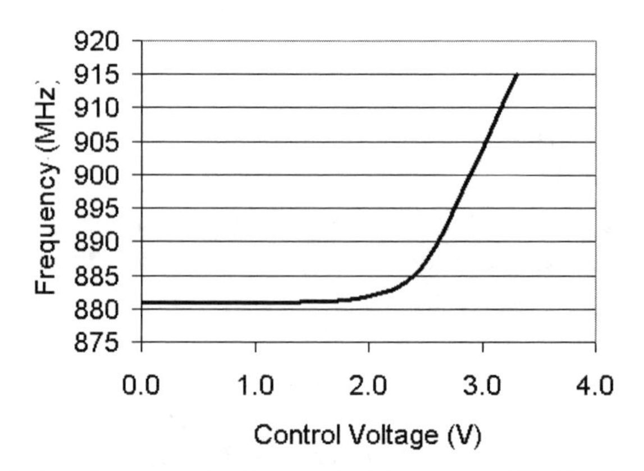

Figure 6.29 Simulated tuning range for the *p*MOSFET varactor VCO. The first pass design yields a 26.5-MHz/V VCO gain in the tuning range of 2.0 to 3.3V.

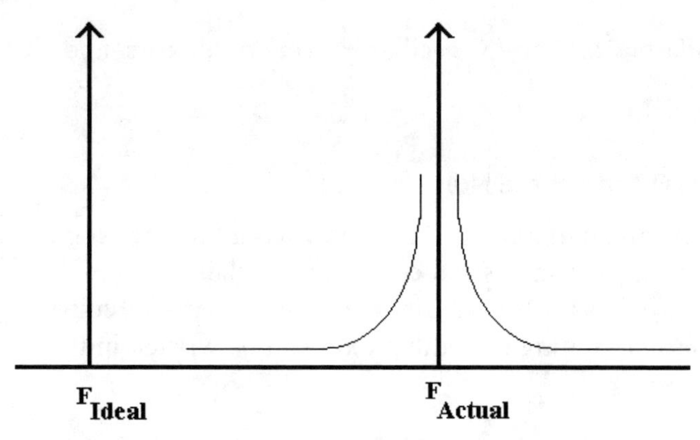

Figure 6.30 Ideal oscillator signal (F_{Ideal}) and oscillator signal with broadened spectrum due to oscillator noise (F_{Actual}).

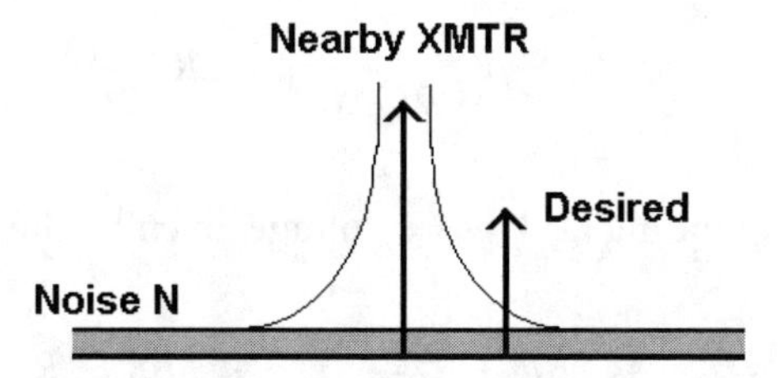

Figure 6.31 A noisy oscillator can degrade the SNR of an adjacent channel.

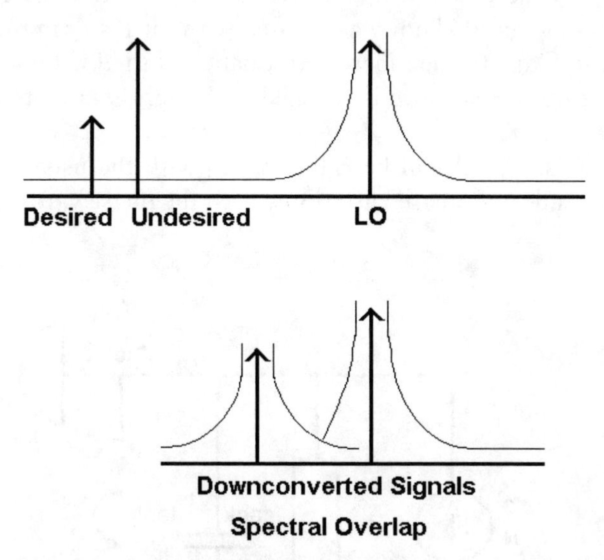

Figure 6.32 Oscillator noise can be downconverted and adds to the noise in the desired channel, decreasing the SNR.

This section covers the origin of noise introduced by the oscillator tank with an extension of this theory to Leeson's general oscillator noise theory [22]. The impact of oscillator noise on LC oscillators, ring oscillators, and VCO control lines is also covered.

6.5.1 LC Tank Phase Noise

The tank circuit of a real LC oscillator has a finite resistance R_p that is related to the tank circuit Q. This resistance introduces thermal noise current into the resonator and eventually into the remainder of the oscillator circuitry (Figure 6.33). The resistive thermal exciting the tank circuit can be written in the usual form for any pure resistance:

$$\bar{i}_N^2 = 4kTB/R_p \tag{6.42}$$

The total impedance of the tank circuit at an *offset frequency* Δf close to the oscillator frequency f_0 can be written as:

$$Z = -jR_p \frac{f_0}{2Q\Delta f}; |Z| = R_p \frac{f_0}{2Q|\Delta f|} \tag{6.43}$$

with the corresponding tank noise voltage given by the product of (6.42) and (6.43):

$$\overline{V}_N^2 = \bar{i}_N^2 |Z|^2 = 4kTBR_p \left(\frac{f_0}{2Q\Delta f}\right)^2 = 4kTBR_{eq}; R_{eq} = R_p \left(\frac{f_0}{2Q\Delta f}\right)^2 \tag{6.44}$$

where R_{eq} is the equivalent noise resistance of the tank circuit. Equation (6.44) shows that the thermal noise associated with the losses in the tank circuit is concentrated near f_0 due to the inherent resonance of the LC tank circuit impedance. The equivalent noise resistance is reduced with high-Q circuits or at large offsets from the carrier.

This noise signal will be riding along with the usual tank signal, introducing both amplitude and phase variations, with the phase variations most acute near the

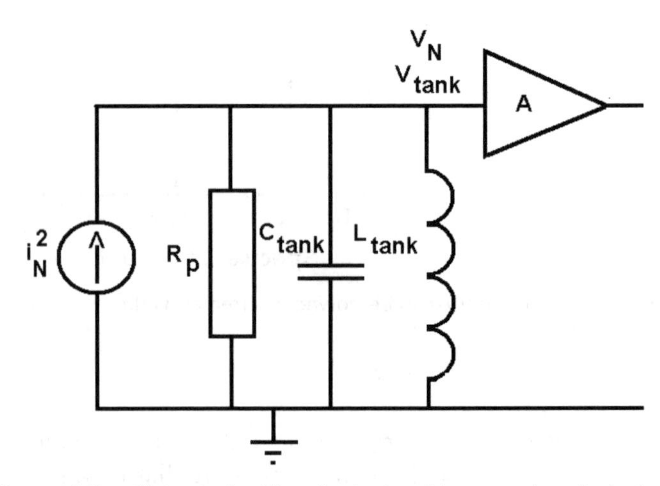

Figure 6.33 General LC oscillator tank with noisy tank resistance and equivalent noise current.

zero crossings of the signal (Figure 6.34). An approximation often used in oscillator noise analysis is to divide the noise equally between amplitude and phase noise [23], the phase noise component being the more important component. Using this approximation, the ratio of the tank noise power to tank signal power P_{sig} is given by

$$\frac{P_N}{P_{sig}} = PN = \frac{2kT}{P_{sig}}\left(\frac{f_0}{2Q\Delta f}\right)^2 \tag{6.45}$$

which shows that oscillator phase noise (PN) impact is minimized by using as large a tank signal as possible (P_{sig}) or by increasing the tank Q (which is another reason why oscillator design engineers are so concerned with oscillator tank Q). This is the formula derived by Leeson for the close to carrier (or $1/f$) phase noise component of

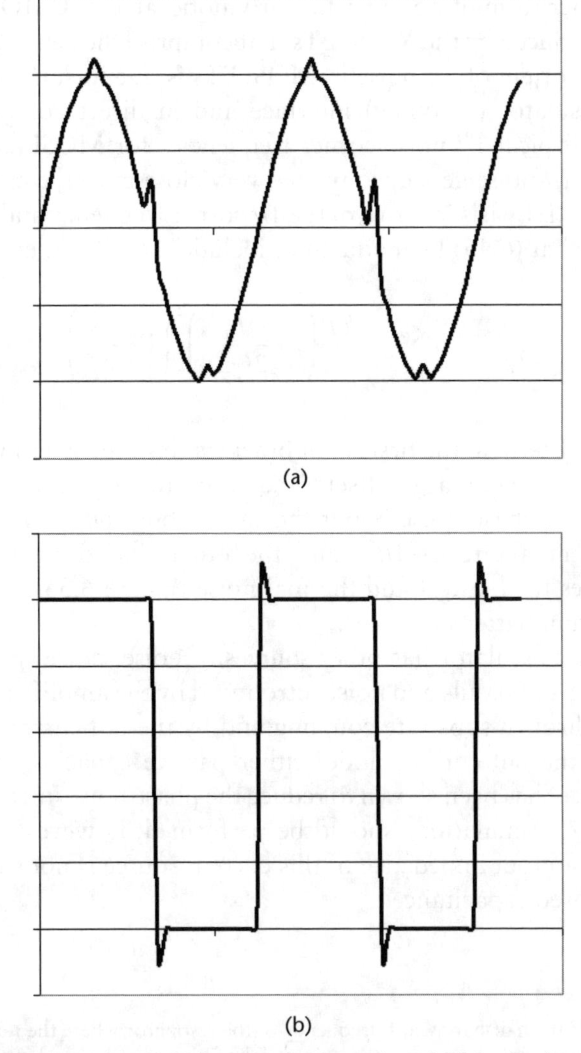

Figure 6.34 The noise introduced (a) by R_p in the LC tank and (b) in the ring oscillator will introduce amplitude and phase noise on the oscillator signal.

the overall oscillator noise spectrum [22]. Other investigators, however, have preferred to keep the more general term *oscillator noise* and not multiply by one-half [24].

Inductor Q is unfortunately a function of frequency and so simulations must be performed to obtain detailed estimates of inductor resistive and reactive components to ensure the proper values are substituted in (6.45) at the design frequency. PN is highest close to the carrier (f_0) and rolls off as $(1/\Delta f)^2$ moves away from the carrier. PN on a per unit bandwidth basis allows (6.45) to be written in decibel terms as follows ($T = 290$K):

$$PN = -116 + 20\log(f_{0-GHz}) - 20\log(\Delta f_{MHz}) - 20\log(Q) - P_{sig-dBm} \text{ dBc/Hz} \qquad (6.46)$$

For example, a 1.0-GHz oscillator signal at a 0-dBm power level at 100-kHz offset with a tank Q of 10 has a phase noise of –116 dBc/Hz. This phase noise decreases by an additional 20 dB at 1.0-MHz offset from the carrier.

Another form of noise particularly noticeable in CMOS RFIC design is the $1/f$ noise introduced by the MOSFETs that comprise the active components of the oscillator. The origin of $1/f$ noise in MOSFETs is due to surface effects at the semiconductor insulator (or oxide) interface and in direct contact with the conducting channel. Typical $1/f$ noise corner frequencies for MOSFETs are around 1.0 MHz [25]. The $1/f$ noise gets upconverted very close in to the carrier (although not on a completely 1:1 basis in terms of the $1/f$ corner frequency and is generally lower [26]), requiring that (6.45) be modified to include the $1/f$ corner frequency $f_{1/f}$:

$$\text{PN} = \frac{2kT}{P_{sig}}\left(1 + \frac{f_0}{2Q\Delta f}\right)^2\left(1 + \frac{f_{1/f}}{\Delta f}\right) \qquad (6.47)$$

where the 1 term in the first set of brackets sets the lower limit for overall oscillator phase noise at very large offsets.[4] Equation (6.47) shows that two distinct regions exist for the phase noise about the carrier but above the thermal noise: one that decreases at a rate of $1/f^3$, and the other that decreases as $1/f^2$ as the noise approaches the background thermal noise (Figure 6.35). Very close to the carrier, the spectrum flattens out [26].

The LC oscillator has other sources of noise: noise spikes coming through the power or ground rails and noise introduced by the amplifier MOSFETs, either internally or through substrate coupling and by the tail current source. A high capacitance on the tail current node (either passive capacitor or large drain substrate capacitance) has been shown to reduce the phase noise introduced by the tail current source [27]. Simulations should be performed, however, to ensure that the high-frequency output impedance of this current source is not substantially degraded by this increased capacitance.

4. The term $f_0/2Q$ is the half-bandwidth frequency of the resonator where the noise tends to flatten out. If the resonator is in series with the output, the noise will be filtered until it reaches the noise floor. (Thanks to the reviewer for pointing out this fact.)

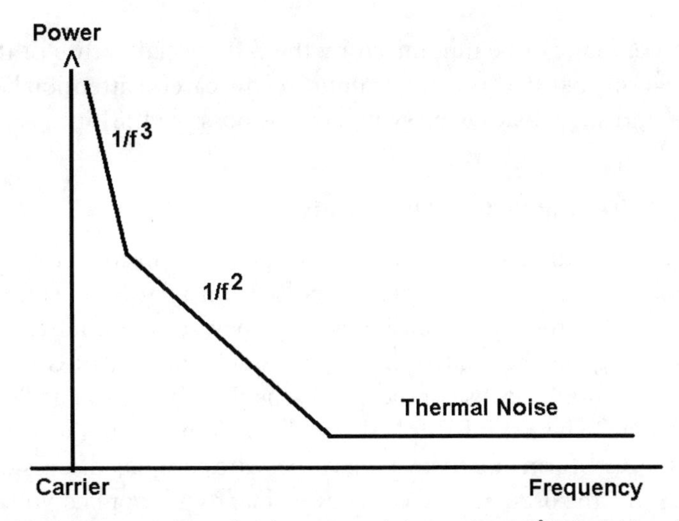

Figure 6.35 Three distinct regions of noise surround the carrier: $1/f^3$ (30-dB/decade roll-off), $1/f^2$ (20-dB/decade roll-off), and thermal noise. The phase noise spectrum also occurs on the reverse side of the carrier.

6.5.2 Ring Oscillator Phase Noise

Ring oscillator phase noise is well known to be poorer than LC resonator-based oscillators [24]. There are several reasons for this poor noise performance. The first reason can be seen from the definition of Q: energy stored per cycle divided by energy dissipated per cycle. In the LC oscillator, the energy loss of R_p is not complete during each cycle. (In other words, the capacitor charge is not completely removed through R_p per cycle.) In the ring oscillator, the load capacitors are pretty much completely charged and discharged once each cycle (from the rail-to-rail switching characteristic of the digital inverter stage), giving a nearly unity Q. Secondly, phase error introduced by noise impulses is highest at the waveform rising and falling edges [Figure 6.34(b)], the same signal phase where energy is being pushed into or pulled from the load capacitors. Analysis has shown that single-ended ring oscillators exhibit a relatively constant amount of phase noise independent of the number of stages N for a given frequency and power dissipation [28]:

$$\mathrm{PN} \cong \frac{16}{3}\frac{kT}{P_{diss}}\left(\frac{f_0}{\Delta f}\right)^2 \tag{6.48}$$

where P_{diss} is power dissipation. This is not the case for differential ring oscillators, which show poorer PN performance as the number of stages N increases [28]:

$$\mathrm{PN} \cong \frac{8}{3}\frac{kT}{P_{diss}}N\left(\frac{V_{DD}}{RI_{tail}}\right)\left(\frac{f_0}{\Delta f}\right)^2 \tag{6.49}$$

where I_{tail} is the differential pair tail current and R the load resistance. However, there are other noise sources, such as common mode noise introduced through the substrate, that can be minimized by the differential nature of this circuit topology. Noise in ring oscillators can be minimized by careful attention being paid to keep the layout and stage loading as symmetric as possible [28].

6.5.3 VCO Control Line Phase Noise

Oscillators, be they LC or ring, will be controlled in some form as part of the VCO. The voltage control line can pick up noise from a variety of sources and translate this noise signal to the VCO, causing a frequency variation in response to the instantaneous change in the control line voltage. The impact of control line noise can be quite substantial because of the VCO sensitivity K_{VCO} and its dependence on control line voltage (whether it is applied directly or comes from the undesired noise source). In its simplest form and to keep this otherwise complex discussion as simple as possible, assume the oscillation frequency is linear with applied voltage (6.39). If a noise source on the control line is assumed sinusoidal in nature with frequency ω_N and amplitude V_N, the VCO control line voltage is the sum of the dc control voltage V_{CTL} and this noise signal. The VCO oscillation frequency can be written as

$$V_{VCO}(t) = V_0 \sin(\omega_{osc} t) = V_0 \sin\left[\omega_0 t + K_{VCO}\left(V_{CTL} + V_N \sin(\omega_N t)\right)\right] \tag{6.50}$$

which, by the way, is also the mathematical form for an FM signal. Equation (6.50) indicates that control line noise shows up in the VCO output spectrum as an FM signal. The noise power-to-signal ratio for this FM noise can be written as

$$N_{FM-CTL} = \left(\frac{K_{VCO} V_N}{2\omega_N}\right)^2 \tag{6.51}$$

and indicates that high noise amplitudes provide high noise power as one might expect, but the high VCO gain K_{VCO} also has a large impact on the noise amplitude for a given noise spike. The equation indicates that low-frequency noise such as $1/f$ noise plays a large role in VCO control line noise because of the $1/f_N$ noise power dependence indicated in (6.51).

Example 6.6: Determine the noise power of a 1-μV, 1-MHz noise signal on the control line of a VCO with a VCO gain as calculated in the example of Section 6.4.4.

The VCO gain of 50 MHz/V will govern the noise power of the 1-μV noise signal on the control line. The noise power from this noise source can be computed as

$$N_{FM-CTL} = \left(\frac{K_{VCO} V_N}{2\omega_N}\right)^2 = \left(\frac{50 \cdot 10^6 \cdot 10^{-6}}{2 \cdot 2\pi \cdot 10^6}\right)^2 = 15.8 \text{ pW/Hz} = -78 \text{ dBm/Hz}$$

6.5.4 PN Example Calculation

A certain communication system has various VCO phase noise requirements for different offsets about the nominal 1000-MHz carrier frequency as shown in Figure

6.36(a). Using the oscillator design example of Section 6.2.6, determine if the phase noise requirements of this system are met. Assume the VCO has a 1/*f* noise corner frequency of 10 kHz.

The example oscillator was defined to have a Q of 10 and a nominal carrier frequency of 1000 MHz, so the only variable left to calculate is the signal power available from the oscillator. The tank voltage was found during simulation to have a 4.3V peak value. With this voltage across the tank equivalent resistance of $R_P = R_s(1 + Q^2) = 1{,}515\Omega$, the signal power P_{sig} is 6.1 mW or 7.8 dBm. The resulting phase noise spectrum is calculated using (6.47) and is shown in Figure 6.36(b). The figure indicates that this oscillator meets the specifications, although the PN at 1.0-kHz offset is approaching the specification limit. Under these circumstances, an oscillator redesign may be in order. A likely candidate for oscillator improvement is to increase the available signal power. This solution, however, requires higher tail current with a subsequent increase in dc power. Improving the inductor Q will obviously help, but the Q of 10 is at the upper limit of easily obtainable on-chip inductor Q values of bulk CMOS processes.

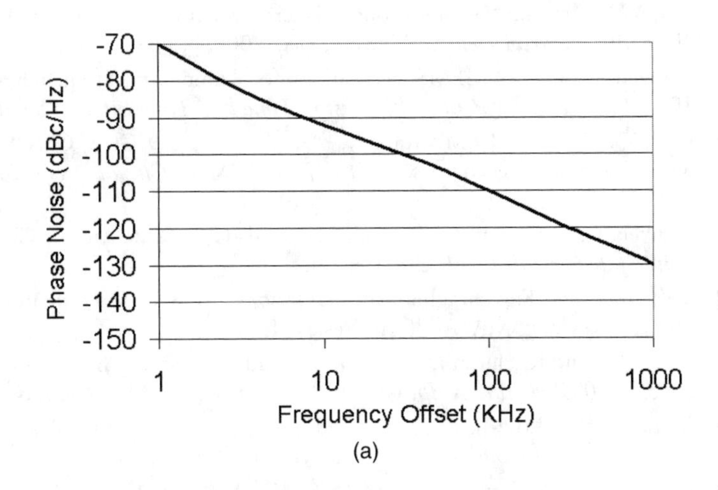

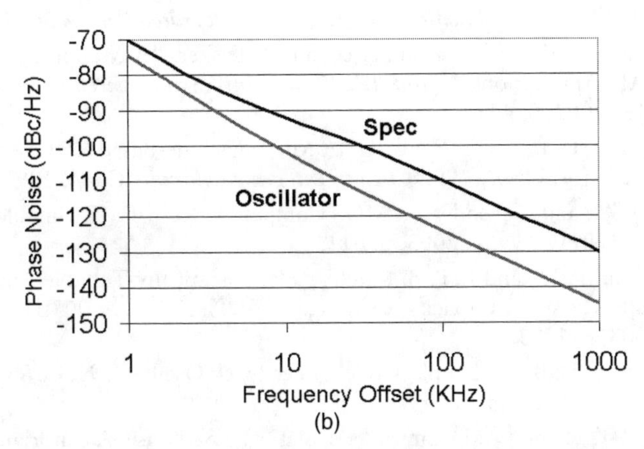

Figure 6.36 (a) Phase noise specification for the PN example. (b) Comparison of phase noise specification and oscillator phase noise based on Leeson's formula.

References

[1] Campbell, R., "RF Oscillators," *2006 IEEE Intl. Microwave Symp. Workshop Notes: Fundamentals of HF Through UHF Design*, 2006.

[2] Nise, N. S., *Control Systems Engineering,* 4th ed., New York: John Wiley and Sons, 2004.

[3] Lee, T. H., *CMOS Radio Frequency Integrated Circuit Design,* 2nd ed., Cambridge, UK: Cambridge University Press, 2004.

[4] Hayward, W., R. Campbell, and B. Larkin, *Experimental Methods in RF Design,* Newington, CT: ARRL, 2003.

[5] Pozar, D., *Microwave Engineering,* 2nd ed., New York: John Wiley and Sons, 1998.

[6] Kral, A., F. Behbahani, and A.A. Abidi, "RF-CMOS Oscillators with Switched Tuning," *Proc. IEEE 1998 Custom Integrated Circuits Conf.,* 1998, p. 555.

[7] van der Tang, J., et al., "A Cost-Effective Multi–Band LC Oscillator for Low–IF FM Radio Receivers," *Proc. 2002 28th European Solid State Circuits Conf., ESSCIRC 2002*, 2002, p. 819.

[8] Yasunaga, T., et al., "A Fully Integrated PLL Frequency Synthesizer LSI for Mobile COMMUNICATION SYSTEM," *2001 IEEE RFIC Symp. Digest*, 2001, p. 65.

[9] Gaddi, R., et al., "Reconfigurable MEMS-Enabled LC-Tank for Multi-Band CMOS Oscillator," *Proc. 2005 IEEE Microwave Symp.*, June 2005, p. 12.

[10] Lin, Y.-W., et al., "Series-Resonant Micromechanical Resonator Oscillator," *Proc. 2003 IEEE Int. Electron Devices Meeting*, Dec. 2003, p. 961.

[11] Nguyen, C. T.-C., "MEMS Technology for Timing and Frequency Control," *Proc. Joint IEEE Int. Frequency Control/Precision Time and Time Interval Symp.*, 2005, p. 1.

[12] Otis, B. P., and J. M. Rabaey, "A 300-μW 1.9-GHz CMOS Oscillator Utilizing Micromachined Resonators," *IEEE J. Solid State Circuits,* Vol. 38, No. 7, July 2003, p. 1271.

[13] Nguyen, C. T.-C., and J. Kitching, "Towards Chip-Scale Atomic Clocks," *Proc. 2005 Int. Solid State Circuits Conf.*, 2005, p. 84.

[14] Weste, N., and K. Eshraghian, *Principles of CMOS VLSI Design: A Systems Perspective,* 2nd ed., Reading, MA: Addison Wesley, 1992.

[15] Liang, D., and R. Harjani, "Comparison and Analysis of Phase Noise in Ring Oscillators," *Proc. 2000 IEEE Int. Symp. Circuits Systems*, Vol. 5, May 2000, p. 77.

[16] Maneatis, J., "Low-Jitter Process-Independent DLL and PLL Based on Self-Biased Techniques," *IEEE J. Solid State Circuits,* Vol. 31, Nov. 1996, p. 1723.

[17] Cheng, K.-H., C. Lai, and Y. Lo, "A CMOS VCO for 1V, 1GHz PLL Applications" *Proc. 2004 IEEE Asia Pacific Conf. Adv. Sys. Integrated Cir.*, Aug, 2004, p. 150.

[18] Bhattacharjee, J., D. Mukherjee, and J. Lasker, "A Monolithic CMOS VCO for Wireless LAN Applications," *Proc. IEEE 2002.Intl. Symp. Circuits Systems ISCAS 2002*, Vol. 3, May 2002, p. 441.

[19] Young, U. J., and B. E. Boser, "A Micromachine-Based RF Low-Noise Voltage-Controlled Oscillator," *Proc. 1997 Custom Integrated Circuits Conf.*, 1997, p. 431.

[20] Oz, A., and G. Fedder, "CMOS Compatible RF MEMS Tunable Capacitors," *Proc. 2003 IEEE RFIC Symp.*, 2003, p. 611.

[21] Peroulis', D., and L. P. B. Katehi, "Electrostatically-Tunable Analog RF MEMS Varactors with Measured Capacitance Range of 300%," *Proc. 2003 IEEE Int. Microwave Symp.*, 2003, p. 1793.

[22] Leeson, D. B., "A Simple Model of Feedback Oscillator Noise Spectrum," *Proc. IEEE*, Vol. 54, Feb. 1966, p. 329.

[23] Lee, T. H., and A. Hajimiri, "Oscillator Phase Noise: A Tutorial," *IEEE J. Solid State Circuits*, Vol. 35, No. 4, March 2000, p. 326.

[24] Kouznetsov, K. A., and R. G. Meyer, "Phase Noise in LC Oscillators," *IEEE J. Solid State Circuits,* Vol. 35, No. 8, Aug. 2000, p. 1244.

[25] Re, V., et al., "Survey of Noise Performances and Scaling Effects in Deep Submicrometer CMOS Devices from Different Foundries," *IEEE Trans. Nuclear Science,* Vol. 52, No. 6, Pt. 2, Dec. 2005, p. 2733.

[26] Edson, W. A., "Noise in Oscillators," *Proc. IRE*, Aug. 1960, pp. 1454–1466.

[27] Hajimiri, A., and T. H. Lee, "Design Issues in CMOS Differential LC Oscillators," *IEEE J. Solid State Circuits*, Vol. 34, No. 5, May 1999, p. 717.

[28] Hajimiri, A., S. Limotyrakis, and T. H. Lee, "Jitter and Phase Noise in Ring Oscillators," *IEEE J. Solid State Circuits*, Vol. 34, June 1999, p. 794.

Selected Bibliography

Goyal, R., *High Frequency Analog Integrated Circuit Design,* New York: John Wiley, 1995.

Ludwig, R., and P. Bretchko, *RF Circuit Design: Theory and Applications,* Upper Saddle River, NJ: Prentice Hall, 2000.

CHAPTER 7

CMOS Mixer Circuits

Up to this point, the circuits we have studied have been linear in nature. In amplifiers, for example, the input and output frequency are the same (assuming no harmonics are generated), only the amplitude has increased (and the phase changed). Even oscillators, while considered nonlinear in nature, are "linear" in the sense that they ideally output a single frequency sinusoid. Linear circuits, however, do not have the ability to translate or convert a signal to a higher frequency (*upconversion*) or to a lower frequency (*downconversion*); another circuit is required. The design of such a frequency converter or *mixer* should consider the following:

- The conversion should be done efficiently and with low loss of signal integrity;
- The conversion should suppress undesired signals and noise;
- The converting circuits should exhibit efficient power consumption.

This chapter describes a number of circuit types that perform these frequency conversion functions. Emphasis is placed on those circuit architectures widely used in CMOS RFIC design, and we also look at circuit enhancements to improve such issues as signal linearity and noise performance. The chapter starts off with a review of frequency conversion concepts and terminology. A number of simple frequency conversion circuits are covered to show both circuit possibilities but to also introduce more advanced concepts. The chapter finishes with several sections on high-performance frequency conversion circuits.

7.1 General Mixer Concepts

This section presents the concept behind frequency conversion and introduces both simple passive and active frequency conversion circuits as well as conceptual circuits for improved circuit performance.

7.1.1 Terminology

From communication theory, if a signal $x(t)$ has as its Fourier transform a spectrum of the form $X(\omega)$, multiplying $x(t)$ by a complex sinusoid of the form $e^{j\omega_0 t}$ *modulates* the spectrum by the amount ω_0:

$$x(t)e^{-j\omega_0 t} \leftrightarrow X(\omega - \omega_0) \tag{7.1}$$

thereby shifting the center frequency of spectrum $X(\omega)$ to ω_0. Equation (7.1) shows that the signal $x(t)$ can be shifted by a multiplicative process, by nature a nonlinear process. Fortunately, *nonlinear devices* provide the capability of performing this signal multiplication and the circuit that performs this multiplication is referred to as a *mixer*. The mixer is fundamentally a three-port device with the signal frequency to translate referred to as the *RF or radio-frequency* signal, the modulating or shifting signal referred to as the *LO or local oscillator* signal, and the result of the mixing process termed the *IF or intermediate frequency* signal. In an ideal mixer, the RF and LO ports are strictly input ports, with the IF being strictly an output port. The mixing efficiency that ultimately produces the IF signal is referred to as the *conversion gain* (or *conversion loss*) and is the ratio of the IF power out to the RF power in (7.2a). The ratio of the IF output voltage to RF input voltage can also be used to define the conversion gain (7.2b).[1]

$$CG = 10\log\left(\frac{P_{IF}}{P_{RF}}\right) \text{dB} \tag{7.2a}$$

$$CG = 20\log\left(\frac{V_{IF}}{V_{RF}}\right) \text{dB} \tag{7.2b}$$

Passive mixers (such as diodes or unbiased transistors) always exhibit negative conversion gain, which is then termed conversion loss. Active mixers, however, can exhibit higher IF power outputs than RF power inputs, and so the term conversion gain directly applies. Determining which term to use (loss or gain) is simply a matter of determining whether the IF power is greater or less than the RF power. In practice, there will be some signal leakage out of the input ports or RF and LO finding their way to the IF port. The level of these nonideal signals is referred to as the *mixer isolation*. Table 7.1 outlines some of these isolation parameters and their importance to mixer performance.

For a single-frequency RF signal and a single-frequency LO signal, the mixer IF occurs at the sum and difference frequencies of these signals, $f_{IF} = |(f_{LO} \pm f_{RF})|$. High side LO injection is defined as an LO frequency f_{LO} *above* the RF frequency, f_{RF}; conversely, low side LO injection is defined as an LO frequency *below* the RF frequency. If a signal resides at a frequency on the other side of the LO from the desired RF signal, this signal may also be mixed down to the same IF with the possibility of interference with the desired signal (especially if this signal is of larger amplitude than the desired one). This *image signal* can be especially problematic in channelized applications where channels are separated by a fixed frequency (Figure 7.1). Various mixer circuit topologies can be employed to reduce these signals including filtering out the image signal prior to the mixer (*image reject filter*). Other circuits that improve mixer performance are presented in the following sections.

Example 7.1: A 430-MHz ISM band signal is to be downconverted to a 25-MHz IF. What are two possible LO signal frequencies and their related image frequencies?

1. The voltage definition of conversion gain is less meaningful because various passive means such as transformers can be used to increase the IF voltage without changing the IF power generated.

Table 7.1 Summary of Mixer Isolation Parameters

Isolation Parameter	*Importance*
LO-RF	The large LO signal appearing at the RF port could actually make it to the antenna and radiate, especially if $\lvert S_{12} \rvert$ of the LNA is poor.
LO-IF	The large LO signal appearing at the IF port may require extensive filtering to ensure that its large amplitude does not overdrive downstream circuits.
RF-IF	One of the isolations of lesser importance; filtering usually can minimize this since the RF signal is small.

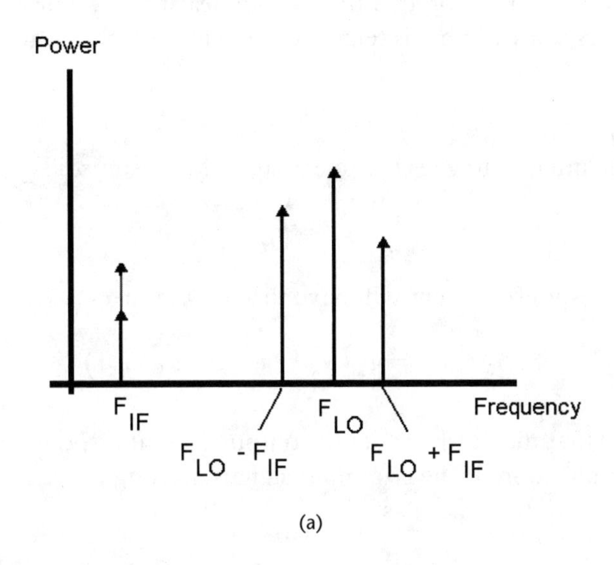

(a)

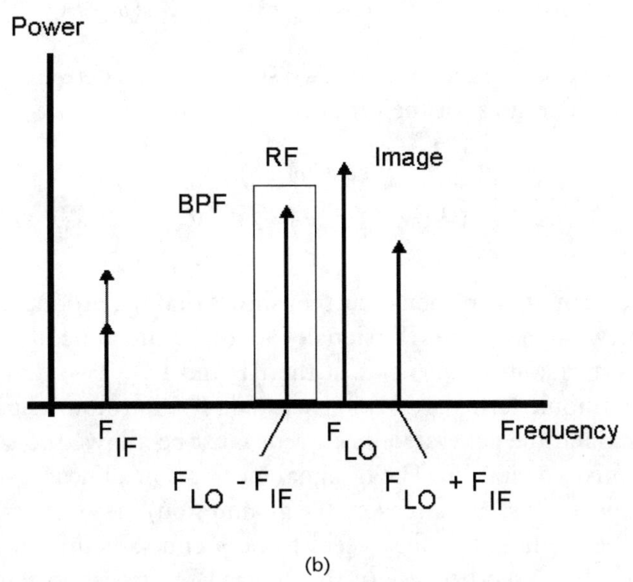

(b)

Figure 7.1 (a) The image signal resides on the opposite side of the LO from the desired RF signal. (b) A bandpass filter centered on the desired RF signal can be used to reduce the image signal amplitude.

Two LO frequencies will give the required downconverted IF signal $f_{IF} = |(f_{LO} - f_{RF})|$: 455 and 405 MHz. For the 455-MHz LO, the RF signal is *below* the LO, and so the corresponding image frequency would be *above* the LO by the IF: 480 MHz. For the 405-MHz LO, the RF signal is above the LO, and so the corresponding image frequency would be 380 MHz.

7.1.2 Ideal Passive Mixers—Weak Nonlinearity

The simplest device nonlinearity that will provide the required signal multiplication is the second-order or quadratic nonlinearity, x^2. For example, if a nonlinear device's response current is related to the square of the input voltage stimulus, then

$$i_{\text{out}} = g_2 v_{\text{in}}^2 \tag{7.3}$$

If the input voltage is the linear sum of two sinusoids

$$v_{\text{in}} = v_{RF} + v_{LO} \tag{7.4}$$

then the response current will have three components:

$$i_{\text{out}} = g_2 \left(v_{RF}^2 + v_{LO}^2 + 2 v_{RF} v_{LO} \right) \tag{7.5}$$

Note that the third term in the output current response provides the desired signal multiplication. If the two input signals are single-frequency sinusoids, the input excitation can be written as

$$v_{\text{in}} = V_{RF} \cos(\omega_{RF} t) + V_{LO} \cos(\omega_{LO} t) \tag{7.6}$$

The mixing product of these two signals can be determined by recalling the trigonometric identities for the product of two sinusoids:

$$i_{\text{out}} = \frac{1}{2} g_2 \begin{bmatrix} V_{RF}^2 + V_{RF}^2 \cos(2\omega_{RF} t) + V_{LO}^2 \\ +V_{LO}^2 \cos(2\omega_{LO} t) + 2 V_{RF} V_{LO} \cos((\omega_{RF} \pm \omega_{LO}) t) \end{bmatrix} \tag{7.7}$$

The form of the response current shows that the product of RF and LO gives rise to a new signal, the IF signal, at the sum (upconversion) and difference (downconversion) frequencies of the RF and LO. Two dc terms and two second-order harmonic terms are also generated. The harmonic signals can usually be filtered out and the generated dc current must be allowed to see a dc return path for proper mixer operation. The IF signal increases in a linear fashion (assuming the LO amplitude is constant, a reasonable assumption) in direct response to the RF signal amplitude. Although it may seem to be a contradiction in terms, the term *linear mixer* is often used because of this linear IF response to the RF amplitude (which also assumes a constant LO signal amplitude).

Real nonlinear devices that exhibit quadratic nonlinearities also exhibit linear and higher order responses as well. These additional responses can be mathematically described by an infinite series representation of the response current:

$$i_{out} = i_0 + g_1 v_{in} + \sum_{n=2}^{\infty} g_n v_{in}^n \tag{7.8}$$

where g_1 is the linear RF response characteristic (in this case, a conductance) and the remaining g_n are the nonlinear RF response coefficients. For $n = 3$ and higher, a host of other signals are generated as part of the overall output current response and potentially appear at the load along with the desired IF signal. These additional (and always unwanted) signals are usually lumped together and referred to as spurious mixer signals or *mixer spurs*. Many of these spurs occur outside of the IF band (*out-of-band spurs*) and are therefore easily filtered. Those mixer spurs that occur within the IF band (*in-band spurs*), however, are not easily removed and can interfere with the desired IF signal. A general expression for the in-band mixer spurs generated by a nonlinear device can be estimated by the following expression:

$$f_{IF} = \left(\begin{matrix}+\\-\end{matrix}\right) m f_{RF} + \left(\begin{matrix}+\\-\end{matrix}\right) p f_{LO} \tag{7.9}$$

where m and p are the various harmonics of the RF and LO signals, respectively. This computation gives only the frequency of the mixer spurs and not their amplitudes. Fortunately, the device nonlinearity coefficients g_n usually decrease rapidly with increasing n and therefore their influence becomes vanishingly small.

Example 7.2: Consider an IF band of 0 to 50 MHz with an RF band of 1,850 to 1,900 MHz and an LO of 1,900 MHz. Determine the mixer spurs if up to third-order harmonics of the RF and LO are considered.

If third-order harmonics of both the RF and LO signal are assumed to exist, the following [m, p] pairs generate in-band mixer spurs: [–3, 3], [–2, 2], and [–1, 1].

7.1.3 Ideal Active Mixers—Switching or Multiplying Mixers

In typical nonlinear devices, even the second-order nonlinearity g_2 is much smaller than the linear coefficient g_1, generating a relatively weak IF response. Another way to generate the frequency conversion is to multiply the RF signal by another signal that is rich in harmonics, with each harmonic providing a frequency translation. A square-wave signal is one such harmonic-rich signal. If a switching signal $S(t)$ is unipolar (0 to +1) at the LO frequency [Figure 7.2(a)], $S(t)$ may be written as a Fourier series as

$$S(t) = \frac{1}{2} + \sum_{k=1}^{\infty} b_k \sin(k\omega_{LO} t);\ b_k = \begin{cases} 0; & k-\text{even} \\ \dfrac{2}{k\pi}; & k-\text{odd} \end{cases} \tag{7.10}$$

This switching signal "chops" the RF signal at the rate of the LO, effectively multiplying the RF signal by $S(t)$:

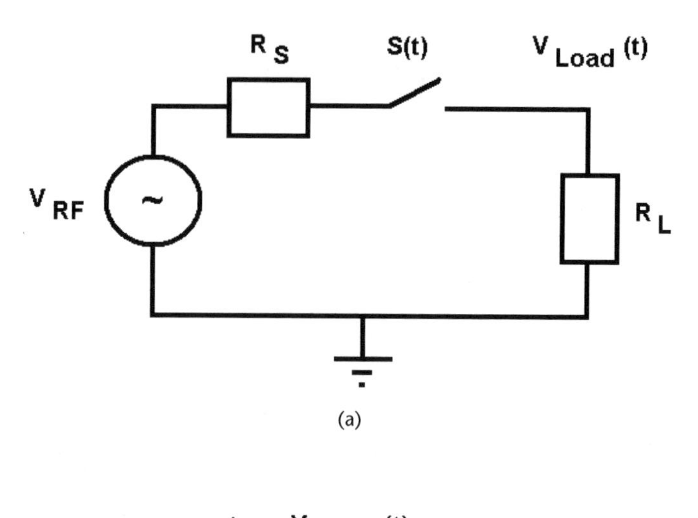

(a)

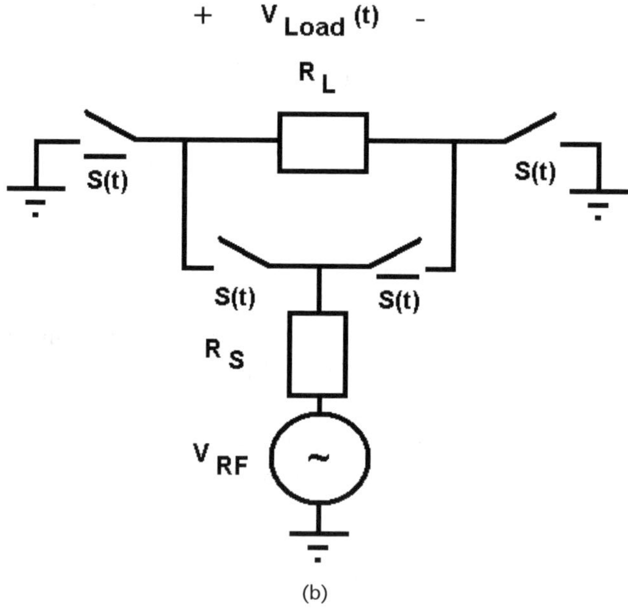

(b)

Figure 7.2 Conceptual circuit of (a) unipolar and (b) bipolar LO switching mixers.

$$S(t)V_{RF}(t) = \left[\frac{1}{2} + \sum_{k=1}^{\infty} b_k \sin(k\omega_{LO}t)\right] V_{RF} \sin(\omega_{RF}t)$$
$$= \frac{1}{2} V_{RF} \sin(\omega_{RF}t) + \frac{1}{2}\sum_{k=1}^{\infty} V_{RF} b_k \sin[(k\omega_{LO} \pm \omega_{RF})t] \quad (7.11)$$

In a pure square wave, b_k for k even equals zero and so these even-order mixer signals are suppressed; only the odd-order harmonics (k odd) remain. The dc component of $S(t)$ generates an RF feed through the signal as shown in (7.11), which can be eliminated by using a bipolar switching scheme for $S(t)$ (−1 to +1). In this bipolar switching case [Figure 7.2(b)], the b_k are twice the value of the unipolar case.

7.1.4 Single- and Double-Balanced Mixers: General Definitions

In the previous circuit examples, significant signal feed-through occurred. Output filtering can improve the isolation, but this filtering is usually performed with on-chip components that may include lossy inductors. A more real estate efficient means of feed-through reduction can be achieved by using circuit techniques yielding in- and out-of-phase signals that when summed provide the desired cancellation. In addition, common mode and noise signals can be reduced by these same signal cancellation techniques. *Balanced mixer* circuits implement this signal cancellation and use in- and out-of-phase components of one or more input RF and LO signals (0° and 180°). Balanced mixers are easily implemented in CMOS RFICs because in many cases both the in- and out-of-phase signals are available from differential amplifiers and oscillators (see Chapters 4 through 6). One such balanced mixer uses two different mixer circuits, each fed with a single RF signal but in- and out-of-phase LO signals (Figure 7.3). The IF output is taken as the difference between the two mixer outputs. The operation of this *single balanced mixer* can be seen by assuming a simple nonlinear I-V relationship for the mixing device:

$$I(V) = \sum_{n=0}^{\infty} g_n V^n \tag{7.12}$$

The mixer inputs assume that the RF and LO signals are ideally combined:

$$V_{TOP} = V_{RF} + V_{LO}; \ V_{BOT} = V_{RF} - V_{LO} \tag{7.13}$$

The mixer IF output is the difference between the two response currents:

$$I_{IF} = I_{TOP} - I_{BOT} = \sum_{n=0}^{\infty} g_n (V_{RF} + V_{LO})^n - \sum_{n=0}^{\infty} g_n (V_{RF} - V_{LO})^n \tag{7.14}$$

Looking at only the first two components of the IF output, the desired quadratic mixing term (n = 2) and an unwanted LO feed-through component (n = 1) are generated:

$$n = 1\text{: } I_{IF} = 2g_1 V_{LO} \quad n = 2\text{: } I_{IF} = 4g_2 V_{RF} V_{LO} \tag{7.15}$$

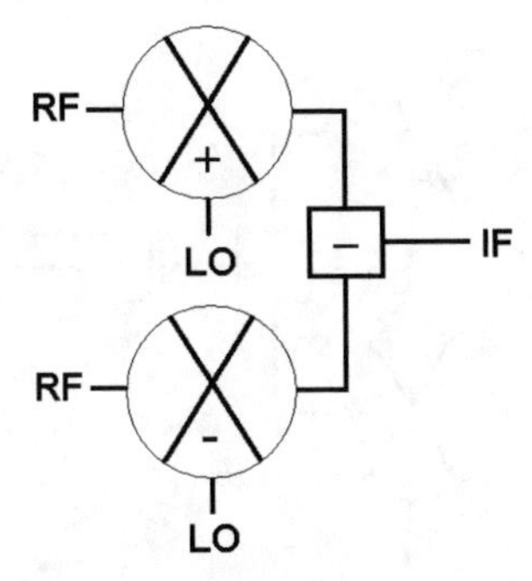

Figure 7.3 Single balanced mixer composed of two mixing elements and an ideal difference block.

No RF signal, however, appears at the IF output and so the RF-IF port isolation is high in this mixer.

If the in- and out-of-phase components are the RF instead of the LO signals, then there is LO cancellation and RF feed-through.

The two unwanted feed-through components can be canceled by including in- and out-of-phase components of both LO and RF. The *double balanced mixer* uses four mixing devices and three ideal difference blocks (Figure 7.4). Using the same analysis procedure as the single balanced mixer gives an IF response current of

$$I_{IF} = \left(\sum_{n=0}^{\infty} g_n (V_{RF} + V_{LO})^n - \sum_{n=0}^{\infty} g_n (V_{RF} - V_{LO})^n\right) - \left(\sum_{n=0}^{\infty} g_n (-V_{RF} + V_{LO})^n - \sum_{n=0}^{\infty} g_n (-V_{RF} - V_{LO})^n\right) \tag{7.16}$$

Looking at only the first two components of the IF output, no unwanted feed-through of either the RF or the LO occurs but the desired quadratic mixing term ($n = 2$) is generated:

$$n = 1: \; I_{IF} = 0 \quad n = 2: \; I_{IF} = 8 g_2 V_{RF} V_{LO} \tag{7.17}$$

Twice the IF response to the RF signal is generated when compared with the single balanced mixer.

7.2 Single MOS Mixer Topologies

This section covers a few basic mixer types: the simple square-law mixer (not really practical but very useful in setting the stage for the analysis of other mixer types), the transconductance mixer, and the switching mixer. A design example is presented as well.

7.2.1 Conceptual MOSFET Mixer $(V_{GS} - V_T)^2$

At first glance, the square-law nature of the MOSFET I-V characteristic in saturation points to the ability of providing the desired signal multiplication and therefore

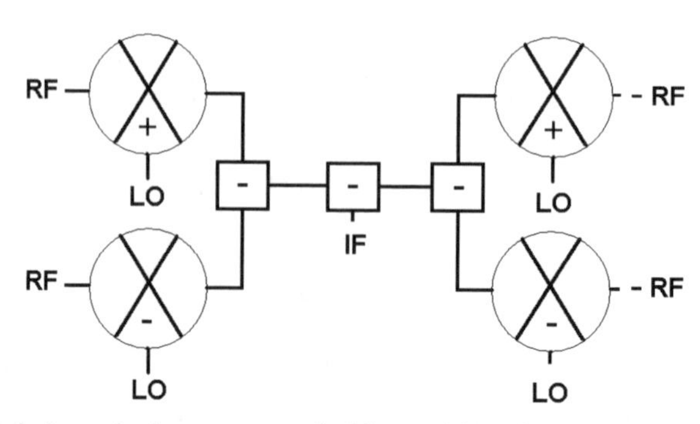

Figure 7.4 Double balanced mixer composed of four mixing elements and three ideal difference blocks.

the desired frequency conversion. Assuming that the RF and LO signals are added at the gate node along with a dc bias (Figure 7.5)

$$V_{GS} = V_{DC} + V_{RF} + V_{LO} \tag{7.18}$$

the resulting drain current can be written as

$$\begin{aligned} I_{DS} &= \frac{1}{2}\mathrm{KP}\frac{W}{L}(V_{GS} - V_T)^2 = \frac{1}{2}\mathrm{KP}\frac{W}{L}(V_{DC} - V_T + V_{RF} + V_{LO})^2 \\ &= \frac{1}{2}\mathrm{KP}\frac{W}{L}(V_{DC} - V_T)^2\left(1 + \frac{V_{RF} + V_{LO}}{V_{DC} - V_T}\right)^2 = I_{DC}\left(1 + \frac{V_{RF} + V_{LO}}{V_{DC} - V_T}\right)^2 \end{aligned} \tag{7.19}$$

By squaring the term in parentheses and combining terms, the multiplication of the RF and LO signals can be derived:

$$I_{DS} = \frac{1}{2}\mathrm{KP}\frac{W}{L}(V_{DC} - V_T)^2\left(1 + 2\frac{V_{RF} + V_{LO}}{V_{DC} - V_T} + \frac{V_{RF}^2 + V_{LO}^2 + 2V_{RF}V_{LO}}{(V_{DC} - V_T)^2}\right) \tag{7.20}$$

This small signal result assumes that neither the RF nor the LO swing are large enough to swing below V_T, driving the MOSFET out of saturation and into cutoff. Some problems with this type of mixer become apparent with a closer look at (7.20). The first problem is that both RF and LO feed-through is relatively high and therefore RF-IF and LO-IF isolation is poor. Lowering either the RF or LO signal amplitude degrades the IF output level as well. A large load resistor provides a large IF output but at the expense of a large dc voltage drop. Increasing the aspect ratio *W/L* of the MOSFET also provides an increase in the mixer response for all components, not just the IF.

A closer look at the IF voltage component shows that if the MOSFET gate is dc biased close to V_T, then the response of the IF increases (as do the feed-through

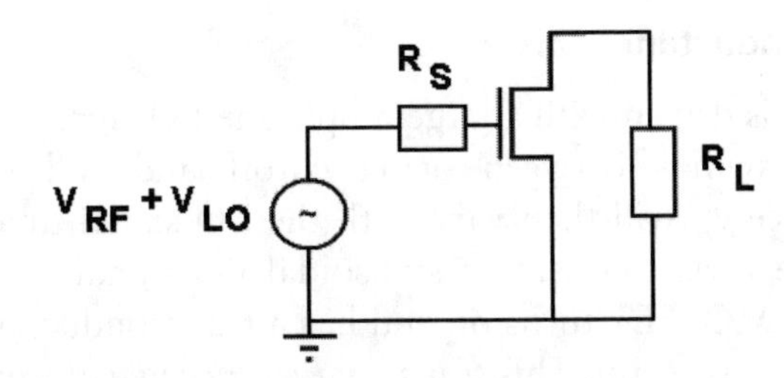

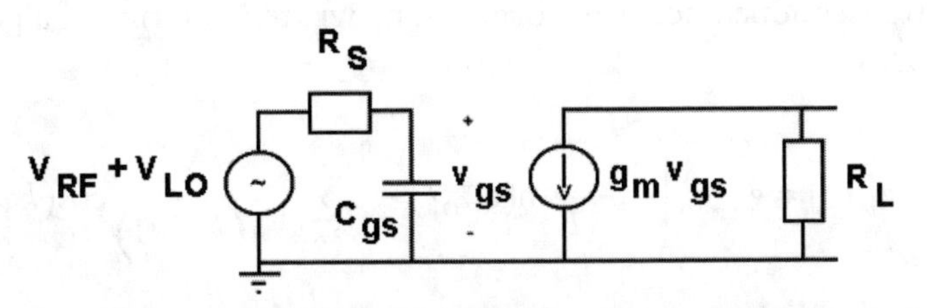

Figure 7.5 A simple mixer based on the MOSFET square-law I-V characteristic.

components but to a lesser extent). This biasing scheme for IF signal enhancement can be problematic in CMOS RFICs because V_T may vary by 50 mV or more from one location on the silicon die due to process variability. The dc bias must be set to make sure that the device will always be above threshold.

Mixer performance then can vary depending on the physical on-chip location of the mixer. In addition to the relatively poor feed-through performance, unless the MOSFET is matched to either the RF or the LO, considerable reflection of energy can occur. If the circuit is matched to the LO frequency, unless the RF and LO frequency are within the matching circuit bandwidth, the input RF is unmatched with a resulting loss in signal. Because the RF is the weaker signal, as much of this signal as possible should be available at the MOSFET gate. A mismatched LO, however, is not without its own problems. If this mixer is preceded by an LNA, any LO reflected energy can find its way to the input of the LNA if $|S_{12}|$ is of even minimal value, with the potential of the LO radiating and causing interference with nearby systems (nothing like a receiver suddenly becoming a transmitter to make good neighbors). If a choice is to be made whether to match the LO or the RF, choose the RF since this is the smaller signal; the LO amplitude can always be increased to compensate for the mismatch.

The form of (7.20) allows a "transconductance" parameter to be defined as the ratio of the IF current output to the RF amplitude [1]:

$$G = \frac{I_{DS-IF}}{V_{RF}} = \mathrm{KP}\frac{W}{L}V_{LO} \tag{7.21}$$

This transconductance term gives some guidance on the important properties of MOSFET mixers even though this circuit would be a rather poor mixer. However, this exercise is useful in that it shows some of the major characteristics of all mixers and serves as an introduction to the general concept of transconductance mixers covered in the next section.

7.2.2 Transconductance Mixer

If the MOSFET is driven with a large-amplitude LO signal on the gate node so that the MOSFET swings between being cutoff and full saturation, the device transconductance g_m will then vary with the LO signal time variation. Figure 7.6 shows this time variation using a sinusoidal LO signal. When the LO amplitude exceeds V_T, the MOSFET turns on and has a transconductance g_m that follows the shape of the LO waveform. This *transconductance waveform* has a maximum value $g_{m,\max}$. The MOSFET turns off and $g_m = 0$ when the LO amplitude falls below V_T. The transconductance waveform can be written in terms of a general Fourier series in ω_{LO} as

$$g_m(t) = g_{m-\max}\left(\frac{1}{\pi} + \frac{1}{2}\sin(\omega_{LO}t) - \sum_{k=2,\,even}^{\infty} \frac{2}{\pi\left(k^2 - 1\right)}\cos(k\omega_{LO}t)\right) \tag{7.22}$$

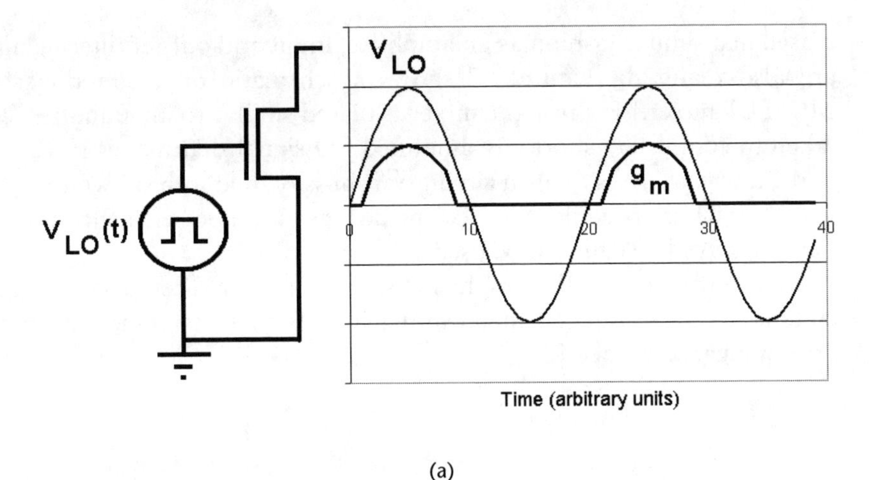

(a)

(b)

Figure 7.6 (a) A MOSFET driven by an LO signal, creating a time varying transconductance waveform, $g_m(t)$ and (b) its RF equivalent circuit.

Equation (7.22) is still an approximation for $g_m(t)$, however, since there are some crossover effects as the LO voltage approaches V_T. If this time-varying transconductance is then multiplied by the RF signal, as is typical with MOSFETs in saturation (i.e., $g_m v_{in}$),

$$V_{RF}(t) = V_{RF}\cos(\omega_{RF}t) \tag{7.23}$$

the resulting load signal contains components at the RF frequency as well as numerous frequency translations of the RF signal. Choosing the $\sin(\omega_{LO}t)$ term only gives the total load voltage as

$$V_{load}(t) = -g_{m-\max}\left[\frac{V_{RF}}{\pi}\sin\omega_{RF}t + \frac{V_{RF}}{2}\sin(\omega_{LO} \pm \omega_{RF})t\right] \tag{7.24}$$

where the second term is the desired IF up- or downconverted signal (depending on which sign is selected). This transconductance waveform also allows some of the RF signal to feed through to the load.

Real transconductance mixers exhibit nonzero values of b_k for k greater than 1, which is the origin for mixer spurs. In practice, the transconductance mixer is dc biased in a similar fashion as an amplifier. Input and output filtering and matching are also required. (Figure 7.7 shows a schematic of a common source single MOSFET mixer.) Because the mixer is biased similar to an amplifier, care must be taken to completely short any amplified LO signal that makes it way to the drain node; the quarter-wave open circuit transmission line stub shown in Figure 7.7 provides this RF short while still allowing dc bias. A lumped element equivalent for this quarter-wave line can also be used.

The power available from the LO source can be written from the equivalent circuit of the single MOSFET mixer shown in Figure 7.6 assuming the input and source are conjugate matched [2]:

$$P_{A\text{-}LO} = \frac{1}{2}(V_G - V_T)^2 \left(\omega_{LO} C_{gs}\right)^2 R_S \tag{7.25}$$

In terms of the RF signal, the load sees the product of the time-varying transconductance $g_m(t)$ and the RF input voltage, giving a drain current at the IF of

$$i_d(t) = \frac{V_{RF}\, g_{m\text{-max}}}{8 j\omega_{RF} C_{gs} R_S} \cos\left[(\omega_{RF} \pm \omega_{LO})t\right] \tag{7.26}$$

The *transducer conversion gain* G_T for the circuit is determined by dividing the IF power developed across R_L by the power available from the RF source ($V_{RF}^2/2R_S$) [2]:

$$G_T = \left(\frac{g_{m\text{-max}}}{C_{gs}}\right)^2 \left(\frac{1}{4\omega_{RF}}\right)^2 \frac{R_L}{R_S} \approx \left(\frac{f_T}{4 f_{RF}}\right)^2 \frac{R_L}{R_S} \tag{7.27}$$

The transducer conversion gain is improved by increasing the load resistance, improving MOSFET g_m by biasing or device sizing, or by lowering C_{gs} by device sizing.

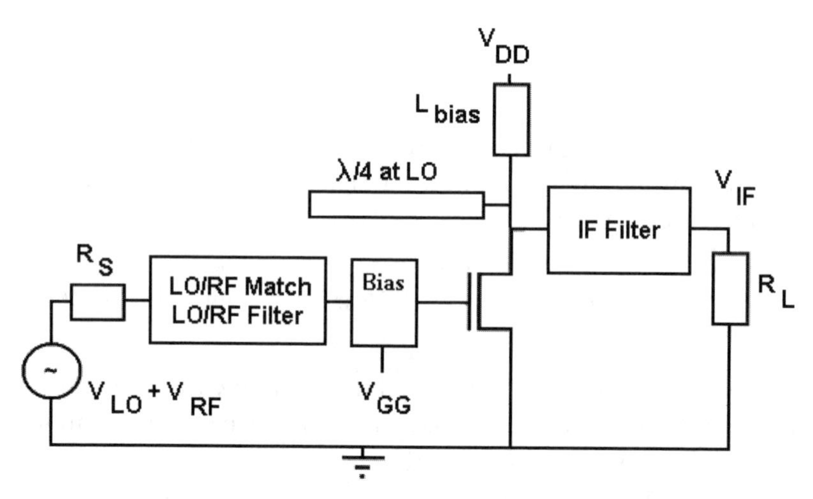

Figure 7.7 Schematic diagram of single MOSFET transconductance mixer.

The design of the single MOSFET mixer begins by making a number of approximations. Once these parameters have been estimated, the designer should go to a circuit simulator to optimize the design. Circuit parameters g_m and C_{gs} can be estimated by assuming a dc gate bias on the MOSFET in weak inversion or slightly above V_T; this operating point provides the greatest variation in transconductance with applied LO voltage as well as low power dissipation [3]. The designer should assume a gate bias voltage near V_T (within 50 and 100 mV) and a mixer bias current consistent with dc power constraints. Biasing near V_T places the MOSFET in a transition region between weak and strong inversion, an area where modeling can be problematic [4]. Because only an estimate of parameters is needed at this stage of the design, one or two iterations between the weak and strong inversion transconductance expressions (7.2a and 2b, respectively) will provide an estimate of MOSFET device size for the assumed bias current:

$$g_{m-weak} \approx \frac{I_{DS}}{kT/q}; \quad g_{m-strong} = \mathrm{KP} \cdot (W/L)(V_{GS} - V_T) = \sqrt{2 \cdot \mathrm{KP} \cdot I_{DS} \cdot (W/L)} \quad (7.28)$$

Because only an estimate of the MOSFET parameters is required at this stage of the design, C_{gs} can be estimated using its strong inversion value:

$$C_{gs} = C_{ox} WL \quad (7.29)$$

Load resistor R_L is sized by looking at both the conversion gain required and the dc voltage drop across R_L ($I_{DS}R_L$) required to keep the MOSFET in saturation during positive LO voltage swings. Active loads can also be designed at this point; these types of loads, however, can introduce the usual MOSFET $1/f$ and thermal noise, increasing the overall level of mixer noise. If the mixer is feeding a purely capacitive load, R_L is assumed to be the output impedance of the MOSFET, r_{ds}. This assumption, however, requires modifying the G_T to include capacitive loading C_L by this stage:

$$|G_T| \approx \left(\frac{f_T}{4f_{RF}}\right)^2 \frac{R_L}{R_S} \frac{1}{\sqrt{1 + (\omega_{RF} R_L C_L)^2}} \quad (7.30)$$

Input and output filtering and matching structures complete the design.

7.2.3 Resistive Mixer

A passive FET mixer structure was advanced by Maas for use with GaAs technology [5]. This concept has been extended for use with CMOS technology. Resistive mixers are noted for their conversion loss like diode mixers, but with a relatively clean IF output due to the relatively weak resistive nonlinearities. Recent work using 90-nm CMOS devices has shown useful resistive mixers at 26 GHz and above [2, 6]. The basic MOSFET resistive mixer, like its GaAs counterpart, has a dc bias on the gate, but the drain remains *unbiased*. In a common source resistive mixer, the LO signal can be applied to the gate and the RF applied to the drain with the IF taken off the drain as well [Figure 7.8(a)]. Alternatively, the LO signal can be applied to the

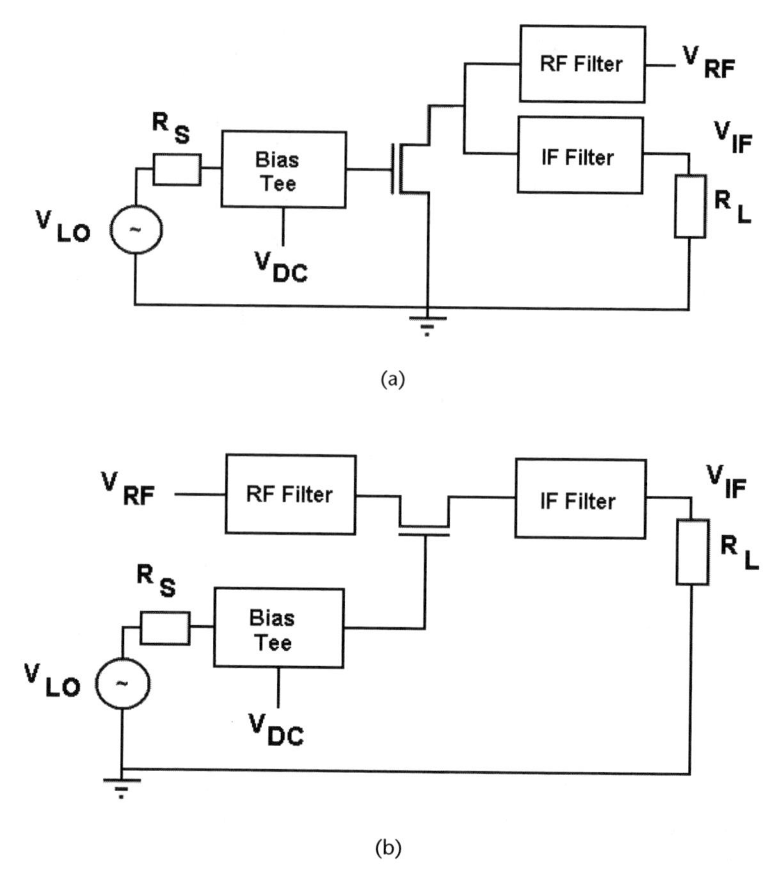

Figure 7.8 (a) Common source and (b) floating passive resistive MOSFET mixers.

gate but the RF applied to the source with the IF taken off the drain [Figure 7.8(b)]. Extensive filtering at all ports is required to make sure that the LO signal is shorted at the drain terminal to prevent the rather large LO signal from self-biasing the MOSFET.

For resistive mixers, the conversion loss is a function of the resistive nonlinearity and can be written as [7]

$$L_{conv} = 10\log\left(\frac{1+\sqrt{1-\varepsilon^2}}{1-\sqrt{1-\varepsilon^2}}\right)\text{dB} \tag{7.31}$$

where ε is the ratio of the first-order resistive nonlinearity Fourier coefficient to the fundamental value:

$$\varepsilon = \frac{R_{DS-1}}{R_{DS-0}} \tag{7.32}$$

For a very strong resistive nonlinearity achieved by strongly pumping the gate node with the LO signal, ε can be as large as 0.5, yielding a conversion loss of 11.4 dB for this type of mixer.

7.2.4 Design Example: Transconductance Mixer

For this example, a 1,000-MHz RF signal needs to be mixed down to 50 MHz with a 1,050-MHz LO signal with 3 dB of conversion gain (gain of 2). A single 0.5-μm gate length MOSFET switching mixer will be used. A 1.0-mA drain bias current with a supply voltage of 3.3V and a gate bias of 100 mV above threshold are assumed.

The MOSFET size can be approximated by using the equations for the weak and strong inversion transconductance. The weak transconductance $g_{m\text{-}weak}$ requirement is approximately 40 mS for 1.0-mA dc bias at room temperature. At 100 mV above threshold, the resulting MOSFET width W (assuming an nMOSFET) can be computed as

$$\begin{aligned} g_{m-strong} &= \mathrm{KP}\cdot(W/L)(V_{GS}-V_T) \Rightarrow \\ W &= L\frac{g_{m-strong}}{\mathrm{KP}(V_{GS}-V_t)} = 0.5\frac{0.04}{56\cdot 10^{-6}(0.1)} = 3{,}571\ \mu\mathrm{m} \end{aligned} \tag{7.33}$$

The gate capacitance C_{gs} can now be estimated using the capacitance relationship give in Chapter 2:

$$C_{gs} = C_{ox}WL = \left(2.56\cdot 10^{-3}\right)\left(3{,}571\cdot 10^{-6}\right)\left(0.5\cdot 10^{-6}\right) = 4.57\ \mathrm{pF}$$

along with the required load resistance (assume a 50Ω source impedance for the first design attempt [2]):

$$R_L = R_S G_T\left(\frac{4\omega_{RF}C_{gs}}{g_{m-\max}}\right)^2 = 50(2)\left(\frac{4\cdot 2\pi\cdot 10^9\cdot 4.57\cdot 10^{-12}}{0.04}\right)^2 = 824\Omega$$

The first simulation was performed used SPICE with BSIM parameters. The resulting mixer output across the 824Ω load resistance is shown in Figure 7.9. The simulation results show a conversion gain of approximately 1.4 dB and a suppressed 1,050-MHz LO signal. However, the LO second-harmonic energy is not shorted since the open-circuited quarter-wave transmission line is a half wavelength at the second harmonic, and so an output IF lowpass filter would be needed to further reduce this unwanted signal energy.

Harmonic balance simulators are also used for mixer designs and this same circuit was entered into one such CAD package to verify the SPICE simulation and to also investigate the conversion gain and linearity of the mixer as a function of LO power to the mixer. Figure 7.10 shows the results of the harmonic balance simulation and shows a maximum in the conversion gain of approximately 9 dB at an LO power 15 dBm. However, this high level of LO power coincides with large mixer nonlinearity as evidenced by the large second-order harmonic generated under these conditions [Figure 7.9(b)]. By backing off the LO power by a few decibels (to 11 or 12 dBm), the linearity of the mixer improves dramatically with only a few-decibel reduction in conversion gain.

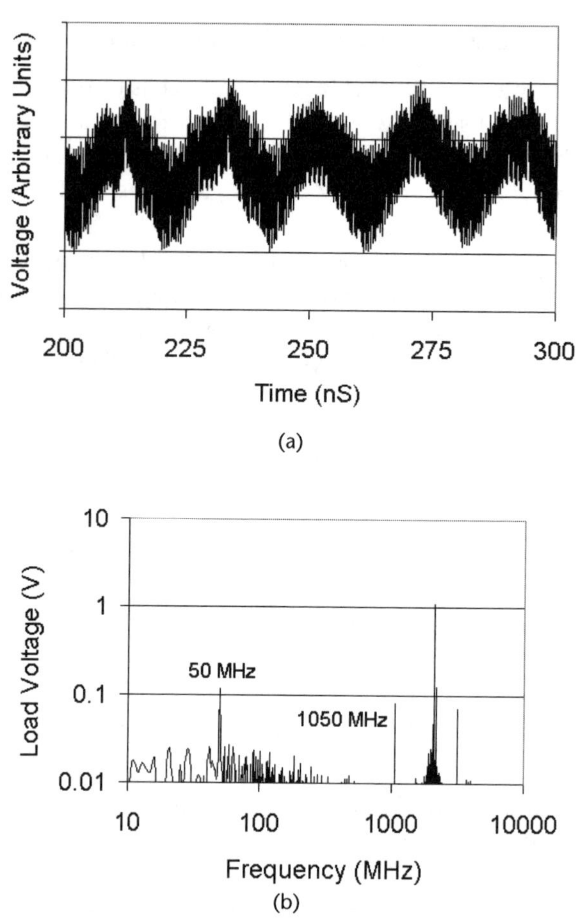

Figure 7.9 Simulated (a) time and (b) frequency spectrum using SPICE showing the 50-MHz IF signal with low conversion gain (+1.4 dB) and a suppressed (but not eliminated) LO feed-through signal (ch7-1.txt).

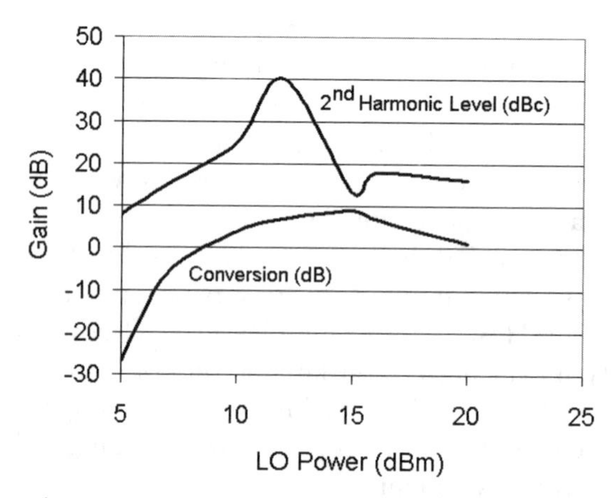

Figure 7.10 A slight reduction in LO power and the resulting small reduction in conversion gain can yield dramatic improvements in mixer linearity.

7.3 Balanced MOSFET Mixers

The switching mixer circuits shown in their conceptual form in Figure 7.2 are the basis for the most widely used mixer topology in CMOS RFIC design. In these switching mixer circuits, the "ideal" switches are replaced with MOSFETs (Figure 7.11) that are switched by the LO signal with each MOSFET configured to have a low on state resistance.

A simple switching mixer based on this concept is shown in Figure 7.12. For the best switching characteristics, the LO switching MOSFETs are dc biased slightly above V_T so that rapid switching occurs when the LO signal exceeds this threshold. For ideal switching, the drain-to-source voltage should be small, which implies a relatively large aspect ratio, *W*/*L*:

$$V_{DS-sat} = \sqrt{\frac{2I_{DS}}{KP(W/L)}} \tag{7.34}$$

Assuming the MOSFET M1 switching is ideal, the RF signal applied to MOSFET M2 is then connected to the load resistance, R_L, with the IF signal taken at the MOSFET drain node. A resistive load R_L has been assumed although active loads are also used. The previous mixer circuits required that the RF and LO signals had to be combined somewhere prior to the mixer. The switching mixer has these as signals on separate ports, improving the RF-LO isolation. The performance of this mixer can be seen by studying the current flow in the circuit (Figure 7.12). M2 current only flows during the on state of the LO switched MOSFET M1. The current

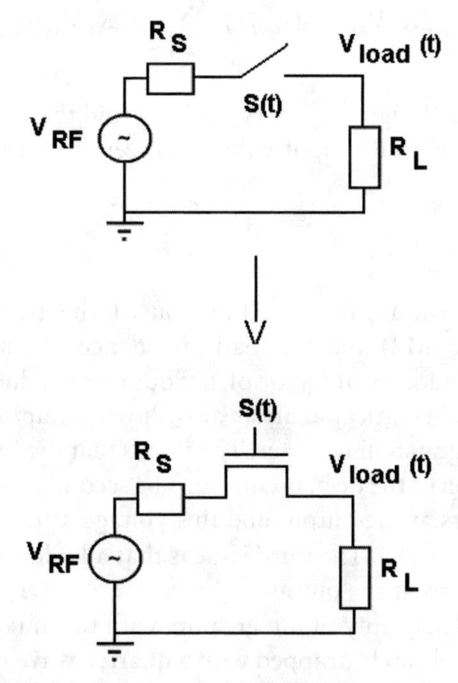

Figure 7.11 The conceptual switching mixer can be created by replacing the ideal switch with a MOSFET switch.

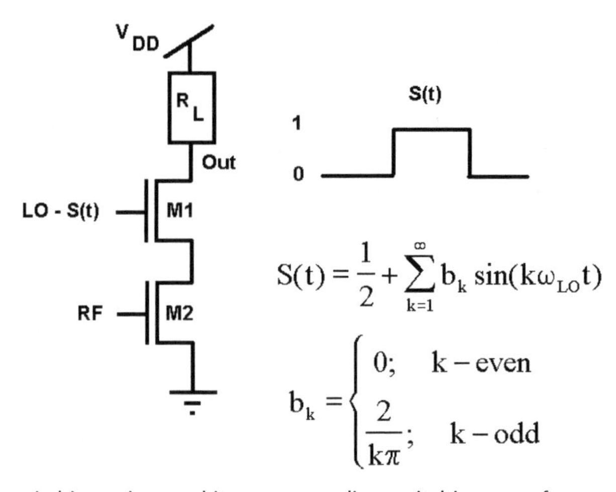

Figure 7.12 The switching mixer and its corresponding switching waveform.

flowing through M2 during the stage can be written as follows using an ideal switching waveform:

$$I_{M1}(t) = \left[I_{DC} - g_m v_{RF}(t)\right]S(t) \tag{7.35}$$

where I_{DC} is the dc bias current and $S(t)$ the LO switching signal. For $S(t)$ shown in (7.10), the resulting voltage at the drain of M1 can be written as

$$\begin{aligned} V_{M1}(1) = V_{DD} - \left(\frac{1}{2}R_L I_{DC}\right) - \sum_{k=1}^{\infty} R_L I_{DC} b_k \sin(k\omega_{LO}t) \\ + \frac{1}{2} g_m R_L V_{RF} \cos(\omega_{RF}t) + \sum_{k=1}^{\infty} \frac{1}{2} g_m R_L V_{RF} b_k \sin\left[(k\omega_{LO} \pm \omega_{RF})t\right] \end{aligned} \tag{7.36}$$

The desired IF signal appears for $k = 1$ and the conversion gain, the ratio of the IF output voltage to RF input voltage, is seen from the appropriate term in (7.36):

$$CG = \frac{1}{\pi} g_m R_L \tag{7.37}$$

This conversion gain is a function of the transconductance g_m of the RF transconductor (M2) and the load impedance, the same as the simple MOSFET amplifier. The additional factor of $1/\pi$ out front is due to the form of the switching signal. The mathematical analysis also shows a number of nonidealities associated with this simple switching mixer. The first is that the LO signal appears at the output and is a function of the dc bias current. The second is that a component of the RF signal also appears at the output and this voltage at the RF frequency is amplified by the gain of M1 ($g_m R_L$). The third issue is that all odd-order harmonics of the LO ($k >$ 1, k-odd) are also at the output.

There will be plenty of mixer spurs with this mixer topology. The large fundamental LO signal can be trapped with a quarter-wave transformer (or lumped equivalent) at the IF output, and the LO harmonics and RF signal can be reduced with an IF filter at the same point. There are other design parameters to consider as well. For example, the previous discussion indicated that M2 could approach a nearly ideal switch for large *W*/*L*. However, large-*W* MOSFETs also exhibit increased capacitance on both the drain and source node($C_{M2\text{-}D}$ and $C_{M2\text{-}S}$), which are relatively high since both area and sidewall capacitances occur at this location. These two capacitances, in addition to the drain capacitance of M1 ($C_{M1\text{-}D}$), capacitively load the output node and can degrade the conversion gain:

$$CG = \frac{1}{\pi} g_m \frac{R_L}{\sqrt{1 + \left[\omega_{RF}\left(C_{M2-D} + C_{M2-S} + C_{M1-D}\right)\right]^2}} \tag{7.38}$$

The design of MOSFET M1 is optimized by choosing a width that is consistent with good switching and good frequency response in the mixer. The width of MOSFET M2 is chosen to provide the transconductance needed for the desired conversion gain.

The output transconductance of M2, g_m, has been considered a constant for this discussion. Any nonlinearity in g_m, however, will generate further mixer spurs. With the strong LO signal pumping the circuit, the drain voltage of M2 can vary and cause nonlinearities in g_m. Several methods have been used to improve the overall transconductance linearity and thereby improve mixer performance. One such technique uses source degeneration on M2, as shown in Figure 7.13. If a general impedance Z_S is used to denote this source term, then an *effective mixer transconductance* expression can be derived:

$$G = \frac{i_d}{V_{RF}} = \frac{g_m}{1 + Z_S\left(g_m + j\omega C_{gs}\right)} \tag{7.39}$$

If inductive degeneration is used ($Z_S = j\omega L_S$), the inductor can be used to resonate C_{gs}, and the effective transconductance *G* becomes

$$G = \frac{i_d}{V_{RF}} = \frac{1}{j\omega L_S} \tag{7.40}$$

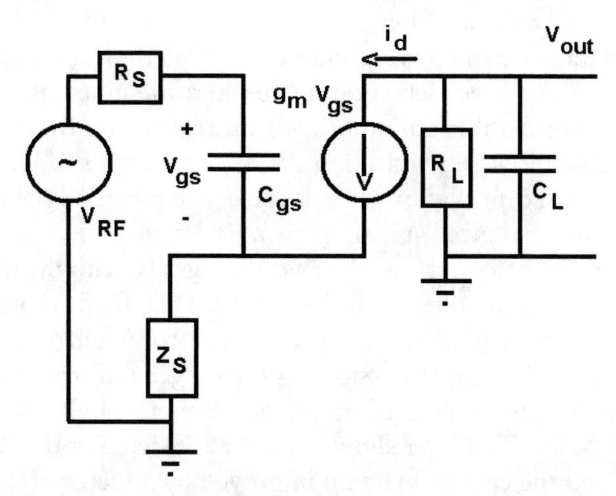

Figure 7.13 Source degeneration (Z_s) is used to improve transconductance linearity.

Inductive degeneration has a number of positive features for the mixer. The effective transconductance, and thereby the conversion gain, are dependent on features external to M2 (R_L, L_S). The conversion gain also rolls off with frequency, improving the harmonic rejection of the mixer. The ideal inductor contains no resistive elements, and so a potential source of noise in the circuit is eliminated. Inductive source degeneration provides an input impedance that is purely resistive with the corresponding improvement in matching [8]:

$$Z_{\text{in}} = \frac{g_m L_S}{C_{gs}} \tag{7.41}$$

Resistive source degeneration can also be used in improving transconductance linearity. Passive polysilicon resistors R_S are often used; however, these resistors introduce resistive thermal noise to the system [9]. For resistive degeneration, the effective transconductance becomes

$$G = \frac{i_d}{V_{RF}} = \frac{g_m}{1 + R_S\left(g_m + j\omega C_{gs}\right)} = \frac{g_m}{1 + g_m R_S} \frac{1}{1 + j\omega \dfrac{R_S}{1 + g_m R_S} C_{gs}} \tag{7.42}$$

For small C_{gs} and large MOSFET transconductance g_m, the effective transconductance G is simply $1/R_S$, with the conversion gain becoming a ratio of two resistances:

$$CG = \frac{1}{\pi} G \cdot R_L \cong \frac{1}{\pi} \frac{R_L}{R_S} \tag{7.43}$$

Resistive degeneration changes the dc bias conditions of the circuit due to the IR drop and reduces voltage headroom. Inductive degeneration may be the better choice in these low-voltage applications.

7.3.1 Single Balanced Mixer

Intermediate frequency filtering is used to reduce LO and RF feed-through components and improve the corresponding port isolation. As discussed in previous chapters, however, lumped element on-chip filters are inherently lossy and the inductors take up a considerable amount of silicon real estate. By using a number of mixers such as those discussed in Section 7.1.4, components of the IF signal can be eliminated by signal cancellation. This is the concept behind the use of *balanced mixers*. The first such balanced mixer, the *single balanced mixer*, contains two identical mixers with both in- and out-of-phase LO signals, with the IF taken as the difference of the drain output voltages. In Figure 7.14, the loads represented by R_L are often some type of active load such as an active resistor or current source. For simplicity of analysis, the following discussion assumes a general resistance value.

An analysis of the currents in the circuit that is similar to that done for the single switching MOSFET mixer shows that the RF signal at the IF output node has been canceled and the conversion gain improved by a factor of 2:

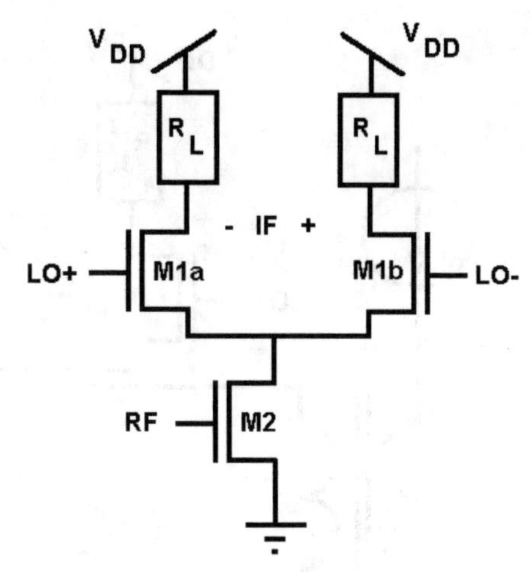

Figure 7.14 The simple single balanced MOSFET mixer.

$$CG = \frac{2}{\pi} g_m R_L \tag{7.44}$$

As with all differential circuits such as the balanced mixer, there is improvement in common mode performance as well as a reduction in common mode noise such as substrate noise. Source degeneration at M2 may help in improving transconductance linearity and hence mixer performance by reducing another contributor to mixer spurs.

The stacking of transistors and mixer loads (passive or active) can cause voltage headroom issues as power supply voltages decrease. The RF stage requires that the V_{DS} be high enough to keep the RF transconductor in saturation for best mixer linearity. One way to improve this headroom is to use passive on-chip transformers to couple the RF signal to the LO switching stage as shown in Figure 7.15 [10]. Here, the IF voltage can be closer to V_{DD} because the LO switches only need a small V_{DS} when in the on state. The RF transconductor then has the full range of V_{DD} to swing.

Some drawbacks to this circuit are lower conversion gain due to transformer losses and relatively low (k = 0.7 or less) transformer coupling. The RF transconductor should be designed with a higher g_m to compensate for these losses. Another technique uses a tuned load at the transconductor with the output feeding the LO stage as shown in Figure 7.15 [11]. The inductor puts the positive rail at the RF transconductor drain but allows the RF within the bandwidth of the LC resonator to transfer into the LO. Here *p*MOSFETs are used as switches with the IF taken across the indicated load resistors. The tuned circuit only peaks at the RF frequency and so provides a degree of g_m linearity to the mixer by filtering out higher harmonics of the transconductance waveform.

7.3.2 Double Balanced Mixer

In the previously discussed single balanced mixer, only a single phase of the RF was used, providing a means for cancellation of the RF signal at the IF port and improving the RF-IF port isolation. This signal cancellation concept can be further

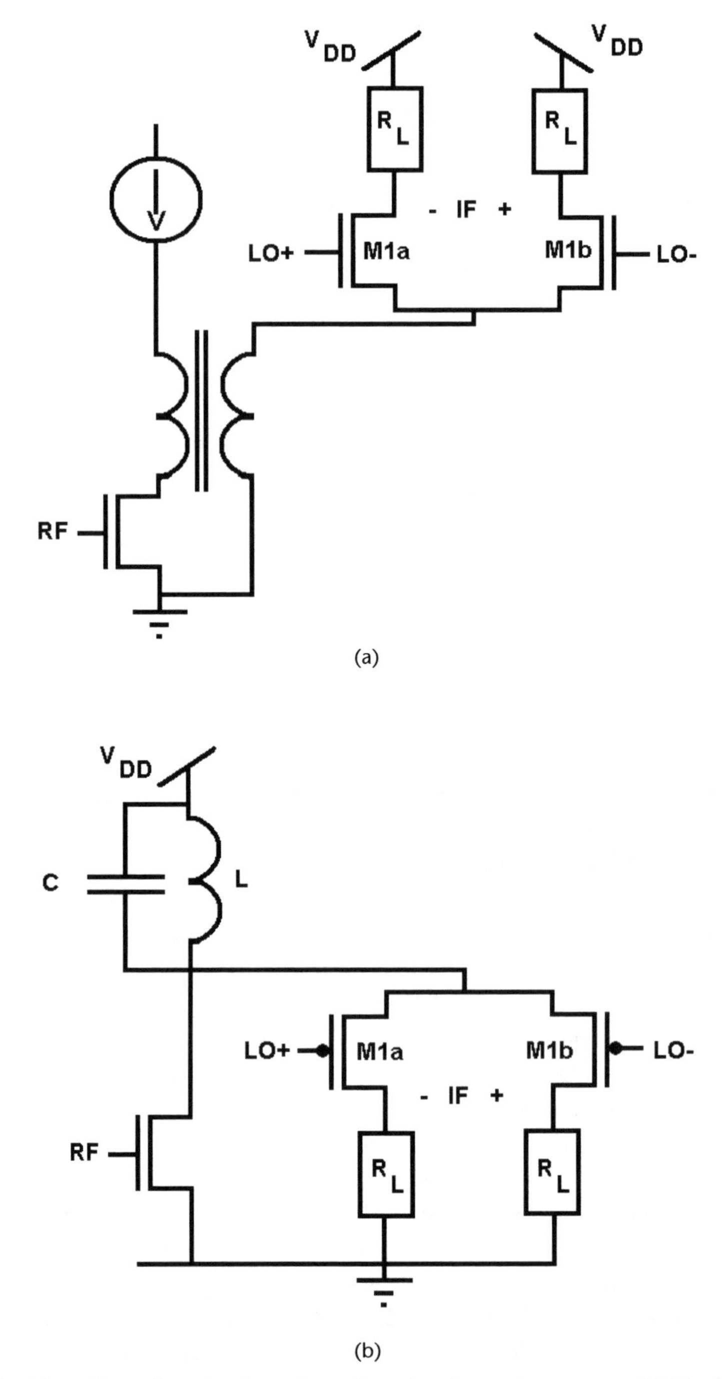

Figure 7.15 Alternative mixer structures for voltage headroom improvement (*After:* [10, 11]): (a) transformer coupling and (b) tuned RF.

extended if both the in-phase and out-of-phase components of the RF signal are available. The conceptual circuit shown in Figure 7.4 provides both RF and LO signal cancellation, implementing what is known as the *double balanced mixer*. The basic CMOS double balanced mixer uses the circuit topology of two single balanced mixers with their IF outputs connected to achieve the proper signal phasing and resulting signal cancellation at the IF ports; this circuit is shown in Figure 7.16. Using a similar current analysis as with the two previous mixer types provides an ideal conversion gain of

$$CG = \frac{4}{\pi} g_m R_L \tag{7.45}$$

This mixer is the MOSFET equivalent to the classic bipolar *Gilbert cell* mixer [12]. This MOSFET double balanced mixer ideally shows a 6-dB increase in conversion voltage gain over the single balanced mixer but at the expense of twice the current draw and hence twice the power consumption. Previously discussed transconductance linearity and matching schemes are often employed with the double balanced mixer.

An alternative transconductance linearization scheme uses a modification of a technique first advanced for bipolar circuits called *multi-tanh linearization* [13]. The *multi-tanh* name comes from the original linearization derivation using the hyperbolic tanh form for the bipolar transistor's I-V relationship. For MOSFET mixers, constant transconductance g_m was defined in a small signal sense as dI_{DS}/dV_{DS}. With large dynamic range mixers, the input RF signal may be anything but "small signal," so g_m can vary significantly over a variety of input signal ranges, and hence the conversion gain will vary as well. The multi-tanh technique takes advantage of the fact that, although g_m is constant over a rather limited range of input voltages, a number of RF transconductors can be paralleled to provide a wider range of constant g_m. Figure 7.17 shows one such multi-tanh circuit with two sections. Each section is dc biased differently (V_i) but the RF is applied to each MOSFET simultaneously. The transconductance g_m will peak around each V_i, and if the V_i are staggered in value down the array, g_m will be more constant for larger input RF voltage swings. For example, a 1.9% variation in g_m over a 1.0V peak-to-peak signal has been achieved using this technique in 0.35-μm CMOSs at

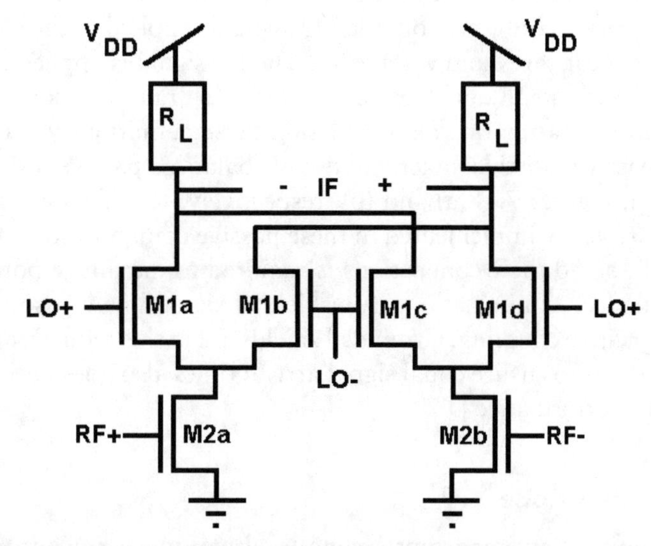

Figure 7.16 Standard double balanced MOSFET mixer.

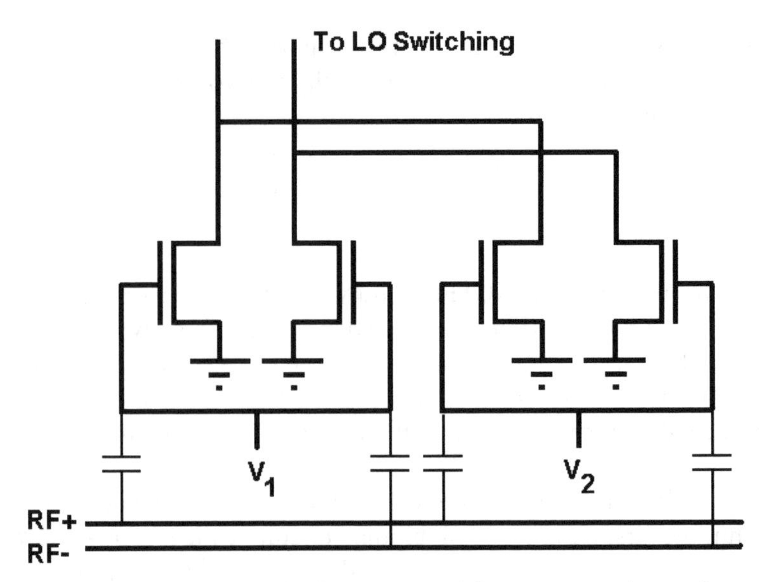

Figure 7.17 Multi-tanh concept to aid in linearization of the RF transconductors for use over a wide input voltage range.

1.9 GHz [14]. Extensive simulations varying bias levels and device widths are employed to optimize the design.

This technique has some drawbacks. Additional transistors require higher bias current draw from the supply, increasing power dissipation. All the drain nodes of the RF transconductors appear as a load capacitance across the mixer load R_L (or its active device equivalent), limiting the overall frequency response of the mixer. Simulations will show the trade-off between the number of stages needed for a given input RF swing and the resulting frequency response. A good first-pass design compromise is to use three multi-tanh blocks and proceed in the design from that point.

The balanced mixer is a general circuit topology not limited to just active mixers; the concept can be extended to use passive resistive mixers as well. In the usual resistive mixer configuration, the LO signal is applied to the gate terminal, which is dc biased near threshold voltage V_T. The RF signal is applied to the unbiased drain terminal with the IF signal taken off either the drain (in a common source circuit) or the source (in a floating circuit). Using these definitions yields possible single balanced passive resistive mixer and double balanced passive resistive mixer circuits as shown in Figures 7.18(a) and (b), respectively. RF matching and LO shorts at the drains are easily implemented in these passive configurations. Careful layout of the MOSFETs and their connections is required to maximize port isolation and noise performance. Minimization of unshielded signal line crossovers reduces crosstalk and capacitive coupling. Grounded shields aid in this reduction. Careful attention to trace lengths to ensure equal signal arrival times also goes far in improving this type of mixer performance [15].

7.3.3 Mixer Noise

Mixer noise is an important system-level parameter since it influences the overall noise figure of the system. Some simple receiver systems such as those utilizing direct

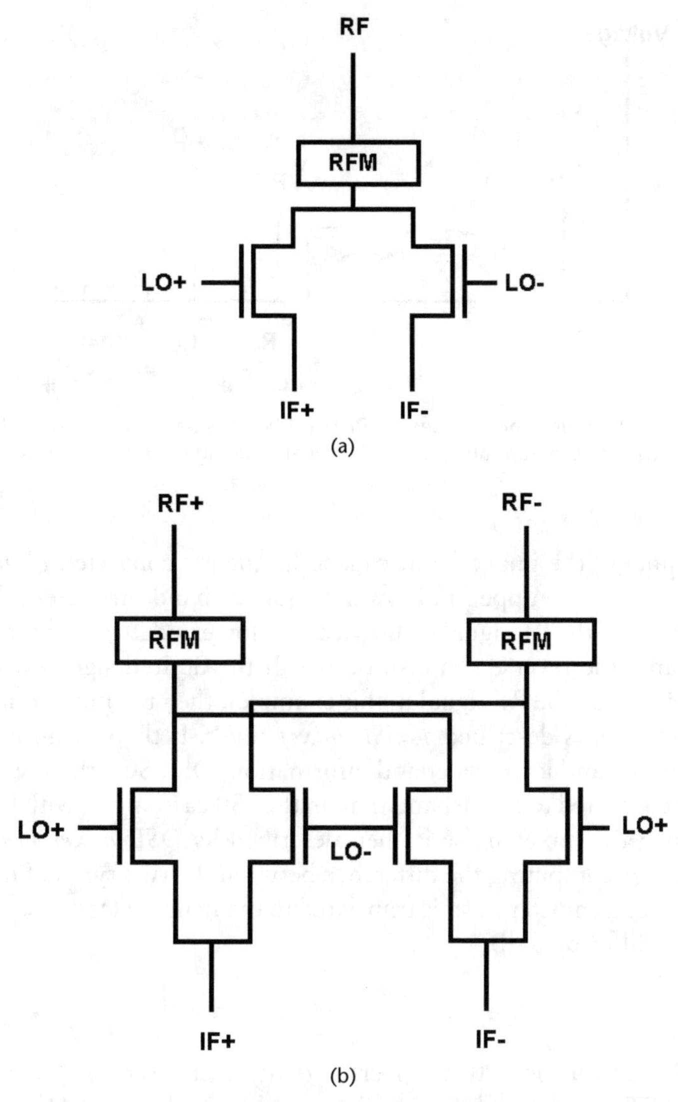

Figure 7.18 (a) Single balanced passive resistive mixer and (b) double balanced passive resistive mixer. RF matching networks (RFMs) are also shown.

conversion have modest RF gain in the LNA stage, and the mixer noise becomes a more substantial portion of the overall system noise figure. Mixers are generally considered "noisy" circuits because both thermal and $1/f$ noise from a number of frequency bands and sources (passive components as well as active ones) are frequency translated to the IF band. Two measures of mixer noise are often used, single-sideband (SSB) and double-sideband (DSB) noise. The difference between the two can be understood as follows. To generate an IF signal, the previous discussions have shown that two signals are required; the RF and the LO. Noise surrounding the RF signal is frequency shifted (in Figure 7.19, downconversion is shown; upconversion has similar characteristics) and the noise in the frequency band of interest comes along with the frequency conversion. However, there is also the image signal lurking in the frequency band opposite that of the LO. The noise

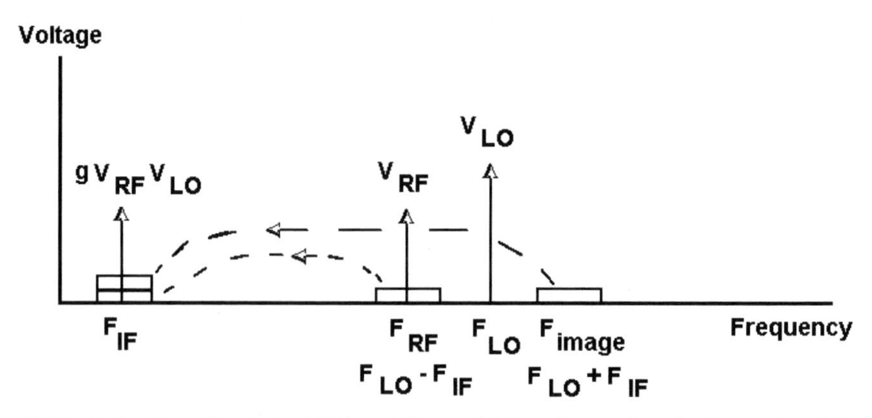

Figure 7.19 Noise from the desired RF band (lower sideband) as well as the image band (upper sideband) are both translated to the IF frequency along with the sideband signal power(s).

surrounding the image band can be frequency converted *even if no actual signal exists there*. The upper and lower frequency bands are referred to as sidebands (in Figure 7.19, the RF signal is shown as the lower sideband and the image as the upper sideband; the reverse can also occur). If the desired signal has power in only one sideband (i.e., the RF signal in this example), then the mixer noise (and the resulting noise figure) is described as *SSB mixer noise*. If the transmission scheme requires both upper and lower sideband information (DSB-SC), then twice as much power is downconverted to the IF band than in the SSB case, along with both noise band components; the mixer noise is then described as *DSB mixer noise*. A rough rule of thumb for computing the difference between the two types of mixer noise is to note that twice as much power is translated to the IF in the DSB case, and so the two noise figures differ by 3 dB:

$$F_{DSB-noise} = F_{SSB-noise} - 3\text{ dB} \tag{7.46}$$

For dc and low IF receivers, $1/f$ noise can seriously degrade the SNR, and MOSFET devices are known to have significant levels of $1/f$ noise. A first approach (and a sensible one too) in minimizing $1/f$ noise at the mixer output is to remove as many sources of $1/f$ noise as possible by, for example, using resistive loading (polysilicon resistors) instead of active resistor or current source loading [15]. If active loading is necessary, then *p*MOSFETs should be used since they have slightly less $1/f$ noise than their *n*MOSFET counterparts [16]. The $1/f$ or flicker noise current for a balanced MOSFET mixer has been shown to be dependent on the tail current capacitance C_{tail}:

$$i_{N-1/f} \cong \frac{2C_{tail}^3}{T}\left(\frac{\omega_{LO}}{g_m}\right)^2 V_n \tag{7.47}$$

Equation (7.47) indicates that large C_{tail} increases the $1/f$ noise and also gives rise to slower LO switching transitions (the term T in (7.47)). To minimize this noise source, a low LO amplitude with as sharp as possible switching transitions is desired [16]. CMOS thermal noise brought down to the IF comes from three major sources:

thermal noise of the load resistances, thermal noise of the LO-switched MOSFETs, and thermal noise of the RF transconductance stage. For balanced MOSFET mixer circuits, the output thermal noise voltage takes the following form [17]:

$$V_{N-IF}^2 = 8kTR_L\left[1 + \frac{\gamma}{\pi}\frac{R_L I_{bias}}{V_{LO-peak}}N + \frac{\gamma}{2}\frac{R_L I_{bias}}{\left(V_{GS-DC} - V_T\right)_{RF}}\right] \tag{7.48}$$

where $N = 1$ for single balanced mixers and $N = 2$ for double balanced mixers (twice as many MOSFETs are contributing to thermal noise in double balanced mixers). Increasing R_L to achieve a high conversion gain or output voltage swing generates increased thermal noise at the IF, so the designer should only attempt modest mixer gains by load adjustment. A large LO amplitude helps in reducing IF thermal noise as does biasing the RF stage well beyond threshold.

7.3.4 Design Example: Single Balanced Mixer

The design flow of a balanced mixer requires that both RF and dc operating points be considered (ch7-2.txt). In this example, a single balanced mixer will be designed subject to the specifications listed in Table 7.2

Because the total current draw is 0.2 mA, the balanced mixer will have 0.1 mA of dc flowing through each branch. With the specified load resistance, the voltage at the drain of the LO switching MOSFETs will be 2.3V. A good LO switch should exhibit as low a voltage drop as possible when switched on; however, the voltage drop is inversely proportional to the device aspect ratio *W*/*L*, so an extremely low voltage drop will require a large area device with the corresponding increase in device capacitance and poorer frequency and switching response. A reasonable design starting point is to assume an LO switch voltage drop of 10% of the drop across the load; in this case, assume a 0.1V drop across the LO switching MOSFETs when they are fully turned on by the LO signal at 1.0V above threshold (1.0V peak LO voltage). This assumption leaves a 2.2V drain-to-source voltage drop across the RF transconductor, which should be operating in saturation. The LO switching MOSFETs are dc biased very close to V_T for good switching characteristics, and since the source of the LO MOSFETs is at 2.2V, the LO dc bias voltage should be $2.2 + V_T$ or 2.86V. The LO MOSFETs will be operating in their linear region, so the linear $I_{DS} - V_{DS}$ expression can be used to determine the aspect ratio of the LO

Table 7.2 Single Balanced Mixer Design Specifications

Parameter	*Specification*
dc power supply/dc current draw	3.3V/0.2 mA
Load resistance	10 kΩ
Intermediate frequency	100 MHz
RF input	1000 MHz
Conversion gain	2
Local oscillator drive	1.0 V_p
Technology	0.5-μm CMOS

switching MOSFETs. With $V_{DS} = 0.1\text{V}$, $V_{GS} - V_T = 1.0\text{V}$ (LO swing above threshold), and $I_{DS} = 0.1$ mA, the gate width W can be computed:

$$I_{DS} = \text{KP}\frac{W}{L}V_{DS}\left(V_{GS} - V_T - 0.5V_{DS}\right) \Rightarrow$$

$$W = L\frac{I_{DS}}{\text{KP}V_{DS}\left(V_{GS} - V_T - 0.5V_{DS}\right)} = 9.4\ \mu\text{m} \Rightarrow 9.5\ \mu\text{m}$$

With this dc bias information, the conversion gain expression can now be employed to provide an estimate of the gate width W for the RF transconductor:

$$CG = \frac{2}{\pi}g_m R_L = \frac{2}{\pi}\left[\sqrt{2I_{DS}\text{KP}\frac{W}{L}}\right]R_L = 2 \Rightarrow W = L\frac{1}{2I_{DS}\text{KP}}\frac{\pi^2}{R_L^2} = 2.2\ \mu\text{m} \Rightarrow 3\ \mu\text{m}$$

The final calculation is to determine the dc bias voltage to apply to the gate of the RF transconductor. Assuming that this device is in saturation, the dc bias voltage can be computed as

$$V_{GS} = V_T + \sqrt{\frac{2I_{DS}}{\text{KP}\frac{W}{L}}} = 0.66 + \sqrt{\frac{2(0.2\text{ mA})}{\text{KP}(6)}} = 1.75\text{V}$$

These results were used as input to SPICE. The first simulation exhibited a total current draw of 0.23 mA, a conversion gain of 1 (0 dB), and dc drain-to-source voltage of the RF transconductor of 1.54, in reasonable agreement with the design process followed above. A few minutes spent with the simulator optimized the circuit performance by changing the RF transconductor width to 5 μm, providing 5 dB of conversion gain. The resulting IF output spectrum is shown in Figure 7.20(a). In addition to the IF and LO signal and its harmonics, no 1000-MHz RF signal is observed, but a number of "mixer spurs" are seen in the 5.0-GHz output of the output spectrum. Many of these are artifacts of the FFT used in obtaining the spectrum. Figure 7.20(b) shows a plot of conversion gain as a function of LO drive voltage. The saturation point at an LO drive of approximately 0.7V is clearly seen in this figure, indicating that LO drive voltages above this level are not required and will actually increase the number of mixer spurs generated; this is borne out by the number of mixer spurs indicated in the output spectrum shown in Figure 7.20(a).

This single balanced mixer design is subject to the nonconstant transconductance in the RF transconductor. If resistive source degeneration is used, the original design must be modified to account for the voltage drop across the source degeneration resistor. The effective transconductance expression for resistive source degeneration gives the required source resistance, in this case assuming a conversion gain of 2:

$$G = \frac{2}{\pi}\frac{R_L}{R_S} = 2 \Rightarrow R_S = 3.15\text{k}\Omega$$

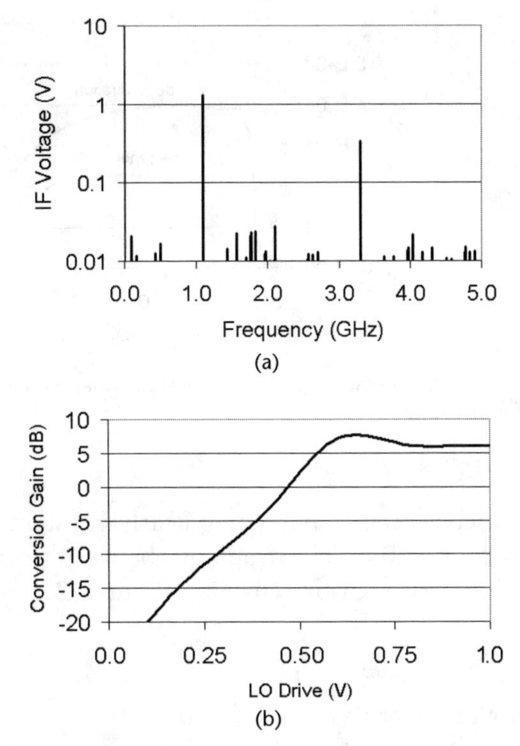

Figure 7.20 (a) Spectrum of the unfiltered IF output showing the desired 100-MHz IF signal and large odd-order harmonics of the LO signal. (b) Conversion gain as a function of LO drive voltage over the range 0.1 to 1.0V.

with an accompanying voltage drop of 0.63V. To maintain the same V_{GS}, the dc voltage on the gate of the RF transconductor needs to be increased by this amount (to 2.38V). The effective transconductance in the preceding expression assumes a large MOSFET transconductance, g_m, as well. Assuming a $g_m R_L$ product of 10 as an initial design parameter and including the increased dc gate bias voltage, we obtain a modified RF transconductor gate width of

$$g_m R_L = \left[\text{KP} \frac{W}{L} (V_{GS} - V_T) \right] R_L = 10 \Rightarrow W = 10L \frac{1}{\text{KP}(V_{GS} - V_T) R_L} = 25 \ \mu\text{m}$$

The bias conditions on the LO switching MOSFETs have not changed, and so the previously derived values can remain the same. Again, these results were used as input to SPICE with a conversion gain of 2 (6 dB) exhibited on the first-pass design (ch7-3.txt).

The main reason to add source degeneration in a mixer is to provide a transconductance that is constant over a wider range of input signals ("linearizing the mixer"). To verify this linearization technique, just the RF transconductor was simulated with and without the source degeneration resistor; no other changes were made to the circuit. Figure 7.21 shows the transconductance versus input voltage both with and without the source degeneration resistor. The top curve (without source degeneration) shows that constant g_m only occurs over a 0.5V voltage range centered on a 1.75V input voltage. The bottom curve (with source degeneration)

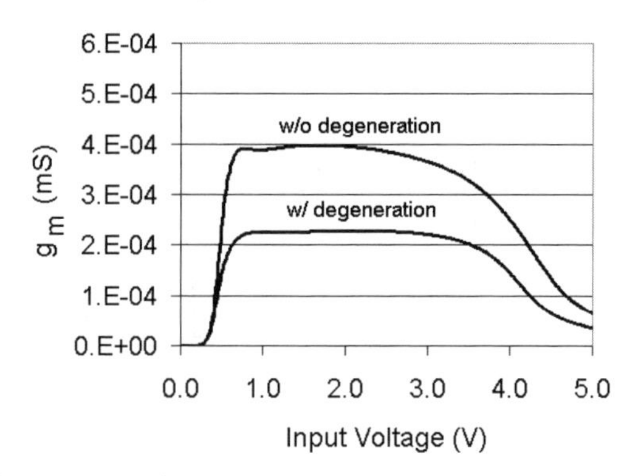

Figure 7.21 Source degeneration (bottom curve) provides a constant transconductance over a wider range of input voltages.

shows the transconductance remaining nearly constant over a 2.0V input voltage range centered on 2.0V. The trade-off between transconductance and transconductance linearity is also seen with about a 2:1 reduction in overall circuit transconductance.

7.3.5 Mixer Nonlinearities

In the preceding discussion, mixer operation has been explained based on the second-order nonlinear response of the mixer device as it operates on a single RF and LO signal to produce the IF. This explanation is sufficient for explaining the basics of small signal mixer operation. As shown in (7.7), the IF output increases with increasing LO signal drive. Increasing the LO signal, however, begins to put the mixer into an operating regime where the third-order (and higher) nonlinearities begin to impact the performance of the mixer. Two such impacts that will be discussed are conversion gain (or loss) compression and spurious signal generation in large LO drive mixers (briefly described earlier in (7.8) and (7.9)).

Equation (1.43) indicates that the odd-order nonlinear components contain a fundamental component that causes an overall reduction in this frequency component at high signal levels; the same is true with the LO input to the mixer. This reduction of effective LO signal level will cause compression in the IF output signal so that any increase in RF power will not create a corresponding change in IF output level. This compression phenomenon is often described in terms of the *mixer 1.0-dB compression point*, measured at a specific LO signal level and by observing the RF input signal level where the linear IF output signal level deviates from its ideal by 1.0 dB.

Another phenomenon that arises from the third- (and higher) order component is the spurious signal components that arise in the IF band when two or more RF signals are incident on the mixer. For "linear" mixer operation (without considering third-order components) and two input signals f_1 and f_2, the resulting IF frequency components may be defined as (assuming downconversion):

$$f_{IF1} = f_{LO} - f_1 \quad f_{IF2} = f_{LO} - f_2 \tag{7.49}$$

A detailed mathematical analysis of the first- through third-order nonlinear responses in a mixer circuit [18] shows that the impact of the third-order component

is to generate a number of mixer products with the most pernicious being the mixer intermodulation products with the IF output band given by

$$\begin{aligned} f_{IM1} &= 2f_{IF1} \pm f_{IF2} = 2(f_{LO} - f_1) \pm (f_{LO} - f_2) = f_{LO}(2 \pm 1) + 2f_1 \pm f_2 \\ f_{IM2} &= 2f_{IF2} \pm f_{IF1} = 2(f_{LO} - f_2) \pm (f_{LO} - f_1) = f_{LO}(2 \pm 1) + 2f_2 \pm f_1 \end{aligned} \tag{7.50}$$

This intermodulation phenomenon can be described in terms of the mixer third-order intercept point MIP3 and the RF input and intermodulation products measured at the IF output:

$$\text{MIP3} = \frac{1}{2}(3P_{RF\text{-in}} - P_{IF\text{-out}}) \tag{7.51}$$

where $P_{RF\text{-in}}$ is the input RF power that gives rise to the specific intermodulation power component at the IF output, $P_{IF\text{-out}}$. MIP3 can be determined from mixer measurements using a technique similar to measurements of amplifier IP3 as described in Chapter 1.

7.3.6 Mixer Summary

The different CMOS RFIC mixer topologies described in this chapter exhibit varying degrees of performance depending on the circuit complexity and phasing of the various input signals (RF or LO). Table 7.3 outlines a few of the important mixer characteristics and shows where one mixer type might have an advantage over another. Note that the information in this table provides only a broad summary of a few of the many performance characteristics of mixer circuit topologies and that careful design, simulation, and layout of the mixer circuit will be the ultimate predictor of mixer performance.

7.4 Image Rejection Circuit Topologies

Earlier in Chapter 7, the image signal was referred to as an "unwanted" signal and one means to remove it was by filtering. Lumped element filters could be employed but take up considerable silicon real estate and have relatively low Q. MEMS filters have high Qs but still require modest real estate and extensive (and expensive) post-processing. Another approach is to take a cue from the balanced mixer concept and construct a mixer architecture that uses signal cancellation to remove the undesired

Table 7.3 Summary of Some of the Mixer Circuits Covered in Chapter 7 and a Performance Comparison

Mixer Type	*Isolation*			*Conversion Gain*	*LO Spurious Rejection*	*Second-order Signal Rejection*	*Noise*
	RF/IF	*LO/RF*	*LO/IF*				
Single ended	Poor	Poor	Poor	Low	Poor	Low	Good
Single balanced (LO)	Good	Good	Poor	Good	Even order	Good	Better
Single balanced (RF)	Poor	Poor	Good	Good	All	Good	Better
Double balanced	Good	Good	Good	Good	All	Good	Better

image signal while retaining the desired signal. This section will look at several of these architectures and provide a figure of merit for the rejection of the image signal and the constraints on the circuit to optimize this rejection.

7.4.1 Architectures

Signal cancellation in balanced mixers was achieved with in- and out-of-phase components of the RF and LO. Placing a 180° phase shift in the mixer unfortunately will not provide the proper phasing since both the desired and image signals will be phase shifted and discerning between the two will be impossible. However, inherent in the desired and image signal structure is the proper phase shift: The desired and image signals are equally separated from the LO, one of them on the upper sideband and the other on the lower sideband. *In-phase and quadrature (I, Q)* techniques can be used to separate these two signals into components that are readily canceled. One such mixer architecture is shown in Figure 7.22(a) where two 90° phase shifts, one positive and one negative, are inserted into one path (top path) of the mixer. The other path (bottom path) has no added phase shifts. In this analysis, the mixers are assumed ideal and identical.

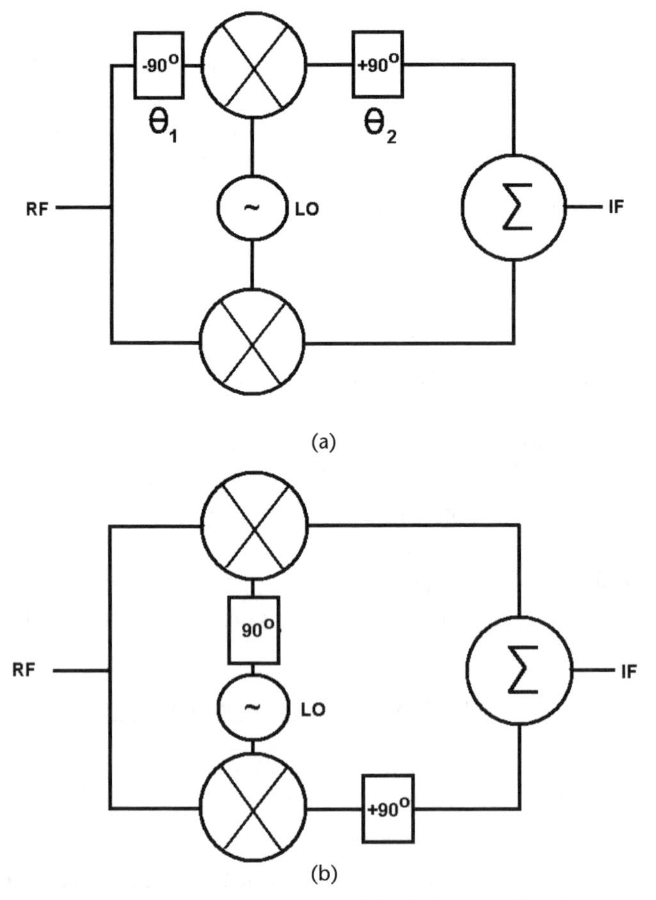

Figure 7.22 (a) Conceptual image reject mixer used for derivation. (b) More practical image reject mixer topology. (Lowpass filters are *not* shown for clarity.)

The applied RF signal has an upper sideband image (amplitude V_{im}) and lower sideband desired signal (amplitude V_{des}) at identical offsets from the LO:

$$\begin{aligned} \omega_{des} &= \omega_{LO} - \Delta\omega, \quad \omega_{im} = \omega_{LO} + \Delta\omega \\ V_{RF} &= \{V_{des}\cos\omega_{des}t + V_{im}\cos\omega_{im}t\} \end{aligned} \tag{7.52}$$

Assuming the two LO signals are identical, the top and bottom path signals prior to the mixer input can be written as follows:

$$\begin{aligned} V_{RF-top} &= \{V_{des}\cos(\omega_{des}t - 90°) + V_{im}\cos(\omega_{im}t - 90°)\} \\ V_{RF-bot} &= \{V_{des}\cos\omega_{des}t + V_{im}\cos\omega_{im}t\} \end{aligned} \tag{7.53}$$

The top and bottom path mixer outputs can be written as follows, where α is a constant describing the output gain of the mixer:

$$\begin{aligned} V_{IF-top} &= V_{RF-top}V_{LO-top} = 0.5\alpha\left[V_{des}\cos(\Delta\omega t - 90°) + V_{im}\cos(\Delta\omega t + 90°)\right] \\ V_{IF-bot} &= V_{RF-bot}V_{LO-bot} = 0.5\alpha\left[V_{des}\cos(\Delta\omega t) + V_{im}\cos(\Delta\omega t)\right] \end{aligned} \tag{7.54}$$

Phase shifting and summing the two IF signals gives the total output IF signal:

$$\begin{aligned} V_{IF} &= V_{IF-bot} + V_{IF-top}(90°) \\ &= 0.5\alpha\begin{bmatrix} V_{des}\cos(\Delta\omega t) + V_{im}\cos(\Delta\omega t) + V_{des}\cos(\Delta\omega t) \\ +V_{im}\cos(\Delta\omega t + 180°) \end{bmatrix} \\ V_{IF} &= \alpha V_{des}\cos(\Delta\omega t) \end{aligned} \tag{7.55}$$

with the result that the desired signal is passed through to the IF output but the image signal is suppressed. Changing the signs on the two phase shifters will cause the lower sideband signal to be suppressed and the upper sideband signal to appear at the IF.

In an ideal world, the unwanted signal will be perfectly canceled. However, phase errors may be associated with the 90° phase shifters. A measure of how imperfect the undesired signal or *image rejection* may be determined for this conceptual circuit by introducing a variable Θ_i that models the phase error in the corresponding phase shifter [Figure 7.22(a)]. The output of the top path mixer can be rewritten to include the effects of the phase error in the shifting process:

$$\begin{aligned} V_{IF-top} = V_{RF-top}V_{LO1} &= 0.5\alpha\left[V_{des}\cos(\Delta\omega t - [90°+\Theta_1])\right. \\ &\left.+ V_{im}\cos(\Delta\omega t + [90°+\Theta_1])\right] \end{aligned} \tag{7.56}$$

Following through the derivation in a similar form as above gives the final form for the IF output voltage:

$$V_{IF} = \frac{1}{2}\alpha\begin{bmatrix} V_{des}\cos(\Delta\omega t + [\Theta_2 - \Theta_1]) + V_{im}\cos(\Delta\omega t) + V_{des}\cos(\Delta\omega t) \\ +V_{im}\cos(\Delta\omega t + 180°+[\Theta_2 + \Theta_1]) \end{bmatrix} \tag{7.57}$$

Using a number of trigonometric identities, the desired and image terms of (7.57) can be cast into a time-varying term (in terms of $\Delta\omega t$ where $\Delta\omega$ is the difference frequency) and the phase errors Θ_i. The term of interest is the time-varying sinusoid's amplitude, which is only related to the phase errors Θ_i. Taking the ratio of the desired signal amplitude to the image signal amplitude and making $V_{des} = V_{im}$, the *image rejection ratio* (IRR) of the mixer can be derived:

$$\text{IRR}_{dB} = 10\log\left[\frac{1+\cos(\Theta_2 - \Theta_1)}{1-\cos(\Theta_2 + \Theta_1)}\right]\text{dB} \tag{7.58}$$

Equation (7.58) shows that perfect image rejection occurs if there are no phase errors in the phase shifters, $\Theta_i = 0$. Any nonzero phase error, however, will upset the cancellation and permit some of the image signal to appear at the IF port. The larger the phase error in the phase shifters, the poorer the IRR. Figure 7.23 shows a plot of IRR versus the absolute value of the phase error. The figure shows how critical phase matching is in ensuring image signal cancellation. Even a 1° error in the phase lowers the IRR to a modest 40 dB. In actual CMOS RFIC designs, phase errors of more than 2.0° can occur, with the best IRR being about 30 dB.

In this analysis, nothing was said about any amplitude errors in the phase shifters or other components or connections in the mixer. As with phase errors, amplitude differences between the two paths in the mixer will also degrade IRR. A nice analysis of both amplitude and phase mismatch was done for a mixer similar to that shown in Figure 7.22(b) with the following result obtained for IRR as a function of both amplitude and phase errors between the two mixer signal paths [19]:

$$\text{IRR}_{dB} = 10\log\left[\frac{1+(1+\Delta A)^2 + 2(1+\Delta A)\cos(\Theta_2 + \Theta_1)}{1+(1+\Delta A)^2 - 2(1+\Delta A)\cos(\Theta_2 - \Theta_1)}\right]\text{dB} \tag{7.59}$$

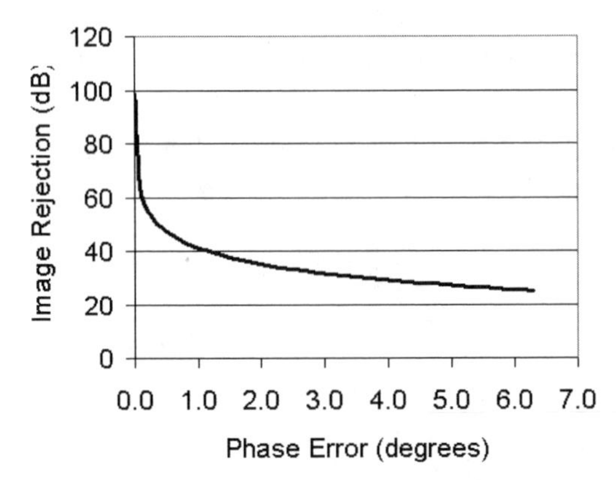

Figure 7.23 Image rejection as a function of error in the 90° phase shift networks. Only 1° of phase error will cause the image rejection to degrade to 40 dB.

where ΔA is the amplitude mismatch between the two paths. An IRR of 25 dB can be achieved with amplitude errors of less than 10% and phase errors of less than 5.7°.

7.4.1.1 Weaver and Hartley Image Reject Mixer Architectures

Difficulties arise when trying to use 90° phase shifters over a wide RF band. A better solution was advanced by Weaver that places the phase shifters in the LO arms where the bandwidth is typically narrower and the designer has more control over signal amplitudes and circuit parameters [20]. Figure 7.24 shows a block diagram of the *Weaver image reject mixer* where the 90° phase shifters in the LO arms are shown. The negative impact of the Weaver is that four mixer circuits and two separate LOs are required, with each mixer configured as a single or double balanced mixer. Not only does this increase circuit complexity, it also increases power dissipation. Phase and amplitude mismatches limit the amount of rejection in this mixer architecture. However, as the name obviously implies, the image rejection is greatly enhanced with this technique.

The image rejection performance of the Weaver architecture can be seen by considering an input signal containing both the desired as well as the image signal:

$$V_{\text{in}} = A_D \cos(\omega_D t) + A_I \cos(\omega_I t) \tag{7.60}$$

where the subscript D denotes the desired signal and the subscript I denotes the image signal. At the outputs of the first mixer and lowpass filter cascade (points A and B in Figure 7.24), the signals have the form:

$$\begin{aligned} &\text{A: } kA_D \cos\left(\omega_1 + \frac{\pi}{2} - \omega_D\right)t + kA_I \cos\left(\omega_1 + \frac{\pi}{2} - \omega_I\right)t \\ &\text{B: } kA_D \cos(\omega_1 - \omega_D)t + kA_I \cos(\omega_1 - \omega_I)t \end{aligned} \tag{7.61}$$

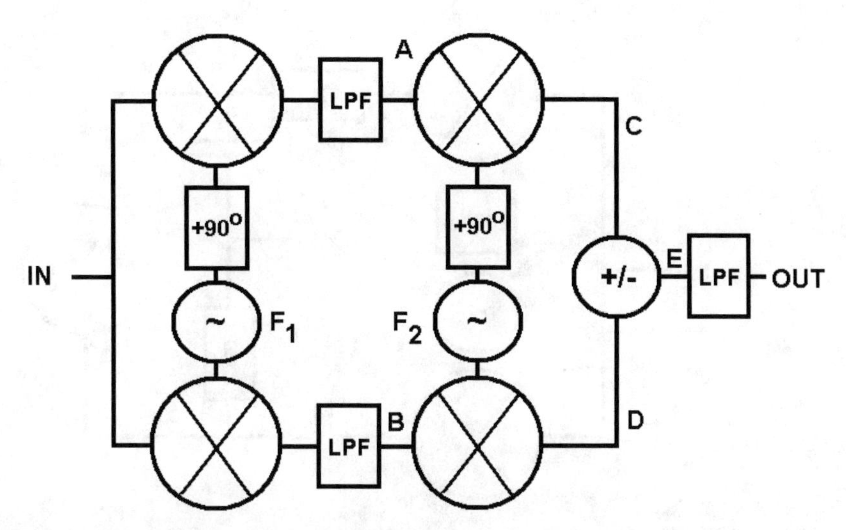

Figure 7.24 Weaver image reject mixer architecture. (*After:* [21].)

and k is some constant describing the frequency conversion process. At the output of the second mixer (points C and D in Figure 7.24), additional frequency components are created by the second frequency conversion process:

$$\begin{aligned}
&\text{C: } kA_D\left[\cos(\omega_2+\omega_1+\pi-\omega_D)t+\cos(\omega_2-\omega_1+\omega_D)t\right]\\
&\qquad +kA_I\left[\cos(\omega_2+\omega_1+\pi-\omega_I)t+\cos(\omega_2-\omega_1+\omega_I)t\right]\\
&\text{D: } kA_D\left[\cos(\omega_2+\omega_1-\omega_D)t+\cos(\omega_2-\omega_1+\omega_D)t\right]\\
&\qquad +kA_I\left[\cos(\omega_2+\omega_1-\omega_I)t+\cos(\omega_2-\omega_1+\omega_I)t\right]
\end{aligned} \tag{7.62}$$

With the result of (7.62), the selection of the appropriate signal can now be discussed. Consider that the desired signal A_D is at a frequency $\Delta\omega$ *above* the first LO frequency ω_1; the image signal A_I then lies at a frequency $\Delta\omega$ *below* ω_1. If the *difference* between the signals at points C and D is taken, then the output at point E becomes:

$$\text{E: } kA_D\cos(\omega_2-\omega_0)t+kA_I\cos(\omega_2+\omega_0)t \tag{7.63}$$

If the output lowpass filter has a corner frequency of ω_2, then the *desired* (or upper) sideband is selected (amplitude A_D component) and the image signal (amplitude A_I component) is rejected. In a similar approach, if the sum of signals at points C and D is performed, the lower sideband is selected with the upper sideband rejected.

Because image rejection really cancels one of the unwanted sidebands, circuit topologies used in *single-sideband suppressed carrier* (SSB-SC or just SSB for short) systems can also be modified to act as image reject mixers. One such classical SSB modulator is the *Hartley image rejection mixer* shown in Figure 7.25 [22]. Two phase shifters are used, with the second phase shifter based on the previously discussed simple RC lowpass/highpass circuit [21]. Phase and amplitude mismatches in the mixers and phase shifters limit the amount of rejection in this mixer architecture as well.

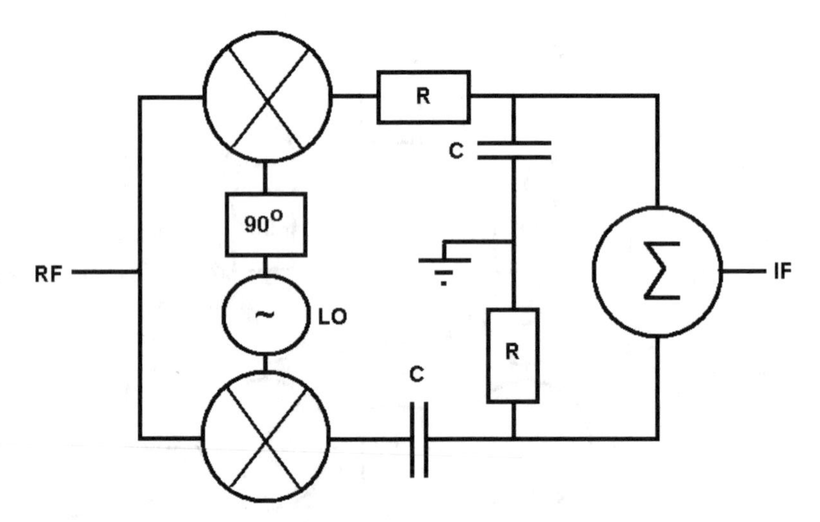

Figure 7.25 Hartley image reject mixer. (*After:* [21].)

7.4.1.2 Phase Shifting Circuits

The quadrature phase shifting circuits in the various image reject mixer architectures can be implemented in a variety of methods. If the RF and/or LO frequencies are high enough (i.e., to the point where on-chip transmission line components become feasible), a branch line coupler or other 90° hybrid can be used. The dependence of IRR on amplitude and phase errors is still subject to the well-studied amplitude and phase behaviors in these circuits (see, for example, [19]). For lower frequencies where real estate issues prohibit distributed element solutions for the phase shifters, lumped element phase shifters are employed. Two such lumped element quadrature phase shifters were discussed in Chapter 5: the simple RC lowpass/highpass phase shifter and the polyphase filter (or in this case, the polyphase phase shifter).

The RC lowpass/highpass phase shifter provides a reasonably well-tracked 90° phase difference between the two output ports over a wide frequency range. The problem with this phase shifter is on the amplitude side; the two amplitudes are identical at only one frequency: $1/2\pi RC$. For lower frequencies, the lowpass output has a higher valued transfer function, whereas the highpass output is higher above this cutoff frequency. The poor amplitude matching limits the simple RF phase shifter to relatively narrowband applications where the output amplitudes of the two ports are similar.

The use of a polyphase filter or phase shifter tends to have better amplitude and phase matching over a wider bandwidth than the simple RC network-based shifter (Figure 7.26). This increase in bandwidth comes from "smoothing out" of the phase variations within the band of interest that the multiple RC elements provide. The passive lumped element components are made with reasonably high-Q capacitors and so no lossy (and real estate intensive) inductors are needed. The design of such a phase shifter needs to be optimized based on the application and frequency range, but a good starting point is to design $(RC)_1$ for the low end of the band and $(RC)_2$ for the high end [1]. Polysilicon resistors and metal-metal or poly-poly capacitors can be used. For the poly-poly capacitors, the resistances associated with those layers

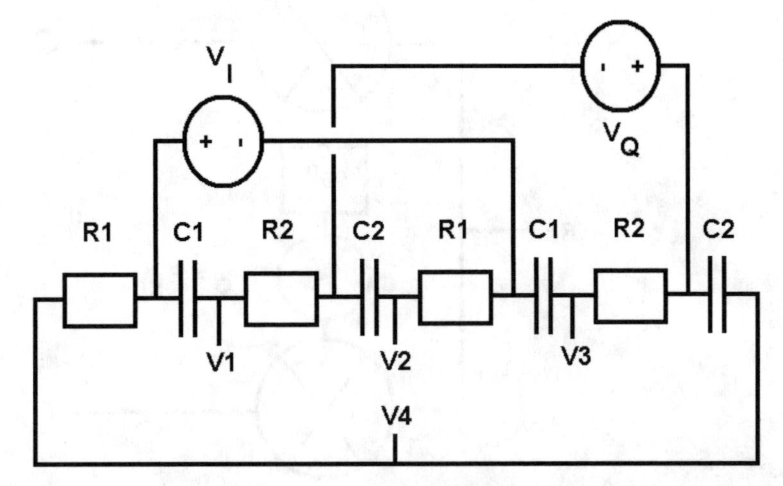

Figure 7.26 The polyphase circuit can be used for quadrature phase shifting in image reject mixers.

can be absorbed into the connecting series resistors. Care must be taken in modeling the parasitic capacitances to ground or substrate of both the resistors and capacitors in the polyphase filter to determine the overall phase shift at the design frequency.

7.5 I/Q Mixer Topologies

In many applications, such as direct conversion (zero IF) receivers (LO at the same frequency as the incoming RF), the emphasis may be on circuit simplicity for cost or power dissipation reasons. In-phase and quadrature (I/Q) mixers can employed in these applications. Information higher than the center of the band and information lower than band center may fold on top of one another in the direct conversion receiving process. However, because these signals are in phase quadrature, no information is lost even though the two spectral components overlap; full processing capability of the transmitted information is now possible.

7.5.1 Architectures

Using only one portion of the previously discussed mixers provides frequency conversion with in phase (I) and quadrature (Q) outputs. The I/Q mixer (Figure 7.27) can provide direct demodulation for angle modulated signals (FM, PM, PSK, etc.). Digital information carried on the RF carrier may be easily demodulated using such a technique. The same conceptual circuit such as shown in Figure 7.27 can be used for downconversion (typically receivers) or upconversion (typically transmitters). For the receive case, the RF appears at the RF port (after an LNA stage to help set the system noise figure) with the baseband demodulated outputs appearing at I/Q (signal flow from left to right). For the transmitter case, the I/Q baseband signals appear at the I/Q ports and are upconverted and combined to create the RF output (signal flow from right to left).

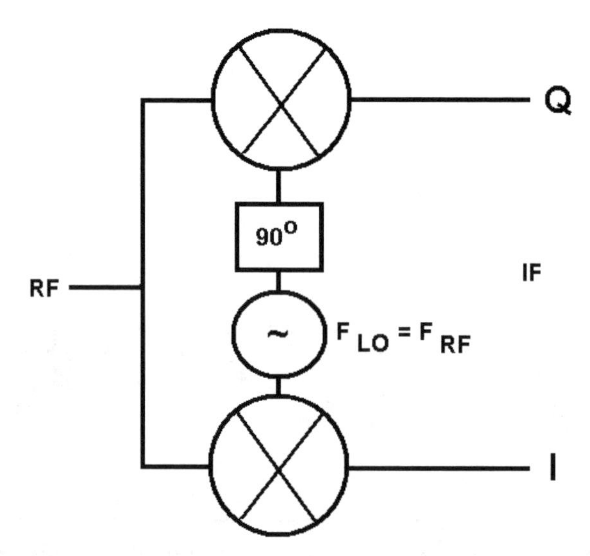

Figure 7.27 I/Q signals can be created with a 90° phase shift in the LO. The same conceptual circuit can be used for downconversion (typically receivers) or upconversion (typically transmitters).

A simple mathematical analysis shows the power of the I/Q demodulator in decoding an RF encoded digital bit stream. Suppose the input RF signal can be written as

$$V_{RF}(t) = V_0 \cos\left[\omega_{RF} t + \phi(t)\right] \tag{7.64}$$

where the function $\phi(t)$ is the bit stream encoded as a time-varying phase shift between 0° and +90°, as shown in Figure 7.28(a). In this figure, no phase shift (or a phase shift of 0°) would be encoded as one logic bit (or bit change according to the actual protocol used) and a phase shift would be encoded as another logic bit (or bit change according to the actual protocol used). Using the mixer mathematics in previous examples and taking just the baseband result (assuming a lowpass filter in both I and Q lines) provides the following signals as those ports:

$$V_Q(t) = \frac{1}{2} V_0 V_{LO-p} \cos\left[\phi(t) - 90°\right];\ V_I(t) = \frac{1}{2} V_0 V_{LO-p} \cos\left[\phi(t)\right] \tag{7.65}$$

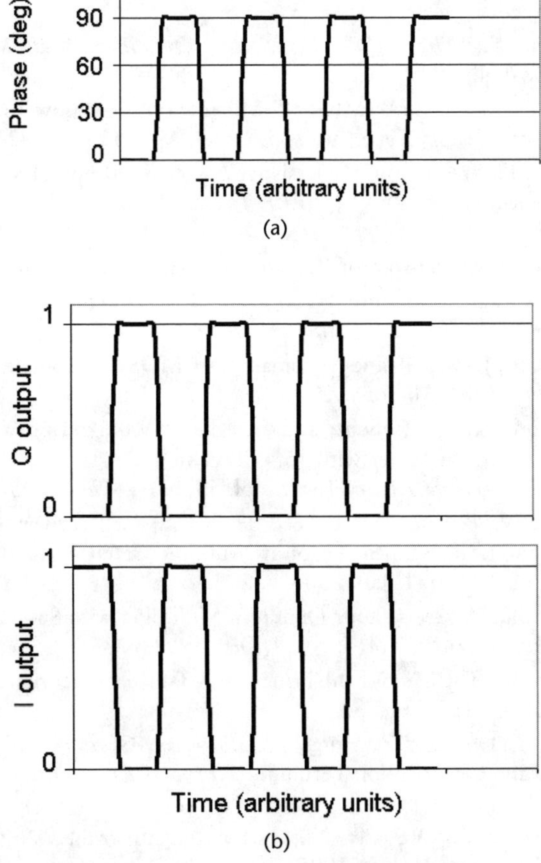

Figure 7.28 Phase modulation example: (a) 90° phase shift modulated signal versus time and (b) Q and I output signals in response to the phase modulated input signal.

The resulting waveforms at the I and Q ports in Figure 7.28(b) clearly show that the $V_Q(t)$ port directly follows the original bit stream $\phi(t)$. Alternatively, if a threshold is set at each of the I and Q ports, a signal above this threshold on the Q port would be interpreted downstream as one logic state and a signal above the threshold on the I port would be interpreted as another logic state. No signal above the threshold on either the I or Q port would indicate no data was being transmitted (i.e., no transmission, signal fading, or other important communication system issue, which is beyond the scope of this book but incredibly interesting nevertheless). Synchronization of the signals is also important, but again beyond the scope of this book. Overviews of the plethora of various modulation schemes (with their associated "alphabet soup" scheme of names) that benefit from I/Q modulation/demodulation techniques can be found in the literature (see, for example, [23]).

References

[1] Lee, T. H., *CMOS Radio Frequency Integrated Circuit Design,* 2nd ed., Cambridge, UK: Cambridge University Press, 2004.

[2] Maas, S. A., *Nonlinear Microwave and RF Circuits,* 2nd ed., Norwood, MA: Artech House, 2003.

[3] Emami, S., et al., "A 60-GHz Down-Converting CMOS Single-Gate Mixer," *Proc. 2005 RFIC Symp.*, June 2005, p. 163.

[4] Tsividis, Y., *The MOS Transistor: Operation and Modeling,* 2nd ed., Boston: McGraw-Hill, 1999.

[5] Maas, S. A., "A GaAs MESFET Mixer with Very Low Intermodulation," *IEEE Trans. Microwave Theory Tech.*, Vol. 35, No. 4, April 1987, p. 425.

[6] Ellinger, F., "26.5–30-GHz Resistive Mixer in 90-nm VLSI SOI CMOS Technology with High Linearity for WLAN," *IEEE Trans. Microwave Theory Tech.*, Vol. 53, No. 8, Aug. 2005, p. 2559.

[7] Saleh, A. A. M., *Theory of Resistive Mixers,* Cambridge, MA: The MIT Press, 1971.

[8] Guan, X., and A. Hajimiri, "A 24-GHz CMOS Front-End," *IEEE J. Solid State Circuits,* Vol. 39, No. 2, Feb. 2004, p. 368.

[9] Rao, K. R., J. Wilson, and M. Ismail, "A CMOS RF Front-End for a Multistandard WLAN Receiver," *IEEE Microwave Wireless Comp. Let.*, Vol. 15, No. 5, May 2005, p. 321.

[10] Hermann, C., M. Tiebout, and H. Klar, "A 0.6V 1.6mW Transformer-Based 2.5GHz Downconversion Mixer with +5.4 dB Gain and –2.8 dBm IIP3 in 0.13μm CMOS," *IEEE Trans. Microwave Theory Tech.*, Vol. 53, No. 2, Feb. 2005, p. 488.

[11] Kim, T. W., B. Kim, and K. Lee, "Highly Linear Receiver Front-End Adopting MOSFET Transconductance Linearization by Multiple Gated Transistors," *IEEE J. Solid State Circuits,* Vol. 39, No. 1, Jan. 2004, p. 223.

[12] Gilbert, B., "A Precise Four-Quadrant Multiplier with Subnanosecond Response," *IEEE J. Solid State Circuits,* Vol. 3, No. 4, Dec. 1968, p. 365.

[13] Gilbert, B., "The Multi-Tanh Principle: A Tutorial Overview," *IEEE J. Solid State Circuits,* Vol. 33, Jan. 1998, p. 2.

[14] Xi, Z., Y. Qin, and Z. Hong, "A 3.3-V, 1.9-GHz, High Linear CMOS Up-Mixer with Multi-Tanh Linearization Technique," *Proc. IEEE 6th Int. Conf. on ASIC: ASICON 2005,* Vol. 1, Oct. 2005, p. 405.

[15] Circa, R., et al., "Resistive Mixers for Reconfigurable Wireless Front Ends," *2005 IEEE RFIC Symp. Digest*, June 2005, p. 513.

[16] Chang, J., A. A. Abidi, and C. R. Viswanathan, "Flicker Noise in CMOS Transistors from Subthreshold to Strong Inversion at Various Temperatures," *IEEE Trans. Electron Devices*, Vol. 41, Nov. 1994, p. 1965.

[17] Darabi, H., and A. A. Abidi, "Noise in RF-CMOS Mixers: A Simple Physical Model," *IEEE Trans. Solid State Circuits*, Vol. 35, No. 1, Jan. 20004, p. 15.

[18] Collin, R., *Foundations for Microwave Engineering,* 2nd ed., New York: McGraw-Hill, 1992.

[19] Rogers, J., and C. Plett, *Radio Frequency Integrated Circuit Design,* Norwood, MA: Artech House, 2003.

[20] Weaver, Jr., D. K., "A Third Method of Generation and Detection of Single-Sideband Signals," *Proc. IRE*, June 1956, p. 1703.

[21] Wu, S., and B. Razavi, "A 900-MHz/1.8-GHz CMOS Receiver for Dual-Band Applications," *IEEE J. Solid State Circuits*, Vol. 33, No. 12, Dec. 1998, p. 2178.

[22] Hartley, R., "Modulation System," U.S. Patent 1 666 206, April 1928.

[23] Hussain, A., *Advanced RF Engineering for Wireless Systems and Networks,* Hoboken, NJ: IEEE Press–John Wiley and Sons, 2005.

Selected Bibliography

Maas, S. A., *Microwave Mixers,* Dedham, MA: Artech House, 1986.

Pozar, D., *Microwave Engineering,* 2nd ed., New York: John Wiley and Sons, 1998.

CHAPTER 8

CMOS PLLs and Frequency Synthesizers

CMOS RFICs are widely used in modern communications systems because of their seamless integration between RF and digital functionality. These communications systems often have significant requirements placed on the on-chip frequency generators: The frequency generators must be highly stable; in channelized systems, the channel frequency must be quickly and accurately set; and the generated noise introduced into the system must be minimized. In GSM systems, for example, the 25-MHz channel bandwidth is divided into 124 separate channels, with a channel spacing of 200 kHz. The GSM system specifications require that the system be frequency agile, with changes occurring on the order of a TDMA time frame of approximately 4.6 ms. All of this needs to be performed within a few parts per million accuracy. On-chip oscillators, both resonator and inverter based, such as those discussed in Chapter 6, do not provide the required stability needed for use in communication systems. The most stable of the oscillator types, those based on piezoelectric crystals, are not feasible for complete on-chip use and have upper frequency ranges of 100 MHz at best. However, by making these oscillators part of a controlled feedback structure with stable reference signals, the stringent communication system requirements can be readily achieved. These feedback control systems are the basis of modern frequency synthesizers.

This chapter covers the basics of frequency synthesis by beginning with a discussion of the basic feedback control structure based on phase lock loop (PLL) principles. The analysis focuses on those components that are most often seen in CMOS RFIC PLL-based frequency synthesizers. The more mathematical discussion of PLL principles is followed by a discussion of behavioral modeling. Behavioral modeling is often used as an intermediate step between the theory-based design and a full circuit simulation of the frequency synthesizer. Examples based on single- and dual-path PLL filters are presented. The chapter concludes with a discussion of the principles of direct digital synthesis, another method of generating a highly accurate signal, albeit one that is often lower in frequency than those based on PLLs.

8.1 Introduction to the Phase Lock Loop

8.1.1 Definitions and Basic Operation

The PLL, as its name implies, compares the output phase of the VCO with some reference phase with the difference (or error signal) filtered and fed back to the VCO control line to lock the VCO to the reference. The term *phase* here is quite a general one. Recall that any time-varying sinusoid has as its argument the phase:

$$x(x) = A\sin[\phi(t)] \tag{8.1}$$

If the phase increases linearly with time, the change in phase with respect to time is referred to as the *frequency,* implying that integrating the frequency over time yields the signal phase:

$$\frac{d\phi(t)}{dt} = \omega;\ \phi(t) - \phi_0 = \int_0^t \omega dt \tag{8.2}$$

where ϕ_0 is an integration constant.

The block diagram of a simple PLL is shown in Figure 8.1 where the VCO *output phase* $\phi_{VCO}(t)$ is returned to the input through a *feedback network* (B) and compared with the *reference phase* $\phi_R(t)$ in a *phase and/or phase frequency detector* (PD/PFD). The PD/PFD output $\phi_e(t)$ is filtered by the *loop filter* (LF) and fed to the VCO control line to control the VCO output phase.

The simplest PD is a mixer or signal multiplier. If the reference and VCO signals are described by their phase expression in the argument as

$$V_R(T) = V_{R0}\sin[\phi_R(t)];\ V_{VCO}(T) = BV_{V0}\cos[\phi_{VCO}(t)] \tag{8.3}$$

then the output voltage of the PD $V_e(t)$ can be written as a function of the two input signals and their respective phases:

$$V_e(t) = \alpha BV_{R0}V_{V0}\cos\phi_{VCO}(t)\sin\phi_R(t) = \frac{1}{2}\alpha BV_{R0}V_{V0}\sin[\phi_R(t) \pm \phi_{VCO}(t)] \tag{8.4}$$

where α is some constant describing the efficiency of the mixing process in the PD. Note that the reference and VCO signals are in phase quadrature with this definition. The resulting phase *difference* term is the one of primary interest because the loop filter will ideally filter out the higher frequency components. For small phase errors, the phase term at the output of the PD can be written as

$$V_e(t) = \frac{1}{2}\alpha V_{R0}V_{V0}[\phi_R(t) - B\phi_{VCO}(t)] = K_{PD}[\phi_R(t) - B\phi_{VCO}(t)] \tag{8.5}$$

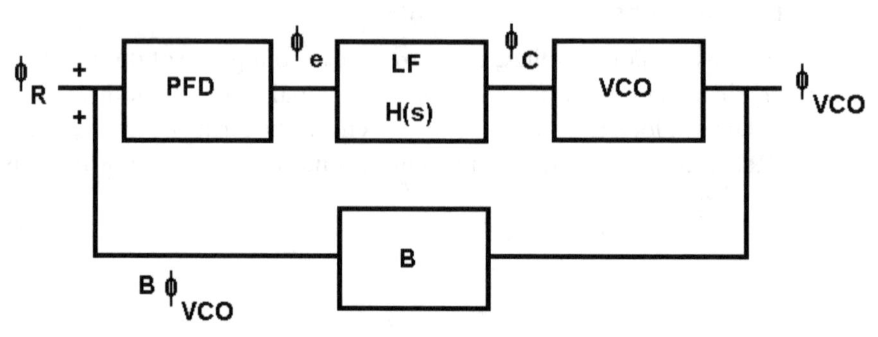

Figure 8.1 Block diagram of the basic PLL.

where K_{PD} is the PD *gain* in units of volts per radian. To keep the mathematics simple, it is better to perform the remaining analysis of the PLL in the frequency domain, which is easily done by taking the Laplace transform of the loop quantities:

$$V_e(s) = K_{PD}\left[\phi_R(s) - B\phi_{VCO}(s)\right] \tag{8.6}$$

If the loop filter LF has a transfer characteristic given by $H(s)$, the output voltage of the LF can be written as the product of the phase error signal and $H(s)$:

$$V_C(s) = K_{PD}H(s)\left[\phi_R(s) - B\phi_{VCO}(s)\right] \tag{8.7}$$

Equation (8.7) shows the control line phase; however, the VCO's frequency is really controlled by the filtered error or the control voltage. Equation (6.39) showed that the oscillation frequency is a function of the product of the VCO gain K_{VCO} and the voltage on the control line with the corresponding phase described by the integration of this phase over time:

$$\omega_{osc} = K_{VCO}V_C(t);\ \phi_{VCO}(t) = \int_0^t \omega_{osc}\,dt = K_{VCO}\int_0^t V_C(t)dt \tag{8.8}$$

Using the form for the integral property of the Laplace transform, the VCO output phase can be written as

$$\phi_{VCO}(s) = \frac{K_{VCO}}{s}\phi_C(s) = \frac{K_{VCO}K_{PD}H(s)}{s}\left[\phi_R(s) - B\phi_{VCO}(s)\right] \tag{8.9}$$

After some algebra, the PLL transfer function can be written in a more recognizable form as

$$\frac{\phi_{VCO}(s)}{\phi_R(s)} = \frac{\dfrac{K_{VCO}K_{PD}H(s)}{s}}{1+\dfrac{K_{VCO}K_{PD}H(s)}{s}B} \tag{8.10}$$

This PLL closed loop transfer function has the same mathematical form as the general feedback system $A/(1 + AB)$ where B is the feedback network term and A is the forward loop gain. Using the same definition as before, the loop gain can then be written as

$$G_L(s) = \frac{K_{VCO}K_{PD}H(s)B}{s} \tag{8.11}$$

One of the main design criteria for a PLL is keeping the loop stable, meaning that adequate gain and phase margins are needed to ensure that $G_L(s)$ does *not* exhibit a phase shift of 180° near the unity gain frequency. For PLLs, the loop filter $H(s)$ is the primary point of interest in PLL design and will be the focus in later sections.

The PLL has several standard definitions describing its operation. The *free running frequency* is the natural oscillation frequency of the VCO with no input on the control line. This does not necessarily mean, however, that zero volts is applied to the control line; the control line is left floating. The *hold range* is the frequency range at which the VCO frequency can be reliably controlled. For LC oscillators, this hold range is a function of the design of the varactor tuning range in the LC tank circuit. In ring oscillators, the hold range is designed into the tail current adjustment or active resistor voltage running range. The *capture range* is the frequency range at which the PLL can be pulled into a locked state. The *capture time* is the time it takes for an unlocked loop to acquire a lock. Figure 8.2 shows these two ranges; the hold and capture ranges may not be exactly centered on the VCO's free running frequency as shown in the figure.

There is an issue, however, with some types of PDs that is called the *dead zone* [1]. This dead zone is a small range in the PD response that does not generate a change in control voltage to the VCO, increasing the system jitter.

8.1.2 Phase Detection and Phase-Frequency Detection

Various circuit schemes have been developed for detecting differences in phase and/or frequency and then providing a single output signal based on this difference. This section looks at basic circuit topologies that provide this detection capability.

8.1.2.1 Mixer Type

The most direct analog method for phase detection is the mixer. Mixer PDs can be used up to very high frequencies and are often used in conjunction with digital PD/PFDs as prescalars. These prescaling circuits perform two major tasks: The circuits mix down a high VCO output frequency to a lower frequency; and the circuits translate the sinusoidal VCO output signal into signal levels compatible with digital circuitry in the feedback loop (B) [2]. The output of the mixer PD was previously described for VCO and reference signals in phase quadrature:

$$V_e(t) = \frac{1}{2}\alpha B V_{R0} V_{V0} \sin\left[\phi_R(t) \pm \phi_{VCO}(t)\right] \tag{8.12}$$

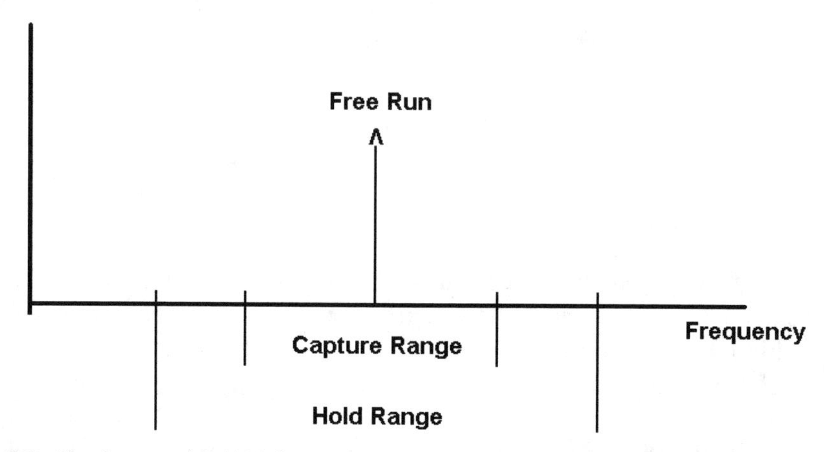

Figure 8.2 The free running VCO frequency is within both the capture and hold ranges of a PLL.

At this stage in the discussion, assume that the feedback term B is unity, which simplifies the expression for the PD output as

$$V_e(s) = K_{PD} \sin[\phi_R(s) - \phi_{VCO}(s)] = K_{PD} \sin[\Delta\phi_R(s)] \tag{8.13}$$

The output voltage error is linearly proportional to the phase difference for small phase errors and follows a sinusoidal shape, with a useful PD range between $-\pi/2$ to $+\pi/2$ (Figure 8.3). Recall, however, that the two phases were already out of phase by 90°; the π range of phase difference is still constant, however; hence, this type of PD is often referred to as a *quadrature phase detector*.

An alternative PD form takes advantage of the relatively large amplitudes of both the VCO and reference signal. Assume these two equal amplitude signals are input to the simple PD circuit composed of two identical *n*MOSFETs in parallel with both drains tied together to a pull-up resistor R as shown in Figure 8.4. If the *n*MOSFETs are dc biased as RF transconductors, the output voltage V_{out} of the circuit can be written as

$$V_{out} = V_{DD} - \frac{1}{2} g_m RV[\cos\phi_R + \cos\phi_{VCO}] \tag{8.14}$$

where the variable V is the amplitude of the reference and VCO signals. If the reference phase is assumed to be 0, then the VCO phase is essentially the phase difference and the output voltage can be simplified to

$$V_{out} = V_{DD} - \frac{1}{2} g_m RV[1 + \cos\Delta\phi] \tag{8.15}$$

The change in V_{out} with phase difference can be used to find the phase detector response:

$$\frac{dV_{out}}{d\Delta\phi} = \frac{1}{2} g_m RV[\sin\Delta\phi] = K_{PD} \sin\Delta\phi \tag{8.16}$$

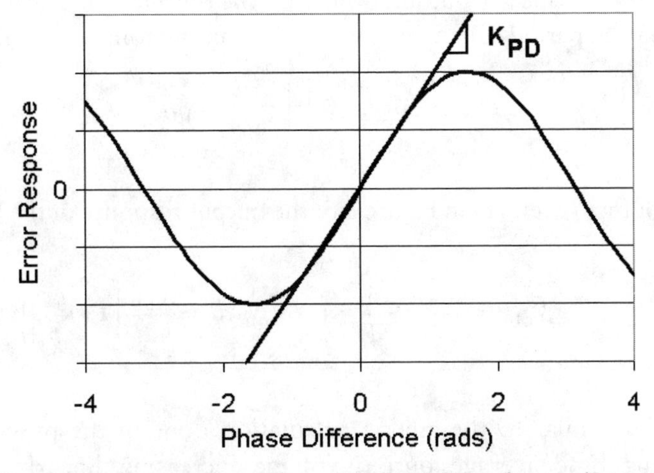

Figure 8.3 Error response for mixer-type phase detectors is linear from $-\pi/2$ to $+\pi/2$ phase difference.

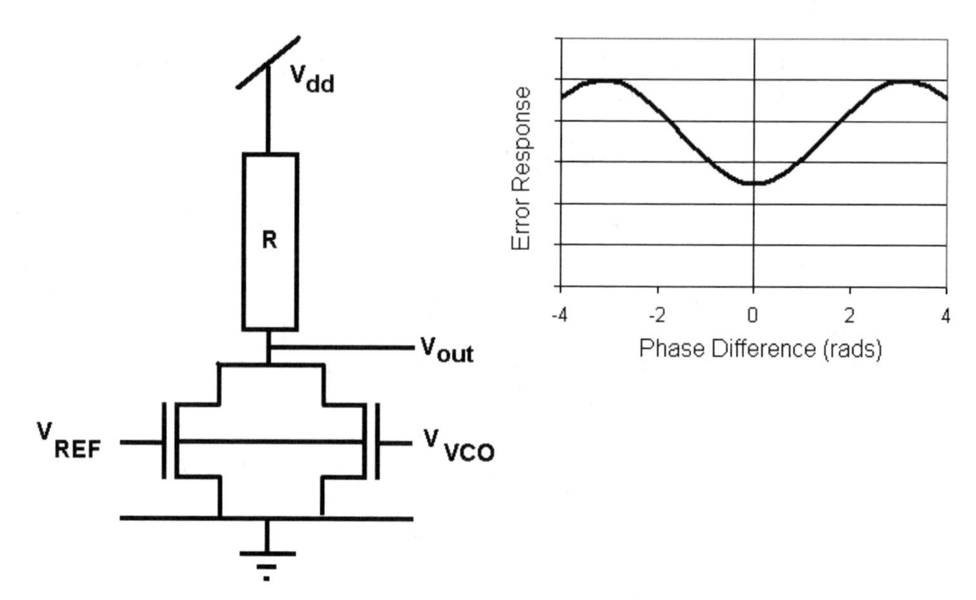

Figure 8.4 An *n*MOSFET-based phase detector with the resulting phase detector response.

A higher gain amplifier provides a higher PD gain. A change of slope occurs at a phase difference of zero but, more importantly, the range of usable phase difference for $\Delta\phi > 0$ has doubled over that of the mixer PD. This type of PD is of interest because it is based on a *digital NOR* circuit topology and acts as an analog introduction to the discussion of digital-based PD and PFD in CMOS RFICs [3]. The digital versions of PD/PFDs are especially attractive in CMOS because of the direct integration with the digital circuitry that will invariably reside on the safe chip.

8.1.2.2 OR Gate Type

If the two input signals to the PD are assumed to be ideal square waves differing only by phase shift $\Delta\phi$, the digital OR function of the two signals produces the waveform shown in Figure 8.5. For the digital OR output, any digital value of 1 (in this case, V_{DD}) will provide a 1 output; otherwise the output is 0. The average output voltage $\overline{V}_o$ over the period T can be computed using the following relationship:

$$\overline{V}_{\text{OUT}} = \frac{1}{T}\int_0^T V_{\text{OUT}}(t)dt \tag{8.17}$$

For the waveform in Figure 8.5, the output response of the PD can be written as

$$\overline{V}_{\text{OUT}} = \frac{V_{DD}}{2} + \frac{1}{T}\int_{\frac{T}{2}}^{\frac{T}{2}-\frac{\Delta\phi}{2\pi}T} V_{DD}\, dt = \frac{V_{DD}}{2}\left[1+\left|\frac{\Delta\phi}{\pi}\right|\right] \tag{8.18}$$

which is similar to the analog calculation done in the previous section with the resulting time average output voltage increasing linearly with $\Delta\phi$ instead of sinusoidally.

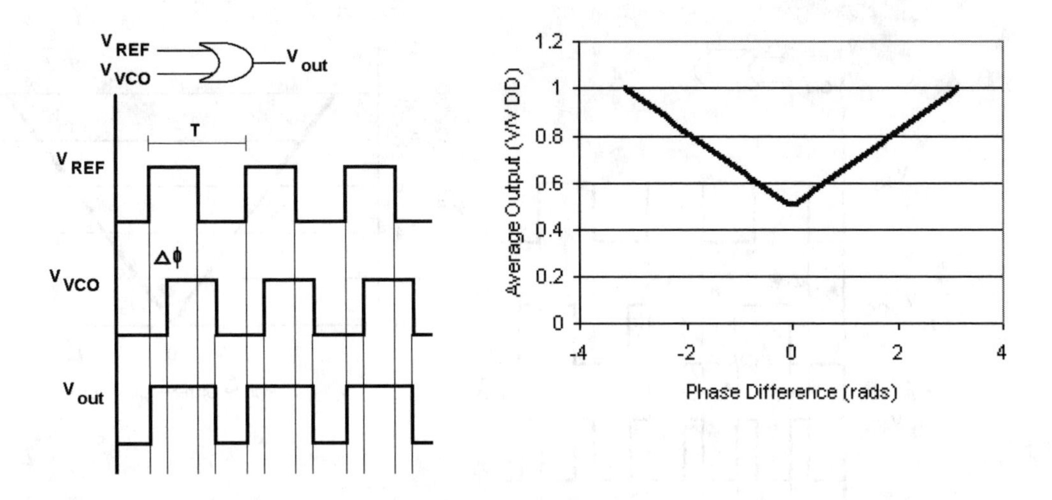

Figure 8.5 Error response for the OR gate-based phase detector.

The fastest response for digital circuits such as these is accomplished with minimum sized MOSFETs. For a CMOS implementation, standard circuit topologies such as that shown in Figure 8.6 can be used [4]. There are two stages in this design; the first is the standard NOR gate and the second is an inverter. The major drawback to the OR-type phase detector can be seen in Figure 8.7 in the next section. A given output voltage can occur with two possible phase differences, centered about zero phase shift. For practical PLLs, only one of these values of phase difference

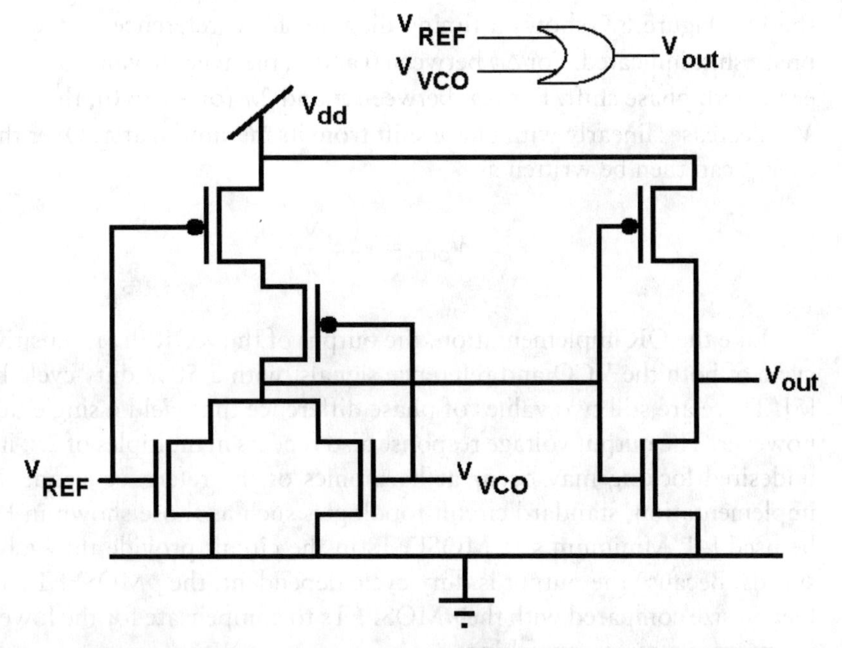

Figure 8.6 CMOS OR-type phase detector. The OR gate is a combination of the classic CMOS NOR gate followed by an inverter stage.

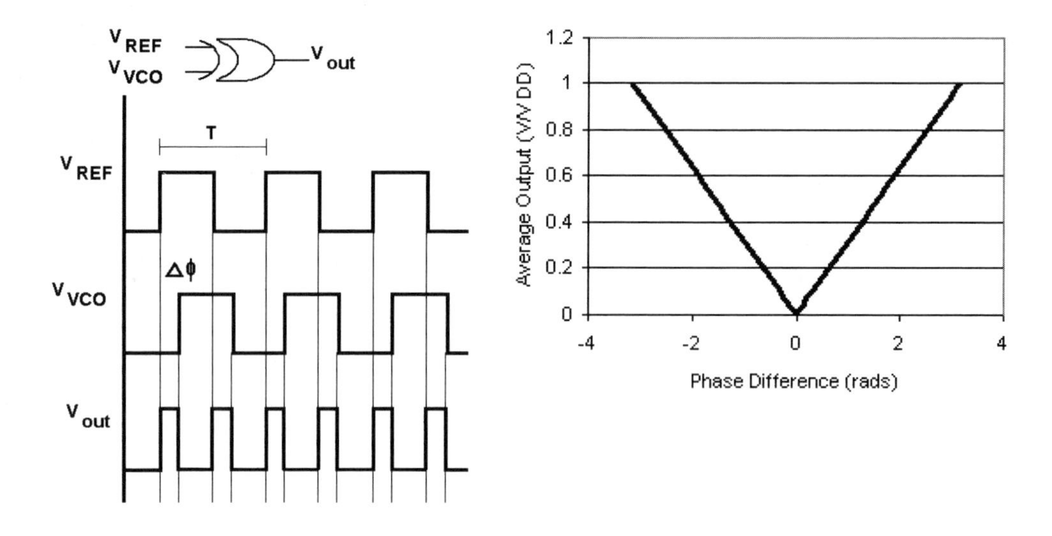

Figure 8.7 XOR phase detector with timing diagram and error response.

results in actual phase lock in the circuit. An improvement to the OR PD that removes this ambiguity is the exclusive OR PD described in the next section.

8.1.2.3 Exclusive OR Type

The digital exclusive OR (XOR) function provides a 1 output when either one or the other input is a logic 1. If the two signals are equal, the output is 0. This definition can be used with the previous definition for computing the average output voltage of the PD. Figure 8.7 shows a timing diagram for a reference and VCO signal with phase shift indicated. For $\Delta\phi$ between 0 and π, the average value of V_{out} increases linearly with phase shift. For $\Delta\phi$ between π and 2π (or $-\pi$ to 0), the average value of V_{out} decreases linearly with phase shift from its maximum at π. Over the range $-\pi$ to π, V_{out} can then be written as

$$\overline{V}_{\text{OUT}} = V_{DD}\left|\frac{\Delta\phi}{\pi}\right| \tag{8.19}$$

Like the OR implementation, the output of the XOR PD is sensitive to the duty cycle of both the VCO and reference signals, with a 50% duty cycle being optimal [5]. There are still two values of phase difference that yield a single output voltage, however. The output voltage response also repeats in multiples of 2π, indicating that undesired locking may occur at harmonics of the reference signal. For a CMOS implementation, standard circuit topologies such as those shown in Figure 8.8 can be used [4]. Minimum size MOSFETs in the circuit provide the highest switching speeds. Because the output is duty cycle dependent, the *p*MOSFETs are often doubled in size compared with the *n*MOSFETs to compensate for the lower KP, providing more symmetric switching.

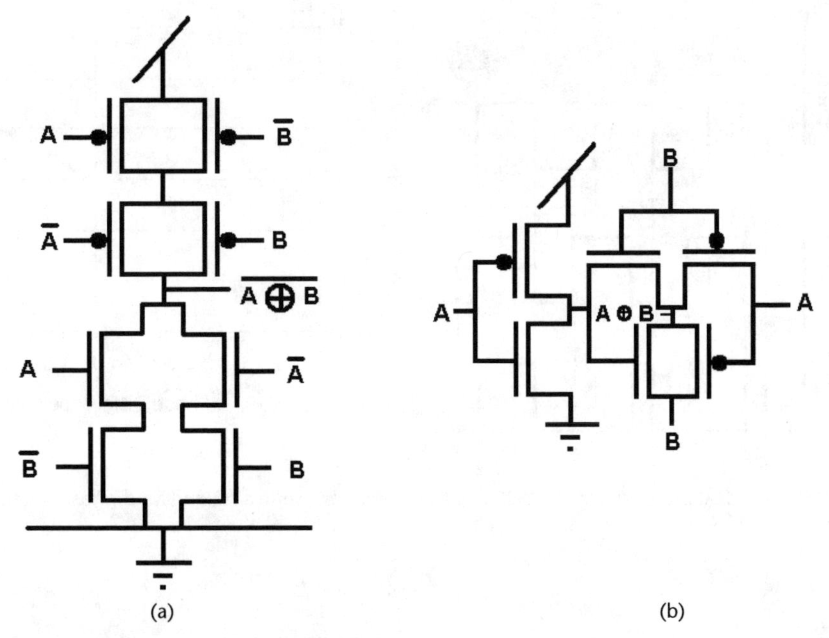

Figure 8.8 Two implementations of XOR functions using CMOS: (a) exclusive NOR PD (follow output with an inverter for XOR) and (b) XOR PD using six CMOSFETs.

8.1.2.4 D-Latch

The two previous digital phase detector implementations provide useful detection for fundamental signals over the range $-\pi < \Delta\phi < \pi$. In the XOR case, Figure 8.9 shows that *two* pulses per reference cycle occur, limiting the overall fundamental detection range. A better phase detection relationship would provide only a single pulse per period, with length dependent on the phase difference, such as that shown in Figure 8.9. This ideal phase difference response also removes the multivalued response of the OR and XOR PDs.

From the figure, the phase detection circuit needs to "remember" the first nonoverlapping signal per reference signal period but not the second, as occurred with the XOR implementation. The use of a memory element in digital circuitry implies the use of a finite state machine or flip-flops in the implementation. A widely used solution for this desired phase detection response uses two D flip-flops (DFF) in configurations variously known as *tristate* or *linear* phase detectors (Figure 8.10) [4, 6].

Because this PD's response covers the entire reference signal period, it can also be used for frequency detection and is therefore termed a *phase frequency detector* (PFD). The conventional DFF-based PFD operates in the following manner (Figure 8.11). When the reference signal V_{REF} first goes high, it sets the top DFF output Q1 high. Assuming that Q2 is low, no signal goes to the clear (CLR) input of the top DFF. When the VCO signal goes high, Q2 goes high and a CLR signal is then sent to both DFFs, driving the outputs of both (Q1 and Q2) to a low. The output Q1 remains low until V_{REF} goes high again at the start of the next reference cycle. For $\Delta\phi > 0$, the operation of the PFD is similar except Q2 is initially set and a CLR signal occurs when Q1 goes high on the rising edge of the reference signal. If the phase

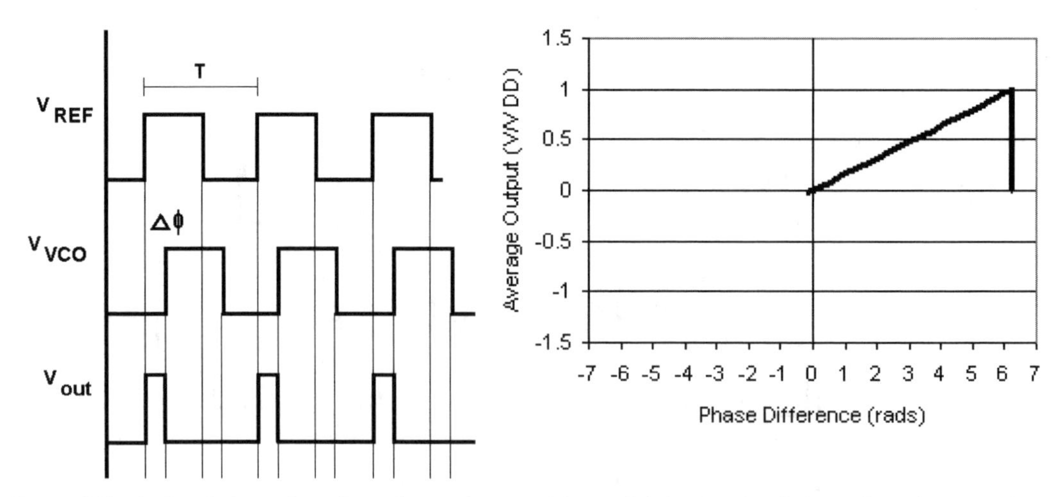

Figure 8.9 An ideal phase detection scheme that provides useful single-valued phase detection over the range $0 < \Delta\phi < 2\pi$.

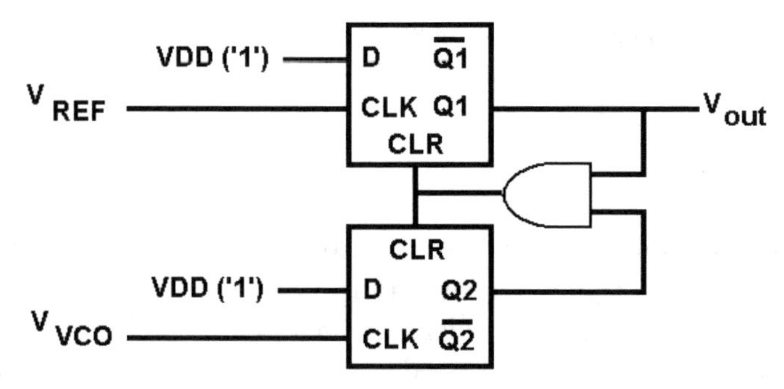

Figure 8.10 Two D flip-flops and an AND gate can be used as a digital phase-frequency detector.

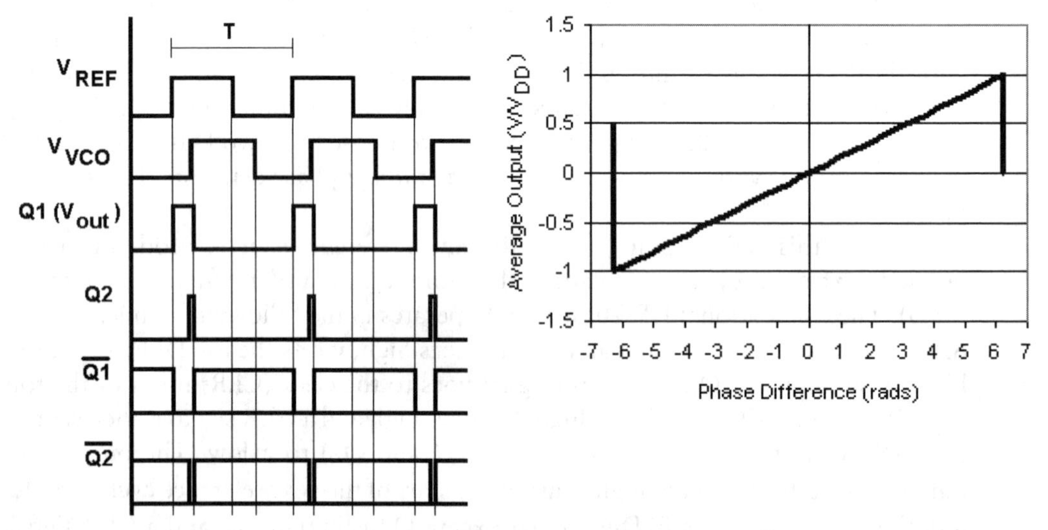

Figure 8.11 Input and output signals for the DFF PFD. Taking the difference between Q1 and Q2 as the output signal doubles the usable phase detection range.

detector output is taken as the difference between Q1 and Q2, the single-valued phase detection range improves to $-2\pi < \Delta\phi < 2\pi$.

Figure 8.12 shows two CMOS implementations of DFFs suitable for use in DFF PFDs. Numerous other DFF-based PFDs are discussed in the literature; these two examples are just meant to show some circuit possibilities.

There is a difference in the PLL operation depending on the phase of ϕ_{VCO} relative to the reference phase ϕ_R. In general, if ϕ_{VCO} lags the reference phase, then the VCO must advance the frequency slightly to that ϕ_{VCO} "catches up" with the

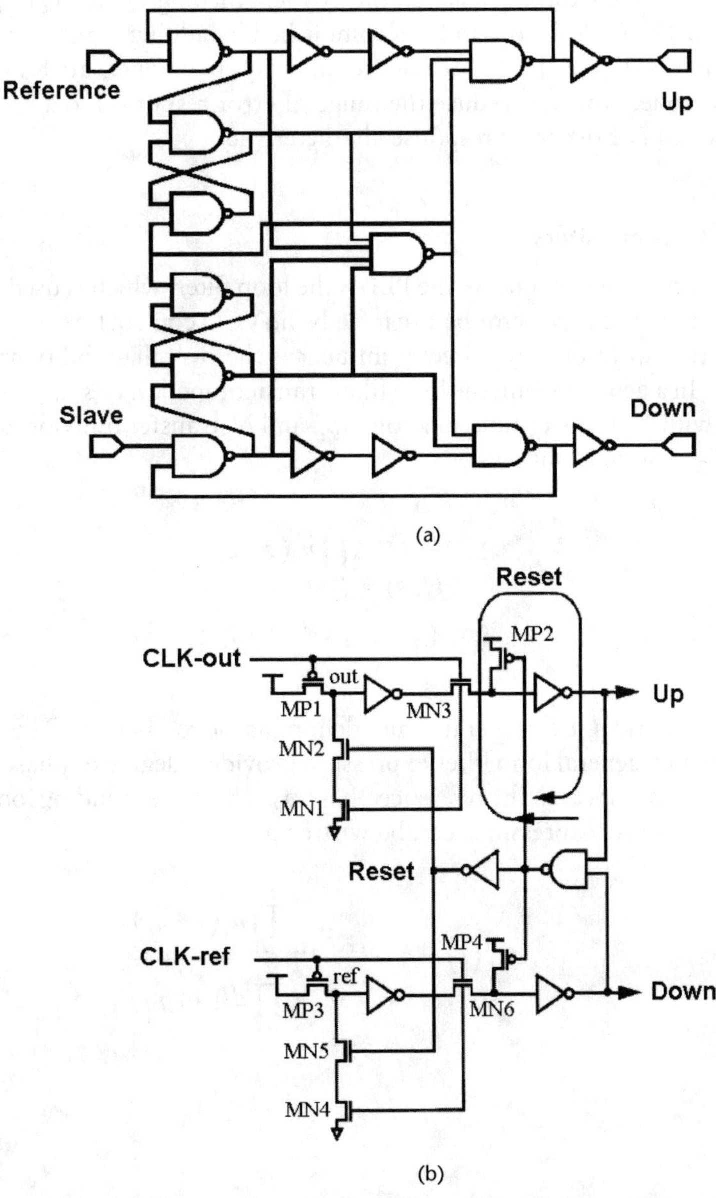

Figure 8.12 Example CMOS implementations for PFDs: (a) conventional PFD (*From:* [1] © 2006 IEEE) and (b) pass transistor PFD. (*From:* [7] © 2003 IEEE.)

reference phase ϕ_R, providing the necessary conditions for the two phases to eventually lock. Conversely, if ϕ_{VCO} leads the reference phase ϕ_R, the VCO must drop its frequency slightly to achieve (or maintain) the locked condition. The loop filter, described in the next section, provides the means to translate the leading or lagging phase into the appropriate VCO control signal.

The DFF PFD has a dead zone zero phase difference due to the inherent gate delay in the clear (CLR) signal to the nearby DFFs. If the gate delay is greater than the time delay due to the input signals' phase difference, Q1 (or Q1-Q2) may have such a narrow pulse width that the average output voltage appears to be zero for a range of small phase differences, creating this *zero phase shift dead zone*. One solution is to delay the CLR signal, thereby lengthening Q1 (or Q1-Q2) and its average value. The CLR delay can be accomplished by adding a number of digital inverters in the AND path [1, 7]. The additional delays to compensate for the zero phase shift dead zone, however, reduce the range of error responses to a value less than 2π on either end of the error response characteristic.

8.1.3 Loop Filters

The second major stage in the PLL is the loop filter, which is used to smooth the output of the phase detector before it feeds the VCO control line. The loop filter, being a function of frequency, directly influences the overall stability and response of the PLL. In a general form, the loop filter transfer function $H(s)$ is a ratio of polynomials in s with N_p transfer function poles (p_i) and N_z transfer function zeros (z_i), most generally written as follows:

$$H(s) = \frac{\prod_{i=1}^{N_z} n_i(s+z_i)}{\prod_{i=1}^{N_p} d_i(s+p_i)} \tag{8.20}$$

where n_i and d_i are numerator and denominator coefficients. The poles (zeros) in the preceding general loop filter expression provide a degree of phase lead (lag) to track ϕ_{VCO} with respect to the reference phase ϕ_R. The corresponding loop gain and overall PLL response expressions can be written as

$$G_L(s) = \frac{BK_{VCO}K_{PD}}{s} \frac{\prod_{i=1}^{N_Z} n_i(s+z_i)}{\prod_{i=1}^{N_p} d_i(s+p_i)} \tag{8.21}$$

and

$$\frac{\phi_{VCO}(s)}{\phi_R(s)} = \frac{\dfrac{K_{VCO}K_{PD}}{s}\dfrac{\prod_{i=1}^{N_z} n_i(s+z_i)}{\prod_{i=1}^{N_p} d_i(s+p_i)}}{1+\dfrac{BK_{VCO}K_{PD}}{s}\dfrac{\prod_{i=1}^{N_z} n_i(s+z_i)}{\prod_{i=1}^{N_p} d_i(s+p_i)}} \tag{8.22}$$

The loop filter transfer function $H(s)$ can be a voltage gain (V/V), current gain (A/A), transconductance (A/V), or transimpedance (V/A) depending on the type of PD preceding it. The typical CMOS RFIC PLL may have VCO gains (K_{VCO}) of 10^8 Hz/V or higher, so the distinct possibility arises that $|G_L|$ may have insufficient gain and/or phase margin to be unconditionally stable. For this reason, a significant amount of work, both theoretical and experimental, has been done in the study of PLL response due to the loop filter [8–10]. Although a large number of possible loop filter transfer functions $H(s)$ have been studied over the years for PLLs, loop filters with a rather small number of poles and zeros are used in on-chip CMOS RFIC PLLs. These loop filters can be classified into three types: passive loop filters, active loop filters (using operational amplifiers), and charge pumps. A further refinement defines the loop filter *Type* as the number of poles located at zero, while the loop filter *Order* is the highest order in s of the loop gain. In general, Type I PLLs have a zero error in phase if the frequency is changed (or ramped) and can only be used if the phase changes. Type II PLLs have zero phase error if the phase or frequency is changed but not swept. Type III PLLs have zero phase error if the phase or frequency is changed or the frequency is swept. Therefore, for a PLL used to change frequency, use a Type II; for sweeping frequency, a Type III is required.

8.1.3.1 Active and Passive Loop Filters

In this discussion on active and passive loop filters, a voltage gain loop filter transfer function $H(s)$ will be assumed. Using this assumption, the simplest loop filter is the single-pole RC lowpass filter, and the loop gain response can be written as a second-order Type I response:

$$G_L(s) = \frac{Bp_1 K_{VCO} K_{PD}}{s(s+p_1)};\ p_1 = \frac{1}{\text{RC}} \tag{8.23}$$

The corresponding closed loop response can then be written for $B = 1$ as

$$\frac{\phi_{VCO}(s)}{\phi_R(s)} = \frac{p_1 K_{VCO} K_{PD}}{s^2 + sp_1 + p_1 K_{VCO} K_{PD}} \tag{8.24}$$

which is a standard form for a second-order feedback system. Second-order feedback systems have step responses with various levels of damping: underdamped, overdamped, and critically damped (Figure 8.13). The underdamped case

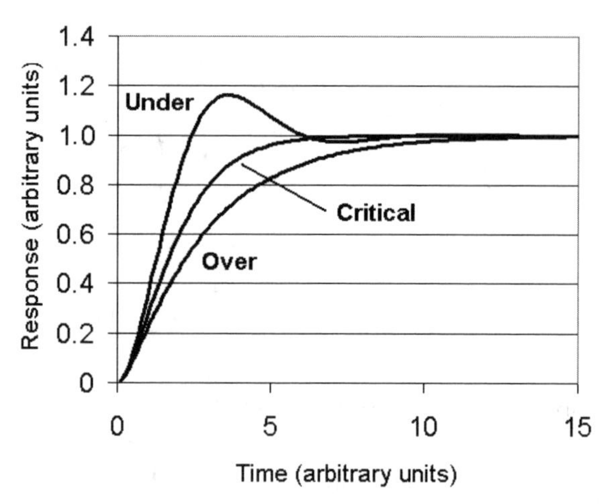

Figure 8.13 Step input response in a second-order system: underdamped, overdamped, and critically damped cases.

is characterized by a slow response to a step input, whereas the overdamped case is characterized by a fast response but with overshoot and ringing occurring during the settling phase. The critically damped case shows the fastest rise time without overshoot. The degree of damping in the system is important for determining PLL settling time on the VCO control line due to phase difference changes at the input of the PD. For the overdamped case, the VCO takes a longer time to settle to the new frequency when a phase difference between the VCO and the reference is detected. Contrast this with the underdamped case where the rise time on the VCO control line is much quicker but then overshoots the final value, taking additional time to settle. The critically damped case is the design goal when it comes to settling time: it results in the fastest settling time with no overshoot occurring. From standard feedback systems theory, the general form that describes these three damping cases is given by the general transfer function form:

$$H(s) = \frac{\omega_0^2}{s^2 + 2\gamma\omega_0 + \omega_0^2} \tag{8.25}$$

where γ is the system damping factor and ω_0 is the natural resonance frequency of the feedback system and is a measure of the loop response time. Values of γ less than 1 (but greater than zero) indicate an underdamped condition, values of $\gamma > 1$ indicate an overdamped condition, and a value of $\gamma = 1$ indicates the critically damped condition. Comparing the second-order general form in (8.25) with the closed loop PLL response in (8.24) shows that both the loop damping factor and the loop resonance frequency are functions of the VCO and PD gains as well as the location of the loop filter pole.

Example 8.1: Determine the location of the loop filter pole for a critically damped system. Assume the product $K_{VCO}K_{PD} = 10^9$ Hz/rad.

Equating the denominators of (8.24) and (8.25) gives expressions for both γ and ω_0 in terms of PLL system parameters. For critical damping, $\gamma = 1$, so the natural resonance frequency of the PLL and the required loop filter pole can be found:

$$\gamma = \frac{\omega_0}{2K_{VCO}K_{PD}} = 1;\ \omega_0 = 2\cdot 10^9 \text{ rps}$$

$$p_1 = \frac{\omega_0^2}{K_{VCO}K_{PD}} = \frac{\left(2\cdot 10^9\right)^2}{10^9} = 4\cdot 10^9 \text{ rps}$$

Suitable values of RC are then chosen to yield this value of pole p_1. An analysis of the loop gain term $G_L(s)$ shows a phase margin of approximately 75° (Figure 8.14).

The single-pole response of Example 8.1 provided a single degree of freedom in the design. Using more complex *lead* and/or *lead-lag filters* provides additional degrees of design freedom. Four examples of simple loop filters often used in

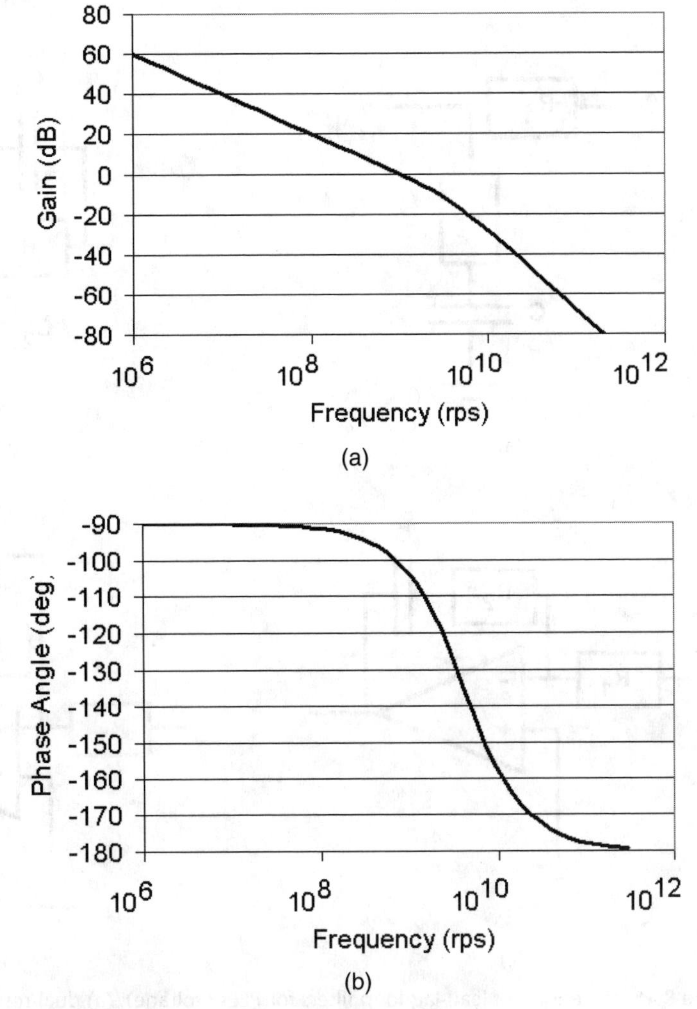

Figure 8.14 (a) Loop gain and (b) phase response for Example 8.1.

on-chip CMOS PLLs are shown in Figure 8.15 with their corresponding transfer function terms and pole zero definitions shown in Table 8.1. The third and fourth circuits shown require the design of an operational amplifier; good discussions of CMOS operational amplifier design procedures can be found in numerous texts [11–13]. In addition, this filter type adds an additional 180° phase shift by virtue of its inverting characteristic.

The design of PLLs with various loop filter configurations can be accomplished by looking more closely at the loop gain term, $G_L(s)$ (8.26), and, more specifically, the phase of $G_L(s)$:

$$G_L(s) = \frac{BK_{VCO}K_{PD}}{s}\frac{n_1(s+z_1)}{d_1(s+p_1)} \tag{8.26}$$

In the relationship for the loop gain in (8.26), the zero adds up to +90° to the overall system phase depending on frequency, with the phase shift at very low and very high frequencies at −90°. In the vicinity of the pole and zero, however, some combination of the two could present a phase shift that approaches −180° with unity

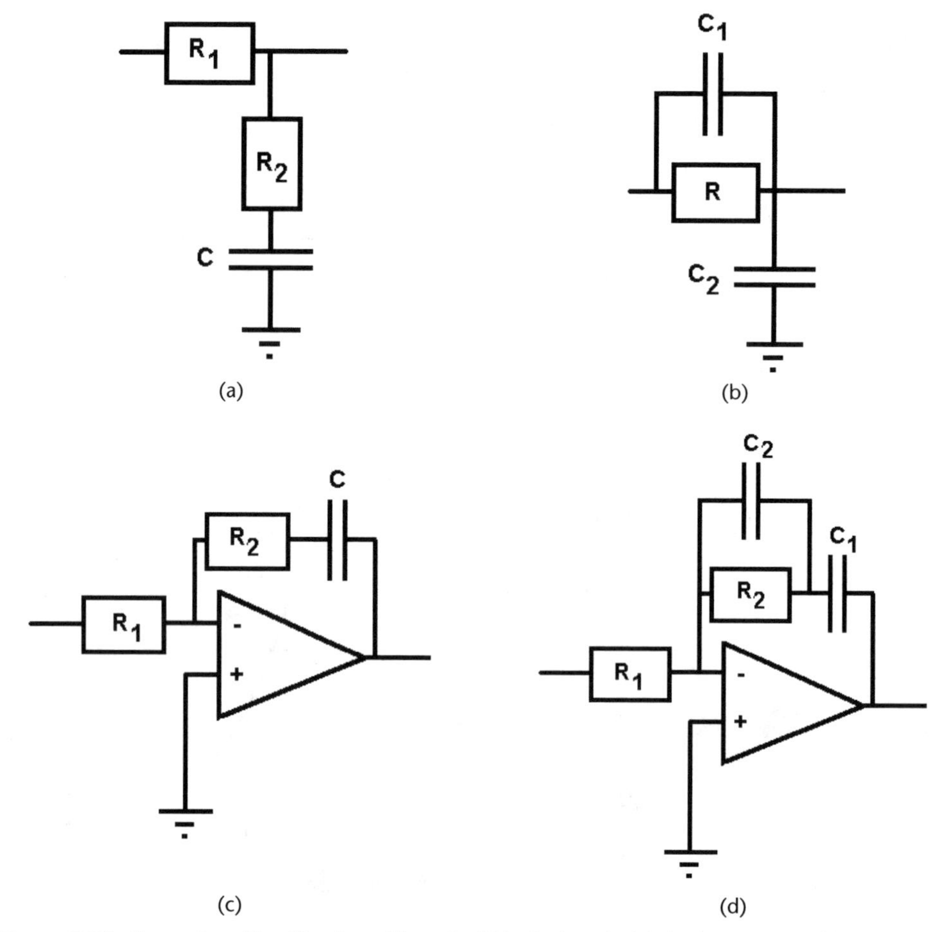

Figure 8.15 Examples of lead-lag loop filters for PLLs (voltage): (a) dual-resistor lead-lag, (b) dual-capacitor lead-lag, (c) integrator with lead, and (d) integrator with lead-lag.

Table 8.1 Gain, Pole, and Zero Locations for the Loop Filter Networks in Figure 8.15

Figure	N_1	d_1	p_1	z_1
8.15(a)	R_2	$R_1 + R_2$	$1/C(R_1 + R_2)$	$1/R_2C$
8.15(b)	C_1	$C_1 + C_2$	$1/R(C_1 + C_2)$	$1/RC_1$
8.15(c)	$-R_2$	R_1	0	$1/R_2C$
8.15(d)	$-(C_1 + C_2)$	$R_1C_1C_2$	$1/R_2C_2$	$1/R_2(C_1 + C_2)$

Note: A second pole is introduced into the loop filter response in Figure 8.15(d) by the second capacitor C_2.

gain ($|G_L| = 1$), running the risk of PLL instability. A loop gain phase margin PM can be defined as the difference between the phase of G_L and -180° (or $-\pi$). This phase margin should be large enough at the unity gain frequency to keep the system stable (typically 60° or more). An expression for this phase margin, PM, for the filters shown in Figures 8.15(a) and (b) can be written as

$$\mathrm{PM}(\omega) = \frac{\pi}{2} + \tan^{-1}\left(\frac{\omega}{z_1}\right) - \tan^{-1}\left(\frac{\omega}{p_1}\right) \tag{8.27}$$

The maximum PM can be found by taking the derivative of (8.27) with respect to frequency, with the result being that the frequency of maximum PM is a function of the pole and zero locations:

$$\omega_0^2 = p_1 z_1 \tag{8.28}$$

The PM for the loop filters shown in Figures 8.15(c) and (d) must include the influence of the two poles at zero (both of which provide -90° phase shift), and so the phase margin is a function of the zero location only:

$$\mathrm{PM}(\omega) = \pi - \tan^{-1}\left(\frac{\omega}{z_1}\right) \tag{8.29}$$

Using a similar algebraic analysis as the previous example, the overall PLL transfer function for the loop filters described in Table 8.1 can be written as follows:

$$\frac{\phi_{VCO}(s)}{\phi_R(s)} = \frac{\dfrac{n_1 K_{VCO} K_{PD}}{d_1} s + \dfrac{n_1 K_{VCO} K_{PD}}{d_1} z_1}{s^2 + s\left(p_1 + \dfrac{n_1 B K_{VCO} K_{PD}}{d_1}\right) + \dfrac{n_1 B K_{VCO} K_{PD}}{d_1} z_1} = \frac{\dfrac{\omega_0^2}{B}\left(\dfrac{s}{z_1} + 1\right)}{s^2 + 2\gamma\omega_0 s + \omega_0^2} \tag{8.30}$$

where the resonance frequency ω_0 and damping factor γ in terms of the pole, zero, and PLL gain terms are given by

$$\gamma = \frac{1}{2}\frac{\omega_0}{z_1}\left(1 + p_1 \frac{d_1}{K_{VCO} K_{PD} n_1}\right); \ \omega_0^2 = B K_{VCO} K_{PD} \frac{n_1}{d_1} z_1 \tag{8.31}$$

For systems where the pole is sufficiently high $\left(p_1 >> \frac{K_{VCO}K_{PD}n_1}{d_1}\right)$, these expressions reduce to

$$\gamma = \frac{1}{2}\frac{p_1}{\omega_0};\ \omega_0^2 = BK_{VCO}K_{PD}\frac{n_1}{d_1}z_1 \tag{8.32}$$

The extra degrees of design freedom come at a cost with lead-lag filters, however. Large values of damping factor γ are necessary to keep the overall loop response from excessive ringing in response to a step phase difference input. Figure 8.16 shows the response of (8.30) versus frequency for several values of damping factor γ. The choice of damping factor γ governs the location of the zero, lowering it as damping factor γ increases. Small values of γ give considerable gain at the resonance frequency, which will also provide noise gain in the PLL bandwidth but will more rapidly attenuate this noise above the filter cutoff. Larger damping factors reduce the noise gain but at the expense of noise and PLL bandwidth.

Example 8.2: Determine the location of the loop filter pole and zero for the loop filter circuit topology shown in Figure 8.15(a) for a loop resonance frequency of $\omega_0 = 10^8$ rps. Assume the product $K_{VCO}K_{PD} = 10^9$ Hz/rad and $B = 1$. Use the simplified pole zero equations ((8.28) and (8.32)).

A damping factor γ must be chosen as the first step. For $\gamma = 0.5$, there is considerable overshoot in the response, whereas for $\gamma > 1$, the response is slow. For this example, choose $\gamma = 0.75$ as a compromise. Using this value of γ and (8.32) allows immediate computation of the pole:

$$p_1 = 2\gamma\omega_0 = 1.5\omega_0 = 1.5 \cdot 10^8 \text{ rps}$$

The zero can be found because the pole is now known; (8.28) can be used since the maximum phase margin at this frequency is required:

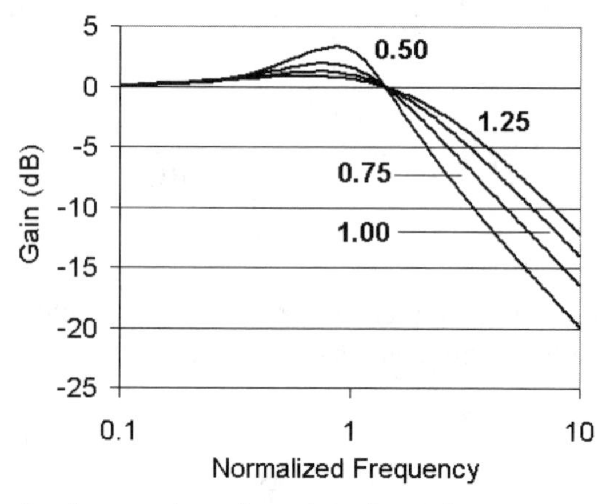

Figure 8.16 The damping factor γ plays a key role in the settling time, bandwidth, and step response overshoot of lead-lag filters in PLLs.

$$\omega_0^2 = z_1 p_1 \Rightarrow z_1 = \frac{\omega_0^2}{p_1} = \frac{\omega_0^2}{1.5\omega_0} = 6.67 \cdot 10^7 \text{ rps}$$

The loop gain (G_L) shows a phase of approximately –75° at the unity gain frequency $\omega_0 = 10^8$ rps with a phase margin of 105°.

8.1.3.2 Charge Pumps

While the loop filters just described are usable with digital PFDs, such as those based on the DFF, a better form of loop filter takes advantage of the control signal outputs of the digital circuits. These signals can be used to control a current source to supply charge to a capacitor, thereby causing the voltage to rise as a function of the amount of charge supplied to it by the current source. These same control signals can also be used to control a current sink to discharge this same capacitor, causing the voltage to drop. For the DFF PFD, an increasing phase difference causes the average output voltage Q1 to rise as Q1's pulse width increases. By using Q1-bar to activate a *p*MOSFET current source, pulses of Q1-bar cause pulses of current to charge up the capacitor; hence, the time average voltage and current are related by the overall impedance seen by the charge pump. In like manner, a discharge signal using Q2 can activate an *n*MOSFET current sink and lower the voltage across the capacitor. Because charge is added or subtracted from the capacitor during each period of the reference signal, the name *charge pump* has been used to describe the circuit operation. A simple charge pump is shown in Figure 8.17 where the control signals UP and DOWN are Q1-bar and Q2, respectively. The control voltage across C can then be written as

$$V_c(s) = Z(s)\frac{I_{pump}}{2\pi}[\phi_R - B\phi_{VCO}] = Z(s)K_{PD}\Delta\phi \tag{8.33}$$

where $Z(s)$ is the impedance seen at the charge pump output.

Because the control signals UP and DOWN are at the rails, the current sourced or sunk by the appropriate MOSFET can be adjusted using the device aspect ratio:

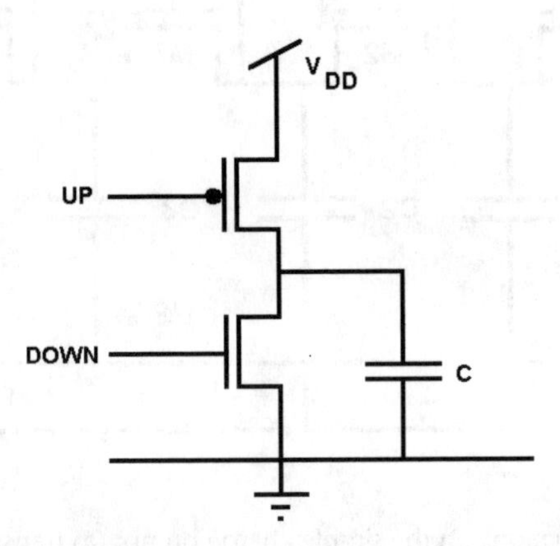

Figure 8.17 Simple charge pump circuit.

$$I_{UP} = \frac{1}{2}\mathrm{KP}_p\left(\frac{W}{L}\right)_p\left(V_{DD} - |V_{Tp}|\right)^2;\ I_{DOWN} = \frac{1}{2}\mathrm{KP}_n\left(\frac{W}{L}\right)_n\left(V_{DD} - |V_{Tn}|\right)^2 \tag{8.34}$$

The charge pump translates voltage pulses to current pulses that charge the capacitor and govern the overall output voltage going to the VCO. Variations on this simple charge pump shown in Figure 8.18 include the use of transmission gates or other circuitry to improve performance [Figures 8.18(a) and (b), respectively] [14].

The transmission gate charge pump [Figure 8.18(a)] helps to minimize control signal feed-through and the subsequent additional charge injection onto C. The diode-connected MOSFETs in the Dickson charge pump [Figure 8.18(b)] allow charge to be pumped from C_1 and to flow only in the direction of current I_o. The clock signals Φ_1 and Φ_2 provide the stimulus and timing of the charge injection into the circuit. The charge pump with a single capacitor load has an impedance function

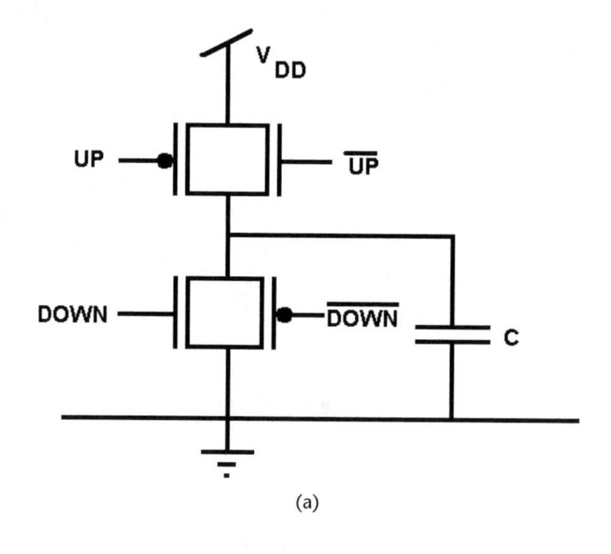

(a)

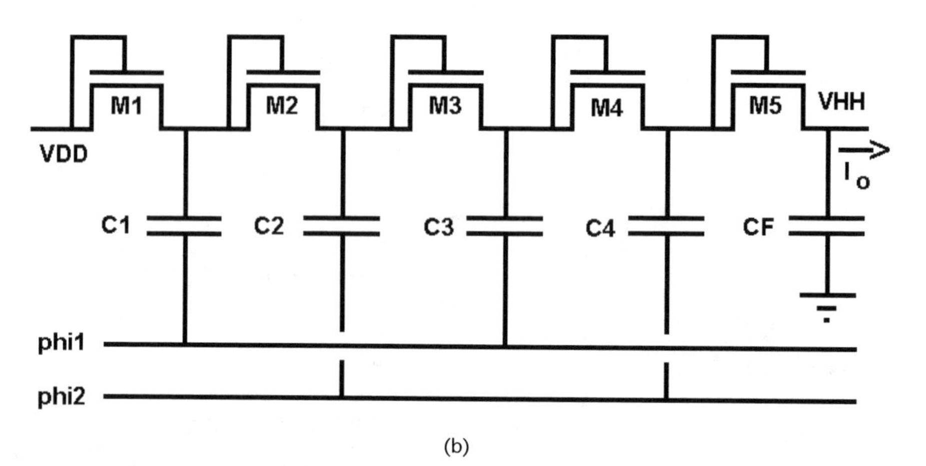

(b)

Figure 8.18 Variations on the simple charge pump: (a) transmission gate-based charge pump and (b) four-stage Dickson charge pump suitable for low-voltage applications. (*After:* [14].)

$Z(s) = (V_c/I_{pump})$ proportional to $1/s$, creating a potentially unstable system since both poles reside on the $j\omega$ axis. The addition of at least one zero in the transfer function, such as adding a resistor in parallel with C, provides a more stable lowpass characteristic to $Z(s)$. Additional filtering sections can be added to increase the degrees of freedom in the PLL design as well as to increase the attenuation above the cutoff frequency. Figure 8.19 shows four such filters. The fourth filter is termed a *dual-path loop filter* since it is fed by two charge pumps [15, 16]. The dual-path loop filter provides additional design options: Both passive element values and charge pump currents can be adjusted for the desired loop performance. Table 8.2 shows the gain and pole zero locations for the loop filters shown in Figure 8.19. The phase margins for the loop gain are also shown. Note that one of the filters exhibits two poles at zero, which adds another –90° phase shift that must be taken into consideration when computing the phase margin for the PLL. The peak PM occurs at the same frequency as before:

$$\omega_0^2 = p_1 z_1 \tag{8.35}$$

The values of resistance and/or capacitance in the loop filter may be quite large and require correspondingly large areas of silicon real estate to implement. Techniques using discrete time filters based on switched capacitor techniques have been used to alleviate this issue [17]. The increase in noise due to switching and other sources common to switched capacitors, however, must be weighed against the benefits (real estate, tighter control of pole and zero locations) obtained using this technique.

The charge pump relies on storage of charge on the pump capacitance. Once the PLL has locked, there are no UP or DOWN pulses to charge/discharge the capacitor. However, this capacitor is connected to the drains of the charge pump MOSFETs, which have small reverse-bias leakage currents to the substrate even when the MOSFETs themselves are completely turned off, so there will be some voltage sag across the capacitor during lock. Typical values of leakage current are 0.01 to 0.1 nA [4], giving a voltage drop over a certain time frame of

$$\Delta V \approx 10^{-4} \frac{\Delta T_{\mu s}}{C_{\mathrm{pF}}} \mathrm{V} \tag{8.36}$$

For example, the voltage across a 1-pF capacitor will sag about 0.1 mV in 1.0 μs, which translates into a 100-kHz frequency change in the VCO output frequency for a VCO gain of $K_{VCO} = 10^9$ Hz/V. However, once any sag is detected as a phase change on the VCO output, the appropriate UP pulse will be sent to charge this capacitor up to keep the loop locked.

Example 8.3: Determine the location of the loop filter pole and zero for the dual-path active loop filter shown in Figure 8.19(d) with a natural loop resonance of 10^8 rad/sec and a phase margin of 60° ($\pi/3$) at unity gain. Assume $K_{VCO} = 10^9$ Hz/V and $B = 1$.

Using the maximum phase margin design specification and (8.35), the pole and zero at the loop resonance frequency can be found:

Table 8.2 Gain, Pole, and Zero Locations for the Charge Pump Loop Filter Networks in Figure 8.19

Figure	n_1	p_1	z_1	*Phase Margin*	*Unity Gain*
8.19(a)	R	0	$1/RC$	$\tan^{-1}\left(\frac{\omega}{z_1}\right)$	$\frac{BK_{VCO}I_pR}{\omega_0^2}\sqrt{\omega_0^2+z_1^2}$
8.19(b)	$1/C_1$	$(C_1+C_2)/RC_1C_2$	$1/RC_2$	$\tan^{-1}\left(\frac{\omega}{z_1}\right)-\tan^{-1}\left(\frac{\omega}{p_1}\right)$	$\frac{BK_{VCO}I_p}{\omega_0^2C_1}\sqrt{\frac{\omega_0^2+z_1^2}{\omega_0^2+p_1^2}}$
8.19(c)	R_1/R_2C_2	$1/R_2C_2$	$1/R_1C_1$	$\tan^{-1}\left(\frac{\omega}{z_1}\right)-\tan^{-1}\left(\frac{\omega}{p_1}\right)$	$\frac{BK_{VCO}I_p}{\omega_0^2}\frac{R_1}{R_2C_2}\sqrt{\frac{\omega_0^2+z_1^2}{\omega_0^2+p_1^2}}$
8.19(d)	k/C_2	$1/RC_2$	$1/kRC_1$	$\tan^{-1}\left(\frac{\omega}{z_1}\right)-\tan^{-1}\left(\frac{\omega}{p_1}\right)$	$\frac{BkK_{VCO}I_p}{\omega_0^2C_2}\sqrt{\frac{\omega_0^2+z_1^2}{\omega_0^2+p_1^2}}$

For the dual path in Figure 8.19(d), the transfer function is defined as $V_{out}(s)/I_2(s)$ and the variable $k = I_1/I_2$.

$$\mathrm{PM}(\omega_0) = \tan^{-1}\left(\frac{\omega_0}{z_1}\right) - \tan^{-1}\left(\frac{\omega_0}{p_1}\right) = \tan^{-1}\left(\frac{p_1}{\omega_0}\right) - \tan^{-1}\left(\frac{\omega_0}{p_1}\right) = \frac{\pi}{3} \tag{8.37}$$

$$\Rightarrow p_1 = 3.72\omega_0 = 3.72\cdot 10^8 \text{ rps} \quad z_1 = \omega_0^2/p_1 = 0.269\cdot 10^8 \text{ rps}$$

where the result of (8.35) was used in the phase margin expression for z_1. At the same frequency, the magnitude of G_L should be unity:

$$|G_L(\omega_0)| = \frac{K_{VCO}I_p}{\omega_0^2}\frac{k}{C_2}\sqrt{\frac{(\omega_0^2+z_1^2)}{(\omega_0^2+p_1^2)}} = 1 \tag{8.38}$$

Assuming that the charge pump current I_p sets the value of I_2 and $I_1 = kI_2$, after a bit of algebraic manipulation, the capacitance C_2 can be written in terms of the pole and charge pump currents as

$$C_2 = \frac{K_{VCO}kI_p}{\omega_0 p_1} = \frac{K_{VCO}kI_2}{\omega_0 p_1} = \frac{K_{VCO}I_1}{\omega_0 p_1} \tag{8.39}$$

For $k = 10$ and a charge pump current $I_1 = 100\ \mu$A, C_2 becomes 2.68 pF. The ratio of z/p yields an expression for $C_1 = C_2\, p/zk$, giving this value of capacitance of 3.73 pF. The corresponding resistance of 1,000Ω completes the loop filter design.

Up to this point in the discussion, the feedback term B has been assumed unity. Reduction in the gain term $K_{VCO}K_{PD}$ by a feedback term $B < 1$ will allow more realistic values of on-chip components to be realized. In addition, $B < 1$ feedback allows the VCO to be referenced to a frequency much lower than its free running value. For example, if B were implemented by a divide-by-N circuit (easily designed digitally using standard finite state machine techniques or microprocessor controlled), then the VCO can be controlled by a reference signal $1/N$ times its free running term (Figure 8.20). A nominal 1,000-MHz VCO followed by a divide-by-10,000 circuit can be locked to a reference signal of 100 kHz. This division technique then allows the use of highly stable, high-Q mechanical resonator-based reference signals

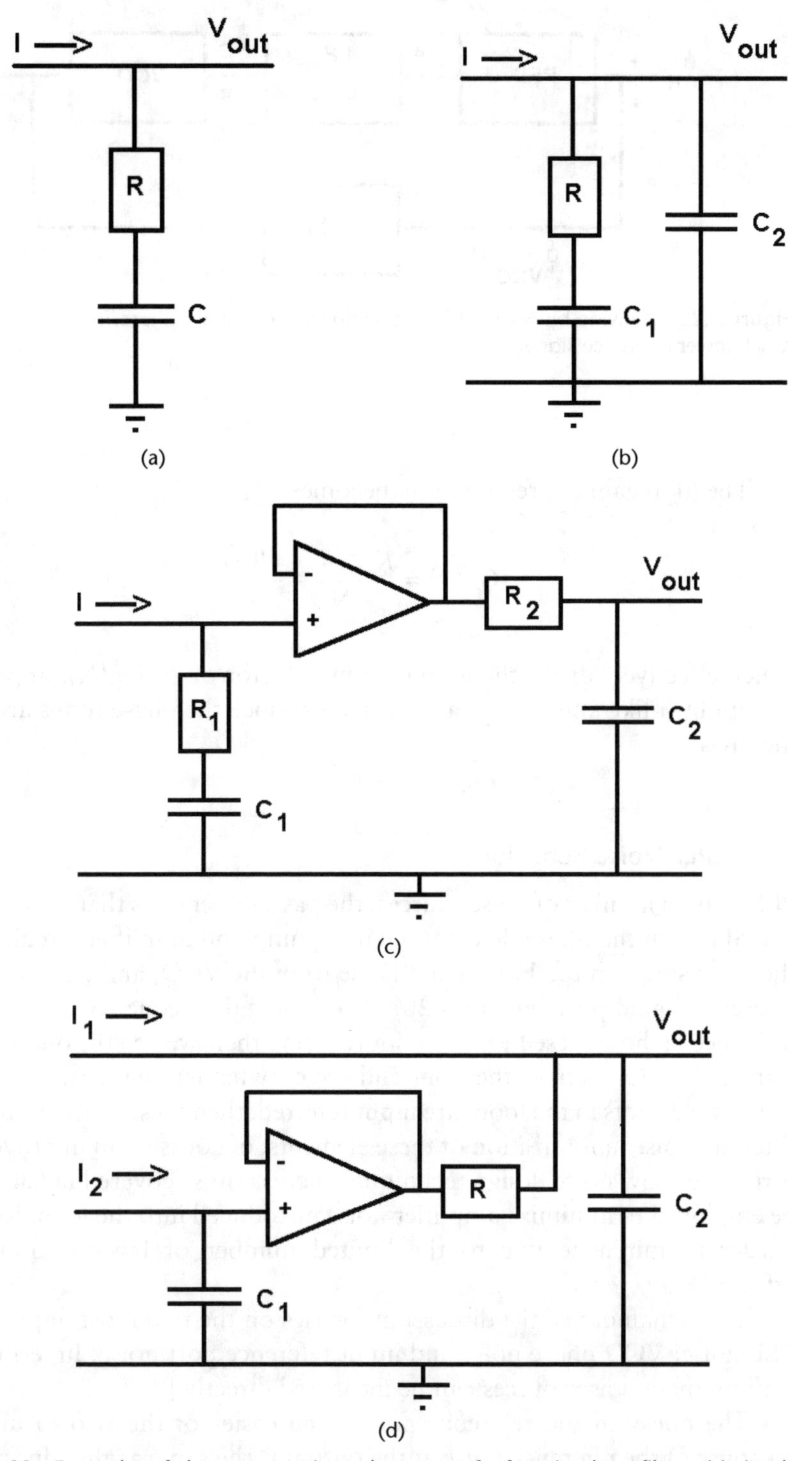

Figure 8.19 Examples of charge pump impedance transfer function loop filters: (a) simple loop filter, (b) loop filter with added pole, (c) loop filter with active element, and (d) dual-path loop filter.

(piezoelectric or MEMS resonator, for example) to accurately control the VCO output frequency.

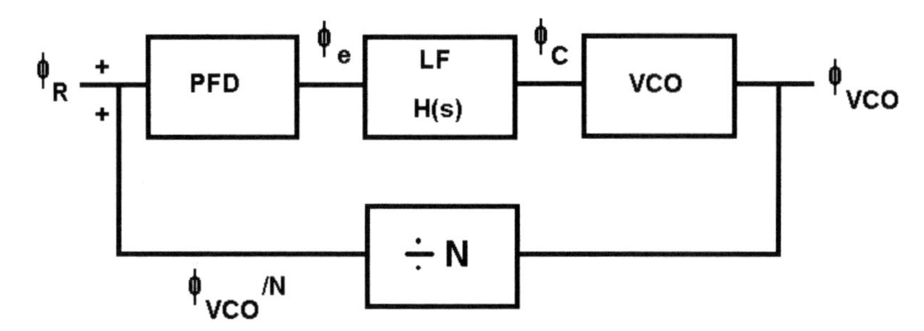

Figure 8.20 A divide-by-*N* circuit in the feedback loop allows phase/frequency referencing to a much lower reference signal.

The loop gain expression now becomes

$$G_L(s) = \frac{K_{VCO} K_{PD}}{N} \frac{H(s)}{s} \tag{8.40}$$

which effectively drops the loop gain by a factor of 20 log(N), improving the gain margin by a like amount for a fixed $H(s)/s$, since the phase terms are unaffected by the division.

8.1.4 PLL Noise Behavior

PLLs have a number of noise sources: the passive elements that make up the PLL, the MOSFETs in the phase detector, charge pump and amplifier circuits that make up the front stages in the PLL, phase noise from the VCO, and noise coming in on the reference signal port (Figure 8.20). The loop gain frequency response has a direct influence on how these PLL noise sources find their way to the output of the system. If the noise sources from the front-end stages (whether they are resistors, MOSFETs, or lossy elements in the loop) are input referred, then these noise sources are lowpass filtered. Noise minimization of these elements, of course, can improve the noise performance. Low noise design techniques such as those covered in Chapter 4 can also be employed to minimize amplifier noise introduced into the loop. Resistive noise is harder to minimize due to the limited number of layer options that CMOS affords [8].

The remainder of the discussion focuses on the two most important sources of PLL noise: VCO phase noise and input reference port noise. In terms of characterization, these noise sources can be measured directly [9].

The noise on the reference port is the easier of the two to understand. The response of the reference noise at the output is the same as that due to the reference signal itself; that is, the same transfer function used in describing the overall PLL response can be applied to describe the reference signal noise:

$$\frac{\phi_{VCO}(s)}{\phi_R(s)} = \frac{\dfrac{K_{VCO}K_{PD}}{s}\dfrac{\prod_{i=1}^{N_z} n_i(s+z_i)}{\prod_{i=1}^{N_p} d_i(s+p_i)}}{1+\dfrac{BK_{VCO}K_{PD}}{s}\dfrac{\prod_{i=1}^{N_z} n_i(s+z_i)}{\prod_{i=1}^{N_p} d_i(s+p_i)}} \tag{8.41}$$

For the previously discussed loop filters, $\phi_{VCO}(s)/\phi_R(s)$ is primarily lowpass in nature, so minimizing the bandwidth of the PLL will tend to minimize the reference noise energy at the output of the PLL. This bandwidth minimization, however, is not conducive to fast loop response time or loop stability, so for that reason, a "quiet" reference signal, such as a piezoelectric or MEMS resonator-based oscillator, is preferred.

The noise situation is a bit more complex for VCO phase noise, which appears at the PLL output and is also part of the overall PLL feedback system. From the discussion of oscillator phase noise in Chapter 6, the VCO phase noise tends to exhibit a $1/f^N$ characteristic depending on the offset frequency (based on Leeson's noise formulation), meaning more noise power occurs at frequencies centered about the carrier. For VCO phase noise, a lowpass loop characteristic would let this low-frequency VCO phase noise be injected into the loop with little or no attenuation. To see the PLL characteristics that can be used to control this noise source, suppose that the VCO phase noise is additive to the output phase of the VCO:

$$\phi_{VCO}(s) = \phi_0(s) + \Delta\phi(s) \tag{8.42}$$

Assuming the PLL reference has locked onto the phase $\phi_0(s)$, the PLL output response to the VCO phase noise term $\Delta\phi(s)$ can be derived as follows, assuming $B = 1/N$:

$$\frac{\phi_{VCO}(s)}{\Delta\phi(s)} = \frac{1}{1+\dfrac{K_{VCO}K_{PD}}{Ns}\dfrac{\prod_{i=1}^{N_z} n_i(s+z_i)}{\prod_{i=1}^{N_p} d_i(s+p_i)}} \tag{8.43}$$

For a simple single-pole loop filter, the PLL response to the VCO phase noise can be written as

$$\frac{\phi_{VCO}(s)}{\Delta\phi(s)} = \frac{s(s+p)}{s^2 + sp + \dfrac{K_{VCO}K_{PD}}{Nd}} \tag{8.44}$$

To show the behavior of (8.44), Figure 8.21 plots the PLL response (magnitude and phase) to additive VCO output phase noise where a pole at unity frequency ($p =$

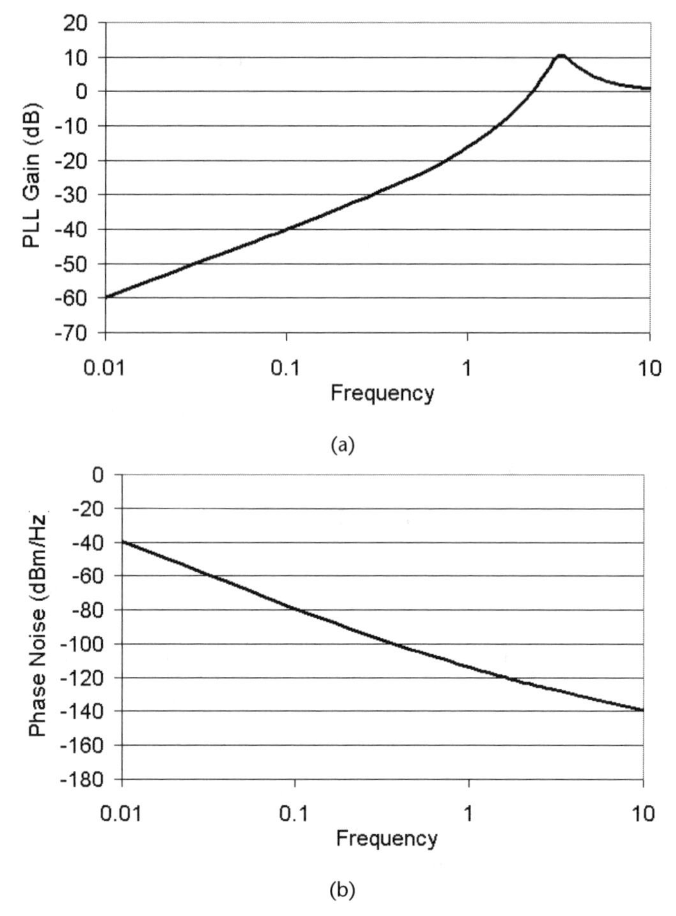

Figure 8.21 (a) PLL response to additive VCO output phase noise and (b) VCO phase noise output spectrum.

1) and $\frac{K_{VCO}K_{PD}}{Nd} = 10$ have been assumed for illustrative purposes. Figure 8.21(a) shows the PLL gain plotted versus the frequency where a highpass characteristic is clearly evident from the shape of the gain curve. With the VCO phase noise response shown in Figure 8.21(b), the best suppression of the VCO phase noise occurs if the PLL has as a *wide bandwidth* so that the suppression of the phase noise components at frequencies about the carrier is higher. Note that this result is contradictory to that of minimum noise on the reference signal port, but a wider loop bandwidth is more consistent with faster PLL response time and stability. A wide bandwidth PLL is often used to "clean up" a VCO with marginal phase noise characteristics, a major reason for its widespread use. Above the pole and zero locations, the PLL output expression shows that

$$\phi_{VCO}(s) = N\Delta\phi(s) \tag{8.45}$$

which shows that large values of N (i.e., large feedback loop division ratios) increase the overall phase noise by 20 log N [9]. The reference signal should be as high as possible to limit the feedback division magnitude to minimize increased output phase noise.

The capture range is increased in wider loop bandwidth PLLs because this capture range is approximately one-half the loop bandwidth [18]. However, if the loop bandwidth is made too large, then any small amount of noise on the reference port becomes as issue. The PLL loop bandwidth should be designed with the minimum loop bandwidth necessary for good phase noise suppression as well as PLL transient response; knowledge of the VCO phase noise properties is crucial for this determination.

8.1.5 PLL Behavioral Modeling

Mathematically based models are a necessary first step in the design of a PLL. An intermediate step in the design process for a PLL between mathematically-based modeling and complex (and time-consuming) circuit-based modeling is to substitute high-level simulation blocks that model the behavior of various components in the PLL. This use of *behavioral modeling* can aid in observing how all of the components of a PLL play together in a form that allows much faster simulation than full circuit analysis. Behavioral modeling also provides an efficient means of seeing the effects of varying PLL component values (VCO gain, loop filter coefficients, charge pump current, feedback division N, etc.) on the final system response. Transient circuit simulators such as SPICE and its variants can be employed for PLL behavioral modeling if translations between the time-varying PLL phase response variables and SPICE circuit responses (voltages or currents) are made [19]. For the remaining discussion, the PLL phase response $\phi(t)$ will be modeled as a node voltage in SPICE.

8.1.5.1 Conceptual PLL Behavior Modeling

The first block in the simple PLL is a phase detector. Because the PLL phase response terms $\phi(t)$ are being modeled as voltages, the system block modeling the behavior of the PD requires that two input voltages be compared with the output and the difference multiplied by the gain. This PD behavioral block can be modeled as a voltage-controlled voltage source (VCVS) with a gain value dependent on the type of PD block being modeled. The controlling nodes for this VCVS are the reference and feedback signals. For example, the XOR-type PD exhibits a gain of $1/\pi$ V/V, whereas the DFF PD exhibits a gain of $1/2\pi$ V/V.

The passive loop filter transfer function $H(s)$ can be easily modeled using the actual resistance and capacitance values. If an active element such as an operational amplifier is required, another VCVS can be used. The VCVS gain is set to be the same as that of the amplifier.

The last block in the PLL is the VCO, with a transfer function discussed previously and given by:

$$\phi_{VCO}(s) = \frac{K_{VCO}}{s} \tag{8.46}$$

For the behavioral modeling, the phases are translated into voltages for simulation purposes. Considering (8.46) in terms of voltages shows that the output voltage from the behavioral block can be modeled using a voltage-controlled current source (VCCS) of gain K_{VCO} driving a unit valued capacitor (1 farad):

$$V_{VCO}(s) = \frac{K_{VCO}}{s} V_{in}(s) = \frac{1}{s}\left[K_{VCO} V_{in}(s)\right] \tag{8.47}$$

Finally, the feedback division-by-N term can be modeled using yet another VCVS with the VCVS gain set as $1/N$. Combining these behavioral blocks gives the behavioral model for the PLL shown in Figure 8.22.

The charge pump in a PLL can also be modeled using behavioral techniques. The basic operation of the charge pump provides the clues needed for modeling its behavior. As long as a control signal (UP or DOWN) is applied, the charge pump will source or sink current. The UP/DOWN control signals are governed by the relative phase position of the VCO output with respect to the reference signal, and because this relative difference is modeled as a voltage, a VCCS can be used for the charge pump. The gain of each VCCS (source and sink) charge pump is the corresponding value of pump current.

8.1.5.2 PLL Behavior Modeling Examples

The first example uses the results of Example 8.1 to show the PLL response to a step input. For a damping factor of unity, the natural resonance frequency was found to be $\omega_0 = 2 \times 10^9$ rps. The inverse of the loop resonance frequency is related to the PLL settling time constant; the settling time T_{settle} is approximately 2π times this value:

$$T_{settle} = \frac{2\pi}{\omega_0} = 3.14 \text{ ns} \tag{8.48}$$

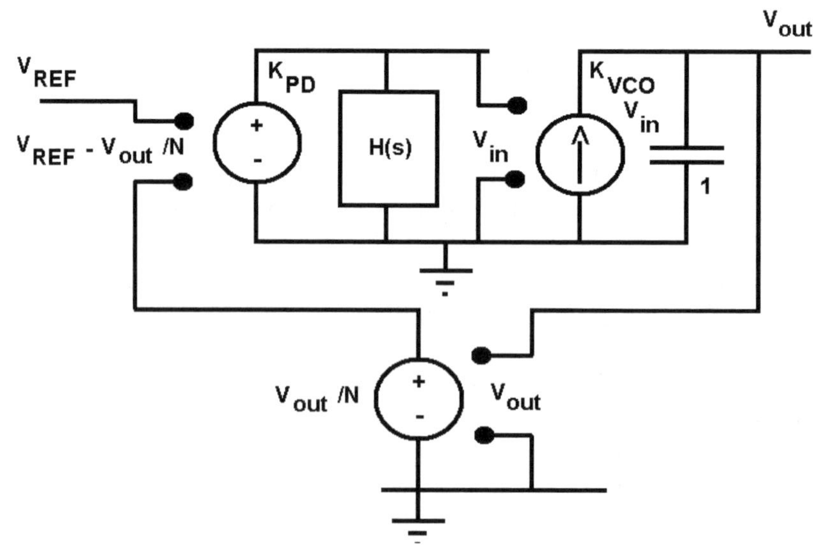

Figure 8.22 Basic PLL behavioral model, in which H(s) is replaced with the loop filter components.

The only design degree of freedom in this example is the location of the loop filter pole, p_1. This pole can be computed as $1/RC$ and, assuming a 1.0-pF on-chip capacitance, the corresponding resistance value is

$$\gamma = \frac{\omega_0}{2K_{VCO}K_{PD}} = 1;\ \omega_0 = 2 \cdot 10^9 \text{ rps}$$
$$p_1 = 4 \cdot 10^9 \text{ rps} = \frac{1}{RC} \Rightarrow R = \frac{1}{p_1 C} = 250\Omega \tag{8.49}$$

A length of polysilicon can be used to implement this modest resistor. (Parasitics should also be modeled to ensure that the parasitic capacitance of R does not unduly affect the PLL response.) A 10% step change in the input reference phase was modeled as a 10% step change in input reference voltage V_{REF}, and this voltage was applied to the PLL behavioral model using the loop filter values calculated earlier and the phase and VCO gains stated in Example 8.1. Figure 8.23(a) shows the SPICE behavioral model circuit used with the following gains: $E1 = 1/2\pi$, $G1 = 2\pi \times 10^9$, and $E2 = 1$. The three $10^{10}\Omega$ resistors are placed in the circuit to aid in bias

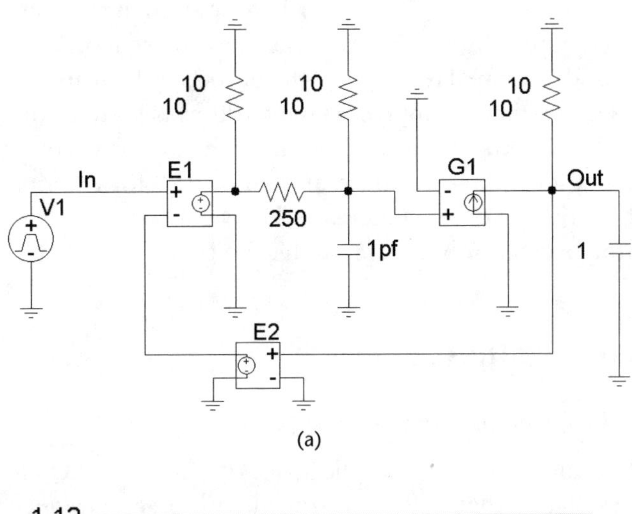

(a)

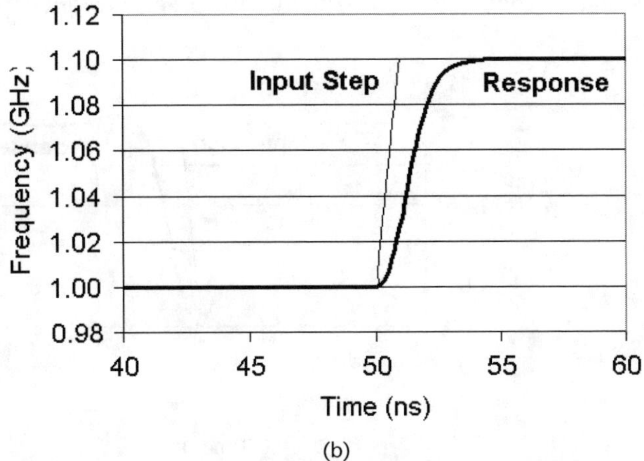

(b)

Figure 8.23 Behavior modeling of the single-pole loop filter PLL of Example 8.1: (a) modeling circuit and (b) output response (ch8-1.txt).

point convergence and do not influence the results. The transient response to an input reference change with a 1.0-ns rise time, shown in Figure 8.23(b), confirms the estimate of the settling time in the PLL of approximately 3.0 ns. The result shows a smooth transition to the step input response indicative of the critically damped design.

The effects of improper loop filter design can be seen if the loop filter resistance is doubled to 500Ω. The behavioral model result (see Figure 8.24) shows an approximately 0.5% overshoot with ringing in the output and a resulting longer settling time.

The second example uses a charge pump as part of the dual-path loop filter and uses the results of Example 8.3. Unity gain in the feedback loop is assumed. Two current sources are required: $I_1 = 100\,\mu$A and $I_2 = 10\,\mu$A, with loop filter capacitance values for C_1 and C_2 of 3.73 and 2.68 pF, respectively. With the pole p_1 found in the example, the loop filter resistance was found to be 1,000Ω. The behavioral model for the dual-loop PLL is shown in Figure 8.25(a). The unity gain amplifier is modeled using VCVS E4 with the charge pump currents I_1 and I_2 modeled as VCCS's G2 and G3, respectively. The loop resonance frequency corresponds to a settling time T_{settle} of 62.8 ns. The large value resistances are used to aid in simulation convergence. The output response to a 10% input change in frequency is modeled and shown in Figure 8.25(b). Note that the phase margin of 60° used in Example 8.3 provides an underdamped response in the output with an approximately 2% overshoot. A loop settling time of approximately 100 ns is seen in the PLL response of Figure 8.25(b). The large settling time is due to the 60° phase margin design requirement. Choosing a less stringent phase margin of 45°, for example, will decrease the rise time of the output but will increase the overshoot in the response. A larger PM will reduce the overshoot but increase the rise time.

8.2 Frequency Synthesis

8.2.1 PLL-Based Synthesizers

One of the most important applications of PLLs in CMOS RFICs is their use in control of oscillators that must be changed with precise frequency steps, and to do this

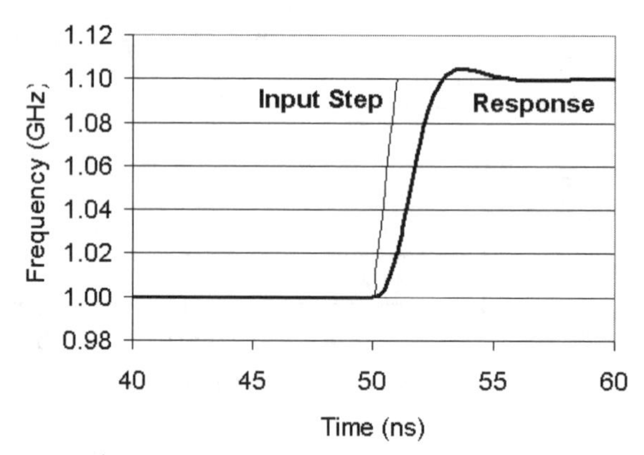

Figure 8.24 Variations in loop filter time constants can cause overshoot in the PLL response. Overdamping of the response is also possible.

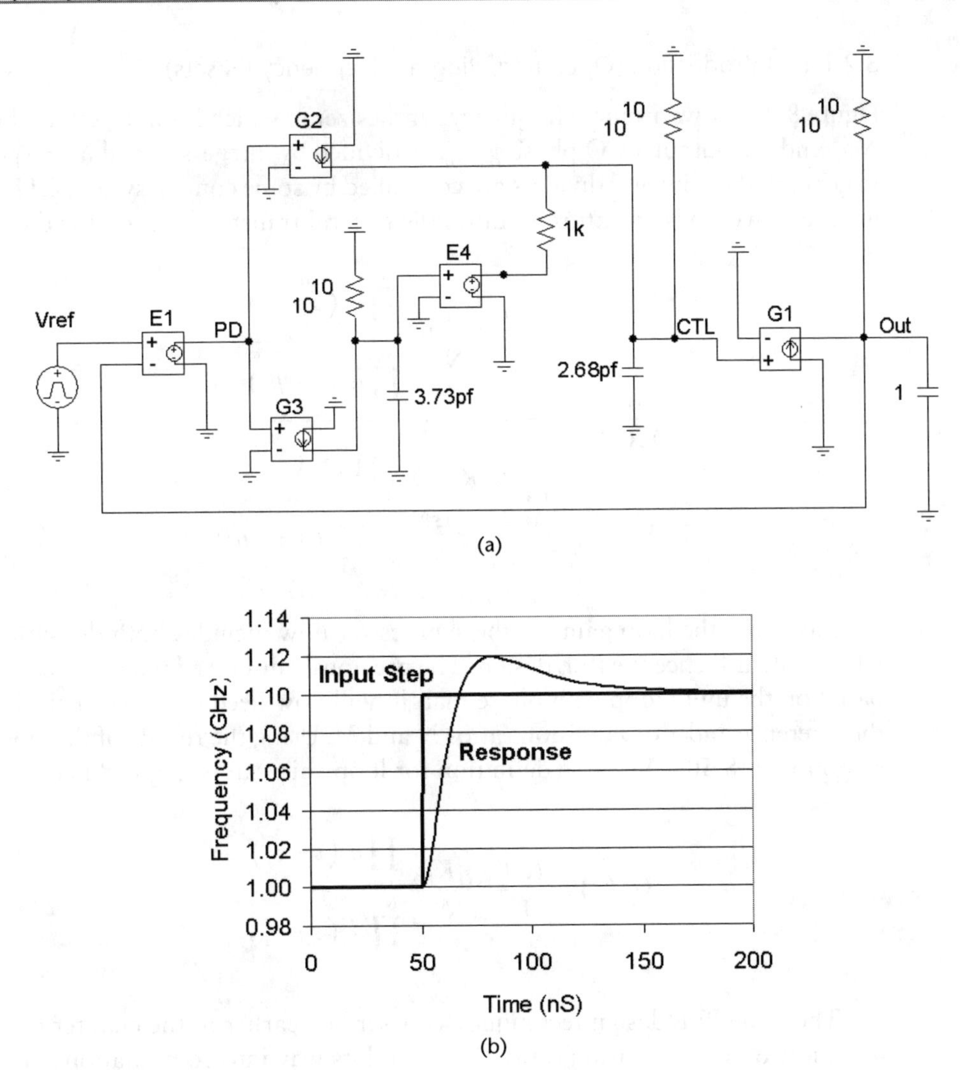

Figure 8.25 Simulation of the PLL response: (a) PLL behavioral model and (b) VCO output response time (ch8-2.txt).

change in a stable and accurate manner. These systems, termed *frequency synthesizers,* require the use of one or more PLLs for channel selection. For example, in GSM applications, the channel spacing is 200 kHz with a nominal carrier frequency of 900 MHz or higher. For frequency hopping systems, the control system varies the channel frequency using a pseudorandom sequence that provides greater security than more deterministic frequency control schemes. These are but a few examples of the many roles available for frequency synthesizers; all, however, require that the frequency change be done in a stable fashion, that the synthesizer reach its final frequency in a reasonable amount of time and with minimal ringing, and that the synthesizer be low noise. A number of excellent books are available that outline in extensive detail frequency synthesizer design; the following discussion is meant to present design fundamentals and to help provide the background for more complex and higher performance designs.

8.2.1.1 Introduction (Operation, Ringing, Frequency Offsets)

Figure 8.26 shows a simple frequency synthesizer in which both the input phase $\phi_{in} = N\phi_R$ and the output VCO phase ϕ_{VCO} are divided by integers N and M, respectively. In general, these integer dividers are controlled by some control system CTL. Including these two division ratios modifies the general transfer function for the system:

$$\frac{\phi_{VCO}(s)}{\phi_{in}(s)} = \frac{\dfrac{K_{VCO}K_{PD}}{sN}\dfrac{\prod_{i=1}^{N_z} n_i(s+z_i)}{\prod_{i=1}^{N_p} d_i(s+p_i)}}{1+\dfrac{N}{M}\dfrac{K_{VCO}K_{PD}}{sN}\dfrac{\prod_{i=1}^{N_z} n_i(s+z_i)}{\prod_{i=1}^{N_p} d_i(s+p_i)}} \tag{8.50}$$

Note that the loop gain for the synthesizer now includes both division factors, which will influence the PLL design. The previously discussed pole-zero calculations based on the unity loop gain phase margin will now need to consider the effects of the reference and VCO division ratios N and M. Using the results of the closed loop response in (8.50), the same definition for loop gain can be applied here:

$$G_L(s) = \frac{N}{M}\frac{K_{VCO}K_{PD}}{sN}\frac{\prod_{i=1}^{N_z} n_i(s+z_i)}{\prod_{i=1}^{N_p} d_i(s+p_i)} \tag{8.51}$$

The same PLL design techniques as described earlier in the chapter can still be used; however, the division ratio M will find its way into computations of the loop filter components since the poles and zeros will now be a function of M. Note that a large feedback loop division factor M will reduce the effective loop gain.

The VCO output frequency is also a function a both division factors as well as the input reference frequency:

$$f_{vco} = \frac{M}{N} f_{in} \tag{8.52}$$

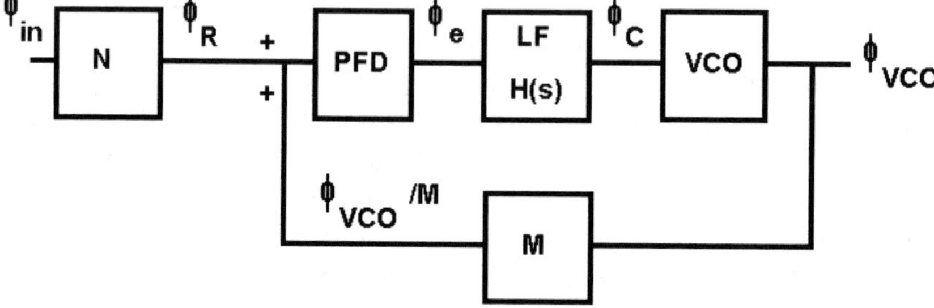

Figure 8.26 General block diagram for a frequency synthesizer.

As an example, if $f_{in} = 1$ MHz and $N = 5$, then the VCO output frequency can be increased by 200 kHz by incrementing M to $M + 1$. If a synthesizer were needed to generate fifty 200-kHz channels over the range of 900 to 910 MHz, then M would need to vary between 4,500 and 4,550, which can be expressed as $f_{VCO} = (4{,}500 + \Delta M) \cdot 2 \cdot 10^5$. In general, the VCO output frequency can be written as:

$$f_{vco} = (M_{\min} + \Delta M)\frac{f_{\text{in}}}{N};\ 0 < \Delta M < M_{\max} - M_{\min} \tag{8.53}$$

where M_{max} and M_{min} are the maximum and minimum division ratios. This is the general operational concept as discussed in Motorola Application Note AN535. In typical applications, the VCO output frequency is in the gigahertz range, so the division factor M may be implemented by several division stages (Figure 8.26). A high-speed, divide-by-2 or divide-by-4 circuit, or *prescalar,* may first be employed to both provide the required division and also shape the sinusoidal VCO output. Slower speed programmable digital circuitry can then be employed for the remaining frequency division blocks. In terms of PLL loop dynamics, a ΔM step change in division causes a PLL response similar to a step response $u(t)$ at the input, with the same issues of loop stability and settling time as described earlier in this chapter. The PLL loop dynamics can cause the VCO output frequency to vary about its final value during the settling period, with consequences such as spectral overlap of adjacent channels, resulting in interference in both the desired channel as well as the adjacent channels.

The output noise from the synthesizer is similar to that of its PLL except for its dependence on the input noise:

$$\phi_{\text{out}} = \frac{M}{N}\phi_{\text{in}} = M\frac{\phi_{\text{in}}}{N} = M\phi_R \tag{8.54}$$

Note that any noise signal on the input reference is *multiplied* by the feedback loop division factor M. For this reason, low-noise input reference signals are an absolute necessity for low-noise frequency synthesizers.

Equation (8.4) showed that the error voltage out of the mixer-type phase detector was a function of the input reference and VCO phases:

$$V_e(T) = \frac{1}{2}\alpha B V_{R0} V_{V0}\left[\sin[\phi_R(t) + B\phi_{VCO}(t)] + \sin[\phi_R(t) - B\phi_{VCO}(t)]\right] \tag{8.55}$$

The assumption was made that the first phase term (sum) was completely removed by filtering. The PLL loop filter is not ideal, however, and a portion of the sum signal does make its way onto the VCO control line. Near phase lock, the two phases are nearly identical $[\phi_R(s) \approx B\phi_{VCO}(s)]$, so the sum term becomes nearly $2\phi_R(s)$. Effectively, this puts energy from the second harmonic of the reference signal (or higher orders in the case of square-wave signals) on the VCO control line, which appears as a relatively low amplitude ripple signal, causing frequency modulation of the VCO output frequency with spurious signals at $Bf_{ref} \pm 2f_{ref}$, $Bf_{ref} \pm 4f_{ref}, \ldots$. These *reference sidebands* or *reference spurs* appear on either side of the main VCO carrier signal. Higher order loop filters can be used to further attenuate these sidebands but at the cost of increased PLL complexity and more detailed analyses to ensure loop stability. In many cases, one or more simple RC single-pole filter stages are added to the loop filter to improve the high-frequency roll-off [20]. An alternative to higher order filters is careful selection of the input reference signal so that the sidebands appear at the far edges of the channel bandwidth [21]. For digital PLLs such as XOR and DFF types, these reference spurs actually appear at $Bf_{ref} \pm f_{ref}$, $Bf_{ref} \pm 2f_{ref}$,... because there are small pulses at the reference signal period even during lock as the PLL tries to maintain its locked state [18, 22].

8.2.1.2 Division-Based Synthesizers

The frequency or channel control in the synthesizer is implemented using digital divider chains in the PLL feedback loop, which can be implemented using standard digital division techniques (see, for example, [4]). A number of division schemes are employed to perform this in an efficient manner since a single divide-by-M stage may not be the best method. In the previous example, a divide-by-4,500 counter was required as the base divisor for a 900-MHz synthesizer. This division may be implemented by several counters of counts N and P, for example. If the channel spacing is called the variable k, then the output frequency may be written as

$$f_{vco} = (NP + k)f_{ref} \tag{8.56}$$

Adding a convenient form of zero to this expression and doing some algebra yields the following result:

$$f_{vco} = (NP + k + kN - kN)f_{ref} = \left[(N+1)k + N(P-k)\right]f_{ref} \tag{8.57}$$

Note that this expression employs what is termed a *dual-modulus counter* (divide by N or $N + 1$ is often written as $N + 1/N$) as part of the feedback loop division chain and includes a counter for k (also sometimes called the *swallow counter*) and a *programmable counter* P–k. The dual-modulus counter counts up to $N + 1$ k times, which then sends a signal to the counter to change to an N counter, which then counts up P–k times. The process then starts over (Figure 8.27) [21]. A variation on this type of divider-based frequency synthesizer is the fractional N divider where the divide ratio can be an integer. Here, one can now use a *higher* reference frequency to control a smaller ($1/N$) frequency (such as the channel spacing). The fractional N divider has lower phase noise and increased bandwidth, allowing for faster settling times (higher speeds in changing frequency). Spurious signals, however, can be a problem in these circuits [22].

Variations on the dual-modulus counters in frequency synthesizers are found throughout the literature (see, for example, [10, 23, 24]). The VCO output does not have to be used directly but rather can be heterodyned to the desired operating frequency. As an example of this process, consider the frequency synthesizer for a 5.0-GHz, 802.11a wireless LAN with a 20-MHz channel bandwidth and a total bandwidth of 300 MHz. The input reference frequency to the synthesizer is 8 MHz with a VCO output of 4 GHz. The feedback loop consists of a dual-modulus divide-by-16/17 counter and a divide-by-32 counter (Figure 8.28). The VCO is further divided by a factor of 4 to provide a 1-GHz output signal. The two output

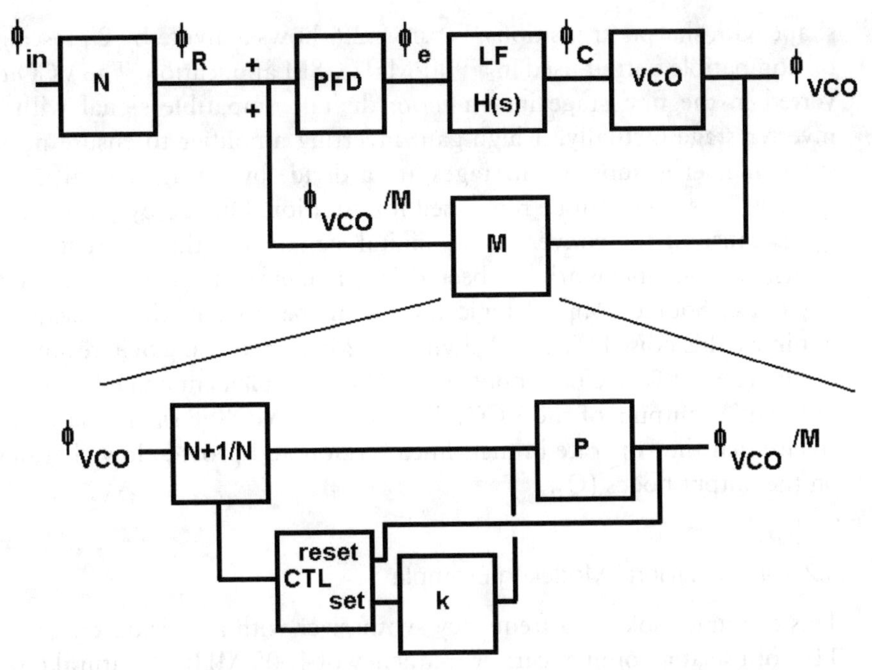

Figure 8.27 Feedback division techniques using dual-modulus prescalar and program and swallow counters. (*After:* [21].)

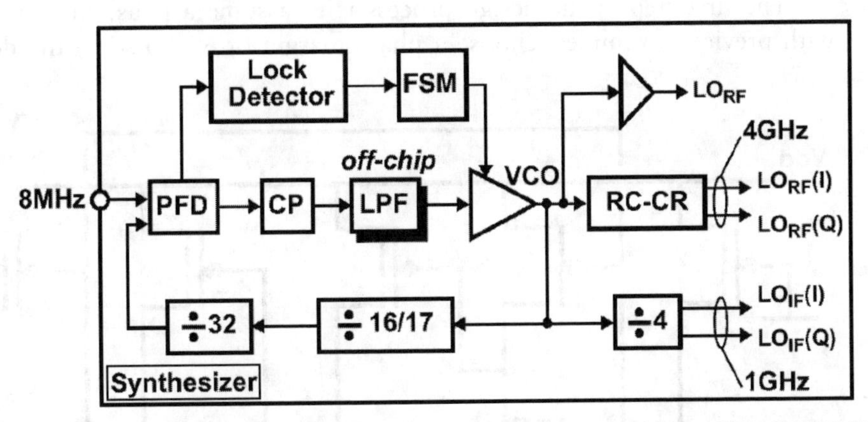

Figure 8.28 Frequency synthesizer for an 802.11a WLAN transceiver. (*From:* [10]. © 2002 IEEE.)

signals are then sent to a mixer stage downstream to be upconverted to the required 5-GHz frequency band. Because the 8-MHz reference controls both the VCO output and its divided counterpart (/4), the final output frequency can be controlled in 10-MHz steps.

8.2.1.3 High-Frequency Prescaling

The mixer types described in Chapter 7 can be used if the application only requires frequency translation. If the ultimate goal is to have a digitally compatible signal, other circuits may need to shape the sinusoidal VCO output to a more square-wave shape with sharper transitions. Figure 8.29 shows a divide-by-2 prescalar with digital compatible output used in a 900-MHz GSM application. The VCO input is converted in the first stage into a more digital-compatible signal with a high-gain inverter stage (actually, a high-gain inverting amplifier to ensure near rail-to-rail switching). The subsequent stages are a divide-by-2 stage to halve the VCO frequency. This stage can be replicated for additional divide-by-2 operations.

Because of the conversion to digital signals and the inherent switching transients, considerable work has been done in an effort to reduce the resulting switching noise. Source coupled logic (SCL) can be used in the prescaling circuits to minimize this noise [25, 26]. A divide-by-2 prescalar using SCL techniques is shown in Figure 8.30. The clock input (*clk* and its complement $\overline{clk}$) are driven by the two differential outputs of the VCO. The pull-up *p*MOSFETs are biased to ground to ensure that they operate in their linear region and provide low charging resistance on the output nodes (Q).

8.2.1.4 Behavioral Modeling Example

This example looks at a frequency synthesizer with a division range of $100 < M < 110$ for use at a nominal carrier frequency of 1000 MHz. A natural loop resonance frequency of 10^7 rps is desired and a 100-μA charge pump loop filter based on Figure 8.19(b) has been chosen. The loop filter components will be designed and the behavioral model response will be presented. Assume $K_{VCO} = 10^9$ Hz/V.

The first step in the design process is to assume a phase margin. In keeping with previous examples, choose a phase margin of 60° to see if the design yields

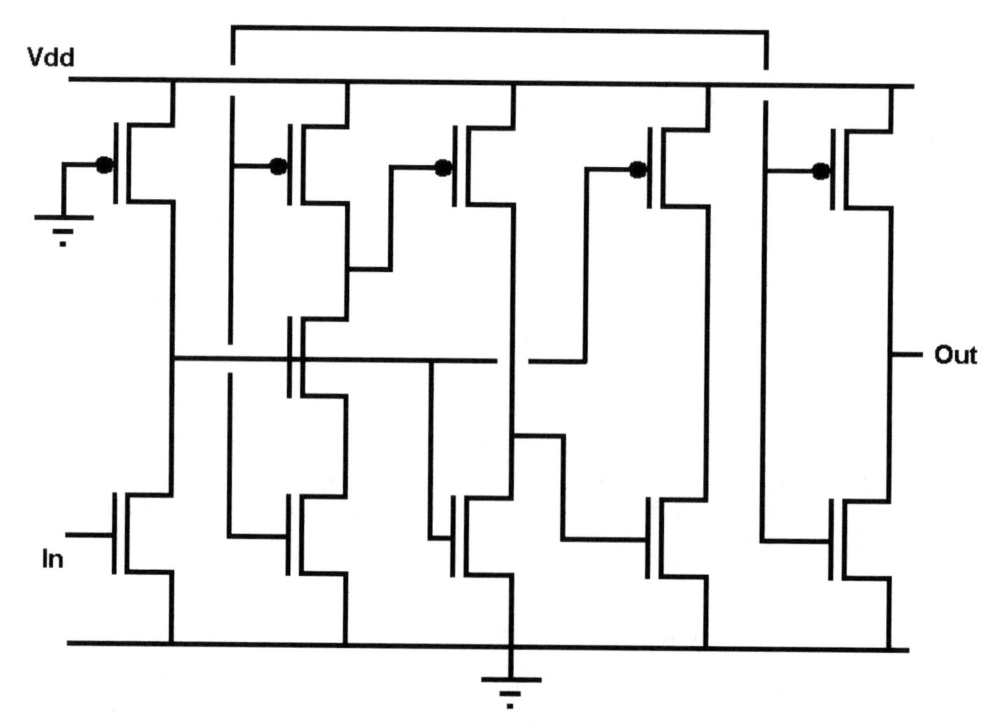

Figure 8.29 Divide-by-2 CMOS prescalar with digital compatible output. (*After:* [2].)

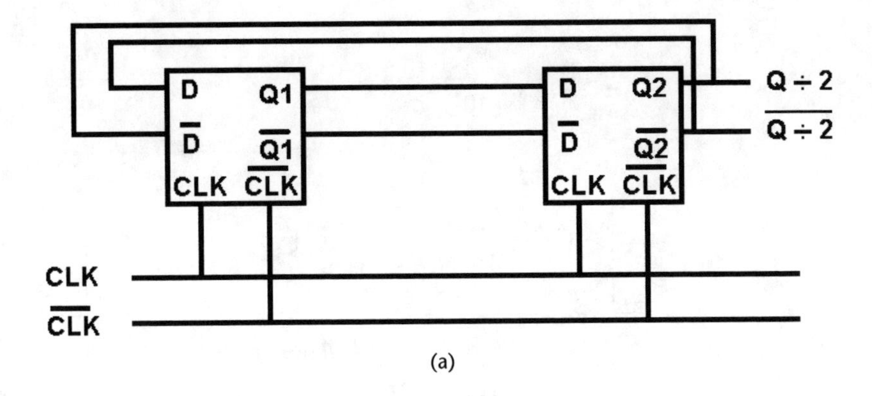

(a)

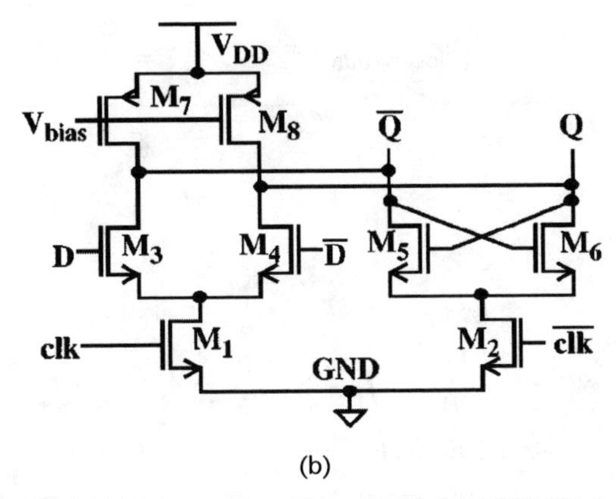

(b)

Figure 8.30 (a) Divide-by-2 CMOS prescalar. (*After:* [26].) (b) MOSFET circuit schematic for the D flip flops using SCL to reduce switching noise. (*From:* [26]. © 2002 IEEE.)

reasonable component values. The resulting pole and zero locations are found to be as follows:

$$p_1 = 3.73\omega_0;\ z_1 = 0.27\omega_0 \tag{8.58}$$

From the unity gain expression, the capacitance C_1 can be computed. A feedback factor midway between the two division ranges ($M = 105$) is a good choice for this first pass in the design:

$$C_1 = \frac{K_{VCO} I_p}{M\omega_0 p_1} = 2.55\ \text{pF} \tag{8.59}$$

The components R and C_2 can then be found from their definitions in Table 8.2:

$$R = \frac{1}{C_1(p_1 - z_1)} = 11.3\ \text{k}\Omega;\ C_2 = \frac{1}{Rz_1} = 33\ \text{pF} \tag{8.60}$$

The results can then be introduced into a behavioral model for the synthesizer as shown in Figure 8.31(a). In this figure, the feedback loop has two branches; the

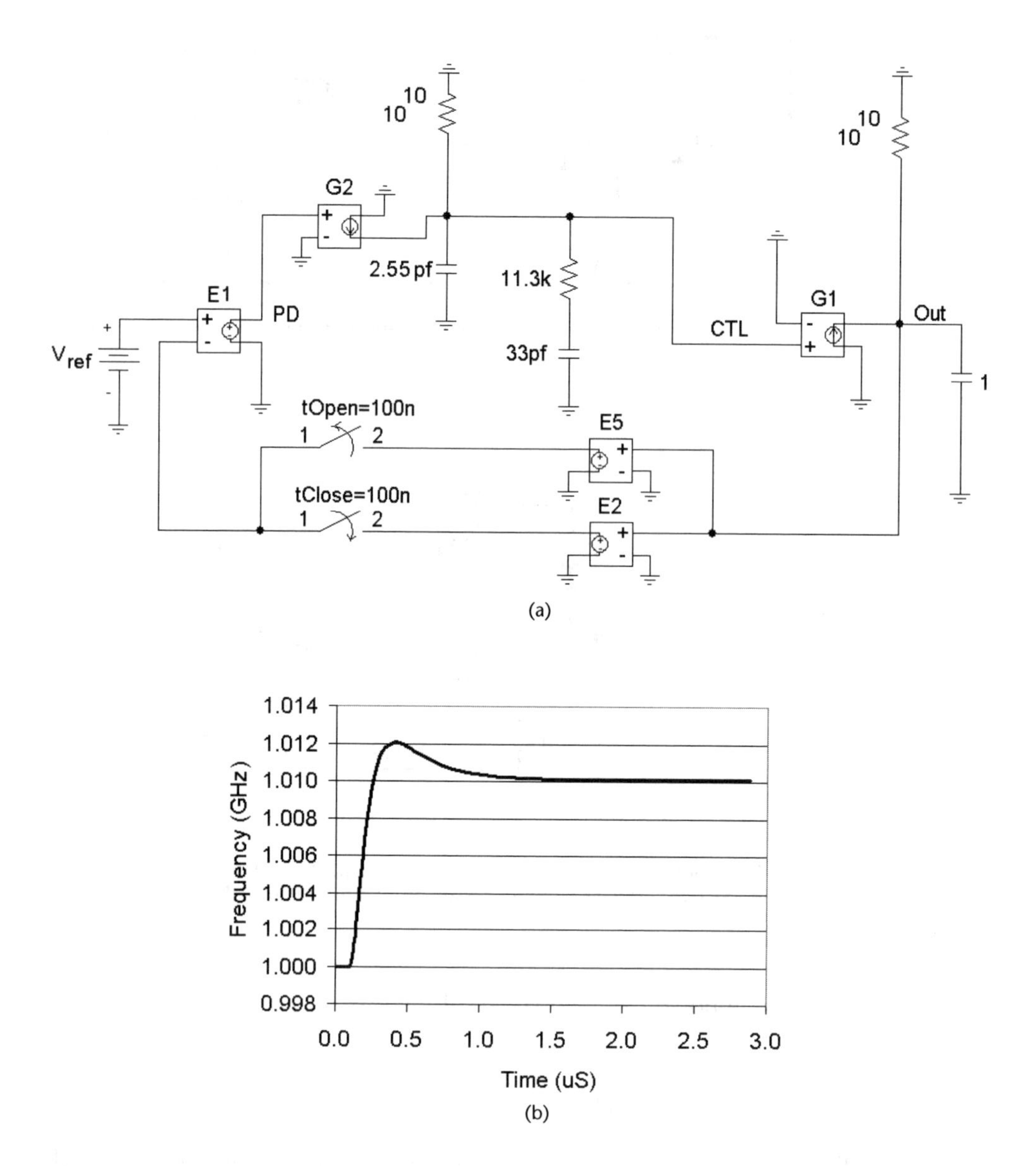

Figure 8.31 Frequency synthesizer behavioral response example: (a) behavioral model and (b) temporal response (ch8-3.txt).

switches are configured to switch between division ratios of 100 and 110 to observe the switching response and settling time over the maximum range of the synthesizer. The switching occurs at time instant 100 ns and shows an approximately 1.25-μs settling time, consistent with the resonance frequency of 10^7 rps chosen for the design.

8.2.2 Direct Digital Synthesis

The advent of high-speed CMOS digital circuitry is currently being exploited to provide highly accurate high-frequency sinusoidal signals directly from digital circuitry instead of using lumped element-based LC oscillators or ring oscillators. This technique is referred to as *direct digital synthesis* (DDS). This section covers the basics of DDS operation, looking at both the conceptually simple *direct waveform synthesis* technique and the more powerful *phase accumulation* technique.

8.2.2.1 Introduction and Basic Operation: Direct Waveform Synthesis

The concept behind the DDS is actually quite straightforward. A digital read-only memory (ROM) or look-up table is programmed with the well-known coefficients of the sine wave (Figure 8.32). Each one of these sine-wave coefficients is then sequentially addressed with a counter clocked at some frequency f_{CLK}, and fed to a DAC to generate the analog output. If there are N sine-wave coefficients, the counter starts at zero and sequentially counts up to $N - 1$, when the counter resets and starts the process again. The frequency of the output sine wave is then directly related to both the clock frequency and the number of bits used to define the square wave, N:

$$f_{\text{out}} = \frac{f_{CLK}}{N} \tag{8.61}$$

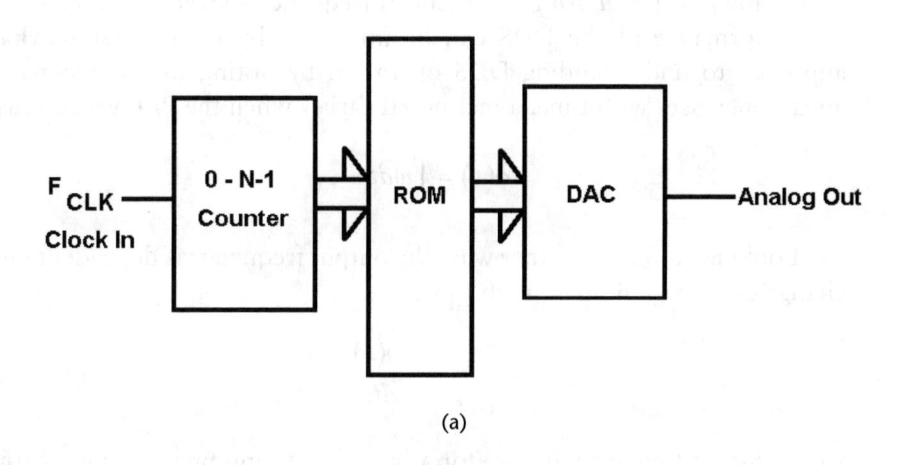

(a)

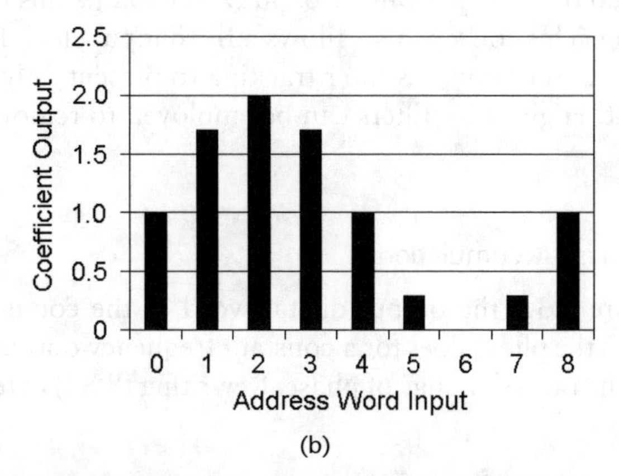

(b)

Figure 8.32 Direct digital synthesis: (a) waveform synthesis block diagram and (b) sine-wave coefficient output versus address word for a 3-bit DDS.

Standard digital design techniques for the counter, ROM, and DAC can be found in the literature for CMOS technology (see, for example, [4, 11]). Any periodic waveform can be synthesized using this technique; the ROM is programmed with the desired waveform coefficients. For waveforms with natural symmetry such as the sine wave, only one-quarter of the actual coefficients need be stored in the ROM; the remaining 270° of signal can be formed by addressing the PAT in the appropriate manner.

Some advantages for DDS are readily apparent. The output signal is easily changed by simply changing the clock rate; this clock signal can be easily controlled by a system control processor. The DAC output is fixed, so that amplitude of the sine-wave output will always be the same regardless of frequency (no AGC or AAC required). There is no need to do a complex loop analysis to determine stability and phase margins since the DDS system is always stable [27]. A number of limitations of this technique are also immediately obvious: The more accurate the sine-wave representation (large N), the lower the output frequency for a given clock rate. A high-speed DAC is required if a high-frequency output is required. The quantized nature of the resulting sine wave requires filtering to be employed to remove the switching signal artifacts. Finally, with a fixed number of samples per output cycle, the clock frequency changes with each requested frequency change, requiring the output lowpass (or *aliasing*) filter cutoff frequency to track as well.

Performance of the DDS can be improved by taking a somewhat different approach to understanding DDS operation by noting that the signal phase $\phi(t)$ always increases with time, a fact noted earlier when the PLL was discussed:

$$\phi(t) = \int_0^t \omega dt + \phi_0 \tag{8.62}$$

Looking at (8.62) another way, the output frequency is dependent on the rate of change of the signal phase with time:

$$\omega = \frac{d\phi(t)}{dt} \tag{8.63}$$

so a constant frequency occurs for a linearly varying phase change with time. The use of digital or *phase accumulation* in DDS exploits this fact and is discussed in the next section. This concept also allows a fixed input clock frequency to be used, simplifying the output lowpass filter tracking to present only a single cutoff frequency for the unit. High-order filters can be employed to removing much of the clocking artifacts.

8.2.2.2 Phase Accumulation

From Figure 8.32, the output digital word of the counter linearly increases with time, just as the phase does for a constant frequency output. The output frequency is related to the rate of change of phase. Rewriting (8.63) in terms of discrete quantities yields

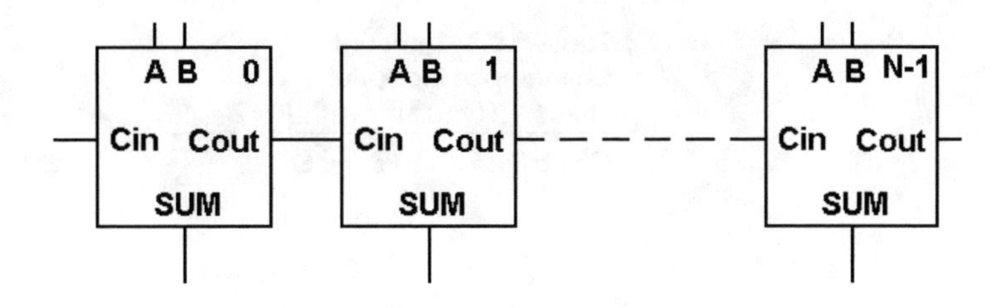

C=Cin

$\text{SUM} = ABC + A\bar{B}\bar{C} + \bar{A}\bar{B}C + \bar{A}B\bar{C}$

$\text{Cout} = AB + AC + BC$

Figure 8.33 High-speed phase accumulator using CMOS full adder technology. The phase accumulation increment is set by the adder inputs. (*After:* [28].)

$$\omega_{\text{out}} = \frac{d\phi(t)}{dt}; f_{\text{out}} = \frac{\Delta\phi}{2\pi}\frac{1}{T_s} \tag{8.64}$$

where T_s is the sampling rate. The term $\Delta\phi/2\pi$ represents the fractional change in phase during a single sampling period T_s. If each sampling period is composed of 2^N possible phase states (N bits) and the term P is that fraction required for the output, then the output frequency is related to the sampling frequency f_s (= $1/T_s$) as

$$f_{\text{out}} = \frac{P}{2^N} f_s \tag{8.65}$$

The change in f_{out} with respect to the fraction change P provides the smallest phase tuning step available:

$$\frac{df_{\text{out}}}{dP} = \frac{f_s}{2^N} \tag{8.66}$$

The code word P describes the number of phase increments to take over any 2π period. This operation takes place in a circuit block called a *phase accumulator* (PA). The PA can be thought of as an integer mathematical integrator:

$$\phi_n = \phi_{n-1} + P\Delta\phi \tag{8.67}$$

As an example of PA operation, suppose a DDS uses a 4-bit PA. For this PA there is a phase resolution of $2\pi/2^4$ or $\pi/8$. If the control word $P = 2$ is input to the PA, this indicates to the PA to jump two phase increments per input clock pulse. Table 8.3 shows the output phase for this control word $P = 2$ for each clock pulse.

A total of eight clock pulses is required to effect a phase change of 2π, providing an output frequency of

$$f_{\text{out}} = \frac{2}{16} f_s = \frac{1}{8} f_s \tag{8.68}$$

Table 8.3 Amount of Phase Change per Clock Pulse in a 4-Bit Phase Accumulator with $P = 2$

Clock Pulse	*Phase, Θ*
1	$2\pi/8$
2	$4\pi/8$
3	$6\pi/8$
4	$8\pi/8$
5	$10\pi/8$
6	$12\pi/8$
7	$14\pi/8$
8	$16\pi/8$

which is also the result of (8.65). Suppose the control word was doubled to $P = 4$; Table 8.4 shows the output phase change during each clock pulse.

This example shows that it takes only four clock pulses to effect a phase change of 2π, effectively doubling the output frequency from the $P = 2$ code word:

$$f_{out} = \frac{4}{16} f_s = \frac{1}{4} f_s \tag{8.69}$$

In both cases, the same clock is used but there are more phase samples defining the lower frequency signal than the higher one. In essence, the higher the value of P, the quicker the rate at which the table is run through with a corresponding higher output frequency. Typical PAs used in modern DDS systems contain 24, 32, or even 64 bits. For a 24-bit PA, a phase resolution of one part in 2^{24} or 16 million will result.

High-speed PAs can take a variety of forms. One such implementation uses a cascade of N full ripple carry digital adders (N-bit PA) with the sum outputs carried back to the adder inputs for accumulation as shown in Figure 8.33 [28]. Also shown in the figure are logic equations for the sum and carry outputs of the full adder; standard CMOS logic can be used to implement this function [4].

The PA is the followed by a *phase-to-amplitude translator* or *generator* (PAT). This PAT is essentially a look-up table containing coefficients of the sine wave or other periodic waveform that are addressed by the PA:

$$\text{PAT} = \frac{1}{2}(1 + \sin\Theta) \tag{8.70}$$

Table 8.4 Amount of Phase Change per Clock Pulse in a 4-Bit Phase Accumulator with $P = 2$

Clock Pulse	*Phase, Θ*
1	$4\pi/8$
2	$8\pi/8$
3	$12\pi/8$
4	$16\pi/8$

For the earlier $P = 4$ example, the PAT has an encoded output of 1, 0.5, 0, and 1. Providing 2^N memory capacity is problematic in terms of the required silicon real estate and access speed, and so a truncated version using only the M most significant bits of the PA is employed in the PAT. A number of compression techniques are available for sine-wave synthesis in DDS (Sunderland, Taylor series, CORIDC) used to reduce the memory requirements (for a review, see [29]).

These coefficients are then sent to a D-bit DAC to provide a sampled voltage or current output waveform. The dynamic range for the DAC is a function of the number of bits [11]:

$$\text{DR} = 6.02\text{D dB} \tag{8.71}$$

For a DAC with a 0 to V_{max} output voltage range, the $P = 4$ example yields an output signal with samples V_{max}, $0.5V_{max}$, 0, and V_{max} as shown in Figure 8.34. An aliasing filter for suppressing clock frequency energy at the output completes the design. This alias filter can be designed with a high filter order since the filter must only remove the constant clock frequency.

The various bit truncations and DAC operations, however, introduce quantization and other noise sources and spurious signals into the DDS output that are not removed by the aliasing filter. The maximum level of the spurious signals below the carrier signal created by this bit truncation in the PAT has been found to be [30]:

$$\text{SPUR}_{max} = -6.02M + 3.92 \text{ dBc} \tag{8.72}$$

The amplitude quantization noise from the DAC is the more dominant of the truncation noise sources. The DAC SNR is a function of the number of bits D [11]:

$$\text{SNR} = 6.02\text{D} + 1.76 \text{ dB} \tag{8.73}$$

An increase in the number of bits improves the SNR but at the expense of DAC complexity. The spurious noise and signals generated by the D-bit DAC can be significantly reduced by replacing this stage with a combination single-bit delta sigma ($\Delta\Sigma$) modulator and single-bit DAC [31].

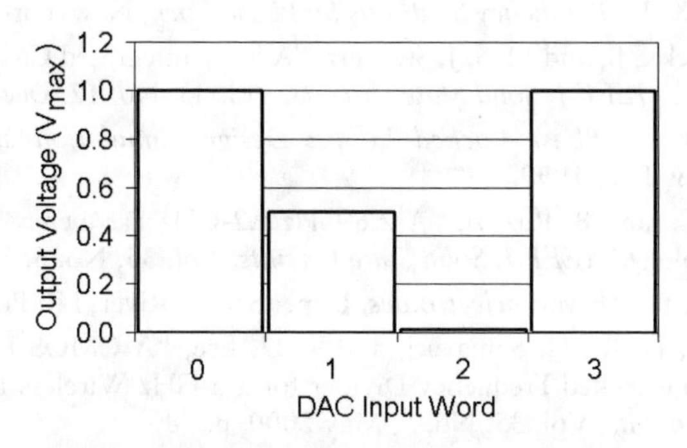

Figure 8.34 The DAC output is a step output waveform.

References

[1] Johansson, H. O., "A Simple Precharged CMOS Phase Frequency Detector," *IEEE. J. Solid State Circuits*, Vol. 33, No. 2, Feb. 1998, p. 2951.

[2] Yan, W., and H. C. Luong, "A 2-V 900-MHz Monolithic CMOS Dual-Loop Frequency Synthesizer for GSM Receivers," *IEEE J. Solid State Circuits*, Vol. 36, No. 2, Feb. 2001, p. 204.

[3] Haskard, M., and I. May, *Analog VLSI: nMOS and CMOS,* Australia: Prentice Hall, 1988.

[4] Weste, N., and K. Eshraghian, *Principles of CMOS VLSI Design: A Systems Perspective,* Reading, MA: Addison Wesley, 1993.

[5] Dayaranta, L., "Theory and Design of Phase Lock Loops," *IEEE 2004 Int. Microwave Symp. Workshop Notes*, 2004.

[6] Seng, L. P., et al., "Design of 2.5V, 900 MHz phase lock loop (PLL) using 0.25 um TSMC CMOS Technology," *Proc. 2004 ICSE*, 2004, p. 431.

[7] Mansuri, M., D. Liu, and C.-K. K. Yang, "Fast Frequency Acquisition Phase-Frequency Detectors for GSamples/s Phase-Locked Loops," *IEEE. J. Solid State Circuits*, Vol. 37, No. 10, Oct. 2002, p. 1331.

[8] Kroupa, V. F., "Noise Properties of PLL Systems," *IEEE Trans. Communications*, Vol. COM-30, Oct. 1982, p. 2244.

[9] O'Leary, M., "Practical Approach Augurs PLL Noise in RF Synthesizers," *Microwaves and RF*, Sept. 1987, p. 185.

[10] Zargari, M., et al., "A 5-GHz CMOS Transceiver for IEEE 802.11a Wireless LAN Systems," *IEEE. J. Solid State Circuits,* Vol. 37, 12, Dec. 2002, p. 1688.

[11] Allen, P., and D. Holberg, *CMOS Analog Circuit Design,* 2nd ed., New York: Oxford University Press, 2002.

[12] Johns, D., and K. Martin, *Analog Integrated Circuit Design,* New York: John Wiley and Sons, 1996.

[13] Gray, P. R., and R. G. Meyer, "MOS Operational Amplifier Design-A Tutorial Overview," *J. Solid State Circuits*, Vol. SC-17, No. 6, June 1981, p. 969.

[14] Wu, J. T., and K.-L. Chang, "MOS Charge Pumps for Low-Voltage Operation," *IEEE. J. Solid State Circuits*, Vol. 33, No. 4, April 1998, p. 592.

[15] Leung, G. C. T., and H. C. Luong, "A 1-V 5.2-GHz CMOS Synthesizer for WLAN Applications," *IEEE J. Solid State Circuits*, Vol. 39, No. 11, Nov. 2004, p. 1873.

[16] Koo, Y., et al., "A Fully Integrated CMOS Frequency Synthesizer with Charge-Averaging Charge Pump and Dual-Path Loop Filter for PCS- and Cellular-CDMA Wireless Systems," *IEEE J. Solid State Circuits,* Vol. 37, No. 5, May 2002, p. 536.

[17] Zhang, B., P. E. Allen, and J. Huard, "A Fast Switching PLL Frequency Synthesizer with an On-Chip Passive Discrete-Time Loop Filter in 0.25-μm CMOS," *IEEE J. Solid State Circuits*, Vol. 38, No. 6, June 2003, p. 855.

[18] Egan, W. F., *Frequency Synthesis by Phase Lock,* New York: John Wiley and Sons, 1981.

[19] Craninckx, J., and M. S. J. Steyaert, "A Fully Integrated CMOS DCS-1800 Frequency Synthesizer," *IEEE J. Solid State Circuits*, Vol. 33, No. 12, Dec. 1998, p. 2054.

[20] Best, R. E., *Phase Locked Loops: Design, Simulation and Applications,* New York: McGraw-Hill, 1999.

[21] Lam, C., and B. Razavi, "A 2.6-GHz/5.2-GHz Frequency Synthesizer in 0.4-μm CMOS Technology," *IEEE J. Solid State Circuits, Vol.* 35, No. 5, May 2000, p. 788.

[22] Razavi, B., *RF Microelectronics,* Upper Saddle River, NJ: Prentice Hall, 1998.

[23] Rategh, H. R., H. Samavati, and T. H. Lee, "A CMOS Frequency Synthesizer with an Injection-Locked Frequency Divider for a 5-GHz Wireless LAN Receiver," *IEEE J. Solid State Circuits*, Vol. 35, No. 5, May 2000, p. 780.

[24] Maeda, T., et al., "Low-Power-Consumption Direct-Conversion CMOS Transceiver for Multi-Standard 5-GHz Wireless LAN Systems with Channel Bandwidths of 5–20 MHz," *IEEE J. Solid State Circuits*, Vol. 41, No. 2, Feb. 2006, p. 375.

[25] Ware, K., et al., "A 200-MHz CMOS Phase-Locked Loop with Dual Phase Detectors," *IEEE J. Solid State Circuits*, Vol. 24, Dec. 1989, p. 1560.

[26] Hung, C.-M., and K. K. O, "A Fully Integrated 1.5-V 5.5-GHz CMOS Phase-Locked Loop" *IEEE J. Solid State Circuits*, Vol. 37, No. 4, April 2002, p. 521.

[27] Cordesses, L., "Direct Digital Synthesis: A Tool for Periodic Waveform Generation (Part 1)," *IEEE Signal Processing Magazine*, July 2004, p. 50.

[28] Turner, S. E., and D. E. Kotecki, "Direct Digital Synthesizer with ROM-Less Architecture at 13-GHz Clock Frequency in InP DHBT Technology," *IEEE Microwave Wireless Comp. Lett.*, Vol. 16, No. 5, May 2006, p. 296.

[29] Essenwanger, K., and V. Reinhardt, "Sine Output DDSs: A Survey of the State of the Art," *Proc. 1998 IEEE Int. Frequency Control Symp.*, 1998, p. 370.

[30] O'Leary, P., and F. Maloberti, "A Direct-Digital Synthesizer with Improved Spectral Performance," *IEEE Trans. Commun.*, Vol. 39, No. 7, July 1991, p. 1046.

[31] Lindeberg, J., et al., "A 1.5-V Direct Digital Synthesizer with Tunable Delta-Sigma Modulator in 0.13-μm CMOS," *IEEE J. Solid State Circuits*, Vol. 40, No. 9, Sept. 2005, p. 1978.

Selected Bibliography

Elliott, K. R., "Direct Digital Synthesis for Enabling Next Generation RF Systems," *Digest 2005 IEEE CSIC Symp.*, Nov. 2005, p. 125.

"Introduction to Direct Digital Synthesis," Intel Application Note 101, June 1991.

Murphy, E., and C. Slattery, "All About Direct Digital Synthesis," Analog Dialogue 38-08, Analog Devices, Aug. 2004 (http://www.analog.com/analogdialogue).

Razavi, B. (Ed.), *Monolithic Phase Locked Loops and Clock Recovery Circuits: Theory and Design,* Piscataway, NJ: IEEE Press, 1996.

CHAPTER 9

CMOS Power Amplifiers

As discussed in detail in Chapter 1 on communication link budgets, the RF power at the receiver terminals is a direct function of the transmitted power. The previous chapters have discussed many on-chip, low-power circuits that allow the generation or detection of wireless communications signals. These low-level signals must be amplified significantly before they are sent to the antenna for transmission. These *power amplifiers* must exhibit decent gain and be able to drive the impedance of an externally connected antenna (usually 50Ω). The PAs must also be thrifty with their power consumption, and convert as much of the dc power drawn into usable RF energy (i.e., *efficiency*). Depending on the transmitted modulation scheme, the amplifiers should exhibit good linearity to prevent odd-order distortion components from broadening the spectral width of the signal (so-called *spectral regrowth*). With the relatively low transconductance of CMOSs, physically large MOSFETs must be constructed with the accompanying large parasitic capacitance values and so the frequency response of the amplifiers becomes a concern.

This chapter presents a review of traditional power amplifiers, which are further divided into so-called *transconductance* amplifiers (Class A through C) and *high-efficiency* amplifiers (Classes E and F, with little mention of Class D).

9.1 Review of Amplifier Terms

A large number of design techniques have been advanced over the years for amplifier designs using discrete amplifying devices. The primary design techniques are based on measured *S*-parameter data available from the manufacturer, measured under specific bias conditions, to design suitable input and output matching networks (Z_{in} and Z_{out}; Figure 9.1) to achieve the desired gain, bandwidth, or other specific system requirement. A variation on this theme is for the designer to obtain discrete devices that may be useful for the application, measure the *S*-parameters using a custom test bed, and continue the design from that point.

Detailed design procedures based on *S*-parameter measurements can be found in numerous references (see, for example, [1, 2]). For the designer of RFIC power amplifiers, a somewhat different design approach is required because the PA devices have not been fabricated and, hence, the *S*-parameters are not available. The RFIC power amplifier design flow must look at the power desired from the amplifier, the amplifier loading, and the desired frequency response so that the active amplifying devices (here, MOSFETs) can be sized and biased accordingly. Design of matching and other passive circuits can then proceed from this point. The discussion in this

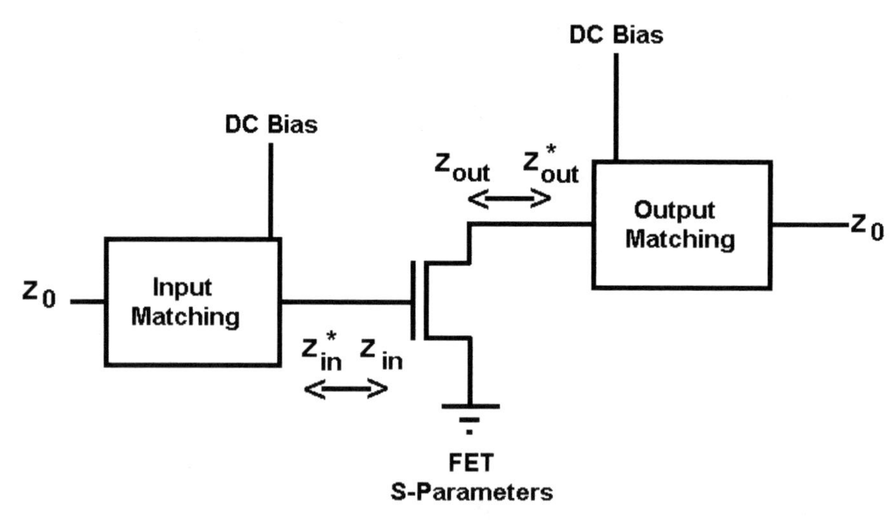

Figure 9.1 Input and output matching networks with conjugate match reflection coefficients.

chapter focuses on some amplifier basics and designs that are applicable to RFIC implementations.

9.1.1 Linear

For PAs, large signal inputs and even larger output signal levels often violate the small signal S-parameter assumptions used a low-power amplifier design. In fact, for PAs, the concept of power amplifier output conditions is often described as "loading the amplifier," rather than designing for a set of specific matching conditions [3]. Here the analysis becomes more problematic, so modifications to the discrete design techniques are required. For RFIC power amplifiers, the design process must include the fact that the transistor layout parameters (primarily sizing), and hence the S-parameters, are not known a priori.[1] However, a few specifications are common to all power amplifiers whether in discrete or integrated form. The first specification is the *power gain* provided by the amplifier, G:

$$G = \frac{P_{RF-\text{out}}}{P_{RF-\text{in}}} \tag{9.1}$$

This equation indicates a linear relationship between input and output power, which is valid up to the point where devices nonlinearities cause the output power to begin to level off or saturate; this is the *1-dB compression point* (P_{1dB}) (Figure 9.2). If a single amplifier stage is insufficient to provide the required output power, two or more amplifiers may be cascaded. In this case, the final amplifier stage is often called the *power amplifier* stage, whereas the preceding stages are often referred to as the *driver amplifiers* or simply the *drivers*.

Another key specification for a PA is how well the amplifier transforms the dc power supplied to the amplifier into useful RF energy. The metric used to describe

1. The results obtained on the test bench after fabrication are often the most important part of the power amplifier design process, regardless of whether discrete [5] or RFIC [4] amplifiers are being discussed.

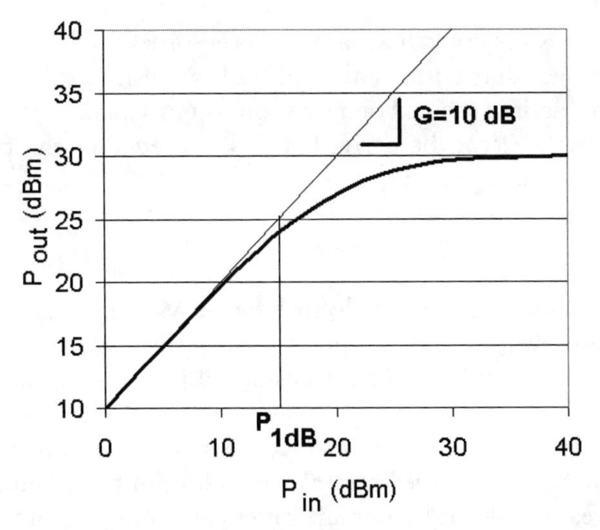

Figure 9.2 Ideal input-output power curve compared with compressed input-output power curve showing the 1-dB compression point.

this transformation is termed the *amplifier efficiency*, η, defined as the ratio of the RF power output with respect to the dc power input:

$$\eta = \frac{P_{RF-\text{out}}}{P_{DC-\text{in}}} \times 100\% \tag{9.2}$$

The amplifier efficiency described in (9.2) takes into account only the output RF power, irrespective of the input drive power. Another efficiency term, the *power added efficiency* (PAE), can be defined as the difference between the output and input RF power; the dc input power remains the same:

$$\eta_{PAE} = \frac{P_{RF-\text{out}} - P_{RF-\text{in}}}{P_{DC-\text{in}}} \times 100\% = \eta\left(1 - \frac{1}{G}\right) \tag{9.3}$$

where the RF amplifier gain G is defined as $P_{RF\text{-out}}/P_{RF\text{-in}}$. Equation 9.3 shows that the gain of the RF amplifier impacts the PAE. Because RF PA devices usually exhibit relatively low gains, significant differences between efficiency and PAE can occur. For power gains of 10 dB or more, η and η_{PAE} are within 10% of each other. For an RF amplifier with a gain of 10 and an efficiency of 40%, the PAE is 36%, a loss of 4% efficiency. If the RF amplifier efficiency is somewhat higher, say, $\eta = 60\%$, then the PAE is 54%, a more serious loss of efficiency of 6%. This effect worsens for lower RF amplifier gains and higher efficiencies [4].

For relatively high gain amplifiers, η and η_{PAE} are nearly the same. PAE will always be less than the initially defined efficiency, η.

9.1.2 Nonlinear

Even before the 1-dB compression point is reached, the amplifier will begin to generate harmonically related distortion components due to device or system nonlinearities. As noted in Chapter 1, even-order nonlinearities usually occur in frequency ranges outside the bandwidth of interest, but the odd-order terms can fall within this desired range. For a two-tone input signal, the third order in band distortion power level can be derived with knowledge of the fundamental power and third-order intercept point, IP3:

$$P_{3D} = 3P_F - 2 \cdot \mathrm{IP3} \tag{9.4}$$

where all quantities are in decibel form. Modern digital communications techniques, however, are not simply composed by a number of individual tones, but instead exhibit a band-limited transmitted power spectrum such as that shown in Figure 9.3.

The ideal PA will multiply all components within this bandwidth by an equal amount and preserve the spectral bandwidth of the original signal. As the power level increases, odd-order nonlinearities cause out-of-band or adjacent channel signal levels to rise. The nonlinear distortion power, occurring both above and below the original signal spectra, causes considerable (and undesirable) broadening of the original signal bandwidth into adjacent frequency bands; this spectral broadening is often termed *spectral regrowth* and is an acute problem in signals with time-varying envelopes such as QAM (Figure 9.4). The power in the adjacent channel(s), referred to as the adjacent channel power (ACP), is specified with respect to the power in the desired signal bandwidth as the adjacent channel power ratio (ACPR) and is measured in decibels below the desired carrier (dBc):

$$\mathrm{ACPR}_{upper} = 10\log \frac{\int_{F_0+F_{OFF}-BW/2}^{F_0+F_{OFF}+BW/2} S(f)df}{\int_{F_0-BW/2}^{F_0+BW/2} S(f)df} \,\mathrm{dBc} \tag{9.5}$$

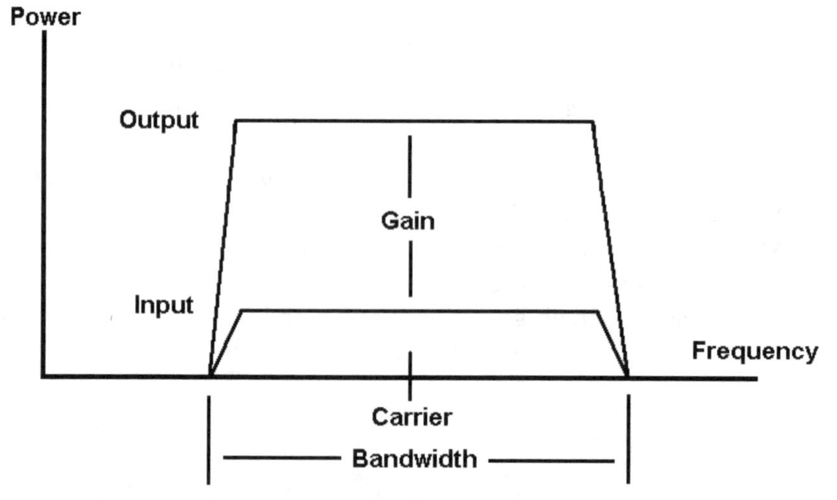

Figure 9.3 The ideal PA will increase the output power without changing the bandwidth of the input signal.

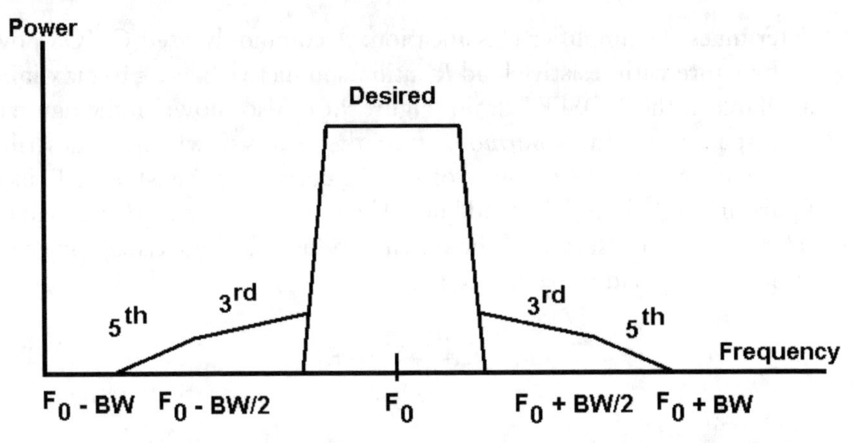

Figure 9.4 Odd-order intermodulation components cause an increase in the spectral bandwidth.

where F_0 is the carrier frequency and F_{OFF} is the offset frequency from the carrier where the measurement is taken.

The results of this measurement over a range of offset frequencies are often referred to as the *spectral emission mask* (see, for example, [6]). Low spectral regrowth is an important aspect in modern communication system design because to use minimum system bandwidth, channels are closely spaced and so any power leaking over from adjacent channels will cause an increase in adjacent channel interference (ACI), as shown in Figure 9.5. If the ACP is large, the interference may spread into a number of neighboring channels.

9.2 Transconductance Amplifiers

The first three major classes of amplifiers to be discussed are termed *transconductance amplifiers* because the output current swing is the result of an input voltage variation. For CMOS implementations, the output dc bias level

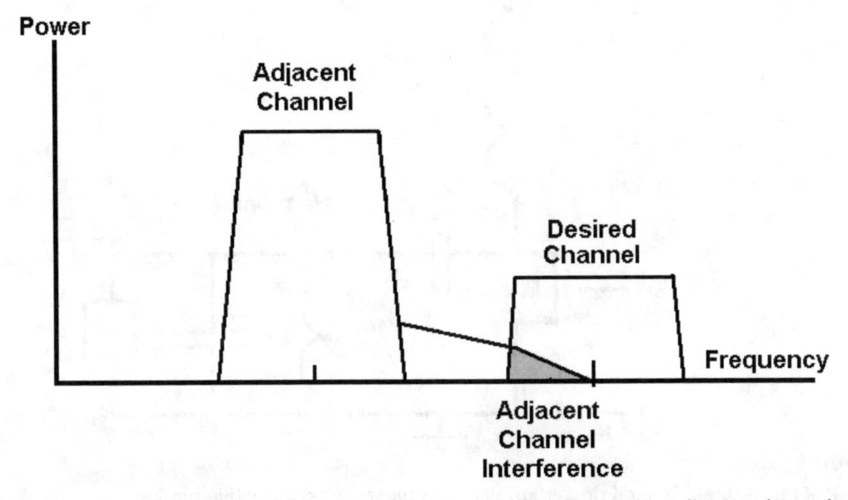

Figure 9.5 Spectral broadening in one channel can cause interference in adjacent channels.

determines the amplifier classification. A commonly used CMOS power amplifier architecture with resistive load R_L utilizes inductive biasing to maximize the voltage available at the MOSFET drain (Figure 9.6). Also shown in the figure is a dc blocking capacitor C_B and a *harmonic trap*, the details of which are described later.

Figure 9.7 shows a family of I_{DS}-V_{DS} curves for the MOSFET; included on the figure are both dc and RF load lines. The dc load line is nearly vertical because of the relatively low resistance of the on-chip inductor. The RF load line's slope is determined by the load resistance R_L:

$$\frac{dV_{DS}}{dI_{DS}} = -R_L \tag{9.6}$$

The intersection of the two load lines at approximately $V_{DS} = V_{DD}$ indicates the range of operating or quiescent Q points of the amplifier. The specific output Q point of the amplifier is set by the input dc bias voltage V_{GS-Q}. A large V_{GS-Q} will shift the Q point to higher levels of output current but will the limit the input voltage swing since the negative-going portion of the RF signal will reach the knee of the I-V curve and a subsequent turning off process of the transistor will begin; the output current waveform will become distorted as a result. Reducing V_{GS-Q} shifts the Q point to a level where the RF input voltage swing is increased; the minimum of the input RF swing just reaches the dotted curve representing the I-V knee. At this point, the input RF voltage swing and output current swing both have the same sinusoidal shape.

The MOSFET PA conducts current during the entire 360° of RF swing. Lowering V_{GS-Q} still further causes a clipping of the negative-going portion of the output RF current; current is flowing only during a portion of the RF voltage swing. This observation provides the basis for defining a PA *conduction angle*, Θ. Knowledge of

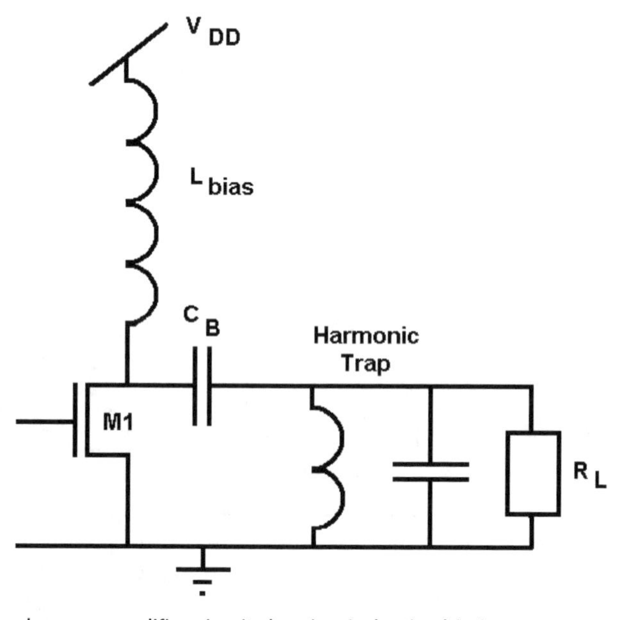

Figure 9.6 General power amplifier circuit showing inductive biasing.

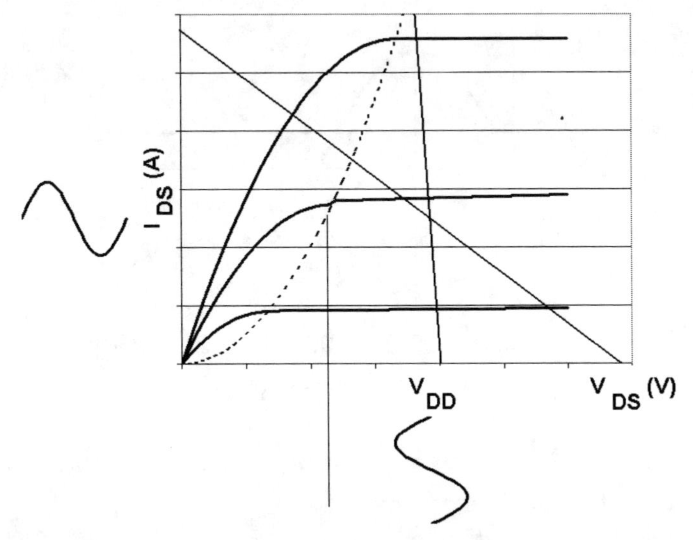

Figure 9.7 Load line analysis of power amplifier showing both dc and RF load lines and the onset of the saturation curve (dashed line).

conduction angle provides some insights into the relationship between efficiency and power output. Because the conduction angle concept is also used in defining the various classifications of power amplifiers, a detailed discussion and derivation are discussed in the next section.

9.2.1 Conduction Angle

The output MOSFET drain current waveform in a PA can vary between a wave shape that is purely sinusoidal to one that is severely clipped on the negative RF peaks. Figure 9.8(a) shows an general example of this waveform with Q point current I_Q and peak current I_P versus phase angle ϕ. The current waveform remains below zero (the clipping point) for all angles except in the range between $-\Theta$ and Θ. A general mathematical expression for this waveform can be written as

$$I(\phi) = \begin{cases} I_Q + I_P \cos\phi & -\Theta < \phi < \Theta \\ 0 & \text{otherwise} \end{cases} \tag{9.7}$$

Using this form, a *conduction angle* term can be defined as $\Theta_c = 2\Theta$. The dc component of this current can be determined by integrating the current $I(\phi)$ over the entire period:

$$I_{DC} = \frac{1}{2\pi}\int_{-\Theta}^{\Theta} I(\phi)\, d\phi = \frac{1}{2\pi}\int_{-\Theta}^{\Theta} \left(I_Q\right) d\phi + \frac{1}{2\pi}\int_{-\Theta}^{\Theta} \left(I_P \cos\phi\right) d\phi \tag{9.8}$$

The result of this integration yields the dc current as a function of the current peak, the Q point current, and the conduction angle:

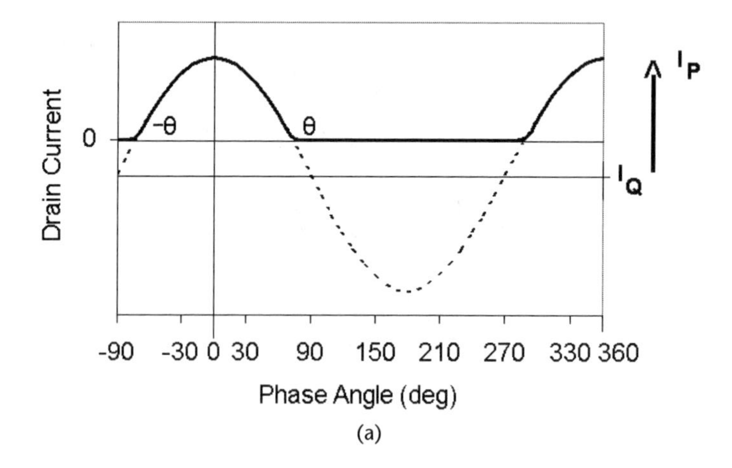

(a)

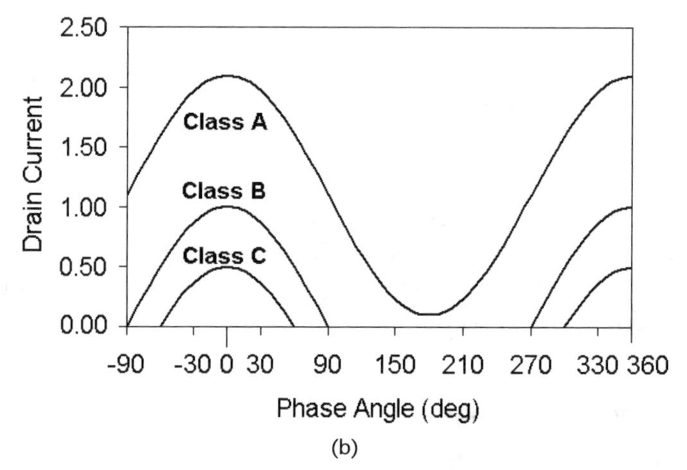

(b)

Figure 9.8 (a) Waveform definition for conduction angle derivation and (b) ideal waveforms for Class A, B, and C amplifier operation versus conduction angle.

$$I_{DC}(\Theta_c) = \frac{2I_P \sin\frac{\Theta_c}{2} + \Theta_c I_Q}{2\pi} \tag{9.9}$$

The clipped version of the output current waveform contains harmonically related components that can be mathematically described by a Fourier series representation. The Fourier coefficients as a function of the harmonic number n can be written as [5]

$$I_n(\Theta_c) = \frac{I_P}{2\pi}\left[\frac{2}{n+1}\sin\left((n+1)\frac{\Theta_c}{2}\right) + \frac{2}{n-1}\sin\left((n-1)\frac{\Theta_c}{2}\right)\right] + \frac{2I_Q}{n\pi}\sin\left(n\frac{\Theta_c}{2}\right) \tag{9.10}$$

The fundamental component ($n = 1$) of the RF signal is the desired component. This selection of fundamental component can be implemented in the PA circuit by the addition of a filter between the amplifier output and the load (the so-called *harmonic trap* [4]). Using a harmonic trap allows the fundamental RF current component to be written as

$$I_{RF}(\Theta_c) = \frac{I_P}{2\pi}\left[\sin(\Theta_c) + \Theta_c\right] + \frac{2I_Q}{\pi}\sin\left(\frac{\Theta_c}{2}\right) \tag{9.11}$$

With no clipping of the RF signal ($\Theta_c = 2\pi$), I_Q and I_P are equal. For conduction only during the positive half-cycle, $I_Q = 0$ and $\Theta_c = \pi$. For a zero conduction angle, $I_Q = -I_P$. A general form for this relationship can then be written as

$$I_Q = -I_P \cos\left(\frac{\Theta_c}{2}\right) \tag{9.12}$$

Combining (9.9), (9.11), and (9.12) gives an expression for the ratio of the fundamental RF current to the dc current:

$$\frac{I_{RF}(\Theta_c)}{I_{DC}(\Theta_c)} = \frac{1}{2}\frac{\Theta_c - \sin\Theta_c}{\sin\frac{\Theta_c}{2} - \frac{\Theta_c}{2}\cos\frac{\Theta_c}{2}} \tag{9.13}$$

One major metric describing amplifier operation is its *drain efficiency* (or simply efficiency), η, defined as the ratio of the fundamental RF power out to the amplifier dc input power. The dc input power P_{DC} is simply the product of the dc current and the power supply voltage $V_{DD} I_{DC}$, where I_{DC} is defined above. Maximum voltage swing is obtained at the output only if a specific load resistance R_{OPT} is chosen [5]. RF power out at the desired fundamental can then be written as

$$P_{RF} = \frac{1}{2}(V_{DD} - V_{DS-sat})I_{RF} = \frac{1}{2}R_{OPT}I_{RF}(\Theta_c)^2 \tag{9.14}$$

where $V_{DD}-V_{DS-sat}$ is the maximum RF voltage swing. Combining (9.9) and (9.13) gives the amplifier efficiency in terms of conduction angle Θ_c:

$$\begin{aligned}\eta &= \frac{P_{RF}}{P_{DC}} \times 100\% = \frac{\frac{1}{2}V_{DD}\left(1 - \frac{V_{DS-sat}}{V_{DD}}\right)I_{RF}}{V_{DD}I_{DC}} \times 100\% \\ &= \frac{1}{4}\left(1 - \frac{V_{DS-sat}}{V_{DD}}\right)\frac{\Theta_c - \sin\Theta_c}{\sin\frac{\Theta_c}{2} - \frac{\Theta_c}{2}\cos\frac{\Theta_c}{2}} \times 100\%\end{aligned} \tag{9.15}$$

For a 360° conduction angle and $V_{DS-sat} = 0$ (voltage swing over the entire available voltage range), the efficiency of the amplifier is only 50%, implying that half of the dc input power is converted to RF output power; the other half is simply dissipated as heat in the amplifying transistor. This is the classical efficiency value for Class A amplifiers, which provide very linear amplification because the output waveform is (ideally) an exact amplified replica of the input. In general, as the conduction angle is reduced, the efficiency improves as shown in Figure 9.9.

For real CMOS Class A amplifiers, V_{DS-sat} is approximately $0.2V_{DD}$, so the actual Class A efficiency for CMOS designs is closer to 40%. In addition, this assumes that the maximum RF swing is occurring. If the Q point is fixed but the

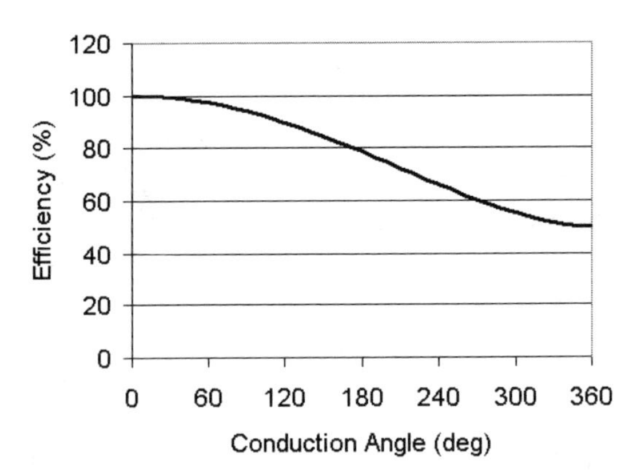

Figure 9.9 Amplifier efficiency improves as output waveform conduction angle is reduced.

peak value of the RF signal is small (such as from a small input signal), then the efficiency is less than 40% since the dc power consumed is still the same (fixed Q point) but the RF power is lower. For a conduction angle of 180°, current conducts only during one-half of the RF swing, but the efficiency improves to 78.5%. This is the classical efficiency value for Class B amplifiers (Figures 9.8 and 9.9). In Class B operation, both the dc and RF current decrease with input power level, resulting in an improvement in efficiency at low power levels when compared with Class A. However, Class B amplifiers (180° conduction angle) have 6-dB less gain compared with Class A amplifiers for equal input drive levels, which may be an issue with low-gain RF amplifier designs. For conduction angles of less than 180°, Class C operation occurs with even better efficiency, approaching 100% for small Θ_c.

All is not what it seems here, however; small conduction angles merely imply that small pulses of output current are occurring during each RF cycle, which will ultimately affect the level of RF output power. The output power under these conditions can be seen by defining the amplifier RF output power normalized to the peak power $\frac{1}{2}V_{DD}I_P$ (Class A operation) as P_N, which can be written in terms of the conduction angle Θ_c as

$$P_N = \frac{1}{2\pi}\left[\sin(\Theta_c) + \Theta_c\right] - \frac{2}{\pi}\sin\left(\frac{\Theta_c}{2}\right)\cos\left(\frac{\Theta_c}{2}\right) \tag{9.16}$$

Figure 9.10 shows the normalized output power P_N plotted versus conduction angle Θ_c. Even though small conduction angles yield the highest amplifier efficiency, the output power drops to low values and is 0 for zero conduction angle.

Equation (9.14) implies that there is an optimum termination resistance for each conduction angle for best efficiency. In practical amplifier configurations, however, the amplifier termination is not variable but is instead a fixed value. If the amplifier termination is assumed fixed for a voltage swing of $V_{DD}-V_{Dsat}$ in Class A, then the amplifier efficiency as a function of conduction angle shown in Figure 9.8 (for an optimum termination) no longer applies; Figure 9.11 shows this efficiency for a fixed termination based on Class A operation. Note that there is a slight increase in

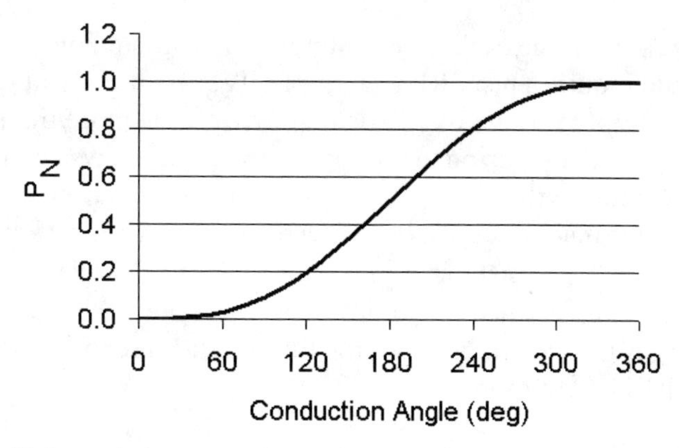

Figure 9.10 The highest relative power output (compared to Class A) occurs at a 360° conduction angle and decreases with this angle.

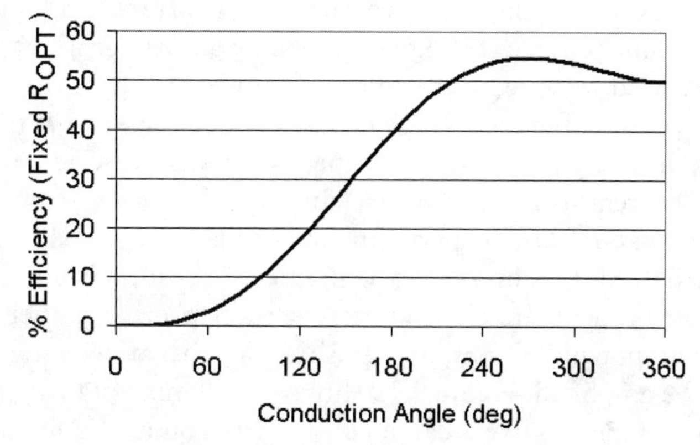

Figure 9.11 The efficiency for a fixed amplifier termination resistance based on Class A operation peaks at a conduction angle of approximately 270°.

efficiency for a conduction angle of approximately 270° in the Class AB region of operation.

The main point in the previous discussion was to show that the output waveform shape determines the efficiency and output power characteristics of the amplifier. From the load line analysis done earlier, the output Q point is determined by the input bias voltage $V_{GS\text{-}Q}$, indicating that both output and input circuit characteristics are needed for proper amplifier design.

At this point in the design of a power amplifier, little mention has been made of the specific power output for a given load. The output power of the PA is directly related to the maximum RF voltage (or current) swing across (through) the load resistor R_L; the RF swing required is then related to the output power and load resistance:

$$P_{\text{OUT}} = \frac{V_{RF-\max}^2}{2R_L};\ V_{RF-\max} = \sqrt{2P_{\text{OUT}}R_L} \tag{9.17}$$

A Q point that maximizes this swing provides the most output power. If the CMOS PA is driving a 50Ω load such as an antenna, the required RF voltage swing is easily found if the desired output power is known. For example, 1W of RF power out would require a peak RF voltage of 10V. Because V_{DS} at the ideal Q point is V_{DD}, a power supply voltage of 10V is required; this is more than five times higher than the 1.8V for RF operation mentioned in the technology road map discussed in Chapter 1. To achieve this voltage swing, either the output power or load resistance, or both in some measure, must be reduced. For a maximum swing of 1.8V and 1W output power, a load resistance of 1.62Ω is required. This value of load resistance, however, will present a severe mismatch with the 50Ω antenna unless impedance matching is performed between the PA and the antenna. For this 1W output power, a peak RF current of more than 1.0A must be delivered on RF voltage swings:

$$I_P = \frac{V_P}{R_L} = \frac{1.8\text{V}}{1.62\Omega} = 1.1\text{A}$$

Needless to say, supplying this large of a current in a battery-operated environment will not provide for extensive operation between battery recharges. In addition, the need to squeeze the most RF power out with this large dc current draw shows that amplifiers with high efficiency are necessary. If the amplifier is Class A, the dc quiescent current equals the peak current, so the actual peak current (RF plus Q point current) is 2.2A, a sizable current!

Note that in Class B operation, half of the RF signal is "wasted" in that only the positive half-cycle is utilized for amplification with harmonic traps and filters used to remove the generated harmonics. If a transformer or other 180° phasing network is used at input and output, both positive- and negative-going portions of the RF signal can be amplified. Figure 9.12 shows an example of this *push–pull* circuit configuration. On the positive RF input swing, M1 conducts and amplifies that portion of the RF signal; on the negative RF input swing, M2 provides the amplification. The output transformer combines the two balanced amplified signals into a single unbalanced output to the load resistance R_L. If center-tapped transformers are used, dc bias at both gate and drain can be applied at points A and B. In most RFIC applications, differential signals from previous stages are already available and so the input transformer can be eliminated from the push–pull circuit. If the PA is driving a balanced load (Figure 9.13), the output transformer can be eliminated as well. Note that in this balanced PA configuration, the optimum load resistance R_{OPT} of each stage combines to yield the balanced load, $R_L = 2\ R_{OPT}$. The RFIC PA designer should try to eliminate as many on-chip transformers as possible because of the power loss associated with these passive networks.

Various methods for biasing MOSFET-based PAs exist. A simple but effective biasing circuit is shown in Figure 9.14. The bias voltage for the amplifying transistor MAMP is set by the current mirror action of M_B according to the standard mirroring expression:

$$V_{DC} = \sqrt{\frac{2I_B}{\text{KP}\left(\dfrac{W}{L}\right)_{MB}}} \tag{9.18}$$

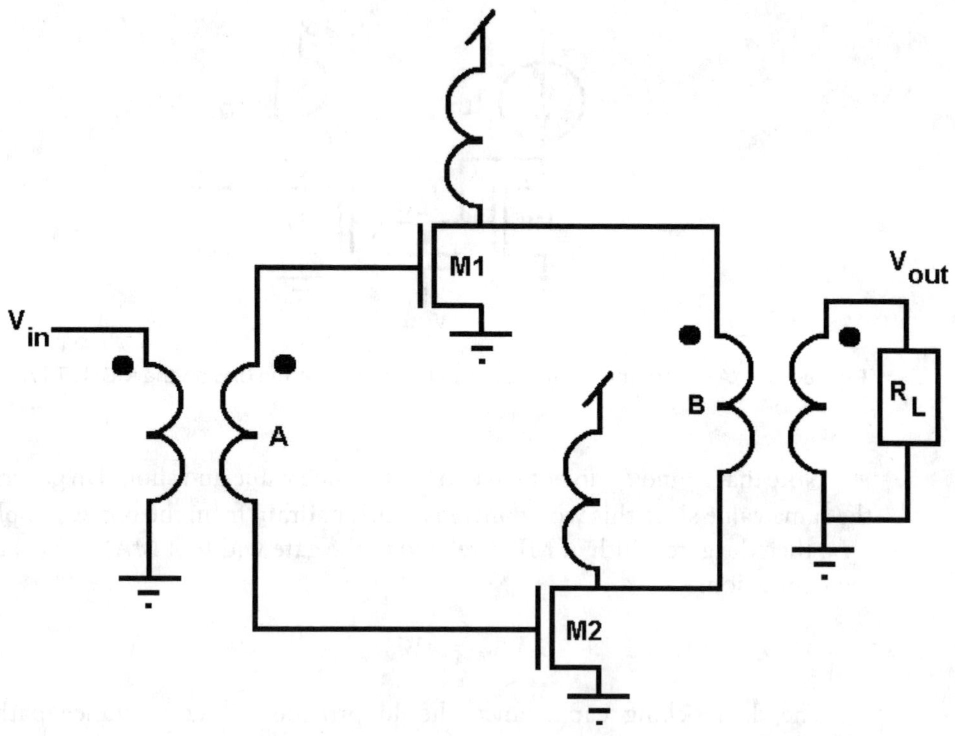

Figure 9.12 The push–pull Class B amplifier configuration uses both negative- and positive-going RF signals. Transformers are required for proper signal phasing and combining.

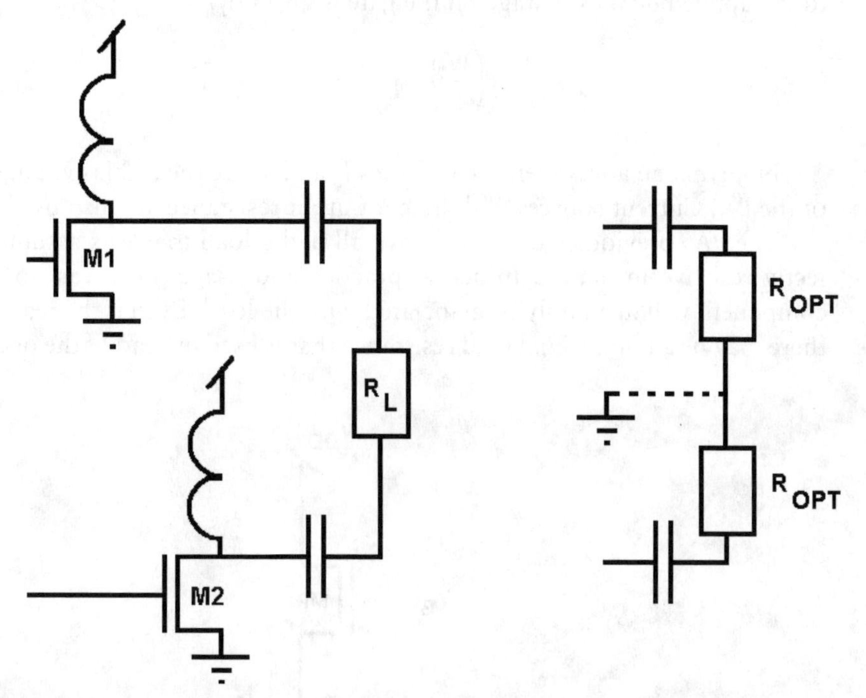

Figure 9.13 A balanced load does not require an output transformer. The dc blocking capacitors are shown.

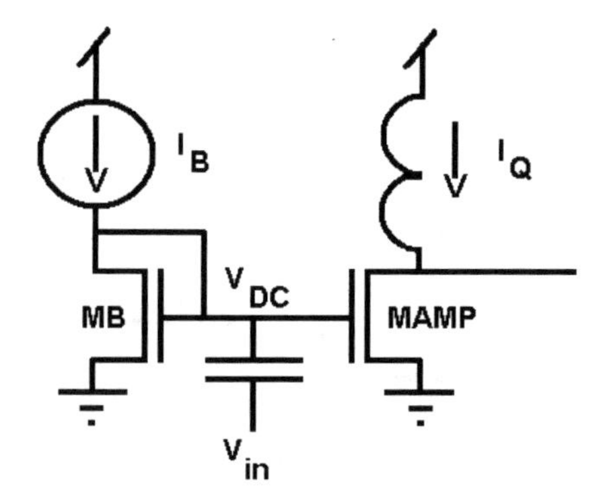

Figure 9.14 A current mirror structure can be used to set the bias on the MOSFET PA.

Note that I_B and I_Q do not have to be the same value and should in general not be the same value since this would increase current drain from the power supply. If $I_B = k\, I_Q$, then the gate width of MB is related to the gate width of MAMP by the following expression:

$$W_{MB} = kW_{MAMP} \tag{9.19}$$

The dc blocking capacitance should provide a low reactance path to the MOSFET gate.

The use of current mirror biasing provides some interesting possibilities for dynamic or adaptive PA biasing. Suppose current source I_B is implemented using a simple *p*MOSFET current source as shown in Figure 9.15. The expression relating I_B to the applied dc bias voltage on the gate is given by

$$I_B = \frac{1}{2}\mathrm{KP}\left(\frac{W}{L}\right)_{MP}\left[V_B - V_{DD} - |V_T|\right]^2 \tag{9.20}$$

Note that an adjustment of V_B varies I_B and hence the dc bias voltage on the gate of the PA. Current sources with higher output resistance are also used.

The PAs previously discussed have all had a load that was assumed to be perfectly resistive in nature. In actual practice and usage, however, some reactance component will inevitably be associated with the load. Even if the reactance is zero, there may be a nonoptimal load resistance that is used instead of the one designed as

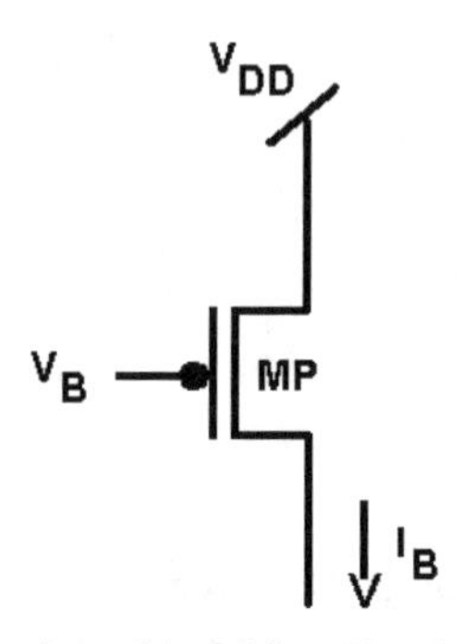

Figure 9.15 Dynamic PA biasing can be achieved by using a variable current source.

optimum. Both of these conditions cause a reduction in PA output power that is termed *load pulling*. Load pull or constant power contours are frequently shown plotted on a Smith chart to show the range of load impedances that yield a given output power (Figure 9.16). The contours are typically defined in terms of the reduction in output power (in decibels) from the optimum.

9.2.2 Class A and B Design Examples

To see how the preceding equations can be used to design a power amplifier, consider the design of a Class A CMOS power amplifier with a desired 100-mW output power. Assume an *n*MOSFET will be used in the design; this device exhibits $V_T = 0.66\text{V}$ and $\text{KP} = 56\ \mu\text{A/V}^2$. Assume a power supply voltage of $V_{DD} = 3.3\text{V}$.

The maximum RF voltage swing can be seen from Figure 9.7 to be:

$$V_{RF-\max} = V_{DD} - V_{DS-sat} = V_{DD} - \left(V_{GS-Q} - V_T\right) = 3.96 - V_{GS-Q}$$

with the resulting RF power output and current swing given by

$$P_{RF} = \frac{1}{2}\frac{\left(V_{DD} - V_{DS-sat}\right)^2}{R_L};\ I_{RF} = \frac{\left(V_{DD} - V_{DS-sat}\right)}{R_L}$$

Assuming the negative RF swing goes down to 1.0V, the peak RF signal swing will be 2.3V:

$$V_{RF-\max} = V_{DD} - V_{DS-sat} = 3.3 - 1.0 = 2.3V$$

which corresponds to an input Q voltage V_{GS-Q} of 1.66V. The load resistance R_L for 100 mW of output power can then be computed as

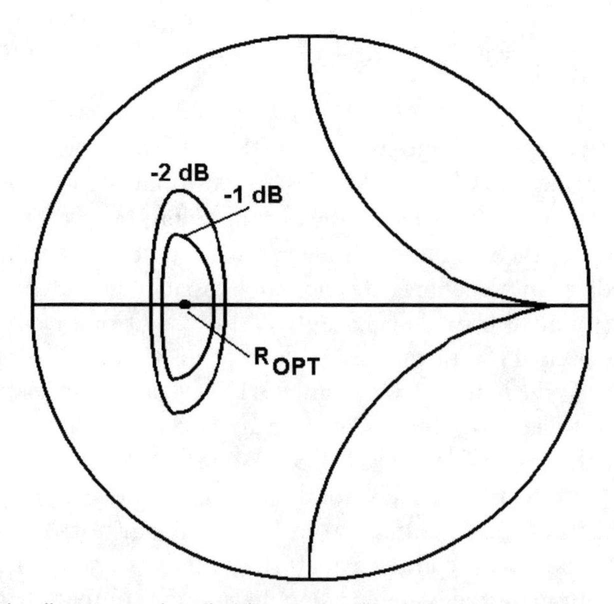

Figure 9.16 Load pull contours describe the range of load impedances corresponding to a reduction in PA output power from its optimum design value.

$$P_{RF} = \frac{1}{2}\frac{\left(V_{DD} - V_{DS-sat}\right)^2}{R_L} \Rightarrow R_L = \frac{1}{2}\frac{\left(V_{DD} - V_{DS-sat}\right)^2}{P_{RF}} = \frac{1}{2}\frac{(2.3)^2}{0.1} = 26.45\Omega$$

with a corresponding RF current swing of

$$I_{RF} = \frac{\left(V_{DD} - V_{DS-sat}\right)}{R_L} = \frac{2.3}{26.45} = 87 \text{ mA}$$

For Class A, the peak RF current is the same as the dc Q point current (9.13) for a 360° conduction angle, which can be used to determine the physical size of the MOSFET:

$$I_{DS} = \frac{1}{2}\text{KP}\frac{\text{W}}{L}V_{DS-sat}^2 \Rightarrow \frac{\text{W}}{L} = \frac{2I_{DS}}{\text{KP}\cdot V_{DS-sat}^2} = \frac{2(0.087)}{56\cdot 10^{-6}(1)^2} = 3{,}107 \Rightarrow 3{,}110$$

or an nMOSFET with a 1,555-μm gate width if 0.5-μm CMOS technology is used. The ideal efficiency of the PA is the ratio of the output RF power to the input dc power:

$$\eta = \frac{P_{\text{RF-out}}}{P_{\text{DC-in}}} \times 100\% = \frac{0.1}{V_{DD}I_{DC}} \times 100\% = \frac{0.1}{(3.3)(0.087)} \times 100\% = 34.8\%$$

which is consistent with the nonideal Class A amplifier efficiency approximation of 40%.

Because this is an amplifier, the maximum input voltage swing must be determined; otherwise the amplifier will be overdriven and the output RF waveform will clip. The voltage gain for the PA can be found from the $g_m R_L$ product and the value of g_m at the Q point:

$$\left|A_V\right| = \frac{V_{\text{out}}}{V_{drive}} = g_m R_L = \left(\text{KP}\frac{\text{W}}{L}V_{DS-sat}\right)R_L = 56\cdot 10^{-6}(3{,}110)(1)(26.45) = 4.6$$

For an output voltage swing of 2.3V, the input voltage swing should be limited to 0.5V peak (2.3V/4.6). This design information was used as input to SPICE (ch9-1.txt); Figure 9.17 shows the output voltage across the 26.45Ω load resistor at 1.0 GHz. For the simulation, a biasing inductance of reactance 10 times that of R_L was used; a low-reactance dc blocking capacitor was chosen to be 100 pF. The first-pass simulation results in SPICE show a Q point current of 97 mA, within approximately 11% of the design current of 87 mA. The peak RF output voltage swing across the load is 2.1V (Figure 9.17), close to the design value of 2.3V. SPICE shows an efficiency of 34.5%, in close agreement with the theoretical value and consistent with typical Class A CMOS PA operation. Small amounts of distortion are seen on the waveforms; the second harmonic value was shown to be more than 25 dB down from the fundamental in SPICE. A harmonic trap at the output would reduce this nonlinear content even further but is not really required for Class A operation because of the inherent low harmonic content. More restrictive spectral

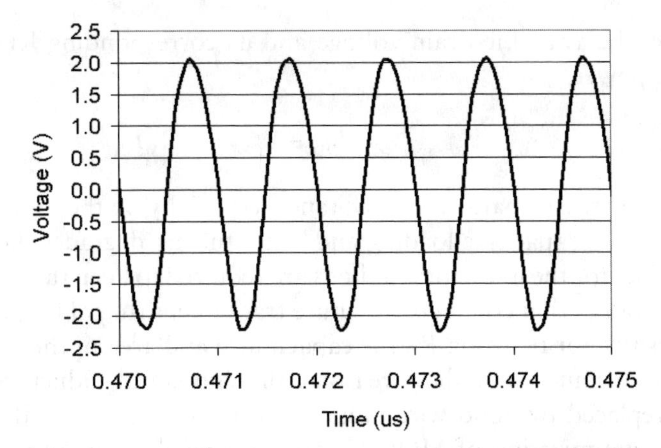

Figure 9.17 Output voltage across load resistance R_L for Class A power amplifier.

requirements may require additional filtering, however. The MOSFET drain current shown in Figure 9.18 shows the 360° conduction angle and swings from 22 mA to a peak of 180 mA.

This procedure showed that simple equations can be used in the design of the Class A power amplifier with good agreement compared with SPICE simulations. The PA design can be modified from this point to increase the output voltage swing and slightly lower the Q point current to bring these parameters into better agreement with the design specifications.

In an actual CMOS PA implementation, the parasitics of M1, primarily the output capacitance, and the parasitics and Q of the biasing inductor must be considered. M1's output capacitance will influence the frequency response of the amplifier by increasing the capacitive loading on the output drain node. The biasing inductor should present a high reactance at the amplifier design frequency (in the previous example, a value of $10R_L$ was used). High-value on-chip inductors, as mentioned previously in Chapter 3, require large areas and many turns, which increases both the parasitic capacitance to ground as well as the series resistance (and degraded Q). The series resistance R_S will create an undesired dc voltage drop across the inductor,

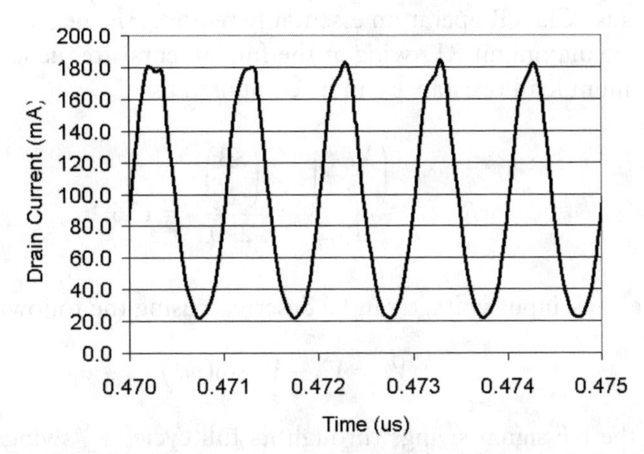

Figure 9.18 The MOSFET drain current conducts for the entire RF cycle, indicative of Class A operation.

limiting the available drain voltage and its corresponding RF swing and therefore output power:

$$V_{RF-swing} = V_{DD} - V_{DS-sat} - R_S I_{DC} \tag{9.21}$$

The increased parasitic capacitance, especially on the drain side of the inductor, increases the capacitive loading and hence further degrades the frequency response. Wide lines for the inductor windings are required to keep the current densities below critical values that could create future trace opens (metal migration). Although that is a positive for reducing R_S, the capacitance and area of the inductor increase for a given inductance. For these reasons, on-chip spiral inductors for PA biasing are often replaced by bond wire inductors [7]. More complex design procedures that include optimization of MOSFET and passive elements in the PA chain have also been developed [8]. In spite of these parasitic limitations, Class A CMOS RFIC amplifiers have been fabricated and show good linearity results for low-power Bluetooth-type applications; efficiencies, however, are still quite low [9].

Note that for the Class A amplifier, there is a reasonably faithful reproduction of the input sinusoidal waveform at the output. Class A amplifiers, if not overdriven, are very linear and are often used in digital transmission of envelope sensitive modulation schemes such as band-limited QPSK to preserve the narrow bandwidth [5].

A second design looks at designing a power amplifier with identical specifications as the previous Class A design, but utilizing higher efficiency Class B operation. For Class B, the output voltage waveform is cutoff for one-half of the RF cycle (180° conduction angle), which implies that the input voltage swing must be driven below the threshold voltage V_T for one-half of the input RF cycle. For this amplifier type, the concept of transconductance becomes problematic because the device is only on for one-half of the RF cycle, violating the "small signal" definition on which the derivation of g_m is based. The design of the Class B amplifier begins with determining the optimum load resistance, R_{OPT}:

$$R_{OPT} = \frac{\left(V_{RF-swing}\right)^2}{2P_{OUT}} \tag{9.22}$$

Because Class B operation essentially removes the negative-going half of the RF signal, the maximum RF swing at the fundamental frequency is $V_{DD}/2$ (1.65V), so the optimum load resistance can be computed as

$$R_{OPT} = \frac{\left(\frac{V_{DD}}{2}\right)^2}{2P_{OUT}} = \frac{\left(\frac{3.3}{2}\right)^2}{2 \cdot 0.1} = 13.6\Omega$$

The peak input voltage can be described using the following equation:

$$V_{in} = V_T + V_T \sin(\omega t)$$

As the RF signal swings through its full cycle, V_{in} swings between 0 and $2V_T$ volts. However, the MOSFET will only conduct when the input drive voltage V_{in} is greater than the threshold voltage V_T, providing the proper input signal drive for the 180° current conduction angle required for Class B operation. This implies that $V_{GS\text{–}Q} = V_T = 0.66\text{V}$ and that the input RF swing is 0.66V (the input voltage is $2V_T = 1.32\text{V}$ at input RF peaks). The peak MOSFET drain current I_P occurs at the peak input voltage value of $2V_T$, so the MOSFET must be able to pass this current during RF peaks; this information aids in determining the MOSFET size:

$$I_P = \frac{V_{DD}}{R_{OPT}} = \frac{3.3}{13.6} = 242 \text{ mA}$$

$$I_P = \frac{1}{2}\text{KP}\frac{W}{L}\left(V_{GS-peak} - V_T\right)^2 = \frac{1}{2}\text{KP}\frac{W}{L}(2V_T - V_T)^2 = \frac{1}{2}\text{KP}\frac{W}{L}(V_T)^2$$

$$\frac{W}{L} = \frac{2I_P}{\text{KP}(V_T)^2} = \frac{2(0.242)}{56 \cdot 10^{-6}(0.66)^2} = 19{,}841 \Rightarrow 20{,}000$$

or a gate width of 10,000 μm in 0.5-μm CMOS technology. Because of the inherently high harmonic content of Class B operation, a harmonic trap is necessary.

The preceding design parameters were used as input for SPICE simulations (ch9-2.txt) to verify the design and operation. A 1.0-GHz signal was used for the analysis. A harmonic trap was implemented using a 10-pF capacitor in parallel with a 2.5-nH inductor, both in shunt with the load resistance; a low-reactance dc blocking capacitor couples the RF energy from the MOSFET drain to the harmonic trap and load. The MOSFET drain current, shown in Figure 9.19, exhibits nearly ideal Class B operation: The current goes to zero for one-half of the RF cycle, corresponding to V_{in} below the MOSFET threshold voltage, V_T. In addition, the current peaks are approximately 240 mA, in good agreement with the design value of 242 mA. The voltage at the load has a value of 1.3V peak, providing a power output of approximately 65 mW, somewhat less than the required 100 mW. With a dc current of 58 mA, the efficiency of this PA is approximately 40%, well below the theoretical maximum of 78.5% but an improvement over the 35% efficiency with the Class A amplifier. Increasing the input drive signal by 100 mV boosts the output voltage

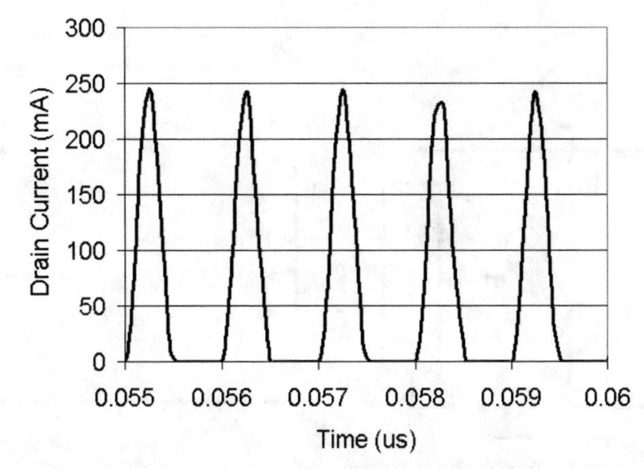

Figure 9.19 Drain current showing 180° conduction angle indicative of Class B amplifier operation.

closer to the 1.65V required for 100-mW operation. The drain current also begins to show the classic Class B bifurcation at this higher input drive level [4].

Both amplifiers used a load resistance that was different from the typical 50Ω output impedance that PAs will typically drive. A simple LC impedance transformation network can be used to transform the 26.45Ω or 13.6Ω optimum amplifier load to the required 50Ω value. Broadband matching is typically performed using transformers.

9.3 Switching Amplifiers

Figure 9.11 shows that the best efficiency for a fixed load termination occurs for Class AB operation at a conduction angle of approximately 250° and a corresponding theoretical efficiency of 60%, although it is reduced somewhat in practical amplifiers. The three classes of amplifiers (A through C) are termed transconductance amplifiers because MOSFETs are used to provide gain to the input signal waveform. A second broad classification of amplifiers with higher theoretical efficiency makes use of the excellent switching characteristics of the MOSFET. (It is this switching action that also makes the MOSFET an ideal device for digital circuit implementation.) These so-called *zero volt switching (ZVS) amplifiers,* or just *switching amplifiers* for short) are the basis for Class D through F amplifier operation. The concept behind the switching amplifier, first advanced by Sokal and Sokal [10], can be observed by looking at an ideal switch that is controlled by an input signal V_{in} (Figure 9.20). When the switch is open (infinite resistance), current will flow through the biasing inductor and into the load resistance R_L with a

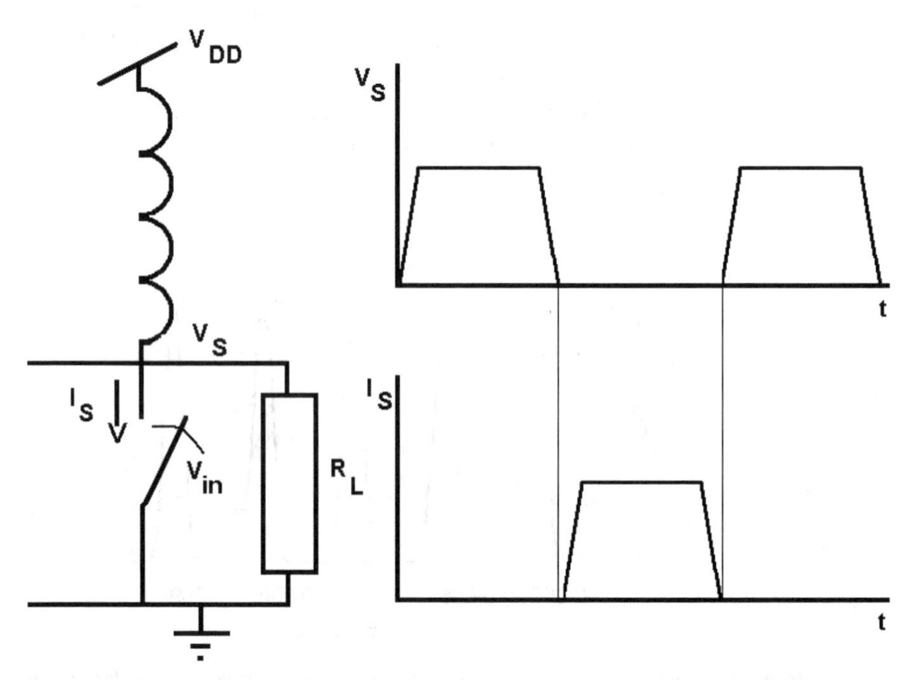

Figure 9.20 Voltage and current waveforms for the idealized switching amplifier circuit. No overlap between the voltage or current waveforms is assumed in the ideal amplifier.

corresponding voltage $V_L = V_{DD}$ developed across the load (the bias inductor is assumed to have zero resistance and high inductive reactance at the design frequency). No current flows through the switch, so therefore no power is dissipated by the switch since the VI product is zero. When the switch is closed, current flows into the switch but with zero voltage drop (the ideal switch having an on-state resistance of 0Ω); again, no power is dissipated by the switch since the VI product is still zero. If no power is dissipated by the switching element during the entire RF cycle, all of the power supplied by the dc supply goes to the load and the efficiency is theoretically 100%.

Note that this 100% efficiency requires that the switch have an infinite off-state resistance, zero on-state resistance, and instantaneous switching, something that is impossible to actually achieve in practice. However, MOSFET switches can be designed to have extremely low on-state resistances (significantly less than 1.0Ω) and high off-state resistances, so actual efficiencies in this class of amplifiers can still be quite high; efficiencies for switching amplifiers of 80% to 90% are not uncommon [4]. The simplest ZVS amplifiers based on this concept are the Class D amplifiers. The instantaneous switching requirement, however, limits the use of Class D amplifiers to high-power, low-frequency applications. To achieve high power, switching devices with low on-state resistances are required. Switching MOSFETs (and indeed, any high-speed switching element) must exhibit large gate widths W with a corresponding large device area that necessarily creates large capacitances. Of particular interest is the large capacitance at the output drain connection. This large capacitance prevents the required instantaneous switching requirement from being achieved in the simple switching amplifier topology shown in Figure 9.20 (the drain capacitance will be across termination R_L and cause significant rise and fall times on the output waveforms). For low-frequency applications, however, Class D amplifiers work quite well. The influence of the output capacitance on the output waveforms, however, can be mitigated by the use of additional circuitry in Class E and Class F ZVS amplifiers. The next two sections look at these two common ZVS amplifier classes, as seen in RFIC applications.

9.3.1 Class E Amplifiers

The classical Class E amplifier shown in Figure 9.21 is a modification of that shown in Figure 9.20 and includes a capacitance C_0 in shunt with the switch, an inductance L_X (for reasons discussed later), and a filter in series with the load. The filter tunes the amplifier to the desired frequency and removes any harmonic components that are an artifact of the square-wave switching process. The shunt capacitance, C_0, is used as a charge storage device in this circuit and is typically a combination of the MOSFET switch drain source capacitance and (possibly) an externally added capacitor. Parasitic capacitance of the circuit inductors can also be absorbed into the design value of C_0.

A number of detailed analyses of the Class E amplifier's current and voltage waveforms (Figure 9.22) have been advanced [4, 11]. The waveform expressions are derived by assuming that when the switch is on (closed) for the second half-cycle, the current flowing through C_0 equals zero:

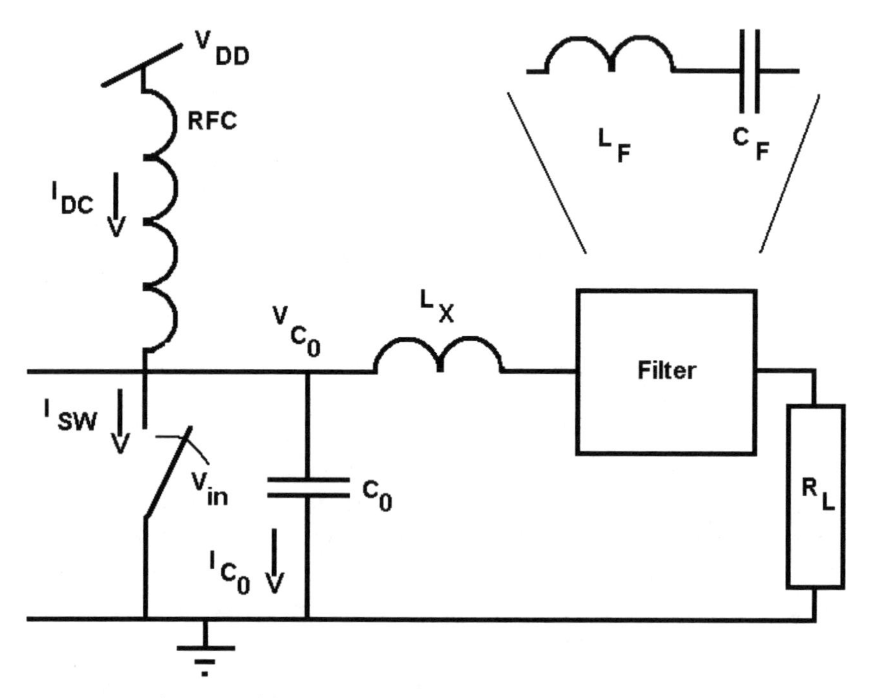

Figure 9.21 Basic Class E amplifier configuration showing shunt capacitance, inductance, filter, and load resistance.

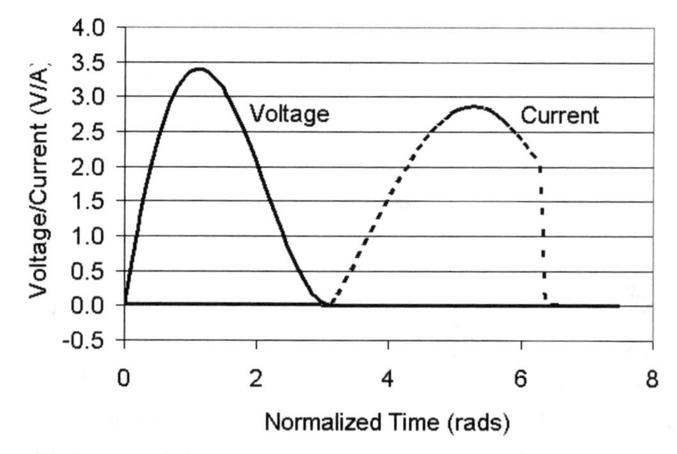

Figure 9.22 Ideal voltage and current waveforms showing no voltage or current overlap, leading to zero power dissipation in the switch and a theoretical 100% efficiency.

$$I_{C_0} = \omega C_0 \frac{dV_{C_0}}{dt} = 0 \tag{9.23}$$

while the current flowing through the switch is the sum of I_{DC} and the sinusoidal load current I_L:

$$I_{SW}(t) = I_{DC} + I_L\left[a \cdot \sin(\omega t + \varphi)\right] \tag{9.24}$$

Here, $\omega = 2\pi/T$ is the switching frequency, I_{DC} is the dc current through the bias inductor, and a and φ are constants that have yet to be determined. When the switch is off (open) in the first half-cycle, the current in (9.24) now flows through the capacitor C_0, and the voltage across C_0 is of the form:

$$V_{C_0}(t) = \frac{1}{\omega C_0}\int_0^{T/2} I_{SW}(t)\; dt = \frac{I_{DC}}{\omega C_0}\left[\omega t + a\cdot\cos(\omega t + \varphi) - a\cdot\cos(\varphi)\right] \tag{9.25}$$

For high amplifier efficiency, the power dissipation in the switch must be zero, which requires that two conditions be satisfied: The current and the voltage must both be zero at the switch transition point, $T/2$ [11]:

$$I_{C_0} = C_0\frac{dV_{C_0}}{dt} = 0 \Rightarrow \frac{dV_{C_0}(t)}{dt}\Big|_{T/2} = 0;\; V_{C_0}(t)\Big|_{T/2} = 0 \tag{9.26}$$

The solution to the preceding set of equations with the specified boundary conditions occurs if a = 1.86 and $\varphi = -0.567$ rad (−32.5°). Using these results with the general form of the capacitor voltage in (9.25), the voltage across capacitor C_0 (and also the switch) as a function of time can be written as follows:

$$V_{C_0}(t)\begin{cases}\dfrac{I_{DC}}{\omega C_0}\left[\omega t + 1.86\cos(\omega t - 0.567) - 1.86\cos(-0.567)\right] & 0 < \omega t < \pi \\ 0 & \pi < \omega t < 2\pi\end{cases} \tag{9.27}$$

while the current through the capacitor is [12]

$$I_{C_0}(t) = \begin{cases} 0 & 0 < \omega t < \pi \\ I_{DC}\left[1 - 1.86\cos(\omega t - 0.567)\right] & \pi < \omega t < 2\pi\end{cases} \tag{9.28}$$

Figure 9.22 shows that both the voltage and current are truncated sinusoids and so are rich in harmonic content. The amplifier's response at the fundamental frequency is the important design parameter and the choice of load on the amplifier optimizes the waveform for low-power dissipation. Performing a Fourier series analysis of both (9.27) and (9.28) and taking the ratio of the fundamental voltage and current components (V_1/I_1) gives an expression for the optimal load impedance that the switching amplifier must see [10, 13, 14]:

$$Z_E = R_L(1 + j1.152);\; R_L = \frac{1}{5.447\omega C_0} \tag{9.29}$$

The inductor L_X required to create the $1.152R_L$ reactive term should have a relatively high Q so that the inductor loss is only a small fraction of the overall resistive term in Z_E:

$$L_X = \frac{1.152R_L}{\omega} \tag{9.30}$$

The harmonic filter is designed to pass the desired fundamental frequency and can be a simple series LC filter (Figure 9.21). The inductance required for the proper termination impedance Z_E and the filter inductance are often combined into a single inductor. The filter inductor should also have low loss (high Q) so that its resistance is only a small fraction of the overall resistive term in Z_E as well:

$$Q_{L_F} = \frac{\omega L_F}{R_L}; C_F = \frac{1}{\omega^2 L_F} \tag{9.31}$$

The choice of R_L directly affects the output power for a given supply voltage V_{DD}. For Class E amplifiers, a good starting point for choosing R_L is the following expression [15]:

$$P_{\text{OUT}} = 0.577\frac{V_{DD}^2}{R_L} \Rightarrow R_L = 0.577\frac{V_{DD}^2}{P_{\text{OUT}}} \tag{9.32}$$

More likely than not, the R_L given by (9.32) is not the desired impedance for a particular application (such as a 50Ω antenna); LC networks or a transformer can be used to transform R_L to the desired impedance.

A further improvement in the Class E amplifier can be done by including both the fundamental and second-order harmonic in the voltage and current waveforms [11]. Design equations for the output circuit of this improved Class E amplifier circuit (Figure 9.23) have been derived [13]:

$$\begin{aligned} C_1 &= \frac{1}{\omega R_L}\sqrt{\frac{R_L}{R_E} - 1} \\ L_2 &= \frac{3}{4}\frac{R_E}{\omega}\left(1.152 + \sqrt{\frac{R_L}{R_E} - 1}\right) \\ C_2 &= \frac{1}{4\omega^2 L_2} \end{aligned} \tag{9.33}$$

These design equations include both proper second-harmonic termination for the Class E amplifier as well as the impedance transformation from optimized R_E to the arbitrary load resistance R_L.

To observe how these equations can be used to design a Class E switching amplifier, consider the design of a 500-mW Class E amplifier using a 3.3V power supply for use at 1.0 GHz (ch9-3.txt). The optimum load resistance to provide the necessary 500 mW is computed using (9.32):

$$R_L = 0.577\frac{V_{DD}^2}{P_{\text{OUT}}} = 0.577\frac{3.3^2}{0.5} = 12.56\Omega$$

The shunting capacitance across the switch can now be computed as:

$$R_L = \frac{1}{5.447\omega C_0} C_0 \Rightarrow \frac{1}{5.447\omega R_L} = \frac{1}{5.447\left(2\pi \cdot 10^9\right)12.56} = 2.32 \text{ pF}$$

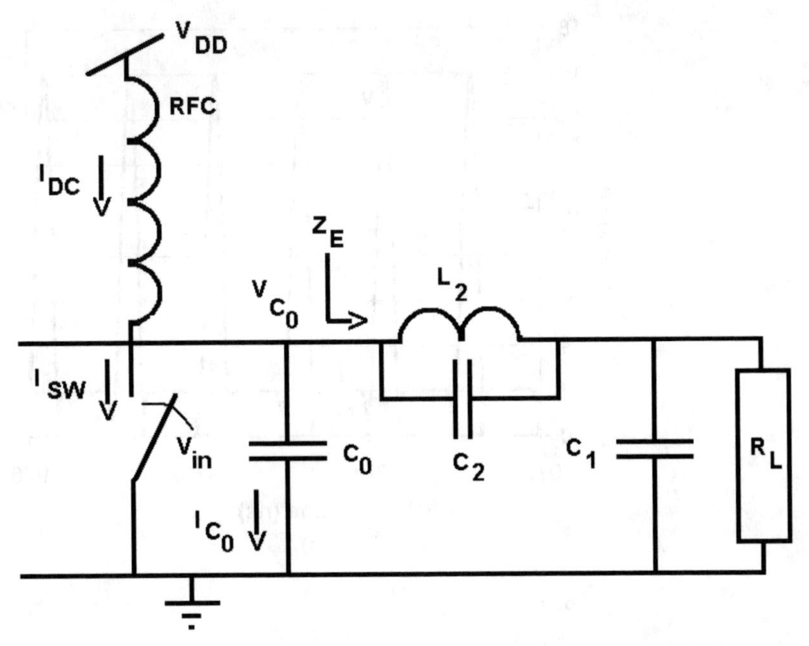

Figure 9.23 Modified Class E amplifier that includes the second harmonic in the switching waveform.

The phasing inductor L_X that provides the proper reactive component for the optimum Class E impedance Z_E can then be computed using (9.30):

$$L_X = \frac{1.152\,R_L}{\omega} = \frac{1.152(12.56)}{2\pi \cdot 10^9} = 2.3 \text{ nH}$$

The results of a SPICE simulation using an ideal switch show an extremely small amount of overlap between the switch voltage and switch current [Figure 9.24(a)]. This voltage current overlap will result in some power dissipation in the switch and is caused by the nonideal switch resistance (0.1Ω on resistance and 10^9Ω off resistance). The load voltage is shown in Figure 9.24(b) with a 1.0-GHz peak voltage component of 3.85V, corresponding to an RF output power of 590 mW. The dc current drawn from the 3.3V power supply is approximately 211 mA, corresponding to a dc power of 690 mW, yielding an amplifier efficiency of 85.5%. This reduction in efficiency from the theoretical 100% is primarily due to the noninstantaneous switching transitions associated with the pulsed input signal (and its corresponding 0.1-ns rise and fall times) and the nonzero on resistance.

The output voltage as shown in Figure 9.24(b) is not a pure sinusoid and has significant harmonic content. Assuming a harmonic trap filter Q of 10, values of L_F = 20 nH and C_F = 1.26 pF can be calculated. Both L_F and L_X can be combined into a single inductor of value 22.3 nH for simulation. The same Class E amplifier simulation of the output voltage but this time with a harmonic trap added is shown in Figure 9.25 (ch9-3a.txt).

In a realistic implementation of the Class E amplifier in CMOS RFIC applications, the ideal switch in Figure 9.21 is replaced by a MOSFET. This replacement

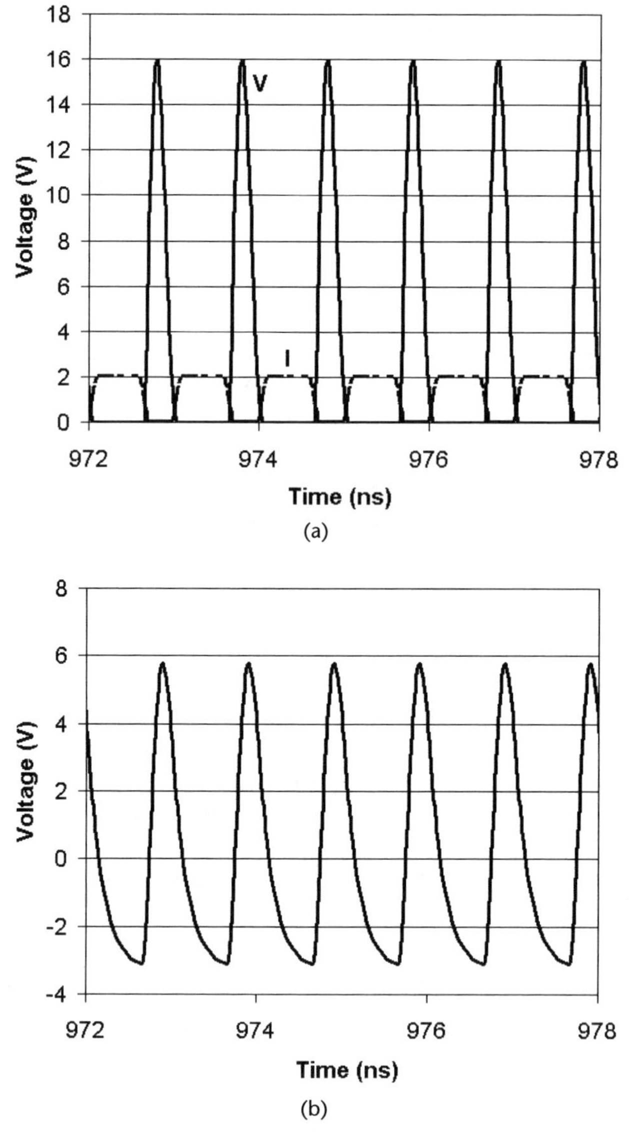

Figure 9.24 Waveforms from Class E amplifier design example: (a) capacitor voltage and switch current and (b) output load voltage waveform.

causes a number of nonidealities to arise. The first nonideality is the finite on-state resistance inherent to the MOSFET. This on-state resistance can be reduced by using large aspect ratio MOSFETs:

$$R_{ON} = \frac{1}{\mathrm{KP}\frac{W}{L}\left(V_{GS} - V_T\right)} \tag{9.34}$$

The smaller the on-state resistance, the more the MOSFET looks like an ideal switch; however, as the MOSFET size increases, both its input gate capacitance C_G and output drain capacitance C_D increase with increasing gate width W. These two

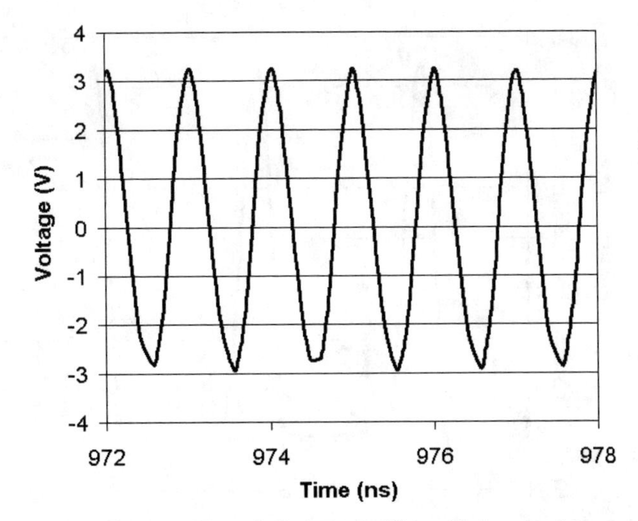

Figure 9.25 Output voltage waveform of Class E amplifier with harmonic trap.

capacitance values can be estimated using the MOSFET capacitance expressions outlined in Chapter 2 and repeated here for convenience:

$$C_G = W\left(C_{gx0} + \frac{\varepsilon_{ox}}{t_{ox}} \cdot L + C_J \cdot L\right) \tag{9.35a}$$

$$C_D = (ab) \cdot C_J + 2(a+b) \cdot C_{JSW} \tag{9.35b}$$

where a and b are the dimensions of the drain diffusion region. The input capacitance C_G influences how quickly the MOSFET can be driven since it is a major contributor to the slewing rate of the driver stage [16]. The large output drain capacitance, however, may not necessarily be a problem for the Class E amplifier; the ideal Class E requires a shunt capacitance C_0. As long as the MOSFET output capacitance C_D does not exceed C_0, the Class E design conditions are the same, only a smaller external capacitance value is added to increase the total output capacitance to C_0. The Class E amplifier waveforms shown in Figure 9.24 indicate that the output voltage swing on the MOSFET drain can be several times that of the power supply voltage, V_{DD}. Care must be taken to ensure that the anticipated peak RF voltage swing does not exceed the breakdown voltage of the switching MOSFET.

To observe how replacing the ideal switch with a low on-resistance MOSFET influences the operation of a Class E switching amplifier, consider the redesign of a 500-mW Class E amplifier with a MOSFET having an on-state resistance of 0.5Ω (ch9-4.txt). Assuming the MOSFET parameters provided in Chapter 2, the required aspect ratio of the switching MOSFET can be found using (9.34):

$$\frac{W}{L} = \frac{1}{\text{KP} \cdot R_{ON}(V_{GS} - V_T)} = \frac{1}{56 \cdot 10^{-6}(0.5)(3.3 - 0.66)} = 13{,}528 \Rightarrow 13{,}600$$

which is a 6,800-μm-wide MOSFET in 0.5-μm technology. Figure 9.26 shows the results of a SPICE simulation of this more realistic implementation of the Class E

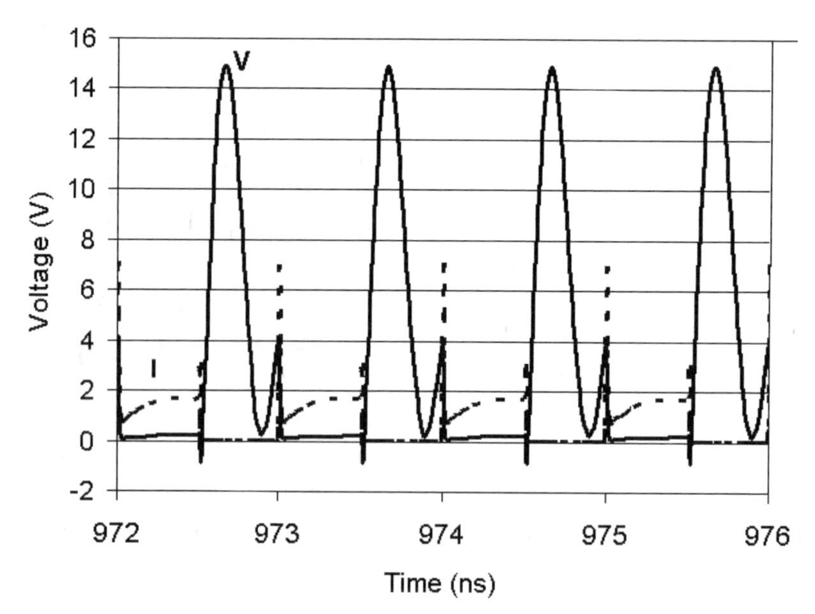

Figure 9.26 Class E amplifier voltage and current waveforms for the realistic implementation using an *n*-channel MOSFET as the switching element (current not to scale). A small decrease in output voltage and some voltage and current spiking are observed in this implementation.

amplifier. These results show similar nearly nonoverlapping capacitor voltage and MOSFET current waveforms as the ideal case with the exception of marked voltage and current spikes during the switching transitions. These spikes are caused by the finite on-state resistance of the MOSFET and are inherent in this actual implementation [4]. The output power delivered to the load is 510 mW with 195 mA pulled from the dc source, resulting in a more realistic amplifier efficiency of approximately 79.2%.

A number of issues in Class E amplifier design must be addressed by the CMOS RFIC designer. The first has been mentioned earlier: the requirement for a high inductive reactance bias inductor. A low-Q inductor will exhibit a relatively large series resistance that will cause a relatively large dc voltage drop with corresponding reduction in output voltage swing and output power. For example, even a 2.0Ω series resistance in the biasing inductor at 195 mA will result in 390 mV of voltage reduction at the drain of the MOSFET. With this significant current drain, wide inductor lines must also be used. This will result in large parasitic capacitances on the drain node, although these may be absorbed into the output capacitance C_0. A number of design procedures have been developed to look at the interactions of the various elements and their parasitics that will optimize Class E amplifier designs [4]. An alternative suggestion has been advanced to make the bias inductor part of the overall load network without assuming high inductive reactance (Figure 9.27) [17]. The following element values for the circuit elements yield the necessary phase shifting required for nonoverlapping voltage and current waveforms:

$$R_L = 0.056\frac{V_{DD}^2}{P_{out}};\ C_0 = \frac{0.071}{\omega R_L};\ L = \frac{3.534R_L}{\omega};\ C_X = \frac{0.204}{\omega R_L} \quad (9.36)$$

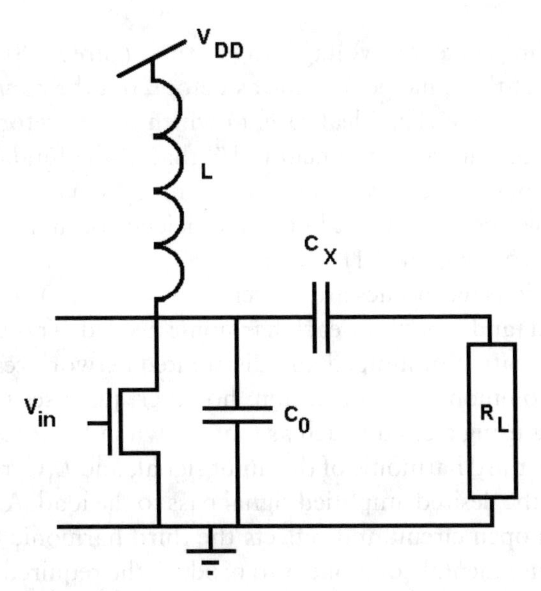

Figure 9.27 Alternate Class E amplifier circuit topology for bias inductor with low inductive reactance.

9.3.2 Class F Amplifiers

A second class of ZVS circuits for high-efficiency amplifiers is the Class F, which actually consists of two subtypes, the traditional Class F and the inverse Class F (F^{-1}). The operation of the Class F-style amplifier is best explained by considering the switching action itself. In a bit of a contrast to Class E amplifiers, Class F amplifiers are driven a little less strongly so that while the amplifier is switching, there are also some times during the input RF cycle at which the signal remains somewhat sinusoidal; the result is that the amplifier output has lower harmonic content than does a Class E amplifier [18]. For Class F amplifiers, these harmonics are actually used to further shape the switching waveform to provide the nonoverlap between voltage and current. The Class F amplifier uses harmonic traps to shape the voltage waveform to a more square-wave shape while leaving the current waveform somewhat sinusoidal. A Class F^{-1} amplifier uses harmonic traps to square up the current waveform, leaving the voltage waveform more sinusoidal. In some respects, a Class F amplifier may be thought of as a moderately overdriven Class B or AB amplifier with resonators used to shape the waveform for higher efficiency [4]. As in the case of the Class E amplifier, the losses in the inductors need to be minimized. Detailed computer modeling of the device and passive characteristics using such optimization schemes as simulated annealing [5] are often used to find the component and component parasitics that optimize amplifier performance. For the harmonic traps, smaller inductor values are preferred, subject to area restrictions on the resonating capacitance value. However, extremely small inductor values are to be avoided because the precision of these inductors decreases with value. Bond wire inductors have similar precision problems but have much higher Qs than their on-chip counterparts.

An alternative view of the Class F/F^{-1} amplifier is that the odd-order harmonics generated by the switching action tend to actually suppress the peak of the corresponding current or voltage waveform (Figure 9.28). The fundamental component is still at the same peak value as before, but the *combined* RF waveform has a lower peak voltage. The ideal case, in which the waveform is flat, occurs when the third-order harmonic component is 1/9 that of the fundamental [4]. Increasing the number of harmonics gives the waveform a "squarer" look and boosts the efficiency. However, the biggest boost in efficiency occurs with just the addition of the third-order component [11].

The various harmonics are chosen by a series of LC resonators with one resonator required (and tuned) for each harmonic needed. Transmission line resonators or some combination of lumped and distributed network resonators can also be used. The most common implementation, however, just uses third-order harmonic voltage peaking using a circuit such as that shown in Figure 9.29. In this case, L_3C_3 are tuned to the third harmonic of the input signal, and L_1C_1 resonates at the fundamental and lets the desired amplified signal pass to the load. At resonance, the L_3C_3 filter presents an open circuit and reflects the third harmonic back to drain to combine with the fundamental component to produce the required waveform.[2] Note that the traditional Class F amplifier does not exhibit the shunt output capacitance at the switch nodes (C_0) as part of the basic circuit topology, making this amplifier more sensitive to output capacitance [18]. If this capacitance is too high, then the higher order harmonics are actually shorted out and will have insufficient amplitude to aid in the construction of the desired waveform. The differences in circuit topology between Class F and Class F^{-1} amplifiers are summed up by the following statement

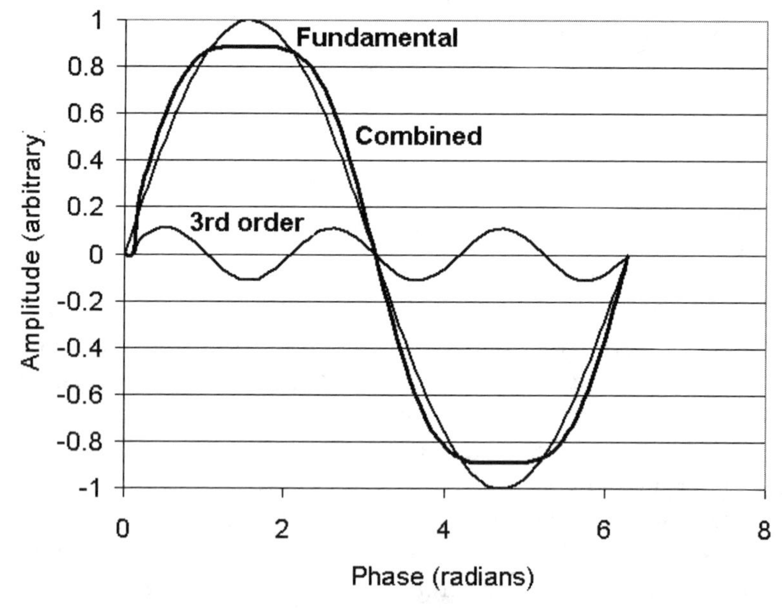

Figure 9.28 A third-order component of 1/9 the amplitude of the fundamental yields a nearly flat maximum in the combined waveform.

2. This type of waveform engineering for improvement in efficiency of RF power amplifiers has been studied by a number of authors [5, 11, 18].

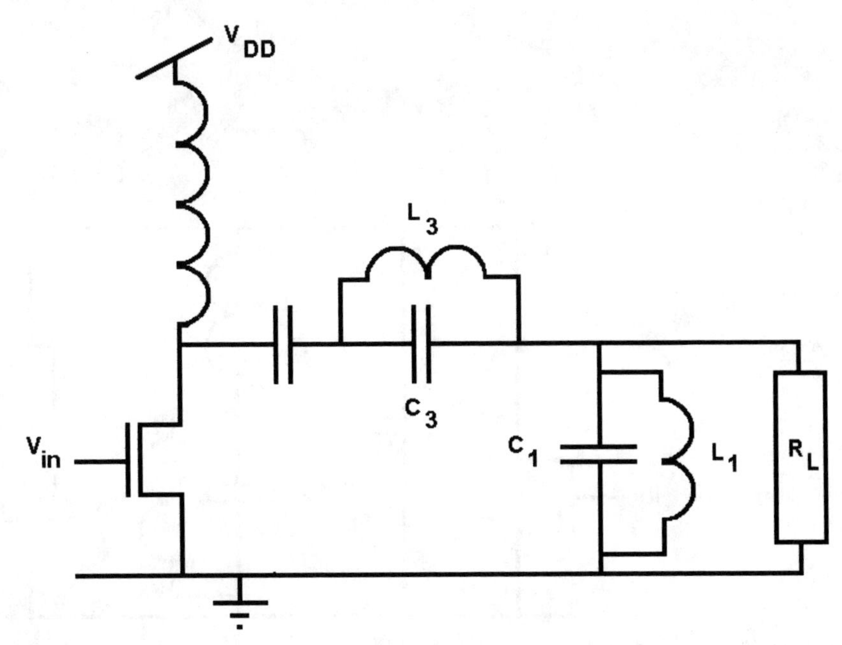

Figure 9.29 Basic Class F amplifier circuit topology forms third-order harmonic peaking. A dc blocking capacitor is also shown.

[18]: "Where Class-F short circuits even harmonics and open circuits odd harmonics, Class F^{-1} open circuits even harmonics and short circuits odd harmonics." For Class F^{-1} operation, the L_3C_3 parallel resonant circuit becomes a series resonant one.

A unified design approach for selection of the various circuit components shown in Figure 9.29 has been advanced for the Class F amplifier [19]. For the circuit shown in Figure 9.29 and with a amplifier bandwidth B, the following design equations can be used:

$$C_1 = \frac{1 - \frac{\pi B}{\omega}}{\omega R_L \left[1 - \left(1 - \frac{\pi B}{\omega}\right)^2\right]}; \; L_1 = \frac{1}{\omega^2 C_1}$$
$$L_3 = \frac{160 L_1 R_L^2}{81\left[9R_L^2 + (2\omega L_1)^2\right]}; \; C_3 = \frac{1}{9\omega^2 L_3} \tag{9.37}$$

The dc blocking capacitor should be approximately eight times C_3. An interesting modification to this design approach uses the circuit shown in Figure 9.30 to provide the proper terminations for the desired waveforms. Note that in this circuit topology, there is a capacitance C_1 in shunt across the switching MOSFET; MOSFET and bias inductor parasitics may be absorbed into this capacitance. The design equations for this modified circuit are indicated by (9.38) [17, 19]:

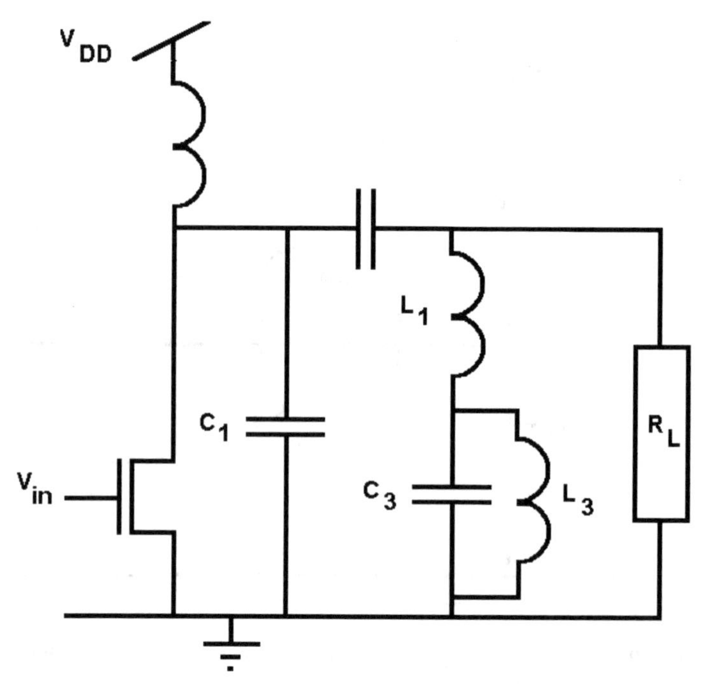

Figure 9.30 Modified Class F amplifier circuit places a capacitance in shunt with the switching transistor.

$$C_1 = \frac{1}{R_L[\omega + \pi B]};\ L_1 = \frac{1}{6\omega^2 C_1}$$
$$C_3 = \frac{12}{5}C_1;\ L_3 = \frac{5}{3}L_1 \tag{9.38}$$

These simple design equations give the CMOS RFIC design a starting point to begin the overall on-chip PA design process. Layout and device parasitics must be considered for a successful final design; several design iterations are usually required before reaching the final design.

To observe how these equations can be used to design a Class F power amplifier, consider the design of a 500-mW amplifier using a 3.3V power supply for use at 1.0 GHz (ch9-5.txt), similar to the previous Class E example. Using the design equations in (9.38) for the Class F amplifier with narrow bandwidth ($B = 0$) yields the following circuit component values:

$$R_L = \frac{V_{DD}^2}{2P_{OUT}} = \frac{3.3^2}{2(0.5)} = 10.9\Omega$$
$$C_1 = \frac{1}{\omega R_L} = 14.6\text{ pF};\ L_1 = \frac{1}{6\omega^2 C_1} = 0.29\text{ nH}$$
$$L_3 = \frac{5}{3}L_1 = 0.48\text{ nH};\ C_3 = \frac{12}{5}C_1 = 35\text{ pF}$$

The same 0.5Ω on-resistance of the MOSFET has been carried over from the Class E amplifier design as well. The results of a SPICE simulation with these design values is shown in Figure 9.31 and shows the voltage across and the current through the MOSFET plotted with $2.0V_p$ input signal drive. The drain voltage waveform shows a "squaring" up of the sinusoid due to the addition of the third-order component. The current waveform also shows a nearly one-half sinusoidal shape (off during voltage peaks) although there is a small spike of current occurring at the negative transition of the MOSFET voltage. The simulated output power is 660 mW at the load with a power supply current of 251 mA, corresponding to an efficiency of approximately 80%.

9.4 Other Amplifiers

The drain and gate capacitances that are inherent to the MOSFET structure are a major limiting factor in the ultimate frequency response of the amplifiers created with these devices. For Class E and F amplifiers, these large capacitances can be used as part of the output loading network to shape the waveforms to achieve high overall amplifier efficiency. Another technique that exploits these parasitic MOSFET capacitances is to use them as part of a phasing network reminiscent of a transmission line; this structure leads to what is called a *distributed amplifier*. A completely different approach, based on a more systems-level observation of amplifier operation, can be used in improving PA efficiency as well. These system-level approaches are many, but two popular types are based on amplifier circuit topologies first described by Doherty and Kahn, and hence bear their names to this day. The next section looks at the operation and design of these alternate PA circuits.

9.4.1 Distributed Amplifiers

The term *distributed amplifier* (DA) is an accurate description for the actual circuit construction of this multiple device amplifier. The MOSFET drain and gate capacitances, in shunt with the drain and gate nodes, can be made part of an overall

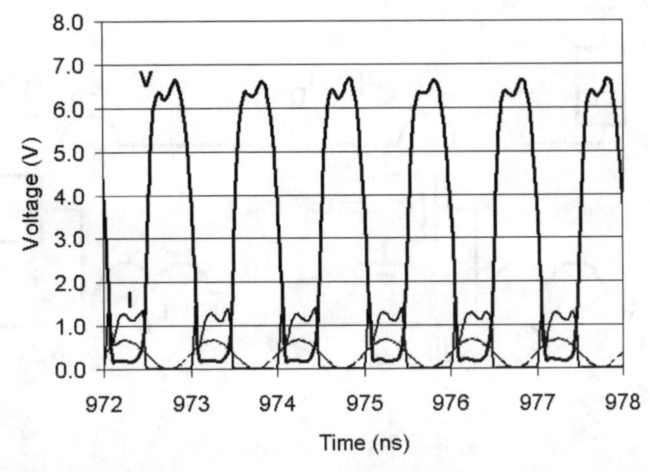

Figure 9.31 SPICE simulation results on the Class F amplifier showing MOSFET drain voltage and current (current not to scale). The dotted line is the input driving voltage (reduced in magnitude and shifted in scale for clarity).

phasing network by adding a series inductor between each MOSFET in the amplifier cascade (Figure 9.32). The drain portion of the DA is connected to the load resistance R_L at the output and drain impedance Z_D at the opposite end. The gate portion of the DA is connected to the source impedance R_S at the input and a gate impedance Z_G at the opposite end. The cascade of a series connection of series L shunt C has an identical look to that of the lossless transmission line model; in fact, transmission line model terminology is used to describe the DA operation. For the gate lead, the input voltage V_{in} changes in phase by an amount $e^{-j\beta_G l_G}$ at each gate node point as the input travels down to the gate impedance Z_G (for this reason, the DA is also sometimes referred to as a *traveling-wave amplifier* or *TWA*) [20].

The amplitude of the voltage at each MOSFET gate is the same if the gate line is lossless; only the phase is different and accumulates by $e^{-j\beta_G l_G}$ as one moves down the DA chain. The phase term β_G is a function of the series inductor L_G added to the gate lead and the inherent gate capacitance C_{GS} of the MOSFET:

$$\beta_G = \omega\sqrt{L_G C_{GS}} \tag{9.39}$$

Along the same avenue as transmission line theory, there will be no reflections on the input line if the phasing network parameters match those of gate impedance Z_G:

$$Z_G = \sqrt{\frac{L_G}{C_{GS}}} \tag{9.40}$$

The drain line circuit is similar to the gate line except for the added effect of the current source originating from the transconductance of the MOSFET (Figure 9.33). At each current source node, one-half of the total current supplied by each transconductance current source heads toward the load and the other half toward

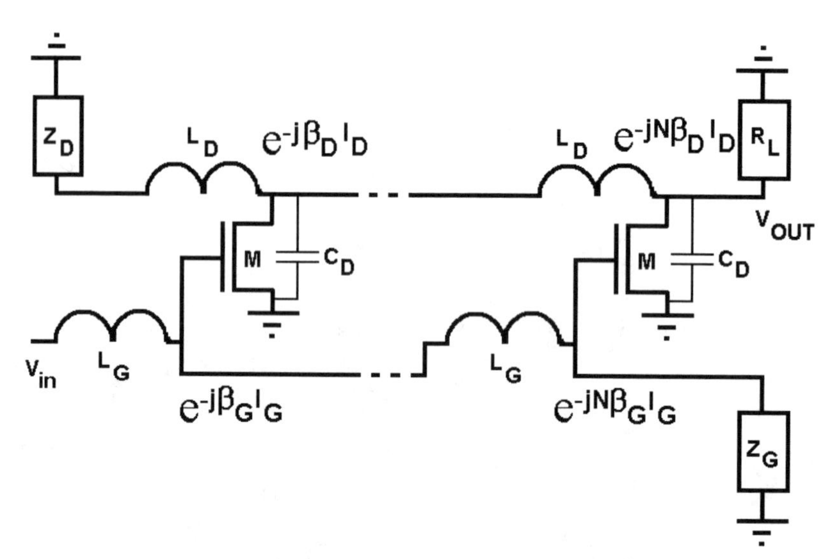

Figure 9.32 General distributed amplifier showing added drain and gate lead inductances for the drain and gate phasing networks.

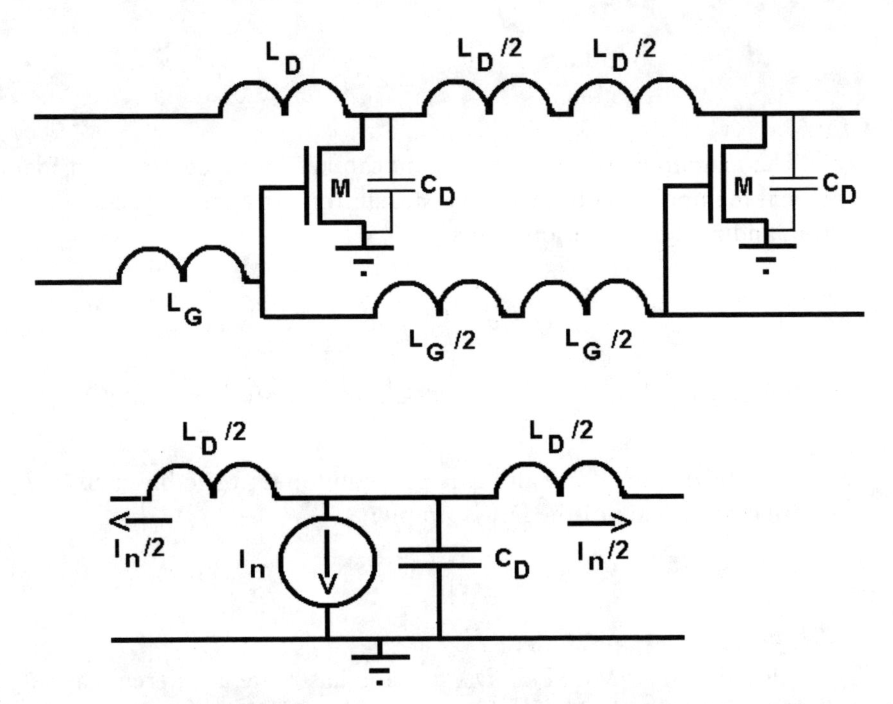

Figure 9.33 Subset of distributed amplifier circuit, showing the drain line RF equivalent circuit, which includes the MOSFET transconductance current source, I_n.

Z_D. Similar to the gate line, the phase of each current element advances by $e^{-j\beta_D l_D}$ as it moves toward the load,

$$\beta_D = \omega\sqrt{L_D C_D};\ Z_D = \sqrt{\frac{L_D}{C_D}} \tag{9.41}$$

where β_D is the drain line phase constant and Z_D is the impedance of the drain line.

For an N-stage DA, the total voltage at the load termination R_L is the sum of each current; keeping track of the phase of each current contribution allows the output voltage to be written in general form as

$$V_{OUT} = R_L \sum_{n=1}^{N} \frac{I_n}{2} e^{-j\beta_D l_D [N-n+1]} \tag{9.42}$$

where I_n is the nth transconductance current source contribution. Each I_n originates from its respective input gate voltage drive and the MOSFET transconductance g_m. Assuming identical MOSFETs with identical transconductances, a general expression for I_n can be written in terms of the original input signal V_{in} as

$$I_n = -g_m V_{\text{in}} e^{-j\beta_G l_G} \tag{9.43}$$

and the resulting output voltage gain can be written as

$$A_V = \frac{V_{\text{OUT}}}{V_{\text{in}}} = -\frac{1}{2} g_m R_L \sum_{n=1}^{N} e^{-j\beta_D l_D [N-n+1]} e^{-j\beta_G l_G} \tag{9.44}$$

The magnitude of the voltage gain can be simplified after considerable mathematical manipulation to provide a straightforward gain expression in terms of the drain and gate propagation terms:

$$|A_V| = \frac{g_m R_L}{2} \frac{\sin\left[\frac{N}{2}(\beta_D l_D - \beta_G l_G)\right]}{\sin\left[\frac{1}{2}(\beta_D l_D - \beta_G l_G)\right]} \tag{9.45}$$

By adjusting the drain and gate propagation terms to be equal ($\beta_D l_D = \beta_G l_G$), the overall cascade gain of the DA is simply:

$$|A_V| = \frac{g_m R_L N}{2} \tag{9.46}$$

Note that the *n*MOSFET DA cascade has quite a different cascade gain result when compared with the simple amplifier cascade. For the simple $g_m R_L$ cascade, the gain increases exponentially with the number of stages N:

$$|A_V| = (g_m R_L)^N \tag{9.47}$$

indicating that the DA gain will be significantly lower than its simple cascade counterpart. However, the advantage of the DA over the typical amplifier cascade is that the frequency-limiting capacitances have been made part of the phasing network, so the frequency response of the DA can actually be greater than that of the individual amplifiers themselves. In fact, the DA gain bandwidth product (GBW) is approximately $Nf_T/2$ where f_T is the unity current gain frequency of the individual MOSFETs [20]. Of course, any losses in the on-chip inductors (which will always be the case since these are relatively low Q devices) will reduce this gain. Differential DA circuit topologies can also be used by replacing the individual MOSFET transconductors with a differential MOSFET transconductor; common mode rejection is improved, and some immunity to substrate injected noise can be achieved with this circuit [21].

A distributed amplifier design technique has been advanced that provides good first-pass results [22]. In this procedure, drain and gate cutoff frequencies are defined based on the phasing networks in those lines, and a "staggering technique" in defining the drain and gate line components is used. The result yields expressions for the phasing network inductance and capacitance for a given line impedance Z_0 (Z_D assumed equal to Z_G):

$$L_G = 0.7\ L_D;\ C_{GS} = 0.7\ C_D \tag{9.48}$$

The factor of 0.7 is the optimum staggering factor and is used to minimize gain peaking near the drain and gate line cutoff frequencies [22]. The gate capacitance

C_{GS} is approximately WLC_{ox}, and so the required transconductance g_m and corresponding MOSFET aspect ratio W/L can be found for $R_L = Z_0$:

$$g_m = \frac{2|A|}{NZ_0} = \frac{2|A|}{N}\sqrt{\frac{C_D}{L_D}};\ \frac{W}{L} = \frac{g_m}{\mathrm{KP}(V_{GS} - V_T)} \tag{9.49}$$

This design procedure only gives an initial starting point for DA design. More detailed computer simulations using such techniques as simulated annealing or other computer optimization techniques are needed to fine-tune the result in the presence of losses (primarily inductor losses) and phase differences due to process variations and interconnections [8]. In fact, the addition of losses in the general DA actually limits the number of stages (N) for optimal DA performance. A number of authors have shown that the optimum value for N can be computed as [1, 23]

$$N_{OPT} = \frac{\ln(\alpha_G/\alpha_D)}{(\alpha_G - \alpha_D)} \tag{9.50}$$

where α_G and α_D are the attenuation factors in the gate and drain line, respectively. For this reason, MOSFET distributed amplifiers are usually composed of three to five stages.

To observe how these equations can be used to design an ideal distributed amplifier, consider the design of a three-stage ($N = 3$) 9.54-dB distributed amplifier (gain of 3) using a 3.3V power supply for use out to 1.0 GHz (ch9-6.txt). Assume 50Ω impedances for all loads and distributed elements. Further, assume that the drain capacitance is known a priori to be 1.0 pF. Using a 0.7 drain and source staggering factor (9.48), the associated gate capacitance can be calculated to be 0.7 pF. For a 50Ω system, the corresponding drain inductance is 2.5 nH (gate inductance of 1.75 nH). For a gain of three, the required transconductance and associated aspect ratio of each transistor is

$$g_m = \frac{2|A|}{NZ_0} = \frac{2(3)}{3(50)} = 0.04\mathrm{S}$$

$$\frac{W}{L} = \frac{g_m}{\mathrm{KP}(V_{GS} - V_T)} = \frac{0.04}{56\cdot 10^{-6}(3.3 - 0.66)} = 270.5$$

To complete the circuit, a 100-nH bias choke is used. SPICE simulation results of the DA gain (Figure 9.34) show the required 9.54-dB gain. The gain is 9.54 dB ± 1.0 dB out to beyond 2.0 GHz. The small gain peaking near the highest operating frequency is due to impedance variations of the distributed LC networks over frequency [22].

To see the effects of inductor parasitics on this circuit, the SPICE simulation was run again but this time with each inductor exhibiting a nominal inductor Q of 3.5 (ch9-7.txt). Parasitic capacitances of the inductors were not included in the simulation. Figure 9.34 shows the output voltage gain over frequency for this DA, which now includes losses in the inductors. The low-frequency gain of both amplifiers is similar but the gain begins a monotonic roll-off at a lower frequency with a

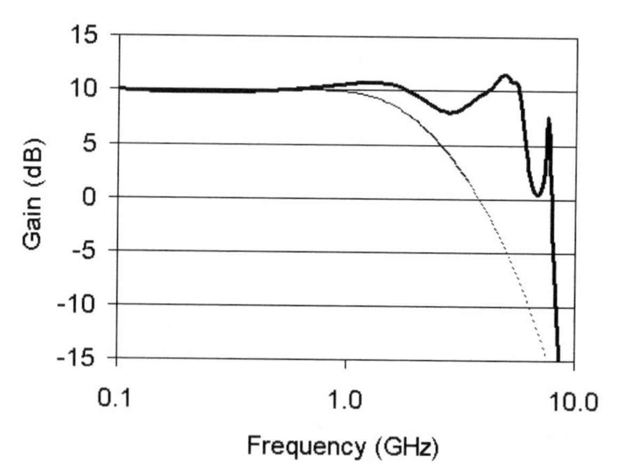

Figure 9.34 SPICE simulation results for the three-stage 1.0-GHz ideal (bold line) and nonideal (dotted line) distributed amplifier.

corresponding reduction in DA bandwidth. Additional device and component parasitics will degrade the performance of the DA even further.

9.4.2 Kahn and Doherty Structures

A systems-level approach for high-power amplifiers has been used in an effort to boost efficiency and linearity. Two of the most commonly used types are those first described by Doherty and Kahn for use in vacuum tube amplifiers decades ago [24, 25]. These two amplifier types are discussed separately.

9.4.2.1 Doherty Amplifier

Operation of the Doherty amplifier can be seen by looking at the block diagram of the basic system as shown in Figure 9.35. There are two amplifiers in the basic Doherty scheme, the *main amplifier* (MAIN) running usually in Class B and the *auxiliary amplifier* (AUX) running in a more efficient Class C. During periods of operation when the output power is low (usually below one-half the maximum output power), only the main amplifier is used and the auxiliary amplifier is off. (The Class C bias is set low enough that the input RF signal swing does not turn this amplifier on.) When higher levels of power are required, both amplifiers contribute to the total output power. The two 90° lines (or lumped element counterparts) are used for impedance matching of the amplifier outputs during all operating regimes [15].

The use of the auxiliary amplifier is illustrated in Figure 9.36. At low power levels, the auxiliary amplifier does not contribute to the total output power; only when the main amplifier starts to compress at its P_{1dB} level does the auxiliary amplifier "kick in," compensating for the main amplifier and keeping the drive level and output power relationship more linear. For this reason, the auxiliary amplifier is often referred to as a *peaking amplifier*.

Because the two amplifiers engage at different power levels, the efficiency of the Doherty amplifier varies throughout the RF output power range. Figure 9.37 shows this efficiency as a function of power level below maximum (so-called back-off

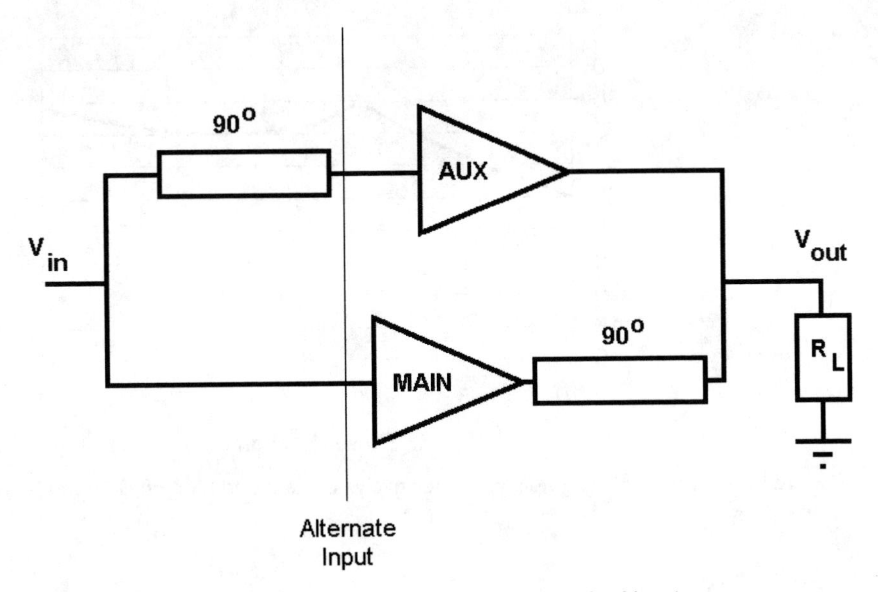

Figure 9.35 Basic Doherty amplifier ("alternate input" is described later).

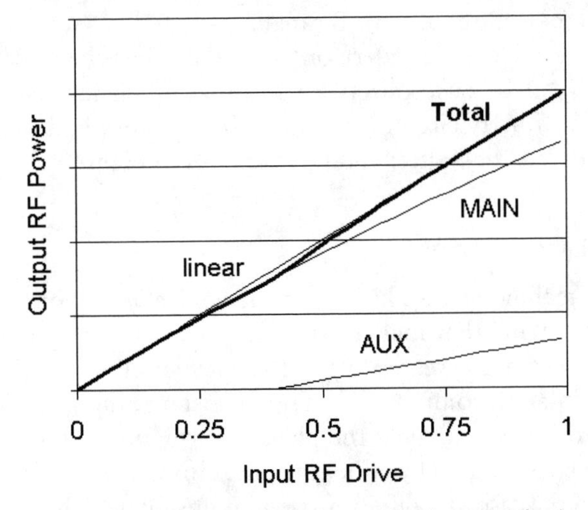

Figure 9.36 The total output power in the Doherty amplifier has contributions from the MAIN and AUX amplifiers.

power from P_{MAX}). At low power levels, the efficiency is typical of that of the Class B amplifier. Once the output power reaches to within 6 dB of P_{MAX}, the efficiency reaches the theoretical 78.5% maximum. At this point, the auxiliary amplifier begins to provide power to the total output; there is a region of somewhat lower efficiency during this range, but the overall efficiency is around 70% or better throughout the high-power range [4]. The high efficiency of the Doherty amplifier comes from the interaction between each amplifier. As noted in the discussion of transconductance amplifiers at the beginning of this chapter, efficiency typically degrades with decreasing output power for a fixed load termination; however,

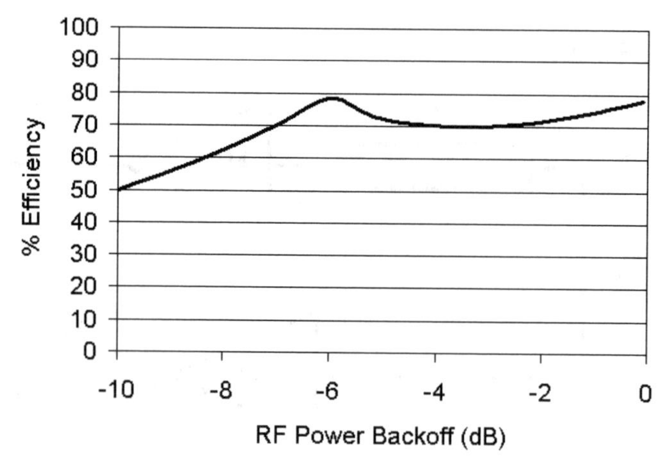

Figure 9.37 The Doherty amplifier has relatively constant and high efficiency at high output power levels.

being able to vary the load termination (R_{OPT}) improves the amplifier efficiency. This is what happens with the Doherty amplifier. The 90° phasing line at the output "pulls" the output resistance (so-called *active load pull* [4]) of each amplifier to a higher value over about a 6-dB range.

With the advent of modern on-chip DSPs, the level of RF drive signals to the two amplifiers can be customized for optimal efficiency over practically any power range. The Doherty circuit is modified from its original to use the DSP to its fullest (the DSP drives the "alternate input" nodes in Figure 9.35).

9.4.2.2 Kahn Technique

The Kahn technique (also known as *envelope elimination and restoration* or EER) is another "system" that makes use of the characteristics of a phase- and amplitude-modulated signal to create a high-efficiency amplifier (Figure 9.38) [15]. The signal phase modulation component can be a constant amplitude signal since only the zero crossings of the signal yield the phase information. This constant amplitude phase signal can be generated by either the programmed output of the on-chip DSP or with a limiter in the case of a purely analog input signal. This phase portion of the RF signal can be amplified by a relatively nonlinear but highly efficient power amplifier (such as Class C, E, or F). The amplitude modulation (or envelope) gets restored at this point by varying the power output of the PA as the envelope varies. The amplitude of the input signal can be obtained by passing the signal through an envelope detector. Because the amplitude modulation is usually at a much lower frequency than the RF signal, the envelope restorer can be both high power and efficient. The EER technique dates back to vacuum tube days, using a Class S power supply modulator as the envelope restorer; the same circuit technique can be used today. The concept of power supply modulation is easy to grasp if one considers the basic definition of the PA output power:

$$P_{\text{OUT}} = \frac{V_{DD}^2}{2R_L} \tag{9.51}$$

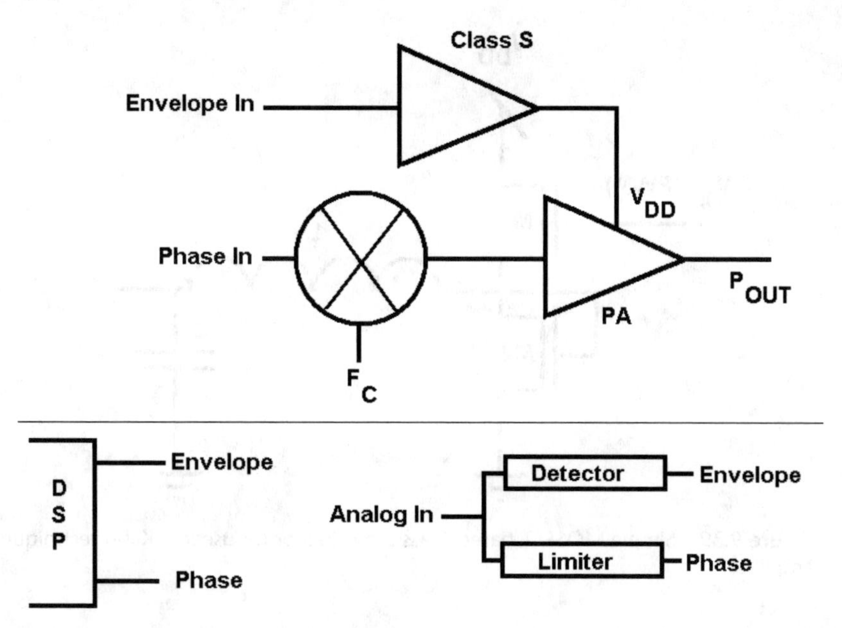

Figure 9.38 Circuit diagram for a Kahn EER amplifying technique. The inputs to the Kahn amplifier may come from a DSP or by using envelope and phase detectors in the input signal. The DSP approach is often called a *polar modulator* because the I and Q signals output from the DSP are in polar form (amplitude or envelope and phase).

If the voltage V_{DD} is made time varying by the signal amplitude envelope $[V_{DD}(t)]$, the output power will follow that envelope $[P_{OUT}(t)]$ and amplitude modulate the output amplifier. The envelope bandwidth should be at least twice the RF bandwidth of the modulating signal [15].

A Class S modulator, shown in Figure 9.39, is a type of switching circuit but the input signal V_{in} is a pulse width modulated (PWM) signal. The pulse width is proportional to the output signal level; the LC filter removes all but the fundamental envelope component desired since the switching process necessarily generates harmonically related signals based on the PWM signal shape. The diode connected MOSFET (M2) may be replaced with a PN junction diode if that is supported by the technology.

9.5 Amplifier Linearizers

The linearity of power amplifiers is becoming increasingly important as more channels and users and being forced to compete for finite spectrum space, with adjacent channel interference and tight spectral masks being major design and system issues. In addition, linear PAs have their best efficiency at power levels approaching the 1-dB compression point, which is also the onset of significant nonlinear operation. The use of linearization techniques can be used to reduce the ACI and delay the onset of P_{1dB} by a few decibels and improve amplifier efficiency. A number of quite complex techniques exploiting the power of on-chip DSPs provide a virtually unlimited number of waveforms that can yield the required amplifier linearization.

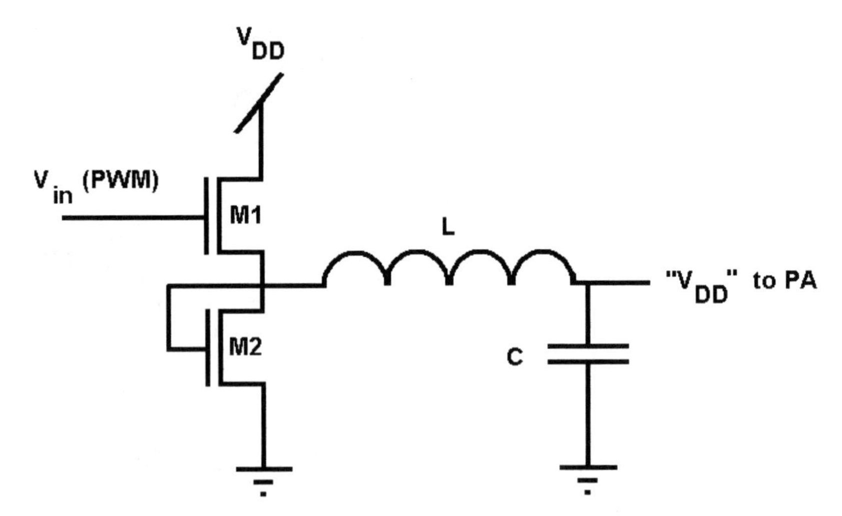

Figure 9.39 Simple MOSFET-based Class S modulator for use in a Kahn technique high-efficiency amplifier.

Alternatively, a number of relatively simple analog techniques are available that can provide linearization over a more modest range but that take the linearization load off the DSPs so that they can pay attention to other tasks. This section discusses the concept behind amplifier linearization and shows various techniques that can be employed to improve amplifier performance.

9.5.1 Basic Amplifier Linearization

The basic amplifier linearizer seeks to reduce odd-order components because it is these intermodulation components that are generated within the desired frequency band. For example, in (1.43), the in-band third-order IMD term has an amplitude of:

$$I_{d-3} = g_3 \frac{3}{4} AB^2 \tag{9.52}$$

If a signal could be applied to the amplifier in such a way that its third-order IMD term is 180° out of phase with (9.52), the distortion is effectively removed. Although this technique works ideally on paper, in actual practice, the circuit used to generate the third-order compensation signal itself introduces nonlinearities and its higher order terms (fifth, seventh, etc.) may actually enhance overall amplifier distortion and increase the output spectral width.[3]

Two major of types of linearizers that perform the nonlinear signal cancellation are delineated by their circuit location. If the cancellation signal is injected at the input of the amplifier, a *predistortion* (PD) type of linearizer is obtained. If the cancellation signal is injected into the amplifier output, a *feed-forward* (FF) type of linearizer is obtained. The operation of these two types of linearizers is sufficiently different that each will be discussed in turn.

3. Murphy's law will almost certainly guarantee this will be the case.

9.5.2 Predistortion Linearizers

As the name implies, the PD linearizer applies a signal of appropriate phase and amplitude *prior* to amplification so that the desired and cancellation signals are operated on by the amplifier so that the nonlinear output component is ideally eliminated. The cancellation signal may be generated by a DSP, which samples the PA output and adaptively changes the input amplitude and phase to yield the desired cancellation signal. Alternatively, a nonlinear device can be placed at the input to the amplifier and by carefully controlling the nonlinear characteristics of this device, the appropriate cancellation signal can be injected into the amplifier (Figure 9.40).

Consider a PA that has a nonlinear gain characteristic out to third order given by the following mathematical relationship:

$$V_{\text{OUT}} = G_1 V_{\text{in}} + G_2 V_{\text{in}}^2 + G_3 V_{\text{in}}^3 \tag{9.53}$$

Consider at the input a square-law predistortion signal that consists of a scaled input voltage and a scaled quadratic version of the input:

$$V_{in} = \alpha V_{RF} + \beta V_{RF}^2 \tag{9.54}$$

Substituting (9.54) into (9.53) and looking only at V_{RF}^3 components yields the following cancellation relationship:

$$2G_2 \alpha\beta V_{RF}^3 + G_3 \alpha^3 V_{RF}^3 = 0 \Rightarrow \beta = -\frac{1}{2}\alpha^2 \frac{G_3}{G_2} \tag{9.55}$$

Equation 9.55 indicates that for proper PD cancellation, the nonlinear gain coefficients of the PA, G_2 and G_3, must be known, which requires detailed knowledge of

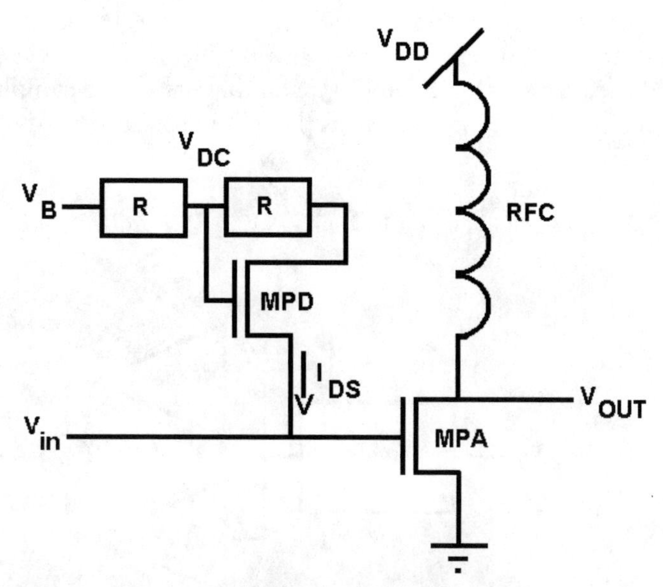

Figure 9.40 A diode-connected MOSFET (MPD) may be used to inject a cancellation signal and act as a predistorter for the MOSFET PA (MPA). (*After:* [27].)

the actual amplifier operation. These coefficients are retrieved from values stored in memory. In actual practice, these (and higher) coefficients change with time, and so adaptive techniques are required that sample the output and compute new values of G_2 and G_3 periodically. Using DSP techniques, ACI improvements of 25 dB are possible [5].

A major drawback to DSP predistorters in low-power applications is the increased amount of dc power necessary for the operation of the DSP. In theory, any device that generates a nonlinear signal that can be added to the desired input signal may be used as a predistorter. A CMOS implementation of this type of PD is based on the use of diode nonlinearities to provide the desired cancellation signal [26]. Rather than using a PN junction diode for the linearizer, a diode-connected MOSFET can be employed as shown in Figure 9.40 [27]. In the figure, MPA is the PA transistor. The square-law characteristic of the input signal is noted by observing that the source node of the MOSFET MPD is at voltage V_{in} and that a dc bias voltage V_{DC} is developed at the junction of the two resistors:

$$I_{DS} = \frac{1}{2}\text{KP}\frac{W}{L}(V_{GS} - V_T)^2 = \frac{1}{2}\text{KP}\frac{W}{L}(V_{DC} - V_{in} - V_T)^2 \tag{9.56}$$

The linear portion of V_{in} comes from the direct connection to the gate of the MOSFET MPA. Extensive simulations must be performed to determine the aspect ratio W/L and bias voltage V_B for the MOSFET MPD to ensure that the appropriate cancellation signal amplitude and phase are injected into the PA MOSFET.

9.5.3 Feed-Forward Linearizers

The converse of the predistortion linearizer, the feed-forward (FF) linearizer, provides the nonlinear correction signal at the output of the power amplifier system. A basic FF linearizer is shown in Figure 9.41 where the MAIN amplifier handles the high-power chores, while the AUX amplifier is designed to provide the nonlinear signal distortion cancellation [28, 29]. Consider a system where the input voltage is divided between the MAIN and AUX amplifiers with a sampling factor of α. If the output of both of the amplifiers has the following set of third-order voltage response characteristics,

$$V_M = (1-\alpha)G_{M1}V_{in} + (1+\alpha)^3 G_{M3}V_{in}^3;\ V_A = \alpha G_{A1}V_{in} + \alpha^3 G_{A3}V_{in}^3 \tag{9.57}$$

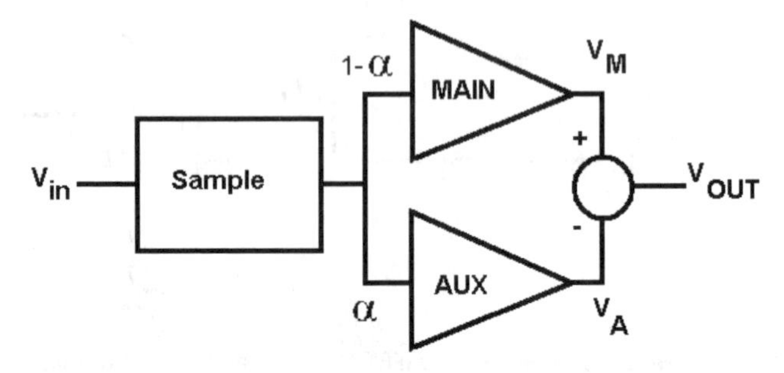

Figure 9.41 Basic feed-forward amplifier and linearizer circuit.

the output voltage will be the difference of the two responses:

$$V_O = V_M - V_A = \left[(1-\alpha)G_{M1} + \alpha G_{A1}\right]V_{\text{in}} + \left[(1-\alpha)^3 G_{M3} - \alpha^3 G_{A3}\right]V_{\text{in}}^3 \quad (9.58)$$

For cancellation of the output third-order signal, the following relationship must hold for sampling factor α:

$$\alpha = \frac{1}{1+\left(\dfrac{G_{A3}}{G_{M3}}\right)^{1/3}} \quad (9.59)$$

This form for α is strongly dependent on the nonlinear properties of each of the amplifiers. To maintain as much of the input signal as possible going into the MAIN amplifier, α should be small, which implies that the factor

$$\left(\frac{G_{A3}}{G_{M3}}\right)^{1/3} \quad (9.60)$$

should be maximized. A reasonably linear MAIN amplifier with a highly nonlinear AUX amplifier provides the necessary nonlinear signal voltage amplitude for third-order cancellation. The MOSFET simplified schematic shown in Figure 9.42 indicates one possible RFIC implementation for the FF linearizer. The MAIN

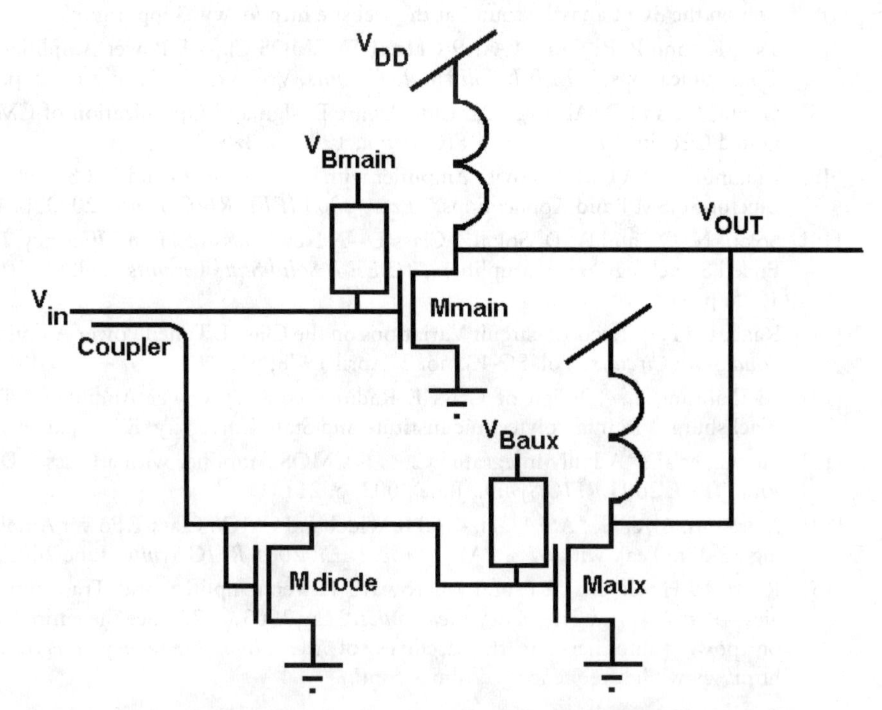

Figure 9.42 MOSFET version of a feed-forward linearizer. MOSFET Mdiode is used as a square-law generator to feed the auxiliary amplifier with a signal rich in second-order components. dc blocking and bypass capacitors not shown.

amplifier bias is set for the desired class of operation. The AUX amplifier bias is adjusted to provide the required cancellation signal amplitude and phase. Alternatively, a diode-connected MOSFET can be used at the input to generate the required nonlinear components to be amplified by the AUX amplifier. This third-order improvement will also provide some forward gain correction, although in actual practice a number of nonidealities in the system will limit this effect. As in the PD linearizer, the time-varying gain characteristics of the amplifiers also limit the overall reduction of third-order components at the output. Digital predistortion is currently being used to replace the more inefficient FF linearization schemes.[4]

References

[1] Pozar, D., *Microwave Engineering,* 2nd ed., New York: John Wiley and Sons, 1998.

[2] Vendelin, G., A. M. Pavio, and U. L. Rhode, *Microwave Circuit Design Using Linear and Nonlinear Techniques,* New York: John Wiley and Sons, 1990.

[3] Raab, F., Statement made during "HF, VHF, UHF Technology Workshop," *2006 IEEE Int. Microwave Symp.,* 2006.

[4] Cripps, S., *RF Power Amplifiers for Wireless Communications,* 2nd ed., Norwood, MA: Artech House, 2006.

[5] Gupta, R., B. Ballweber, and D. Allstot, "Design and Optimization of CMOS RF Power Amplifiers," *IEEE J. Solid State Circuits*, Vol. 36, No. 2, Feb. 2001, p. 166. (See also Hajimiri, A., "Fully Integrated RF CMOS Power Amplifiers—A Prelude to Full Radio Integration," *Proc. 2005 IEEE RFIC Symp.,* June 2005, p. 439.)

[6] 3rd Generation Partnership Project, "3GPP Technical Specification TS25.101, V7.4.0, UE Radio Transmission and Reception," TS25.101, V7.4.0, June 2006. (Up-to-date information on the 3GPP may be found at the website http://www.3gpp.org/.)

[7] Tsai, K., and P. R. Gray, "A 1.9-GHz, 1-W CMOS Class-E Power Amplifier for Wireless Communications," *IEEE J. Solid State Circuits*, Vol. 34, No. 7, July 1999, p. 962.

[8] Gupta, R., and D. Allstot, "Parasitic Aware Design and Optimization of CMOS RF Integrated Circuits," *IEEE 1998 RFIC Symp.*, 1998, p. 325.

[9] Khannur, P., "A CMOS Power Amplifier with Power Control and T/R Switch for 2.45 GHz Bluetooth/ISM Band Applications," *Proc. 2003 IEEE RFIC Symp.*, 2003, p. 145.

[10] Sokal, N. O., and A. D. Sokal, "Class E—A New Class of High Efficiency Tuned Single-Ended Switching Power Amplifiers," *IEEE J. Solid State Circuits*, Vol. SC-10, No. 3, June 1975, p. 168.

[11] Raab, F. H., "Effects of Circuit Variations on the Class E Tuned Power Amplifier," *IEEE J. Solid State Circuits*, Vol. SC-13, no. 2, April 1978, p. 239.

[12] Al-Shahrani, S., "Design of Class E Radio Frequency Power Amplifier," Ph.D. Thesis, Blacksburg: Virginia Polytechnic Institute and State University, EE Department, 2001.

[13] Ho, C., et al., "A Fully Integrated Class-E CMOS Amplifier with a Class-F Driver Stage," *Proc. IEEE 2003 RFIC Symp.*, June 2003, p. 211.

[14] Mazzanti, A., et al, "A 1.4 GHz-2 GHz Wideband CMOS Class-E Power Amplifier Delivering 23 dBm Peak with 67% PAE," *Proc. IEEE 2005 RFIC Symp.*, June 2005, p. 12.

[15] Raab, F. H., et al., "RF and Microwave Power Amplifier and Transmitter Technologies—Part 2," *High Frequency Electronics*, May 2003, p. 22. (See the entire five-part series on power amplifiers in the archives of the *High Frequency Electronics* website, http://www.highfrequencyelectronics.com/.)

4. Thanks to the reviewer for pointing out this fact.

[16] Allen, P., and D. Holberg, *CMOS Analog Circuit Design,* 2nd ed., New Ork: Oxford University Press, 2002.

[17] Grebennikov, A., "Load Network Design Technique for Switched-Mode Tuned Class E Power Amplifiers," *High Frequency Electronics*, July 2004, p. 19.

[18] Kee, S. D., et al., "The Class-E/F Family of ZVS Switching Amplifiers," *IEEE Trans. Microwave Theory Tech.*, Vol. 51, No. 6, June 2003, p. 1677.

[19] Grebennikov, A., "Circuit Design Technique for High-Efficiency Class-F Amplifiers," *Int. Microwave Symp. Digest*, Boston, MA, June 13–15, 2000, Vol. 2, pp. 771–774. (See also Trask, C., "Class-F Amplifier Loading Networks: A Unified Design Approach," *Proc. 1999 IEEE Int. Microwave Symp.*, June 1999, p. 351.)

[20] Zhang, F., and P. R. Kinget, "Low-Power Programmable Gain CMOS Distributed LNA," *IEEE. J. Solid State Çircuits*, Vol. 41, No. 6, June 2006, p. 1333.

[21] Ahn, H., and D. J. Allstot, "A 0.5–8.5 GHz Fully Differential CMOS Distributed Amplifier," *IEEE. J. Solid State Circuits*, Vol. 41, No. 6, Aug. 2002, p. 985.

[22] Ballweber, B. M., R. Gupta, and D. J. Allstot, "A Fully Integrated 0.5–5.5 GHz CMOS Distributed Amplifier," *IEEE. J. Solid State Circuits*, Vol. 35, No. 2, Feb. 2000, p. 231.

[23] Amaya, R., and C. Plett "Design of High Gain Fully-Integrated Distributed Amplifiers in 0.35μm CMOS," *Proc. 29th ESSCIRC Solid State Circuits Conf.*, Sept. 2003, p. 145.

[24] Doherty, W. H., "A New High Efficiency Power Amplifier for Modulated Waves," *Proc. IRE*, Vol. 24, No. 9, Sept. 1935, pp. 1163–1182.

[25] Kahn, L. R., "Single Sideband Transmission by Envelope Elimination and Restoration," *Proc. IRE*, Vol. 40, No. 7, July 1952, pp. 803–806.

[26] Kazuhisa, Y., et al., "A Novel Series Diode Linearizer for Mobile Radio Power Amplifiers," *Proc. 1996 IEEE Int. Microwave Symp.*, June 1996, p. 831.

[27] Yen, C., and H. Chuang, "A 0.25-μm 20-dBm 2.4-GHz CMOS Power Amplifier with an Integrated Diode Linearizer," *IEEE Microwave Wireless Components Lett.*, Vol. 13, No. 2, Feb. 2003, p. 45.

[28] Yu, S., and P. Roblin, "Analysis and Simulation of Low-Frequency Feed-Forward Linearization," *Proc. 44th IEEE MWSCAS*, Vol. 2, Aug. 2001, p. 789.

[29] Yang, Y., and B. Kim, "A New Linear Amplifier Using Low-Frequency Second-Order Intermodulation Component Feedforwarding," *IEEE Microwave Guided Wave Lett.*, Vol. 9, No. 10, Oct. 1999, p. 419.

Selected Bibliography

Aoki, I., et al., "Fully Integrated CMOS Power Amplifier Design Using the Distributed Active-Transformer Architecture," *IEEE J. Solid State Circuits*, Vol. 37, No. 3, March 2002, p. 371.

Aoki, I., et al., "A Fully-Integrated 1.8-V, 2.8-W, 1.9-GHz, CMOS Power Amplifier," *Proc. 2003 IEEE RFIC Symp.*, June 2003, p. 1999.

Inoue, A., et al., "Analysis of Class-F and Inverse Class-F Amplifiers," *Proc. 2000 IEEE Int. Microwave Symp.*, Vol. 2, June 2000, p. 775.

Lu, L. H., T. Chen, and Y. Lin, "A 32-GHz Non-Uniform Distributed Amplifier in 0.18-μm CMOS," *IEEE Microwave Wireless Comp. Lett.*, Vol. 15, No. 11, Nov. 2005, p. 745.

Raab, F. H., "Maximum Efficiency and Output of Class-F Power Amplifiers," *IEEE Trans. Microwave Theory Tech.*, Vol. 47, No. 6, June 2001, pp. 1162–1166.

APPENDIX A

Sample SPICE-3 Parameters

These sample SPICE-3 parameters are available on the CD as file: X:spice3parms.txt (where X: is your CD drive). These parameters are for example purposes only in this text.

```
.MODEL CMOSN1 NMOS LEVEL = 3
+TOX=1.38E-8     VTO=0.66        KP=56.30e-6   U0=450       PHI=0.7
+LAMBDA=0.01     GAMMA=0.5       NSUB=1.50E15  XJ=0.5E-6
+CGDO=2.18E-10   CGSO=2.18E-10   CGBO=1E-9
+CJ=4.27E-4      PB=0.9          MJ=0.43
+CJSW=3.05E-10   PBSW=0.8        MJSW=0.21
```

```
.MODEL CMOSP1 PMOS LEVEL = 3
+TOX=1.38E-8      VTO=-0.91       KP=1.91E-5    U0=152     PHI=0.7
+LAMBDA=0.01      GAMMA=0.5       NSUB=1E17     XJ=2E-7
+CGDO=1.87E-10    CGSO=1.87E-10   CGBO=1E-10
+CJ=3.04E-4       PB=0.9          MJ=0.44
+CJSW=1.69E-10    PBSW=0.8        MJSW=0.10
```

APPENDIX B

Sample SPICE BSIM Parameters

These sample SPICE BSIM parameters for a 0.5-μm double poly triple metal CMOS process are available on the CD as file: X:bsimparms.txt (where X: is your CD drive). (Courtesy of MOSIS, http://www.mosis.com. Used with permission.)

```
*n-channel MOSFET parametrs
*SPICE 3f5 Level 8, Star-HSPICE Level 49, UTMOST Level 8
*Temperature_parameters=Default
.MODEL CMOSN NMOS(                                    LEVEL =49
+VERSION = 3.1          TNOM =27                      TOX =1.38E-8
+XJ =1.5E-7             NCH =1.7E17                   VTH0 =0.6559455
+K1 =0.86934            K2 =-0.0988243                K3 =25.6950929
+K3B =-7.7863337        W0 =1E-8                      NLX =1E-9
+DVT0W =0               DVT1W =0                      DVT2W =0
+DVT0 =3.4348179        DVT1 =0.3738587               DVT2 =-0.0888276
+U0 =456.5594281        UA =2.474049E-12              UB =1.487E-18
+UC =4.911677E-12       VSAT =1.900944E5              A0 =0.554448
+AGS =0.1241863         B0 =2.736758E-6               B1 =5E-6
+KETA =-3.522862E-3     A1 =9.530913E-6               A2 =0.3098765
+RDSW =1.185055E3       PRWG =0.0758314               PRWB =0.0183752
+WR =1                  WINT =2.386617E-7             LINT =7.69339E-8
+XL =1E-7               XW =0                         DWG =-5.133E-9
+DWB =3.839878E-8       VOFF =0                       NFACTOR =0.5292434
+CIT =0                 CDSC =2.4E-4                  CDSCD =0
+CDSCB =0               ETA0 =2.048817E-3             ETAB =-2.0867E-4
+DSUB =0.0683907        PCLM =2.547959                PDIBLC1 =1
+PDIBLC2 =2.466948E-3   PDIBLCB =-2.660252E-3         DROUT =0.9718917
+PSCBE1 =6.382393E8     PSCBE2 =2.111041E-4           PVAG =9.95724E-3
+DELTA =0.01            RSH =82.6                     MOBMOD =1
+PRT =0                 UTE =-1.5                     KT1 =-0.11
+KT1L =0                KT2 =0.022                    UA1 =4.31E-9
+UB1 =-7.61E-18         UC1 =-5.6E-11                 AT =3.3E4
+WL =0                  WLN =1                        WW =0
+WWN =1                 WWL =0                        LL =0
+LLN =1                 LW =0                         LWN =1
+LWL =0                 CAPMOD =2                     XPART =0.5
+CGDO =2.18E-10         CGSO =2.18E-10                CGBO =1E-9
+CJ =4.276473E-4        PB =0.9144082                 MJ =0.4304161
+CJSW =3.05451E-10      PBSW =0.8                     MJSW =0.2124257
+CJSWG =1.64E-10        PBSWG =0.8                    MJSWG =0.2124257
+CF =0                  PVTH0 =0.1217063              PRDSW =-183.54194
+PK2 =-0.0283047        WKETA =-0.017074              LKETA =4.0323E-3)
```

```
* p-channel MOSFET parameters
* SPICE 3f5 Level 8, Star-HSPICE Level 49, UTMOST Level 8
* Temperature_parameters=Default
.MODEL CMOSP PMOS (                                     LEVEL =49
+VERSION =3.1          TNOM =27                 TOX =1.38E-8
+XJ =1.5E-7            NCH =1.7E17              VTH0 =-0.9479551
+K1 =0.5428151         K2 =8.112852E-3          K3 =8.2586198
+K3B =-0.8301998       W0 =1E-8                 NLX =4.89505E-8
+DVT0W =0              DVT1W =0                 DVT2W =0
+DVT0 =2.4704748       DVT1 =0.4435188          DVT2 =-0.0848431
+U0 =216.5840726       UA =2.970391E-9          UB =1.6159E-21
+UC =-5.85255E-11      VSAT =1.774764E5         A0 =0.8022077
+AGS =0.1453224        B0 =1.269528E-6          B1 =5E-6
+KETA =-2.485273E-3    A1 =5.751071E-4          A2 =0.3010313
+RDSW =2.918751E3      PRWG =-0.042716          PRWB =-0.0213817
+WR =1                 WINT =2.688789E-7        LINT =1.02652E-7
+XL =1E-7              XW =0                    DWG =-1.3444E-8
+DWB =1.896646E-8      VOFF =-0.0583573         NFACTOR =0.7322289
+CIT =0                CDSC =2.4E-4             CDSCD =0
+CDSCB =0              ETA0 =0.5122687          ETAB =-0.0755207
+DSUB =0.9878092       PCLM =2.2234682          PDIBLC1 =0.0366688
+PDIBLC2 =3.166899E-3  PDIBLCB =-0.0462103      DROUT =0.1847743
+PSCBE1 =5.816939E9    PSCBE2 =5.495972E-10     PVAG =5.52722E-3
+DELTA =0.01           RSH =104.3               MOBMOD =1
+PRT =0                UTE =-1.5                KT1 =-0.11
+KT1L =0               KT2 =0.022               UA1 =4.31E-9
+UB1 =-7.61E-18        UC1 =-5.6E-11            AT =3.3E4
+WL =0                 WLN =1                   WW =0
+WWN =1                WWL =0                   LL =0
+LLN =1                LW =0                    LWN =1
+LWL =0                CAPMOD =2                XPART =0.5
+CGDO =2.38E-10        CGSO =2.38E-10           CGBO =1E-9
+CJ =7.238338E-4       PB =0.965516             MJ =0.4984422
+CJSW =2.831971E-10    PBSW =0.99               MJSW =0.3097831
+CJSWG =6.4E-11        PBSWG =0.99              MJSWG =0.3097831
+CF =0                 PVTH0 =5.98016E-3        PRDSW =14.8598424
+PK2 =3.73981E-3       WKETA =4.760239E-3       LKETA =-5.025E-3)

*
```

APPENDIX C

Y-Parameters of the MOSFET Model

The RF equivalent circuit shown in Figure C.1 for the MOSFET model is used extensively in determining the impact of the various equivalent circuit parameters on device performance. The Y-parameters of the MOSFET RF equivalent circuit shown in the figure are presented in (C.1). The drain channel noise current source (i_{dN}), gate resistance noise voltage source (v_{gN}), and bulk noise voltage source (v_{bN}) are also shown. The corresponding S-parameters for this RF equivalent circuit (minus noise sources) can be obtained using the Y-parameter–to–S-parameter conversion relationships presented in Appendix D. Note in Figure C.1 that the MOSFET source contact is RF grounded, eliminating both the body effect and the influence of the source-bulk diffused region; this equivalent circuit makes the expressions for the Y-parameters more tractable and is used to model many of circuit types described in this book. The drain-bulk diffused region remains in the equivalent circuit since, in the majority of cases, this is at the output, and hence any impedances here have a direct influence on the output characteristics of the device.

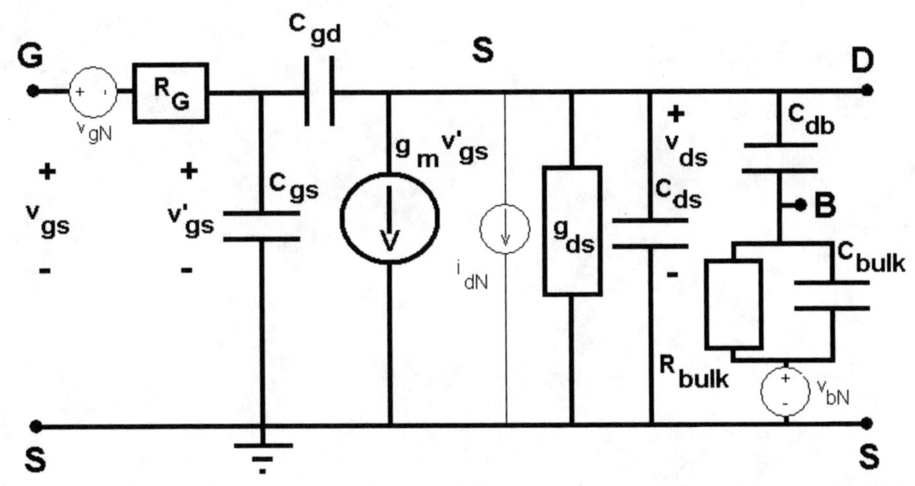

Figure C.1 RF equivalent circuit for a MOSFET. Noise sources and the drain-bulk effect are included.

$$
\begin{aligned}
Y_{11} &= \left.\frac{I_1}{V_1}\right|_{V_2=0} = \frac{j\omega\left(C_{gs}+C_{gd}\right)}{1+j\omega\left(C_{gs}+C_{gd}\right)R_G} \\
Y_{12} &= \left.\frac{I_1}{V_2}\right|_{V_1=0} = \frac{-j\omega C_{gd}}{1+j\omega\left(C_{gs}+C_{gd}\right)R_G} \\
Y_{21} &= \left.\frac{I_2}{V_1}\right|_{V_2=0} = \frac{g_m - j\omega C_{gd}}{1+j\omega\left(C_{gs}+C_{gd}\right)R_G} \\
Y_{22} &= \left.\frac{I_2}{V_2}\right|_{V_1=0} = g_{ds} + j\omega C_{ds} + j\omega\left(C_{gs}+C_{gd}\right) + \frac{j\omega C_{db}\left(1+j\omega C_{bulk}R_{bulk}\right)}{1+j\omega R_{bulk}\left(C_{db}+C_{bulk}\right)} \\
&\quad + \frac{\left[g_m - j\omega C_{gd}\right] j\omega C_{gd} R_g}{1+j\omega\left(C_{gs}+C_{gd}\right)R_G}
\end{aligned}
\tag{C.1}
$$

APPENDIX D

Parameter Conversion Equations for Two-Port Networks

S-Parameters to *Y*-Parameters:

$$Y_{11} = Y_0 \frac{(1 - S_{11})(1 + S_{22}) + S_{12} S_{21}}{(1 + S_{11})(1 + S_{22}) - S_{12} S_{21}}$$

$$Y_{21} = Y_0 \frac{-2S_{12}}{(1 + S_{11})(1 + S_{22}) - S_{12} S_{21}}$$

$$Y_{21} = Y_0 \frac{-2S_{21}}{(1 + S_{11})(1 + S_{22}) - S_{12} S_{21}}$$

$$Y_{22} = Y_0 \frac{(1 + S_{11})(1 - S_{22}) + S_{12} S_{21}}{(1 + S_{11})(1 + S_{22}) - S_{12} S_{21}}$$

Y-Parameters to *S*-Parameters:

$$S_{11} = \frac{(Y_0 - Y_{11})(Y_0 + Y_{22}) + Y_{12} Y_{21}}{\Delta Y}$$

$$S_{12} = \frac{-2Y_{12} Y_0}{\Delta Y}$$

$$S_{21} = \frac{-2Y_{21} Y_0}{\Delta Y}$$

$$S_{22} = \frac{(Y_0 + Y_{11})(Y_0 - Y_{22}) + Y_{12} Y_{21}}{\Delta Y}$$

where

$$\Delta Y = (Y_{11} + Y_0)(Y_{22} + Y_0) - Y_{12} Y_{21}$$

and Z_0 is the system reference impedance ($Y_0 = 1/Z_0$).

APPENDIX E

Constants and Some Properties of Silicon and CMOS-Related Materials

Tables E.1 and E.2 show constants and some properties of silicon and materials related to CMOS circuits at a temperature of 290K and metric unit parameters.

Table E.1 Important Physical Constants

Parameter	*Symbol*	*Value*
Free space permittivity	ε_0	8.85×10^{-12} F/m = 8.85 pF/m
Free space permeability	μ_0	$400\pi \times 10^{-9}$ H/m = 400π nH/m
Relative dielectric constants		
Silicon	ε_{Si}	11.8
Silicon dioxide (SiO_2)	ε_{ox}	3.9
Bulk mobility (channel mobility from SPICE BSIM parameter UO)		
Electron	μ_n	1,400 cm^2/V-s
Hole	μ_p	400 cm^2/V-s
Intrinsic carrier concentration	n_i	1.5×10^{16} m^{-3}
Boltzmann's constant	k	1.38×10^{-23} J/K
Electronic charge	q	1.6×10^{-19} C
Thermal voltage (300K)	kT/q	0.0258V

Table E.2 Metric Units

Parameter	*Symbol*	*Multiplier*
atto	a	10^{-18}
femto	f	10^{-15}
pico	p	10^{-12}
nano	n	10^{-9}
micro	μ	10^{-6}
milli	m	10^{-3}
centi	c	10^{-2}
deci	—	10^{-1}
deca (deka)	—	10^{1}
kilo	K	10^{3}
mega	M	10^{6}
giga	G	10^{9}
tera	T	10^{12}
peta	P	10^{15}

About the Author

Robert H. Caverly was born in Cincinnati, Ohio, in 1954. He received his Ph.D. in electrical engineering from the Johns Hopkins University, Baltimore, Maryland, in 1983. He received the M.S.E.E and B.S.E.E degrees from North Carolina State University, Raleigh, in 1978 and 1976, respectively.

Dr. Caverly has been a faculty member at Villanova University, Villanova, Pennsylvania, in the Department of Electrical and Computer Engineering since 1997. Previously, he was employed for more than 14 years at the University of Massachusetts, Dartmouth (formerly Southeastern Massachusetts University). In 1990, with support from the National Science Foundation, he was a visiting research fellow with the Microwave Solid-State Group at the University of Leeds in the United Kingdom. Dr. Caverly's research interests are in silicon RFIC design, as well as characterizing semiconductor devices, such as PIN diodes and FETs in the microwave and RF control environment.

Index

Y

Z

Recent Titles in the Artech House Microwave Library

Active Filters for Integrated-Circuit Applications, Fred H. Irons

Advanced Techniques in RF Power Amplifier Design, Steve C. Cripps

Automated Smith Chart, Version 4.0: Software and User's Manual, Leonard M. Schwab

Behavioral Modeling of Nonlinear RF and Microwave Devices, Thomas R. Turlington

Broadband Microwave Amplifiers, Bal S. Virdee, Avtar S. Virdee, and Ben Y. Banyamin

CMOS RFIC Design Principles, Robert Caverly

Computer-Aided Analysis of Nonlinear Microwave Circuits, Paulo J. C. Rodrigues

Design of FET Frequency Multipliers and Harmonic Oscillators, Edmar Camargo

Design of Linear RF Outphasing Power Amplifiers, Xuejun Zhang, Lawrence E. Larson, and Peter M. Asbeck

Design of RF and Microwave Amplifiers and Oscillators, Pieter L. D. Abrie

Digital Filter Design Solutions, Jolyon M. De Freitas

Distortion in RF Power Amplifiers, Joel Vuolevi and Timo Rahkonen

EMPLAN: Electromagnetic Analysis of Printed Structures in Planarly Layered Media, Software and User's Manual, Noyan Kinayman and M. I. Aksun

Essentials of RF and Microwave Grounding, Eric Holzman

FAST: Fast Amplifier Synthesis Tool—Software and User's Guide, Dale D. Henkes

Feedforward Linear Power Amplifiers, Nick Pothecary

Foundations of Oscillator Circuit Design, Guillermo Gonzalez

Fundamentals of Nonlinear Behavioral Modeling for RF and Microwave Design, John Wood and David E. Root, editors

Generalized Filter Design by Computer Optimization, Djuradj Budimir

High-Linearity RF Amplifier Design, Peter B. Kenington

High-Speed Circuit Board Signal Integrity, Stephen C. Thierauf

Intermodulation Distortion in Microwave and Wireless Circuits, José Carlos Pedro and Nuno Borges Carvalho

Introduction to Modeling HBTs, Matthias Rudolph

Lumped Elements for RF and Microwave Circuits, Inder Bahl

Lumped Element Quadrature Hybrids, David Andrews

Microwave Circuit Modeling Using Electromagnetic Field Simulation, Daniel G. Swanson, Jr., and Wolfgang J. R. Hoefer

Microwave Component Mechanics, Harri Eskelinen and Pekka Eskelinen

Microwave Differential Circuit Design Using Mixed-Mode S-Parameters, William R. Eisenstadt, Robert Stengel, and Bruce M. Thompson

Microwave Engineers' Handbook, Two Volumes, Theodore Saad, editor

Microwave Filters, Impedance-Matching Networks, and Coupling Structures, George L. Matthaei, Leo Young, and E.M.T. Jones

Microwave Materials and Fabrication Techniques, Second Edition, Thomas S. Laverghetta

Microwave Mixers, Second Edition, Stephen A. Maas

Microwave Radio Transmission Design Guide, Trevor Manning

Microwaves and Wireless Simplified, Third Edition, Thomas S. Laverghetta

Modern Microwave Circuits, Noyan Kinayman and M. I. Aksun

Modern Microwave Measurements and Techniques, Second Edition, Thomas S. Laverghetta

Neural Networks for RF and Microwave Design, Q. J. Zhang and K. C. Gupta

Noise in Linear and Nonlinear Circuits, Stephen A. Maas

Nonlinear Microwave and RF Circuits, Second Edition, Stephen A. Maas

QMATCH: Lumped-Element Impedance Matching, Software and User's Guide, Pieter L. D. Abrie

Phase-Locked Loop Engineering Handbook for Integrated Circuits, Stanley Goldman

Practical Analog and Digital Filter Design, Les Thede

Practical Microstrip Design and Applications, Günter Kompa

Practical RF Circuit Design for Modern Wireless Systems, Volume I: Passive Circuits and Systems, Les Besser and Rowan Gilmore

Practical RF Circuit Design for Modern Wireless Systems, Volume II: Active Circuits and Systems, Rowan Gilmore and Les Besser

Production Testing of RF and System-on-a-Chip Devices for Wireless Communications, Keith B. Schaub and Joe Kelly

Radio Frequency Integrated Circuit Design, John Rogers and Calvin Plett

RF Design Guide: Systems, Circuits, and Equations, Peter Vizmuller

RF Measurements of Die and Packages, Scott A. Wartenberg

The RF and Microwave Circuit Design Handbook, Stephen A. Maas

RF and Microwave Coupled-Line Circuits, Rajesh Mongia, Inder Bahl, and Prakash Bhartia

RF and Microwave Oscillator Design, Michal Odyniec, editor

RF Power Amplifiers for Wireless Communications, Second Edition, Steve C. Cripps

RF Systems, Components, and Circuits Handbook, Ferril A. Losee

Stability Analysis of Nonlinear Microwave Circuits, Almudena Suárez and Raymond Quéré

System-in-Package RF Design and Applications, Michael P. Gaynor

TRAVIS 2.0: Transmission Line Visualization Software and User's Guide, Version 2.0, Robert G. Kaires and Barton T. Hickman

Understanding Microwave Heating Cavities, Tse V. Chow Ting Chan and Howard C. Reader

For further information on these and other Artech House titles, including previously considered out-of-print books now available through our In-Print-Forever® (IPF®) program, contact:

Artech House
685 Canton Street
Norwood, MA 02062
Phone: 781-769-9750
Fax: 781-769-6334
e-mail: artech@artechhouse.com

Artech House
46 Gillingham Street
London SW1V 1AH UK
Phone: +44 (0)20 7596-8750
Fax: +44 (0)20 7630 0166
e-mail: artech-uk@artechhouse.com

Find us on the World Wide Web at: www.artechhouse.com